Ionel Haiduc, Luminiţa Silaghi-Dumitrescu
Organometallic Chemistry

Also of Interest

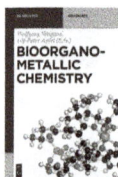

Bioorganometallic Chemistry
Wolfgang Weigand, Ulf-Peter Apfel (Eds.), 2020
ISBN 978-3-11-049650-5, e-ISBN 978-3-11-049657-4

Selenium and Tellurium Reagents.
In Chemistry and Materials Science
Risto Laitinen, Raija Oilunkaniemi (Eds.), 2019
ISBN 978-3-11-052794-0, e-ISBN 978-3-11-052934-0

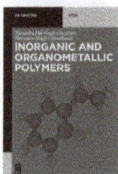

Inorganic and Organometallic Polymers
Narendra Pal Singh Chauhan, Narendra Singh Chundawat, 2019
ISBN 978-1-5015-1866-9, e-ISBN 978-1-5015-1460-9

Organometallic Reagents in Organic Synthesis
Narendra Pal Singh Chauhan, Narendra Singh Chundawat,
Divya Singh, planned 2022
ISBN 978-1-5015-1916-1, e-ISBN 978-1-5015-1919-2

Ionel Haiduc, Luminiţa Silaghi-Dumitrescu

Organometallic Chemistry

Fundamentals and Applications of Organometallic Compounds

2nd Edition

DE GRUYTER

Authors
Prof. Dr. Ionel Haiduc
Faculty of Chemistry and
Chemical Engineering
Babes-Bolyai University
Arany Janos Str. 11
400028 CLUJ-NAPOCA
Rumania
ihaiduc@acad.ro

Prof. Luminiţa Silaghi-Dumitrescu
Faculty of Chemistry and
Chemical Engineering
Babes-Bolyai University
Arany Janos Str. 11
400028 CLUJ-NAPOCA
Rumania
luminita.silaghi@ubbcluj.ro

ISBN 978-3-11-069526-7
e-ISBN (PDF) 978-3-11-069527-4
e-ISBN (EPUB) 978-3-11-069544-1

Library of Congress Control Number: 2022930569

Bibliographic information published by the Deutsche Nationalbibliothek
The Deutsche Nationalbibliothek lists this publication in the Deutsche Nationalbibliografie;
detailed bibliographic data are available on the Internet at http://dnb.dnb.de.

© 2022 Walter de Gruyter GmbH, Berlin/Boston
Cover image: Customdesigner/iStock/Getty Images Plus
Typesetting: Integra Software Services Pvt. Ltd.
Printing and binding: CPI books GmbH, Leck

www.degruyter.com

Foreword

This book was initiated as a new edition of *Basic Organometallic Chemistry*, by Ionel Haiduc and J.J. Zukerman, published by Walter de Gruyter, Berlin, 1985. The present volume is a heavily revised and updated, with significant changes from the previous book. While many chemical diagrams were reproduced here, the text was rewriten and updated, some sections were eliminated and new chapters were added.

The book is not a monograph; it is intended as an introductory textbook. Therefore, literature references are not given for each information and are limited to some general reviews (recommended as further reading) and to specific or recent data. The information which is by now classical (first edition) is not referenced. References are provided only for new work, added in this second edition.

Parts I–III have been prepared by IH and Part IV by LS-D.

https://doi.org/10.1515/9783110695274-202

Contents

Part III: Organometallic compounds of transition metals
General — 159

Part IV: **Application of organometallics in organic synthesis**

Part I: **General**

1 The scope of organometallic chemistry

Organometallic chemistry is the discipline dealing with *compounds containing at least one direct metal–carbon bond*. It should be added that organometallic chemistry deals with compounds in which an organic group is attached to an atom which is less electronegative than carbon (electronegativity $X = 2.50$). On this basis, the organic derivatives of some non-metals (with the electronegativities shown in brackets), namely boron ($X = 2.01$), silicon ($X = 1.74$) and arsenic ($X = 2.20$) are traditionally included in organometallic chemistry, although these elements are not metals (but often described as metalloids). In this volume, the organic derivatives of B, Si, Ge and As are not included, assuming that their metallic character is not predominant.

All elements, except the noble gases (other than xenon), form compounds with element–carbon bonds. Therefore, organometallic chemistry embraces the organic derivatives of the alkali and alkaline earth metals, the non-transition metals (Main groups 13–15), the transition metals (d-block elements, plus lanthanides and actinides) and some nonmetals (or metalloids) such as boron, silicon, antimony and tellurium.

It should also be mentioned that several classes of compounds which contain a metal and carbon in their composition, are not described as organometallic if a direct metal–carbon bond is absent. Thus, compounds such as metal alkoxides (with M–OR bonds), metal amides (with M–N bonds), chelate complexes (e.g., acetylacetonates) or the metal salts of carboxylic acids, are not considered organometallic. Often, such compounds are described as *metal organic*.

Perhaps a classification as described may seem somewhat arbitrary but it is practical and generally accepted.

https://doi.org/10.1515/9783110695274-001

2 Organometallic molecules: the nature of metal–carbon bonds

The stability and reactivity of various organometallic compounds are very different. Some are very sensitive and react spontaneously with oxygen and water or are thermally unstable, whereas others are perfectly stable and can be handled in open atmosphere at room temperature without any special precautions. The chemical properties of organometallic compounds are determined by the nature of the metal–carbon bonds, and they can differ very much between various metals.

The classical bond types, covalent and ionic, are present in many organometallic compounds, but there are some metal–carbon interactions discovered and present only in the metal–carbon compounds.

2.1 Sigma-covalent (bicentric bielectronic) metal–carbon bonds

These are the classical covalent bonds formed by pairing of two electrons of opposite spin and are possible for all elements.

These are typical for all main group (nontransition) elements but also occur in transition metal derivatives. Because of the electronegativity differences between carbon and metals, the covalent metal–carbon bonds are polar $M^{\delta+}-C^{\delta-}$ covalent bonds, and the degree of polarity (percent of ionic character) depends on the electronegativity difference. The stability of polar covalent bonds is influenced by the nature of organic substituents. Electron-attracting substituents in the organic group (e.g., fluorine) increase the stability of the M–C bonds. This is reflected in the fact that the $M-CF_3$ and $M-C_6F_5$ derivatives are significantly thermally more stable than the nonfluorinated analogues, especially in the case of transition metals.

The stability of σ-covalent organometallic compounds is determined by thermodynamic and kinetic factors.

The main group elements form homoleptic compounds of MR_n type, where n is the typical valence of the metal, which are in general thermally stable. The transition metals show less tendency to form stable homoleptic compounds; their low stability is of kinetic origin and is due to incomplete occupation of the d orbitals. Stability is gained by adding π-acceptor ligands like CO, PR_3 and $\pi\text{-}C_5H_5$, which form additional dative bonds to increase the kinetic stability. Thus, $Ti(CH_3)_4$ is unstable at room temperature but the cyclopentadienyl derivative $(\pi\text{-}C_5H_5)_2Ti(CH_3)_2$ is stable.

The low stability of some σ-bonded organometallic compounds is caused by the tendency to eliminate the organic group as an olefin with the formation of a metal hydride. This is called β-elimination:

https://doi.org/10.1515/9783110695274-002

$$M\text{-}CH_2CH_2\text{-}R \rightarrow H_2C{=}CH\text{-}R + M\text{-}H$$

When the structure of the organic group, for example, $CH_2\text{-}SiMe_3$, $CH_2\text{-}CMe_3$ and $CH_2\text{-}C_6H_5$, makes the β-elimination impossible, the σ-covalent compounds are more stable.

The thermodynamic stability can be measured by the bond energies (or thermal dissociation energies). The values of thermal dissociation energies of $M\text{-}CH_3$ bonds in metal and nonmetal methyl compounds are given for comparison in Tab. 2.1.

Tab. 2.1: Thermal dissociation energies of E–C bonds (kcal)*.

	BMe_3	CMe_4	NMe_3	OMe_2	FMe
	87.5	85.3	71.5	84.4	ca. 120
	$AlMe_3$	$SiMe_4$	PMe_3	SMe_2	ClMe
	62.9	72.1	65.3	70.3	82.1
$ZnMe_2$	$GaMe_3$	$GeMe_4$	$AsMe_3$	$SeMe_2$	BrMe
41.5	57.5	n.a.	51.5	n.a	68.9
$CdMe_2$	$InMe_3$	$SnMe_4$	$SbMe_3$	$TeMe_2$	IMe
32.8	n.a.	53.3	47.9	n.a.	54.4
$HgMe_2$	$TlMe_3$	$PbMe_4$	$BiMe_3$		
29.5	n.a.	35.9	33.7		

*Data from I.H. Long, https://citeseecx.ist.psu.edu).

Some tendencies can be noted. The M–C bond energies decrease on descending in a group, due to more diffuse character of the s and p orbitals of the heavier elements, which causes a less efficient overlap with the carbon sp^3 hybrid orbitals.

The compounds with weak M–C bonds (e.g., those of Cd, Hg, Pb and Bi) decompose thermally to deposit the metal. These compounds are thermodynamically unstable with respect to their decomposition to metals and hydrocarbon. Their isolation is possible due to kinetic stability, that is, the lack of a decomposition mechanism with a low activation energy.

It should also be noted that all organometallic compounds are thermodynamically unstable with respect to oxidation, due to the formation of very stable compounds such as metal oxides, carbon dioxide and water. The oxidative stability (e.g., in open atmosphere) of compounds such as SiR_4, SnR_4 and HgR_2 is due to kinetic factors, that is, the absence of oxidation reaction mechanisms with low activation energy. This is true for metals without vacant orbitals of low energy, but those with vacant orbitals (such as AlR_3, ZnR_2, LiR and NaR) are very reactive and spontaneously flammable in air. These considerations should be taken into account when handling the organometallic compounds which require special conditions (e.g., AlR_3, ZnR_2 and LiR). This behavior is determined by the polarity of the M–C bonds and is favored by the presence of low-energy vacant orbitals at the metal. This is the situation with compounds such as AlR_3, ZnR_2, MgR_2 and LiR, which have more polar M–C bonds. Similar considerations are valid for transition metal organometallics.

2.2 Highly polar and ionic metal–carbon bonds

The organometallic compounds of alkali metals and rare earths are often described as ionic and it is often stated that in these compounds the metal is present in cationic form, as M^{n+}, and the organic group is a carbanion. This concept should be considered with some caution. All metal–carbon bonds are polarized as $M^{\delta+}-C^{\delta-}$ and exhibit a certain degree of ionic character. This can be roughly calculated with the aid of Pauling's equation:

$$i = 1 - e^{-0.25(X_A - X_B)^2}$$

where i is the percentage of ionic character, and X_A and X_B are the electronegativities of elements A and B. The value of i is never larger than 0.80 (i.e., for francium, a metal not encountered in organometallic chemistry) and is 49% for cesium, 46% for potassium, 48% for rubidium, 46% for potassium, 42% for sodium and 44% for lithium. Therefore, the corresponding metal–carbon bonds are obviously very polar but retain a certain degree of covalent character. The values of i for rare earths (Ca, Sr and Ba) are in the same range of values. Keeping this in mind, for practical purposes and simplicity, the description as ionic may be acceptable for alkali metal and rare earth organometallic compounds, as frequently found in the literature, but it is preferable to describe their metal–carbon bond as polar (or strongly polar).

Genuine ionic character, with cation–anion separation, occurs in compounds in which the negative charge is delocalized in a ring, like cyclopentadienyl in sodium cyclopentadienide $Na^+[C_5H_5]^-$ in a delocalized ring fragment or moiety, like in the benzyl carbanion, or at the terminal carbon of an acetylenic carbanion. This leads to stabilization of the carbanion and a decrease in the reactivity.

A particular (but rare) case is the formation of organometallic anions by addition of an electron from an alkali metal to an aromatic hydrocarbon, without substitution of a hydrogen. Thus, naphthalene can accept an electron into a vacant antibonding molecular orbital with the formation of a colored, reactive radical anion $(C_{10}H_8^{-\cdot})$.

The polar and ionic organometallic compounds are very sensitive to moisture. The carbanion abstracts a proton from water and produces a hydrocarbon and the alkali metal hydroxide:

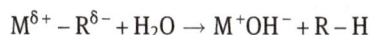

$$M^{\delta+} - R^{\delta-} + H_2O \rightarrow M^+OH^- + R - H$$

Other reagents containing mobile hydrogen (acids, alcohols, amines and thiols) react in a similar way. In all these cases, the kinetic stability (or reactivity) plays the leading role.

The strongly polar and ionic organometallics are also very sensitive to oxygen.

2.3 Electron-deficient (localized tricentric bielectronic) metal–carbon bonds

This is a particular type of chemical bonds that manifests in $M \cdots CH_3 \cdots M$ bridges (e.g., with M = Be, Al) in which an electron pair connects three atoms, forming a tricentric bielectronic bonding system. These are weaker than the common bicentric bielectronic bonds and are formed when enough valence electrons are not available to fill all the bonding orbitals with electron pairs. The formation of these bonds in the dimeric trimethylaluminum is illustrated in Fig. 2.1.

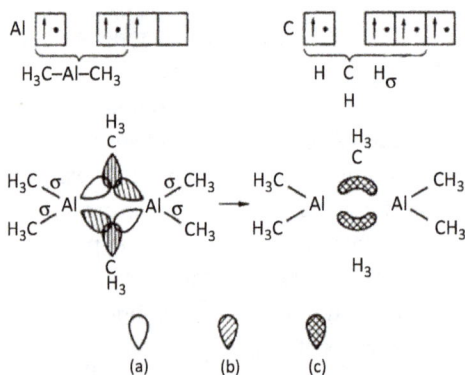

Fig. 2.1: Formation of tricentric bielectronic systems in dimeric trimethylaluminum.

Tricentric bielectronic bonds explain the supramolecular association of dimethylberyllium in solid state (Fig. 2.2).

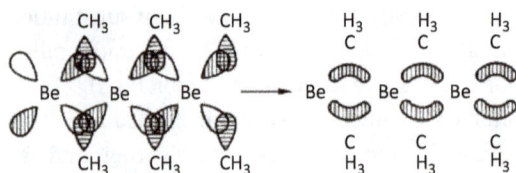

Fig. 2.2: The formation of three-center bonds in $[Be(CH_3)_2]_x$.

2.4 Delocalized bonds in polynuclear systems

Lithium forms some polynuclear compounds $(LiR)_n$ based upon Li_4 tetrahedra and Li_6 polyhedra with delocalized metal–metal bonds (Fig. 2.3). Here the organic group is attached simultaneously to several metal atoms, usually three, on a polyhedral triangular face. The formation of this type of bonding can be explained in terms of molecular orbital theory. Similar interactions are present in polymetallic metal carbonyl clusters, where the metal atoms are held together by collectivization of the valence orbitals and electrons.

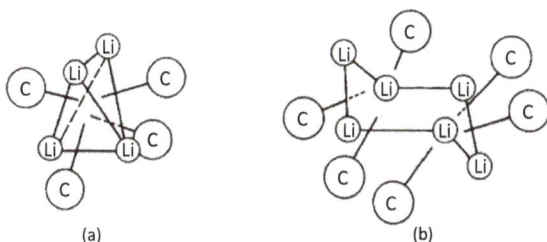

Fig. 2.3: The structure of $(LiR)_4$ and $(LiR)_6$.

2.5 Sigma donor–π-acceptor dative bonds

An important type of bonding in transition metal complexes is the so-called dative bond. In this type, the carbon atom donates an electron pair to the metal atom, and the ligand molecule accepts back some electron density from the metal into a vacant antibonding molecular orbital (a process called *back donation*). The ligands capable of this interaction are molecules possessing an electron pair at carbon, for example, carbon monoxide (:C=O), cyanides (:CN⁻) and isocyanides (:C=N–R). The formation of dative bonds is typical for metal carbonyls and is illustrated in Fig. 2.4. In fact, we have here a metal–carbon double bond, with electron donation in both directions.

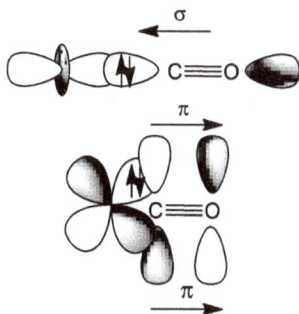

Fig. 2.4: The formation of dative bonds in metal carbonyls.

2.6 π-Bonding of unsaturated molecules to transition metal atoms

A particular type of dative bonding in transition metal organometallics is the so-called π-bonding of unsaturated molecules to transition metal atoms. The unsaturated molecule releases a number of electrons from its π-molecular orbitals into vacant atomic orbitals of the metal and simultaneously accepting electrons from the occupied metal atom orbitals into the π*-antibonding molecular orbitals. Such organic molecules act as ligands and can contribute a variable number of electrons, for example, two (ethylene), three (allyl groups), four (butadiene and cyclobutadiene), five (cyclopentadienyl ring) and six (benzene ring).

The formation of ethylene complexes through interaction of its π-molecular orbitals with the metal s, p and atomic orbitals is illustrated in Fig. 2.5. The ethylene molecule donates electron density from the occupied molecular orbital into the vacant s, p and d orbitals of the metal and accepts back donation from the occupied p and d atomic orbitals of the metal with appropriate orientation.

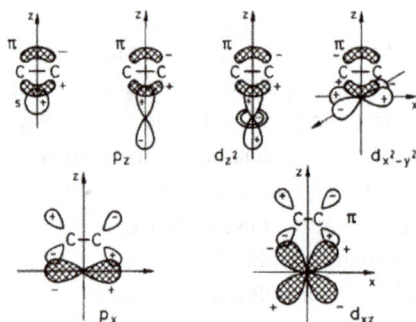

Fig. 2.5: Participation of various orbitals in the formation of π-olefin complexes.

The π-complexes formed by unsaturated molecules donating from two to six π-electrons, with transition metal atoms are represented in Fig. 2.6.

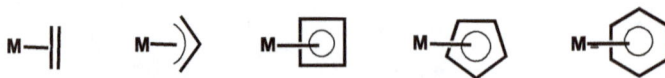

Fig. 2.6: Formation of π-complexes.

Depending on the number of electrons required by the metal atom to satisfy its noble gas configuration, an unsaturated molecule can participate in different bonding modes.

A first example is illustrated here for cyclopentadiene. It can contribute five electrons to form a so-called pentahapto π-complex, η^5-C_5H_5M, which can form a carbanion $C_5H_5^-$ or can contribute only four (η^4), three (η^3) and two (η^2) π-electrons to form a π-complex, and finally, can form organometallic compounds connected by a simple sigma M–C bond (Fig. 2.7).

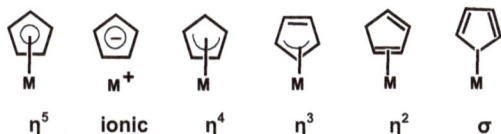

Fig. 2.7: Coordination modes of cyclopentadiene.

In a similar manner, benzene can use all six π-electrons or only some of them in bonding to metal atoms (Fig. 2.8).

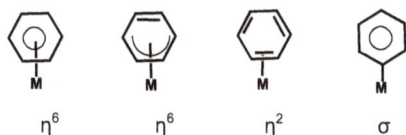

Fig. 2.8: Coordination modes of benzene.

Larger rings can behave similarly, as shown for seven-membered cycloheptatriene and eight-membered cyclooctatetraene rings (Fig. 2.9).

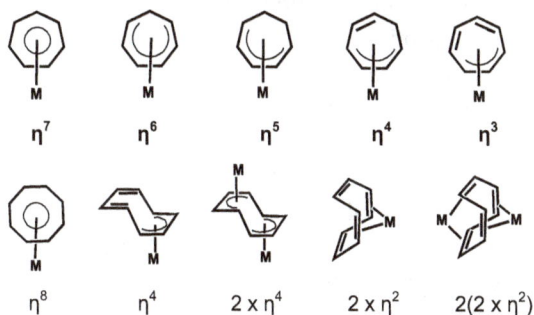

Fig. 2.9: Coordination modes of cycloheptatriene and cyclooctatetraene.

Note: The hapto symbol η, with numerical superscript, provides a description for the bonding of hydrocarbons and other π-electron systems to metals and indicates the connectivity between the ligand and the central atom.

Obviously, such a variety of bonding possibilities will result in a broad variety of organometallic compounds.

2.7 Carbenes, carbynes, carbones, and carbides

2.7.1 Carbenes

The classical carbenes are neutral divalent carbon compounds: CR_2 with a lone pair of electrons which can be donated to a metal atom to form a special type of organometallic compounds. There are two types of carbenes known as Fischer carbenes, in singlet state, and Schrock carbenes, in triplet state. They can be stabilized as transition metal complexes, as shown in Fig. 2.10, by donation of electrons into the d orbitals of the transition metal.

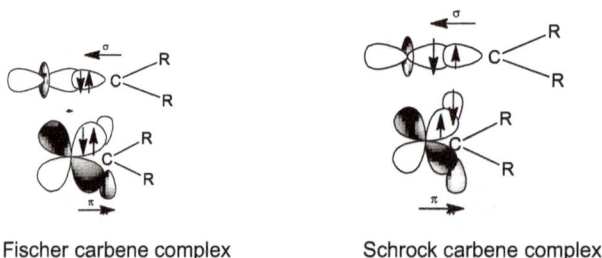

Fischer carbene complex   Schrock carbene complex

Fig. 2.10: Formation of metal–carbon bonds in Fischer and Schrock carbene complexes.

The Fischer carbenes are formed by metals in low oxidation state (e.g., Cr(0), Mo(0) and Fe(0)) and bear alkoxy or alkylamino susbtituents on the carbenoid atom [1].

The Schrock carbenes are formed by metals in a high oxidation state, like Ti(IV) or V(V), with π-donor ligands and bear hydrogen or alkyl substituents at the carbenoid atom [2, 3].

Classical carbenes are not available in free state. They occur usually as transient intermediates and must be stabilized by complexation to a metal.

An important, relatively recent development was the synthesis of N-heterocyclic carbenes which are stable and can be isolated as free ligands and discrete compounds (Arduengo carbenes) [4, 5]. They are stabilized by π-donor substituents, and are excellent σ-donor and poor π-acceptors. Their bond to carbon is regarded as single bonds, while the bonds to carbon of Fischer and Schrock carbenes are double bonds.

Nitrogen heterocyclic carbenes have been much used lately as ligands in numerous organometallic complexes. Many transition metals form organometallic derivatives with these carbenes. Even nontransition metals, for example, mercury and beryllium, have been reported as forming such organometallic compounds (Fig. 2.11).

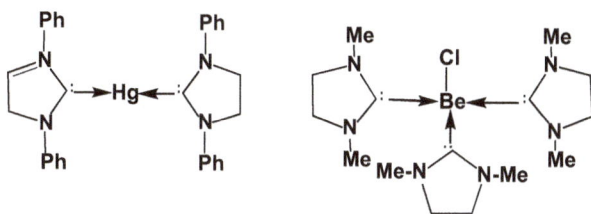

Fig. 2.11: Heterocyclic carbenes as ligands.

2.7.2 Carbynes

The C–R moieties are known as carbynes. They cannot be isolated as discrete compounds but are able to form transition metal organometallic compounds by donation of three electrons to a transition metal atom. Again there are two types of carbyne complexes, as shown in Fig. 2.12, depending on the source of bonding electrons.

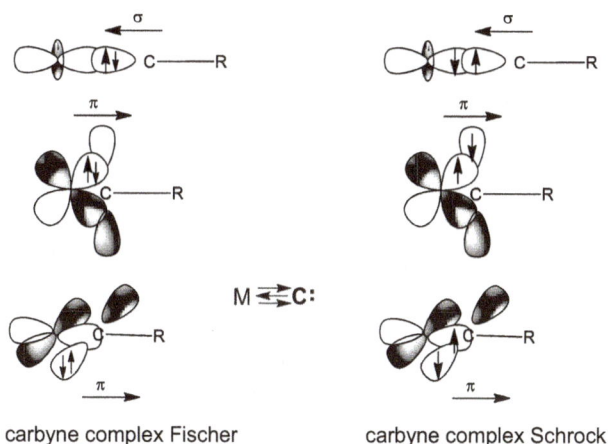

carbyne complex Fischer   carbyne complex Schrock

Fig. 2.12: Formation of carbyne complexes.

2.7.3 Carbones

"Naked" carbon atoms can form transition metal organometallic compounds known as carbone complexes, by donating one or two electron pairs and accepting two electron pairs into the valence orbitals, as shown in the following examples (Fig. 2.13).

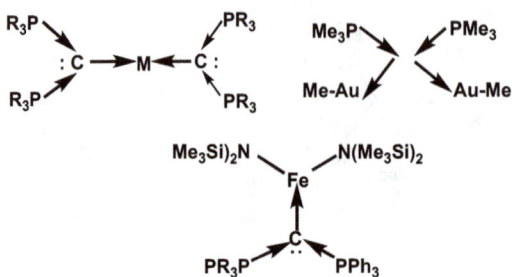

Fig. 2.13: Some carbone complexes.

2.7.4 Carbides

When the naked carbon atoms are terminal ligands, the resulting moieties can donate a pair of electrons and accept two pairs of electrons into the valence orbitals to form transition metal organometallic compounds known as carbide complexes (Fig. 2.14).

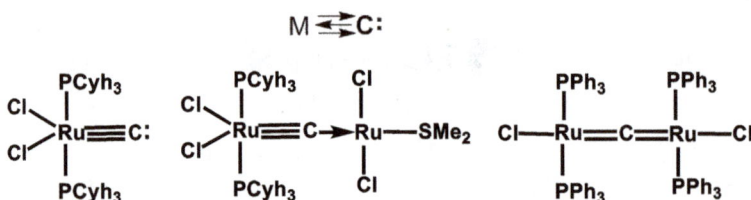

Fig. 2.14: Carbide complexes.

Such compounds could be regarded as carbon-centered inverse coordination complexes.

A number of cluster compounds with a carbon atom as center in polymetal polyhedron are known. One example is a rhenium compound containing $[(\mu_4\text{-C})Re_4(CO)_{15}]^-$ anion, in which the central carbon is embedded in a tetrahedrally distorted square [6] (Fig. 2.15).

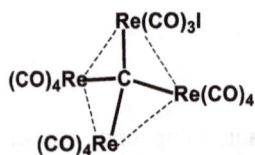

Fig. 2.15: Carbon-centered tetrahedral rhenium compound.

Some surprising carbide complexes are cations with four gold atoms in a square pyramidal geometry with methyl groups in apical position, $[(\mu_4\text{-CCH}_3)Au_4(PR_3)_4]^+$ (R =

Ph, Cyh) [7], the trigonal bipyramidal $[(\mu_5\text{-C})Au_5(PPh_3)_5]^+$ [8] and the octahedral $[(\mu_6\text{-C})Au_6(PPh_3)_6]^+$ [9–11] cations (Fig. 2.16).

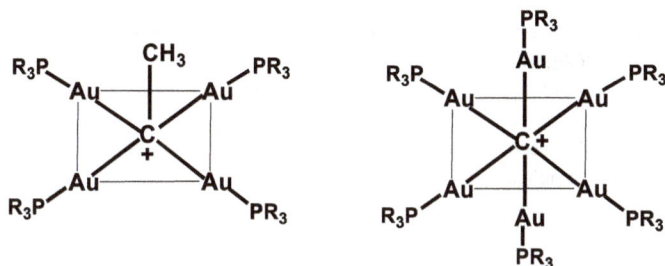

Fig. 2.16: Carbon-centered gold complexes.

Iron also forms carbido inverse coordination complexes with four $[(\mu_4\text{-C})Fe_4(CO)_{13}]^{2-}$, five $[(\mu_5\text{-C})Fe_5(CO)_{15}]^{2-}$ Fig. 2.17. [12] and six metal atoms $[(\mu_6\text{-C})Fe_6(CO)_{16}]^{2-}$ [13]. Nickel $[(\mu_6\text{-C})(NiCp)_6]$ [14] and similar ruthenium carbide complexes $[Ru_6C(CO)_{16}]^{2-}$ [15], and rhodium $[(\mu_6\text{-C})Ru_6(CO)_{16}]^{2-}$ [16, 17] are also known.

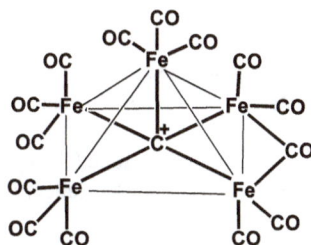

Fig. 2.17: A carbon-centered iron carbonyl complex.

Further reading

Skinner HA The strength of metal-carbon bonds. Adv Organomet Chem 1968, 2, 49–114.

Vidal I, Melchior S, Dobado JA On the nature of metal-carbon bonds. J Chem Phys 2005, 109, 7500–08.

Pe S. M-C bond strengths in transition metal complexes. J Phys Chem 1995, 99, 12723.

Franking GT, Fröhlich NS The nature of bonding in transition metal complexes. Chem Rev 2000,100, 717–74.

Bourissou D, Guerret D, Gabbaï FP, Bertrand G. Stable carbenes. Chem Rev 2000, 100, 39–60.

Hahn FE, Jahnke MC, Heterocyclic carbenes: Synthesis and coordination chemistry. Angew Chem Int Ed 2008, 47, 3122–72.

Nolan SP N-Heterocyclic Carbenes: Effective Tools for Organometallic Synthesis. Wiley-VCH, Weinheim, 2014.

Röthe A, Kretschmer RT Syntheses of bis(N-heterocyclic Carbene)s and their application in carbenes main-group chemistry. J Organomet Chem. 2020, 918, 121289.

Grützmacher H, Marchand CM Heteroatom stabilized carbenium ions. Coord Chem Rev. 1997, 163, 287–344.

Fremont NM, Nolan SP Carbones: Synthesis, properties and organometallic chemistry. Coord Chem Rev 2009, 253, 862–92.

Zhao L, Chai C, Petz W, Franking G Carbones and carbon atoms as ligands in transition metal complexes. Molecules 2020, 25, 4943–48.

Voloshkin VA, Tzouras NV, Nolan SP, Recent advances in the synthesis and derivatization of n-heterocyclic carbene metal complexes. Dalton Trans 2021, 50, 12058–68.

References

[1] Fischer EO, Maasböl A. On the existence of a tungsten carbonyl carbene complex. Angew Chem Int Ed 1964, 3, 580–81.

[2] Schrock RR. First isolable transition metal methylene complex and analogs. characterization, mode of decomposition, and some simple reactions. J Am Chem Soc 1975, 97, 6577–78.

[3] Schrock RR. Multiple metal–carbon bonds for catalytic metathesis reactions (Nobel lecture). Angew Chem Int Ed 2006, 45, 3748–59.

[4] Arduengo AJ. Looking for stable carbenes: The difficulty in starting anew. Acc Chem Res 1976, 32, 913–21.

[5] Arduengo AJ, Harlow RL, Kline M. A stable crystalline carbene. J Am Chem Soc 1991, 113, 361–63.

[6] Beringhelli T, Ciani G, D'Alfonso G, Sironi A, Freni M. A new metallic environment for carbon in a carbido metal cluster: X-ray crystal structure of the anion Re$_4$C(CO)$_{15}$l$^-$. Chem Commun 1985, 978–79.

[7] Steigelmann P, Bissinger H, Schmidbaur H. 1,1,1,1-Tetrakis[triorganylphosphineaurio(I)]-ethanium(+) tetrafluoroborates - hypercoordinated species containing [H$_3$c-c(AuL)$_4$]$^+$ cation. Z Naturforsch B 1993, 48, 72–78.

[8] Scherbaum F, Grohmann A, Müller G, Schmidbaur H. Synthesis, structure, and bonding of the Cation [{(C$_6$H$_5$)$_3$PAu}$_5$C]$^+$. Angew Chem Int Ed 1989, 28, 463–65.

[9] Scherbaum F, Grohmann A, Huber B, Krüger C, Schmidbaur H. "Aurophilicity" as a consequence of relativistic effects: the hexakis(triphenylphosphaneaurio)-methane dication [(Ph$_3$PAu)$_6$C]$^{2+}$. Angew Chem Int Ed 1988, 27, 1544–46.

[10] Gabbaï FP, Schier A, Riede J, Schmidbaur H. Synthesis of the hexakis[(triphenylphosphane)-gold(I)]methanium(2+) cation from trimethylsilyldiazomethane; crystal structure determination of the tetrafluoroborate salt. Chem Ber 1997, 130, 111–14.

[11] Lei Z, Nagata K, Ube H, Shionoya M. Ligand effects on the photophysical properties of N,N′-diisopropylbenzimidazolylidene-protected C-centered hexagold(I) Clusters. J Organomet Chem 2020, 917, 121271.

[12] Kuppuswamy S, Wofford JD, Joseph C, Xie Z-L, Ali AK, Lynch VM, Lindahl PA, Pa, Rose MJ. Structures, interconversions, and spectroscopy of iron carbonyl clusters with an interstitial carbide: localized metal center reduction by overall cluster oxidation. Inorg Chem 2017, 56, 5998–6012.

[13] Hill EW, Bradley JS. Tetrairon carbido carbonyl clusters. Inorg Synth 1990, 27, 182–88.

[14] Buchowicz W, Herbaczyńska B, Jerzykiewicz LB, Lis T, Pasynkiewicz S, Pietrzykowski A. Triple C–H bond activation of a nickel-bound methyl group: Synthesis and X-ray structure of a carbide cluster (NiCp)$_6$(μ$_6$-C). Inorg Chem 2012, 51, 8292–97.

[15] Cariati E, Dragonetti C, Lucenti E, Roberto D. Cluster and polynuclear compounds. Inorg Synth 2004, 34, 210.
[16] Martinengo S, Strumolo SD, Chini DP. Dipotassium μ6-carbido-nona-μ-carbonyl-hexacarbonylhexarhodate(2-) $K_2[Rh_6(CO)_6(\mu\text{-}CO)_9\text{-}\mu\text{-}C]$. Inorg Synth 1980, 20, 212–15.
[17] Muratov DV, Dolgushin FM, Fedi S, Zanello P, Kudinov AR. Octahedral (cyclopentadienyl) rhodium clusters $[Rh_6Cp_6(\mu_6\text{-}C)]^{2+}$ and $[Rh_6Cp_6(\mu_3\text{-}CO)_2]^{2+}$: synthesis, structures and electrochemistry. Inorg Chim Acta 2011, 374, 313–19.

3 Supramolecular organometallic association

The previous chapter was concerned with the structures of distinct organometallic molecules. In the solid state, and quite often in solution and even in vapor state, many organometallic molecules are associated through a variety of noncovalent bonds. Such structures are known as supramolecular.

The concept of supramolecular chemistry was introduced by Jean-Marie Lehn (Nobel Prize 1987) who described the supramolecular chemistry as "the chemistry of intermolecular bond dealing with organized entities of higher complexity that result from the association of two or more chemical species held together by intermolecular forces." There are two types of objects in supramolecular chemistry: *supermolecules*, that is, "well-defined discrete oligomolecular species that result from the intermolecular association of a few components" *and supramolecular assemblies (systems) or supramolecular arrays, that is,* "polymolecular entities that result from the spontaneous association of a large undefined number of components" (J.-M. Lehn).

The intermolecular, noncovalent forces leading to supramolecular association (*self-assembly*) can be very different. The most common are the dative coordinate (donor–acceptor or Lewis acid–base interactions), electrostatic (primarily ionic) interactions, hydrogen bonds and π-bonds. Less common (frequent although usually neglected) are secondary bonds (or "soft–soft" interactions or semibonds). New types were recognized more recently, namely, main group element/lone pair–π-arene and metal carbonyl/lone pair–π-arene interactions. Combination of above types (cooperativity) is also possible.

Association through dative coordinate bonds leads to formation of oligomeric cyclic supermolecules (dimers, trimers and tetramers) and of large polymeric arrays with undefined number of units. Examples are found with the functional derivatives of group 3 organometallics and organotin halides (Fig. 3.1).

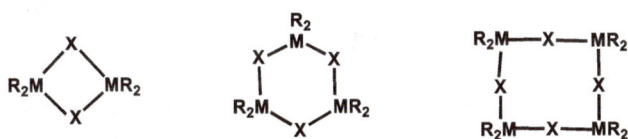

$[R_2M\text{-}X]_n$ M = Al, Ga, In, Tl;   n = 2, 3 or 4; R = alkyl, aryl;
X = halogen, OH, OR, NHR or NR'R", SR, SeR, PR_2, AsR_2, etc.

Fig. 3.1: Supramolecular cyclic self-assembly through dative bonds.

Supramolecular self-assembly occurs with alkali metal strongly polar organometallics, for example, lithium and sodium hexamethyldisilazan (Fig. 3.2).

https://doi.org/10.1515/9783110695274-003

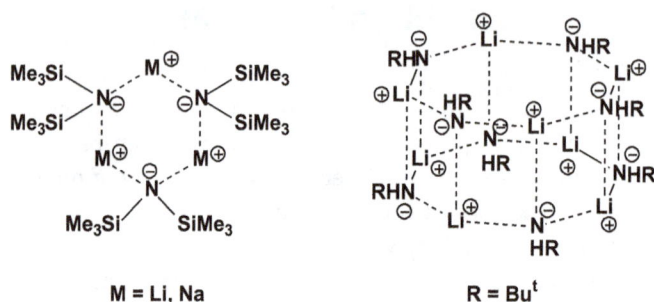

M = Li, Na   R = Bu^t

Fig. 3.2: Supramolecular self-assembly through electrostatic interactions.

The supramolecular association through hydrogen bonds connects molecules with the formation of cyclic oligomers and polymeric arrays (Fig. 3.3).

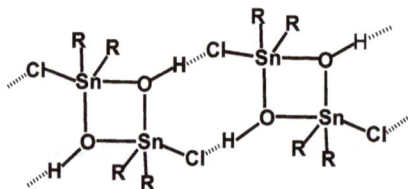

Fig. 3.3: Supramolecular self-assembly through hydrogen bonds.

Secondary bonds are weak interactions, leading to interatomic distances intermediate between single bonds and Van der Waals distances. Such interactions (sometimes also called "semibonds") are observed between soft metals (Hg, Tl, Sn, Pb, Sb, Bi, Te) and soft nonmetals (S, Se, P, As). They are composed of a normal covalent bond A–Y associated with a soft–soft interaction. The secondary bonds have a bond order lower than the normal single bonds between the same atoms. The formation of secondary bond interactions is explained by donation of a lone pair from X into an s* orbital of the A–Y bond (N.W. Alcock).

The supramolecular self-assembly through secondary bond interactions is well illustrated by the case of triphenyl lead dimethyldithiophosphinate, with the parameters indicated in Fig. 3.4.

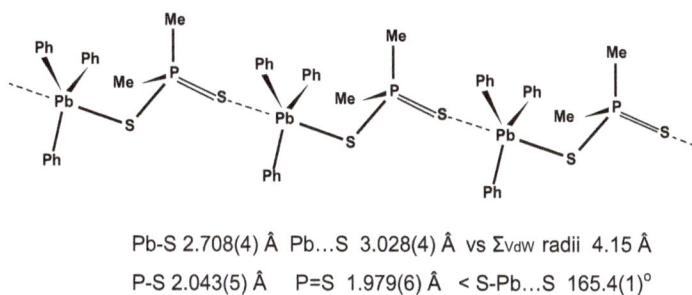

Pb-S 2.708(4) Å Pb...S 3.028(4) Å vs Σvdw radii 4.15 Å

P-S 2.043(5) Å   P=S 1.979(6) Å < S-Pb...S 165.4(1)°

Fig. 3.4: Supramolecular self-assembly through secondary bonds.

The cooperativity of several types of bonding interactions in the formation of supramolecular organometallic assemblies is illustrated by the organotin compound
$[(Me_3Sn)_3(\mu\text{-}OH)_2]^+Br^-$ in which the association of Me_3SnOH molecules occurs with
participation of donor–acceptor O→Sn, secondary Sn · · · Br and hydrogen Br · · · H
bonds (Fig. 3.5).

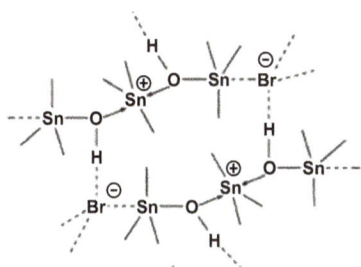

Fig. 3.5: An example of supramolecular cooperative association.

A spectacular case is that of bis(cyclopentadienyl)lead $Pb(C_5H_5)_2$ which forms both
a cyclic oligomer and a supramolecular chain-like array (Fig. 3.6).

Ionic cyclopentadienyl groups also occur as bridges in the structures of alkali
metal cyclopentadienyls, for example, the sodium and potassium compounds
$M^+[C_5H_5]^-$.

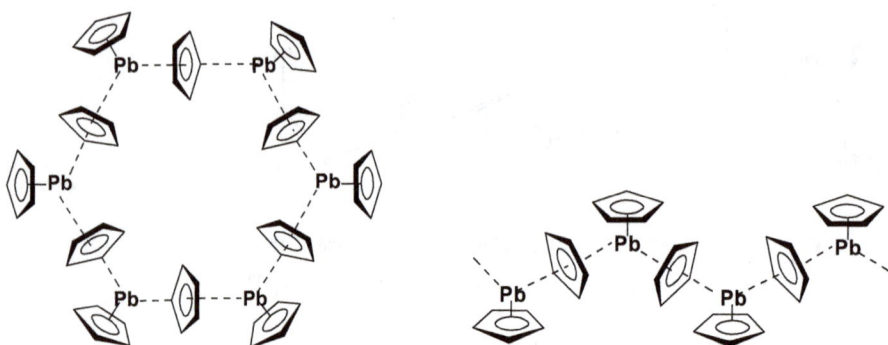

Fig. 3.6: Supramolecular self-assembly of Pb(C$_5$H$_5$)$_2$ through π-bonding.

The organometallic compounds containing carbonyl ligands can sometimes self-assemble into supramolecular structures through carbonyl lone pair–π-arene interactions. An example is the structure of [Mo(CO)$_4$(1,10-phenanthroline)] complex (Fig. 3.7).

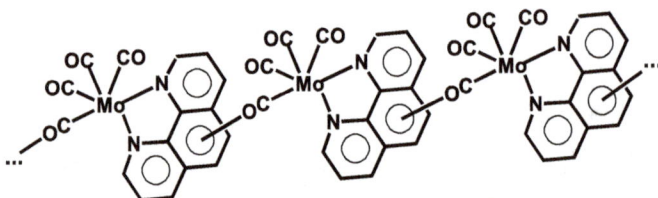

Fig. 3.7: Supramolecular self-assembly through lone pair–arene interactions.

Further reading

Lehn J.-M. Supramolecular Chemistry. Concepts and Perspectives. Weihneim: VCH, 1995.
Lehn J.-M. Supramolecular Chemistry - Scope and Perspectives. Molecules, Supermolecules, and Molecular Devices (Nobel Lecture). Angew Chem Int Ed 1988, 27, 89–112.
Haiduc I, Edelmann FT. Supramolecular Organometallic Chemistry. Weinheim: Wiley-VCH, 2000.
Alexeev YuE, Kharisov BI, Hernandez-Garcia TC, Garnovskii AD. Coordination motifs in modern supramolecular chemistry. Coord Chem Rev 2010, 254, 794.
Chandrasekar V., Boomishanka, R., Nagendrans S. Chem Rev 2004, 104, 5847.
Alcock NW. Bonding and Structure. New York, London: Ellis Horwood, 1993, 195.
Haiduc I. Secondary bonding, in vol. Encyclopedia of Supramolecular Chemistry. Edited by J. Steed and J. Atwood, New York: Marcel Dekker Inc., 2004, 1215.
Mooibroek TJ, Gamez P, Reedijk J Lone pair . . . π-aryl interactions: a new supramolecular bond?. CrystEngComm 2008, 10, 1501.
Zukerman-Schpector J, Haiduc I, Tiekink ERT. The metal carbonyl . . . π(aryl) interaction as a supramolecular synthon. Chem Commun 2011, 47, 12682.

4 Inverse organometallic compounds

Traditionally, most organometallics are compounds in which a metal center is surrounded by a number of M–C bonded organic groups and possibly some additional functional moieties surrounding the metal center. A reversed type of structure is also known, in which an organic molecule acts as centroligand and is connected by two or more metal atoms through metal–carbon bonds. Such compounds can be described as "inverse organometallic compounds."

Inverse organometallics can involve polar metal–carbon bonds (typical for alkali metals) and normal covalent bonds (typical for main group metals). These are illustrated by 1,3,5-trilithium benzene $C_6H_3Li_3$ and the tri-Grignard reagent 1,3,5-trimagnesium tribromide 1,3,5-$C_6H_3(MgBr)_3$ [1] (Fig. 4.1).

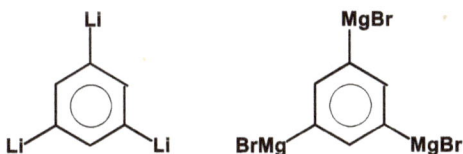

Fig. 4.1: Inverse organometallics with lithium and magnesium.

Other compounds of main group metals can be cited, with covalent metal-carbon bonds, for example, some with bearing three and four organotin substituents [2] (Fig. 4.2).

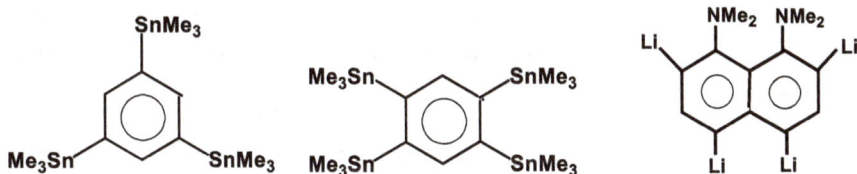

Fig. 4.2: Tri- and tetra-metallic inverse organometallics.

Polymetallated heterocycles, for example, 2,6-dilithiopyridine and 2,6-dimagnesium dibromide, are also known (Fig. 4.3).

Fig. 4.3: Inverse organometallics with central pyridine.

https://doi.org/10.1515/9783110695274-004

Another type of inverse organometallics comprises compounds in which two metal atoms are attached on the two sides of an aromatic ring, connected through π-bonding interactions. These can be described as "inverse sandwich" compounds.

Inverse sandwich compounds with transition metals are formed with cyclopentadienyl, benzene or cyclooctatetraene aromatic groups as centroligands (Fig. 4.4).

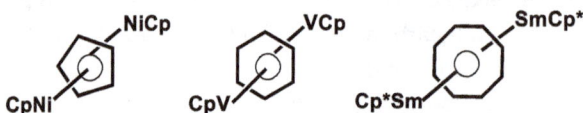

Fig. 4.4: Inverse sandwich compounds.

Fused bicyclic and tricyclic aromatic molecules can also serve as centroligands in inverse sandwich organometallics (Fig. 4.5).

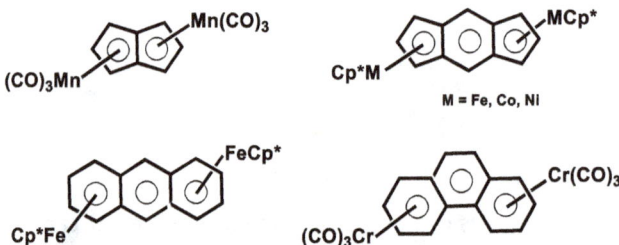

Fig. 4.5: Inverse sandwich complexes with fused aromatic rings.

Based on the principle of inverse π-bond connectivity, a broad variation of molecular structures can be expected, and indeed it is observed throughout transition metal chemistry.

A spectacular inverse sandwich is the iron compound $[(\mu\text{-}C_6H_3)\{FeCp(CO)_2\}_3\{\eta^6\text{-}Cr(CO)_3\}]$, in which there are both sigma- and π-bond metal interactions with the organic moiety [3] (Fig. 4.6).

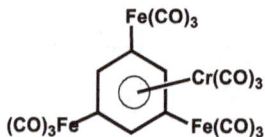

Fig. 4.6: Inverse organometallic compound with mixed bonding types.

References

[1]   Rot N, Bickelhaupt F. Formation of 1,3,5-trilithiobenzene and its conversion to the
      corresponding magnesium, mercury, and tin derivatives. Organometallics 1997, 16, 5027–31.

[2]   (a) Rot N, de Kanter FJJ, Bickelhaupt F, Smeets WJJ, Spek AL. Synthesis of 1,3,5-tri- and
      1,2,4,5-tetrasubstituted tin and mercury derivatives of benzene. J Organomet Chem 2000,
      593/594, 369–79, (b) Yakubenko AA. Karpov VV, Tupikina EYu. Antonov AS. Lithiation of
      2,4,5,7-Tetrabromo-1,8-bis(dimethylamino)naphthalene: Peculiarities of directing groups'
      effects and the possibility of polymetallation. Organometallics 2021, 40, 3627–3636.

[3]   Hunter AD. σ,π-Complexes of benzene. Organometallics 1989, 8, 1118–20.

Part II: Organometallic compounds of main group metals

5 Organometallic compounds of group 1 (alkali metals)

The alkali metals generate monovalent cations by losing their single valence electron and in association with carbanions form ionic organometallic compounds with charge separation. The first element in the group, lithium, behaves somewhat differently by maintaining certain covalent character of the metal carbon bonds and by forming associated $[LiR]_n$ cluster compounds with metal–metal bonds.

In the organometallic compounds of the alkali metals there are vacant orbitals which tend to be occupied wherever possible and this favors strong solvation in donor solvents and formation of complexes with donor molecules (e.g., ethers and amines).

5.1 Organolithium compounds

The organolithium compounds are the most important in the family of alkali metal organometallic compounds due to their extensive use as reagents in preparative chemistry. They are readily prepared, exhibit high chemical reactivity and are soluble in hydrocarbons.

The structure of organolithium compounds is more complex than the simple LiR formula suggests. The organolithium compounds are associated in solutions of non-donor solvents (hydrocarbons), in the solid state and even in the vapor phase (indicated by mass spectra), either as dimers, tetramers or hexamers, depending on the nature of organic group. For example, methyl lithium is a tetramer $[LiCH_3]_4$ made of four lithium atoms, with the methyl groups attached to the faces of the tetrahedron. The oligomeric associated molecule is electron deficient and is formed by delocalization of the valence electrons in the Li_n polyhedron. The compound forms a tetrahydrofuran adduct $[(THF)LiCH_3]_4$ by Li-THF coordination.

The organolithium compounds are prepared under oxygen free and anhydrous conditions.

The best method is the direct reaction of lithium metal and organic halides in hydrocarbon solvents:

$$2\ Li + RX \rightarrow LiR + LiX$$

The reaction of lithium metal with organomercury compounds is now avoided because of the toxicity of the mercury reagents:

$$2\ Li + HgR_2 \rightarrow 2\ LiR + Hg$$

In some cases, the transmetallation (metal–metal exchange) reactions are preferred, for example, for the synthesis of vinyl wlithium:

https://doi.org/10.1515/9783110695274-005

$$Sn(CH=CH_2)_4 + 4\ LiPh \rightarrow 4\ LiCH=CH_2 + SnPh_4$$

$$Pb(CH=CH_2)_4 + 4\ Li \rightarrow 4\ LiCH=CH_2 + Pb$$

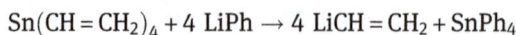

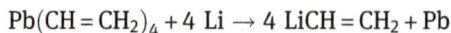

An unusual reactions are the halogen–metal exchange reactions for the synthesis of pentachlorophenyl lithium from hexachlorobenzene and the hydrogen–metal exchange (in tetrahydrofuran at low temperature, bellow −35 °C):

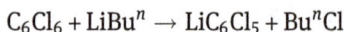

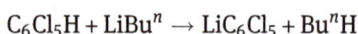

$$C_6Cl_6 + LiBu^n \rightarrow LiC_6Cl_5 + Bu^nCl$$

$$C_6Cl_5H + LiBu^n \rightarrow LiC_6Cl_5 + Bu^nH$$

Donor ligands stabilize the monomeric form by formation of adducts LiR.D (D = donor molecule). In the di- and polyamine complexes of organolithium compounds, the ligand is coordinated to the metal atom, supplying electrons into its vacant orbitals. As a result, $LiCPh_3$.tetramethylethylenediamine (TMEDA) is a monomer and the triethylenediamine complex of benzyl lithium forms a supramolecular chain (Fig. 5.1).

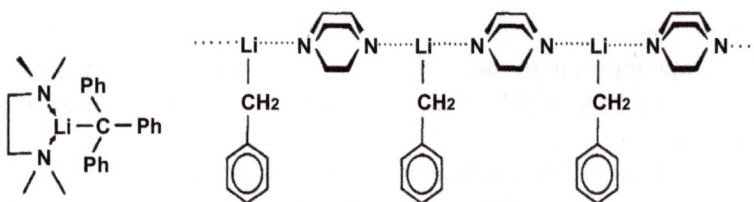

Fig. 5.1: Organolithium diamine adducts.

An inverse organolithium compound, hexalithiobenzene, C_6Li_6, is formed in a reaction of hexachlorobenzene with a large excess of *tert*-butyl lithium at extremely low temperature (−25 °C).

5.2 Organometallic derivatives of sodium and heavier alkali metals

The organometallic compounds of sodium, potassium, rubidium and cesium are very reactive, nonvolatile ionic compounds, nonmelting, insoluble in most organic solvents. They react violently with water, oxygen, carbon dioxide and most organic compounds except saturated hydrocarbons and are spontaneously flammable in air. Such compounds are usually prepared for further use in various reactions without isolation.

Some alkali metal organometallic compounds have been obtained pure in solid state and their structures was determined by X-ray diffractometry. It was found that ethyl sodium displays double layers of isolated Na^+ and $C_2H_5^-$ ions and in methylpotassium K^+ and CH_3^- ions alternate in a layered structure. The sodium cyclopentadienide adduct with TMEDA is a supramolecular chain of alternating cationic $[Na(TMEDA)]^+$ moieties and $C_5H_5^-$ anions (Fig. 5.2).

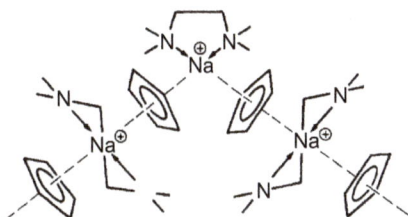

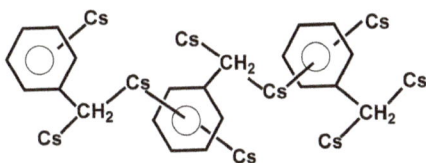

Fig. 5.2: Supramolecular structure of NaC_5H_5. TMEDA.

Benzyl cesium displays a rare supramolecular chain structure formed through Cs-π-arene and $Cs–CH_2$ sigma bonds [1] (Fig. 5.3).

Fig. 5.3: Supramolecular structure of benzylcesium.

A good preparation of organosodium compounds is by using the reaction of organo-lithium reagents with sodium alkoxides:

$$NaOR + LiR' \rightarrow NaR + LiOR'$$

Organosodium compounds were also prepared in a reaction of sodium metal and organomercury derivatives in petroleum ether:

$$2\,Na + HgR_2 \rightarrow 2\,NaR + Hg$$

Organo-zinc, -cadmium and -lead compounds react similarly.The use of dimethyl-mercury must be avoided due to extreme toxicity of this compound (a fatal accident of a researcher using it as NMR standard being known).

The reactive hydrocarbons like triphenylmethane, cyclopentadiene and substituted acetylenes can be metallated with sodium and potassium, or with sodium hydride, in liquid ammonia or tetrahydrofuran. This is a very important reaction for the synthesis of sodium cyclopentadienide $Na[C_5H_5]$, the reagent required for the preparation of metal cyclopentadienyls.

Organo-sodium and -potassium compounds are also formed by cleavage of carbon–carbon bonds of polyarylethanes with the metal amalgams:

$$R_3C-CR_3 + 2\ M/Hg \rightarrow 2\ M^+[CR_3]^- + 2\ Hg\ \ M = Na,\ K;\ R = aryl$$

Naphtalene and other polycyclic aromatic hydrocarbons accept an electron from sodium atoms to form anion radicals, in strong coordinating solvents like tetrahydrofuran or dimethoxyethane, without replacing a hydrogen atom (Fig. 5.4).

Fig. 5.4: Naphtalene-sodium complex.

Organopotassium compounds can be prepared by reacting potassium metoxide
with organolithium reagents and by treating organomercury compounds with potassium. The potassium compounds are very reactive, attack even saturated hydrocarbons and decompose spontaneously with formation of potassium hydride. They
remain mostly as laborarory curiosities.

Further reading

Rappoport Z, Marek I (Eds.) Chemistry of organolithium compounds, Wiley, Chichester, 2007.
Gessner VH, Däschlein C, Strohmann C. Structure, formation principles and reactivity of
        organolithium Compounds, Chem-Eur J 2009, 15, 3320–34.
Reich HJ. What's going on with these lithium reagents, J Org Chem 2012, 77, 5471–91.
Reich HJ. Role of organolithium aggregates and mixed aggregates in organolithium mechanisms.
        Chem Rev 2013, 113, 7130–78.
Seyferth D. Alkyl and aryl derivatives of the alkali metals: Useful synthetic reagents as strong
        bases and potent nucleophiles. 1. Conversion of organic halides to organoalkali-metal
        compounds, Organometallics 2006, 25, 2–24.
Seyferth D. Alkyl and aryl derivatives of the alkali metals: Strong bases and reactive nucleophiles.
        2. Wilhelm Schlenk's organoalkali-metal chemistry Organometallics 2009, 28, 2–33.

Reference

[1]   Orzechowski L, Jansen G, Harder S. Methandiide complexes (R_2CM_2) of the heavier alkali
        metals (M = potassium, rubidium, cesium): Reaching the limit? Angew Chem Int Ed 2009, 48,
        3825–29.

6 Organometallic compounds of group 2 metals (rare earths)

The elements of this group have two valence electrons in the ns orbital and three vacant np orbitals (Fig. 6.1) and this electronic structure dictates the chemical behavior of these elements through the tendency of the metal atoms to make full use of their ns and np orbitals in bonding.

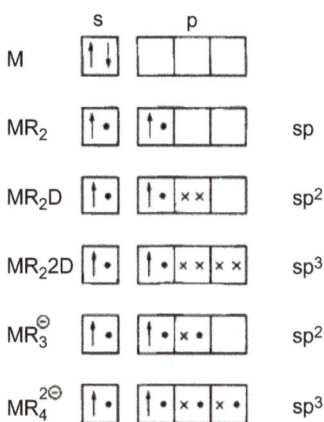

M

MR$_2$ — sp

MR$_2$D — sp^2

MR$_2$2D — sp^3

MR$_3^{\ominus}$ — sp^2

MR$_4^{2\ominus}$ — sp^3

Fig. 6.1: The use of valence orbitals and electrons in the organometallic compounds of group 2.

With the two electrons in the ns atomic orbitals, the elements of group 2 are strongly electropositive and can form dipositive M^{2+} cations, resulting in ionic organometallic compounds. This tendency is increasing for the heavier elements in the order Ca < Sr < Ba < Ra.

The first two elements of the group, beryllium and magnesium, with sp hybridization form two covalent bonds, which are rather polar. The vacant p orbitals tend to accept electrons, and this tendency influences the chemical behavior of these elements. When two electron pairs are accepted from a suitable donor, MR$_2$.D and MR$_2$.2D adducts are formed with sp^2 and sp^3 hybridizations, respectively. By accepting single electrons in the unoccupied p orbitals, these can be paired to form additional M–C bonds in hypervalent anionic [MR$_3$]$^-$ (sp^2 hybridization) and [MR$_4$]$^{2-}$ (sp^3 hybridization). The [MR$_3$]^r anions (e.g., [BePh$_3$]$^-$ are isoelectronic with group 13 trigonal neutral MR$_3$ molecules and the [MR$_4$]$^{2-}$ are isoelectronic with neutral tetrahedral MR$_4$ compounds of the group 14.

Another way to make full use of the p atomic orbitals is the formation of tricentric bielectronic bonds, as mentioned above, for supramolecular structure of dimethylberyllium.

https://doi.org/10.1515/9783110695274-006

6.1 Organoberyllium compounds

6.1.1 Homoleptic compounds BeR$_2$

The diorganoberyllium compounds BeR$_2$ are monomeric only with bulky substituents (R = *tert*-Bu, mesityl) (Fig. 6.2) and are associated as supramolecular dimers (R = Et, *n*-Pr, *iso*-Pr, *n*-Bu) in benzene solution or polymers (R = Me) in the solid state.

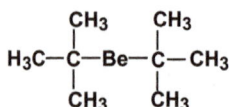

Fig. 6.2: Structure of monomeric di(*tert*-butyl)beryllium.

The linear structure of BeMe$_2$ in the solid state (established by X-ray diffractometry) is in agreement with the sp hybridization (Fig. 6.3).

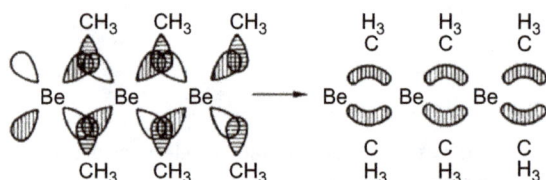

Fig. 6.3: The structure of dimethylberyllium.

Bis(cyclopentadienyl) beryllium (beryllocene) Be(C$_5$H$_5$)$_2$ displays a rare η^1:η^5 slipped sandwich structure, with one η^5 and one η^1 C$_5$H$_5$ rings connected to the metal [1], but bis(pentamethylcyclopenadienyl) beryllium has the normal η^5:η^5 sandwich structure [2].

Diphenylberyllium has an unusual trinuclear structure with four bridging and two terminal phenyl groups [3] (Fig. 6.4).

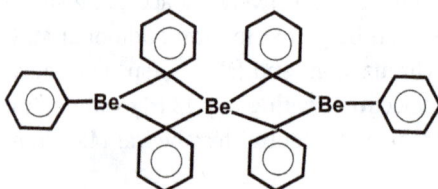

Fig. 6.4: Trimeric structure of BePh$_2$.

The diorganoberyllium compounds are prepared from anhydrous beryllium chloride and phenyl lithium or Grignard reagents. Organolithium reagents can also be used, but an excess can lead to trisubstituted derivatives. A thermal reaction of beryllium metal with diorganomercurials (e.g., diphenylmercury) can be used. *Caution: avoid dimethylmercury!*

$$Be + HgR_2 \rightarrow BeR_2 + Hg$$

6.1.2 Hypervalent species [BeR_3]^- and [BeR_4]^{2-}

In the presence of 12-crown-4 macrocycle diphenylberyllium, BePh$_2$ dissociates into a triphenylberyllium anion [BePh$_3$]$^-$ and [PhBe]$^+$ cation entrapped in the macrocycle (Fig. 6.5).

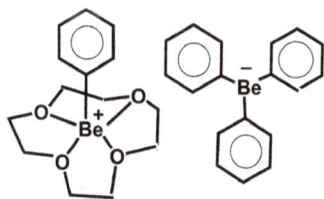

Fig. 6.5: Structure of [PhBe(12-crown-6)]$^+$[BePh$_3$].

With LiPh, diphenylberyllium forms supramolecular chain-like [BePh$_3$Li]$_x$ which in turn can pick up two diethyl ether molecules to form monomeric Li[BePh$_3$].2Et$_2$O] (Fig. 6.6).

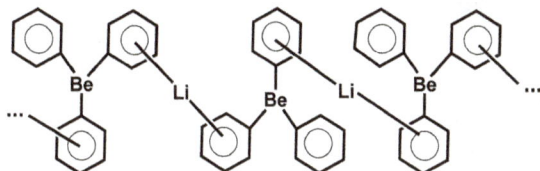

Fig. 6.6: Supramolecular structure of [BePh$_3$Li]$_x$.

A new type of compounds with three beryllium–carbon bonds is represented by a heterocyclic carbene derivative of diphenylberyllium (Fig. 6.7).

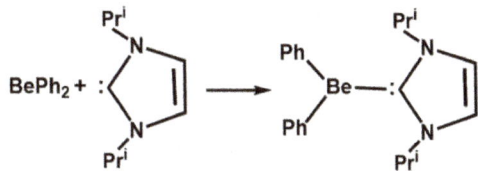

Fig. 6.7: A heterocyclic carbene complex of BePh$_2$.

Compounds with four beryllium–carbon bonds can be illustrated with anionic $[BeR_4]^{2-}$. The tetrahedral tetramethylberyllato anion of $Li_2[BeMe_4]$ is obtained from dimethylberyllium and methyl litium in diethyl ether.

6.1.3 Diorganoberyllium donor adducts $R_2Be.D$

The diorganoberyllium compounds are prepared from anhydrous beryllium chloride and phenyl lithium or Grignard reagents in diethyl ether when ether adducts are obtained [4]:

$$BeCl_2 + 2 \text{ RMgX} + Et_2O \rightarrow BeR_2.OEt_2 + 2 \text{ MgXCl}$$

The ether can be removed with stronger coordination ligands, for example, fluoride, but in this case fluoride bridged, inverse coordination anions are formed:

$$2 \text{ } BeR_2.OEt_2 + KF \rightarrow K^+ [R_2Be–F–BeR_2]^- + 2 \text{ } Et_2O$$

With amines, phosphines, nitrogen heterocycles and other donor molecules, the diorganoberyllium compounds form tri- and tetra-coordinate adducts (Fig. 6.8).

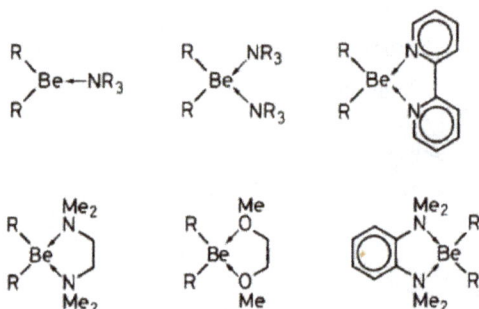

Fig. 6.8: Adducts of BeR_2 compounds with amines.

6.1.4 Organoberyllium halides, RBeX

Organoberyllium halides can be prepared directly from beryllium metal and alkyl halides in the presence of $HgCl_2$ as catalyst (X = CI, Br):

$$Be + RX \rightarrow R–Be–X$$

Organoberyllium halides dimerize through halogen bridges and are monomeric in ethereal solvents, owing to the formation of adducts such as $RBeCl.2Et_2O$. Thus, the

etherate of *tert*-butylberyllium chloride is dimeric in benzene solution but monomeric in diethyl ether (Fig. 6.9).

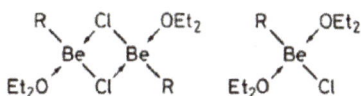

Fig. 6.9: Diethyl ether adducts of organoberyllium
chlorides.

6.1.5 Functional organoberyllium compounds, RBeX

The functional monoorganoberyllium compounds, RBeX, with $X = NR_2$, OR, halogen, SR and SeR, are associated in solution and solid state, through donor–acceptor
bonds, as supramolecular oligomers (dimers, trimers, tetramers) (Fig. 6.10)

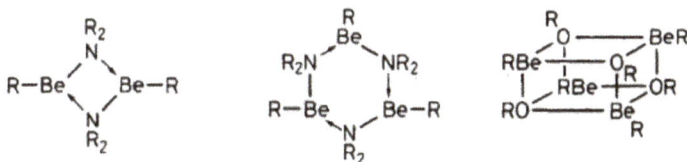

Fig. 6.10: Supramolecular oligomers $[RBeX]_n$.

Sterically encumbered heteroleptic compounds $ArBeX \cdot Et_2O$ ($Ar = C_6H_3$–2,6-Mes_2; X =
Cl, Br, SMes, NHPh, NHSiPh₃, N(SiMe₃)₂) are monomeric [5].

Further reading

Coates GE, Morgan GL. Organoberyllium compounds. Advan Organomet Chem 1971, 195–257.

6.2 Organomagnesium compounds

The organometallic compounds of beryllium and magnesium are similar in many respects related to structure and composition. However, the beryllium compounds are
somewhat exotic while organomagnesium reagents are common and are routinely
used as reagents in many organic and organometallic chemistry laboratories. Cer-
tainly, very few chemists have had in their hands an organoberyllium compound but
probably there is no chemist who has not prepared a Grignard organomagnesium
compound, in his/her undergraduate student years.

6.2.1 Homoleptic compounds, MgR$_2$

Diorgano derivatives, MgR$_2$, are known, but they are much less investigated or used. Dimethylmagnesium is a supramolecular chain-like array built up from three center-two electron bonded bridging methyl groups, like dimethylberylium.

6.2.2 Hypervalent anions [MgR$_3$]$^-$

Derivatives with hypervalent trisubstituted magnesium Li$^+$[MgPh$_3$]$^-$ and tetrasubstituted magnesium Li$^+_2$[MgMe$_4$]$^{2-}$ can be obtained from diorganomagnesium derivatives. The apparently five-coordinated Li$^+_3$[MgMe$_5$]$^{3-}$ may be a LiMe.Li$^+_2$[MgMe$_4$]$^{2-}$ adduct.

6.2.3 Subvalent cations [RMg]$^+$

The subvalent methylmagnesium cation [CH$_3$Mg]$^+$ can be trapped in the cavity of a tetranitrogen macrocycle with formation of a host–guest complex with Mg–N bonds (Fig. 6.11).

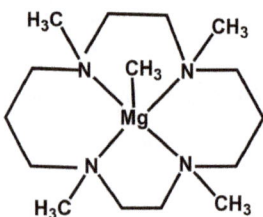

Fig. 6.11: Macroring complex of [MeMg]$^+$ cation.

6.2.4 Organomagnesium donor adducts MgR$_2$.D

Diorganomagnesium compounds form complexes with various nitrogen donors, for example, bidentate and tridentate amines, resulting chelate or inverse coordination complexes (Fig. 6.12).

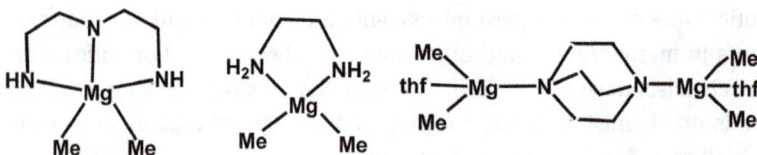

Fig. 6.12: Complexes of MgMe$_2$ with nitrogen donors.

6.2.5 Inverse organomagnesium compounds

Organomagnesium compounds with the metal doubly connected to a central benzene molecule, described as "inverse crown ether " complexes, have been described [6] with a rare benzene core (Fig. 6.13).

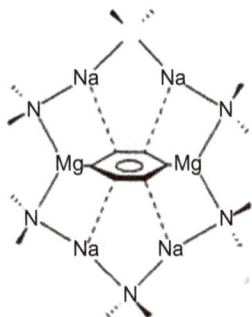

Fig. 6.13: Unusual structure of a magnesium inverse complex.

A typical example is the mixed magnesium complex with sodium and tetramethylpyperidine/tetramethylethylenediamine [7]. This uncommon structure is also displayed by toluene and methoxybenzene [8] complexes. A related inverse complex is known with naphthalene [9] (Fig. 6.14).

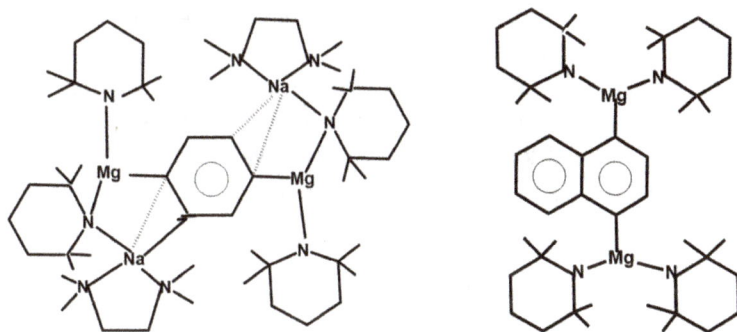

Fig. 6.14: More unusual inverse organomagnesium compounds.

6.2.6 Organomagnesium halides

Organomagnesium halides, RMgX, known as Grignard reagents, are the most important organomagnesium compounds. These are very accessible, reactive compounds.

Their structure is much more complex than indicated by the simple formula RMgX. A redistribution known as Schlenk equilibrium in solution has been demonstrated:

$$MgR_2 + MgX_2 \rightleftharpoons 2\,RMgX$$

In solution, there is also a competition between the coordinating capacity of halogen and ether toward magnesium to give either simple monomeric ether solvates or halogen-bridged dimers (Fig. 6.15).

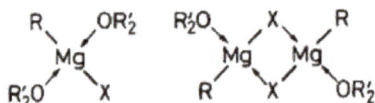

Fig. 6.15: Diethylether adducts of organomagnesium halides.

Monomeric species are predominant in strongly coordinating solvents or dilute solution. Dimerization and polymerization are favored by higher concentrations, weaker donor solvents or non-donating solvents like hydrocarbons. The halides are monomeric as diethyl ether adducts, for example, $R_2MgBr.2Et_2O$ (R = Et, Ph).

The Grignard reagents are readily prepared from organic halides and magnesium metal in anhydrous ether or tetrahydrofuran.

$$RX + Mg \xrightarrow{\text{ether}} RMgX$$

The reactivity of organic halides toward magnesium decreases in the order RI > RBr > RCl > RF. The fluorides are virtually inert except when they are used with activated magnesium freshly prepared by reduction of an anhydrous halide with potassium metal. Usually, the Grignard reagents, being sensitive toward moisture and oxygen, are used in further reactions without isolation, immediately after preparation. However, the solutions can be stored for a long time and are even commercially available. For the use of Grignard reagents in organic synthesis, *vide infra*.

6.2.7 Functional organomagnesium compounds RMgX

Of certain importance are functional organomagnesium compounds, $[RMgX]_n$ where X = NR'$_2$, OR', SR'. The derivatives of secondary amines are supramolecular cyclic dimers $[RMg-NR'_2]_2$ containing a four-membered Mg_2N_2 ring (Fig. 6.16), while the alkoxy derivatives are cyclic dimers $[RMg-OR']_2$ or tetramers $[RMg-OR']_4$ with cubane structures

Further reading

Cossy J Grignard reagents and transition metal catalysts. Berlin, Boston: De Gruyter Berlin, 2016.

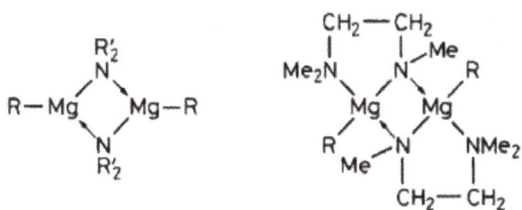

Fig. 6.16: Supramolecular dimers with nitrogen donors.

6.3 Organometallic compounds of calcium

The organometallic compounds of heavier group 2 metals are mostly laboratory curiosities rather than common chemicals. These compounds have more pronounced
ionic character than the beryllium and magnesium analogues and are very reactive
and sensitive to oxygen, water and any reagents with mobile hydrogen.

6.3.1 Homoleptic CaR$_2$ compounds

Dimethylcalcium [CaMe$_2$]$_x$ was obtained by a reaction of calcium bis(trimethylsilyl)-
amide and methyllithium in diethyl ether [10].

A general procedure for the preparation of diorganocalcium compounds uses
the reaction of arylcalcium iodides with potassium *tert*-butoxide [11] (Fig. 6.17).

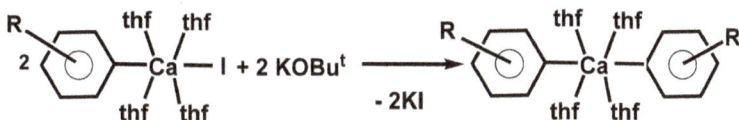

Fig. 6.17: Synthesis of diarylcalcium compounds.

The ionic cyclopentadienyls of alkaline earths can be obtained by reacting the metals or the metal hydrides with cyclopentadiene.

6.3.2 Hypervalent [CaR$_3$]$^-$ anions

Calcium amalgam cleaved hexaphenylethane, Ph$_3$C–CPh$_3$ to form Ca(CPh$_3$)$_2$.nTHF
and reacted with Ph$_3$CCl with formation of hypervalent red [Ca(CPh$_3$)$_3$]$^-$ anion.

A hypervalent organocalcium anion with three metal–carbon bonds [Ca(CHPh₂)₃ (THF)]⁻ has been obtained from Ca{N(SiMe₃)₂}₂(THF)₂ and Ph₂CHLi.TMEDA in ether (Fig. 6.18).

$$
\begin{array}{c}
\text{[Ca\{N(SiMe}_3)_2\}_2\ \text{(thf)}_2] \\
+\ 3\ \text{[Ph}_2\text{CHLi tmeda]}
\end{array}
\xrightarrow[\substack{-\ 2\ \text{LiNi(SiMe}_3)_2 \\ -\ \text{tmeda}}]{\text{Et}_2\text{O,20°C}}
$$

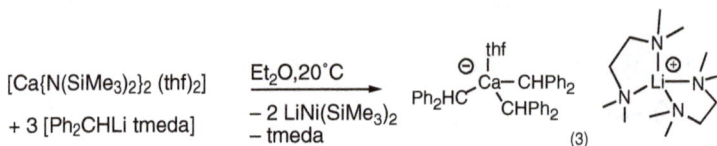

Fig. 6.18: Hypervalent organocalcium anion.

Another compound with three calcium–carbon bonds, the hypervalent anionic [Ca {CH(SiMe₃)₃}₂]⁻, was obtained from the reaction of anhydrous CaI₂ with KCH₂SiMe₃ in benzene (1:3 ratio) (Fig. 6.19).

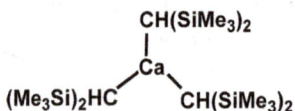

CH(SiMe₃)₂
|
Ca
(Me₃Si)₂HC   CH(SiMe₃)₂   Fig. 6.19: Structure of [Ca{CH(SiMe₃)₂}₃]⁻ anion.

Hypervalent compounds containing three or six calcium–carbon bonds can be prepared with the aid of phenyl substituents bearing two heterocyclic carbene moieties [12] (Fig. 6.20).

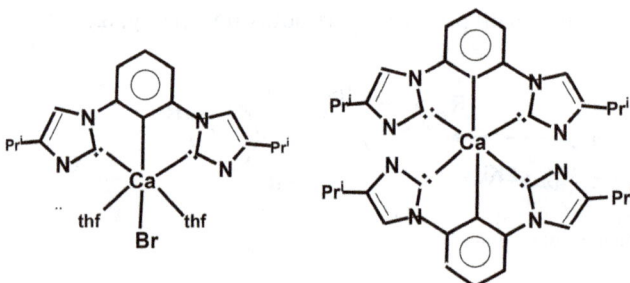

Fig. 6.20: Hypervalent organocalcium compounds.

6.3.3 Other organocalcium compounds

There is evidence that organocalcium analogues of the Grignard reagents, RCaX, are formed from calcium metal and organic halides, in diethyl ether, the reactivity order being again RI > RBr > RCl.

The organocalcium chemistry is recent and rather exotic. This metal was neglected for a long time, and only in the last 20 years, it focused some attention, with
spectacular results. This includes a unique inverse organocalcium sandwich complex [(THF)$_3$Ca{μ-C$_6$H$_3$Ph$_3$}Ca(THF)$_3$] with two calcium atoms at either side of an
arene. It has been obtained by a reaction of activated calcium metal and 1,3,5-
triphenylbenzene [13] (Fig. 6.21).

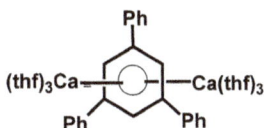

Fig. 6.21: A unique organocalcium inverse sandwich compound.

6.4 Organostrontium and -barium compounds

Organostrontium and organobarium compounds are scarce and very similar.
With strontium, the bis(tetraisopropyl-cyclopentadienyl) Sr(η^5-C$_5$H$_3$Pr$^i_2$)$_2$ [14, 15],
and for barium, bis(η^5-pentamethyl-cyclopentadienyl) Ba(η^5-C$_5$Me$_5$)$_2$ [16] derivatives
can be mentioned as homoleptic diorgano compounds (Fig. 6.22).

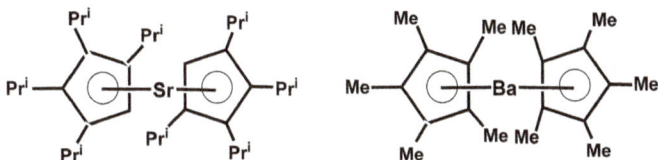

Fig. 6.22: Strontium and barium cyclopentadienyls.

There are several diorganostrontium and -barium donor adducts R$_2$M.D. These include, among others, adducts with tetrahydrofuran M(η^5-C$_5$Me$_5$)$_2$(THF)$_2$ M = Sr, Ba
[17] and bipyridine M(η^5-C$_5$Me$_5$)$_2$.bipy (M = Sr [18], Ba [19].
Crown ether complexes with bis{2-(triphenylsilyl)ethynyl} groups of the type [M
(C≡C-SiPh$_3$)$_2$(18-crown-6)] are formed with both strontium and barium atoms incorporated in the macrocycle [20] (Fig. 6.23)

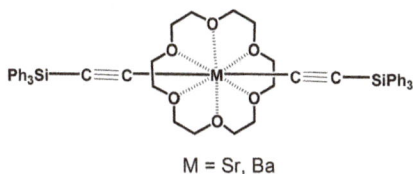

M = Sr, Ba

Fig. 6.23: Organostrontium and -barium crown
ether complexes.

Some exotic organometallic compounds are the inverse sandwich complexes with cyclooctatetraene as centroligand known for both strontium [(μ-C_8H_8){Sr(N$(SiMe_3)_2(THF)_2$}$_2$] [21] and barium [(μ-C_8H_8){Ba(η^5-$C_5HPr^i_4$)$_2$] [22]. (Fig. 6.24).

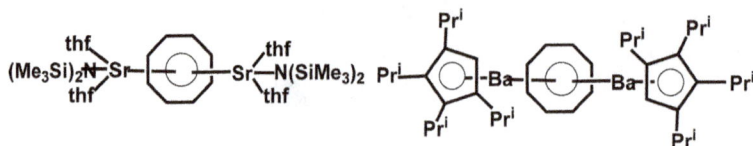

Fig. 6.24: Strontium and barium inverse sandwich compounds with cyclooctatetraene.

Arylstrontium iodides are obtained from strontium metal with aryl iodides, catalyzed by mercury metal.

Further reading

Smith JD Organometallic compounds of the heavier s-block elements - What next? Angew Chem Int Ed 2009, 48, 6597–99
Williams RA, Hanusa TP, Huffman JC Solid state structure of bis(pentamethylcyclopentadienyl) barium, (Me₅C₅)₂Ba; the first X-ray crystal structure of an organobarium complex. J Chem Soc Chem Commun 1988, 1045.

References

[1]   Fernández R, Carmona E Recent developments in the chemistry of beryllocenes. Eur J Inorg Chem. 2005, 3197–206.
[2]   Del Mar Conejo M, Fernández R, Del Río D, Carmona E, Monge A, Ruiz C, Márquez AM, Sanz JF Synthesis, Solid-state structure, and bonding analysis of the beryllocenes [Be(C₅Me₄H)₂], [Be(C₅Me₅)₂], and [Be(C₅Me₅)(C₅Me₄H)], Chem-Eur J 2003, 9, 4452–61.
[3]   Müller M, Buchner MR. Diphenylberyllium reinvestigated: Structure, properties and reactivity of BePh₂, [(12-crown-4)BePh]⁺ and [BePh₃]⁻. Chem-Eur J 2020, 26, 9915–22.
[4]   Coates GE, Roberts PD Propyl- and t-butyl-beryllium complexes. J Chem Soc A, Inorg Phys Theor 1968, 2651–55.
[5]   Ruhlandt-Senge K, Bartlett RA, Olmstead MM, Power PP Synthesis and structural characterization of the beryllium compounds [Be(2,4,6-Me₃C₆H₂)₂(OEt₂)], [Be{O(2,4,6-tert-Bu₃C₆H₂)}₂(OEt₂)], and [Be{S(2,4,6-tert-Bu₃C₆H₂)}₂(THF)].PhMe and determination of the structure of [BeCl₂(OEt₂)₂]. Inorg Chem 1993, 32, 1724–28.
[6]   Armstrong DR, Kennedy AR, Mulvey RE, Rowlings RB Mixed-metal sodium-magnesium macrocyclic amide chemistry: A template reaction for the site selective dideprotonation of arene molecules. Angew Chem Int Ed 1999, 38, 131–33.
[7]   Armstrong DR, Clegg W, Dale S, Graham DV, Hevia E, Hogg LM, Honeyman GW, Kennedy AR, Mulvey RE Dizincation and dimagnesiation of benzene using alkali-metal-mediated metallation. Chem Commun 2007, 598–600.

[8] Martinez-Martinez AJ, Kennedy AR, Mulvey RE, O'Hara CT Directed ortho-meta'- and meta-meta'-dimetallations: A template base approach to deprotonation. Science 2014, 346, 834–37.

[9] Martinez-Martinez AJ, Armstrong DR, Conway B, Fleming BJ, Klett J, Kennedy AR, Mulvey RE,, Robertson SD, O'Hara CT Pre-inverse-crowns: Synthetic, structural and reactivity studies of alkali metal magnesiates primed for inverse crown formation. Cheml Sci 2014, 5, 771–81.

[10] Wolf BM, Stuhl C, Maichle-Mössmer C, Anwander R. Dimethylcalcium. J Am Chem Soc 2018, 140, 2373–83.

[11] Langer J, Krieck S, Görls H, Westerhausen M An Efficient General Synthesis of Halide-Free Diarylcalcium. Angew Chem Int Ed 2009, 48, 5741–44.

[12] Koch A, Krieck S, Görls H, Westerhausen M Directed Ortho Calciation of 1,3-Bis(3-isopropylimidazol-2-ylidene)benzene, Organometallics 2017, 36, 2811–17.

[13] Krieck S, Görls H, Westerhausen M Mechanistic Elucidation of the Formation of the Inverse Ca (I) Sandwich Complex [(thf)$_3$Ca(μ-C$_6$H$_3$-1,3,5-Ph$_3$)Ca(thf)$_3$] and Stability of Aryl-Substituted Phenylcalcium Complexes. J Am Chem Soc 2010, 132, 12492–501.

[14] Westerhausen M, Gärtner M, Fischer R, Langer J, Yu L, Reiher M Heavy Grignard Reagents: Challenges and Possibilities of Aryl Alkaline Earth Metal Compounds. Chem-Eur J 2007, 13, 6292–306.

[15] Williams RA, Hanusa TP, Huffman JC Structures of ionic decamethylmetallocenes: Crystallographic characterization of bis(pentamethylcyclopentadienyl)calcium and -barium and a comparison with related organolanthanide species. Organometallics 1990, 9, 1128–34.

[16] Williams RA, Hanusa TP, Huffman JC Solid state structure of bis (pentamethylcyclopentadienyl)barium, (Me$_5$C$_5$)$_2$Ba; the first X-ray crystal structure of an organobarium complex. J Chem Soc Chem Commun 1988, 1045–46.

[17] Ihanus J, Hänninen T, Hatanpää T, Aaltonen T, Mutikainen I, Sajavaara T, Keinonen J, Ritala M, Leskela M Atomic Layer Deposition of SrS and BaS Thin Films Using Cyclopentadienyl Precursors. Chem Mater 2002, 14, 1937–44.

[18] Kazhdan D, Hu Y-J, Kokai A, Levi Z, Rozenel S (2,2-Bipyridyl)bis(η5-pentamethylcyclopentadienyl)strontium(II). Acta Crystallogr Sect E Struct Reports Online 2008, 64, m1134–m1134.

[19] Kazhdan D, Rozenel S (2,2'-Bipyridyl-к-2N,N')bis(η5-pentamethylcyclopentadienyl)barium. Acta Crystallogr Sect E Struct Reports Online 2013, 69, m429–m429.

[20] Green DC, Englich U, Ruhlandt-Senge K. Calcium, Strontium, and Barium Acetylides - New Evidence for Bending in the Structures of Heavy Alkaline Earth Metal Derivatives. Angew Chem Int Ed 1999, 38, 354–57.

[21] Sroor FM, Vendier L, Etienne M Cyclooctatetraenyl calcium and strontium amido complexes. Dalton Trans 2018, 47, 12587–95.

[22] Walter MD, Wolmershäuser G, Sitzmann H. Calcium, Strontium, Barium, and Ytterbium Complexes with Cyclooctatetraenyl or Cyclononatetraenyl Ligands. J Am Chem Soc 2005, 127, 17494–503.

7 Organometallic compounds of group 13 metals

In spite of the fact that most treatises and textbooks include organoboron compounds in the organometallic section, organoboron compounds will not be discussed here. The nonmetallic character of boron is strongly evident, and its properties are quite different from those of the rest of group 13 elements. Boron chemistry is rather unique, and like carbon, boron deserves to have its own discipline and (although we are boron enthusiasts) we will not include it in this book (maybe to the dismay of some readers, for which we apologize).

7.1 Organoaluminum compounds

The composition and structure of group 13 organometallic compounds can be rationalized on the basis of the electronic structure shown in Fig. 7.1.

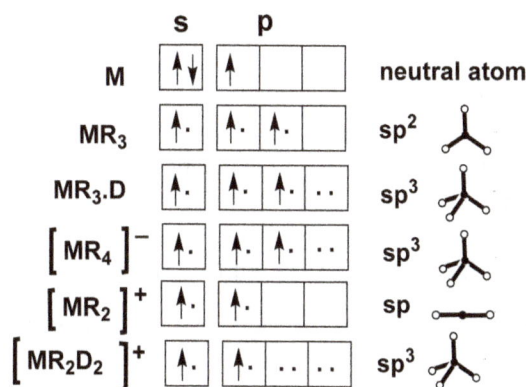

Fig. 7.1: The use of valence orbitals and electrons in the compounds of group 13.

The organometallic chemistry of group 13 elements is determined by the three valence electrons and the vacant p orbitals which tend to be occupied by participation in an sp^2 hybridization to form homoleptic AlR_3 compounds, and by accepting additional electrons to form tetrahedral $[MR_4]^-$ hypervalent anions and $MR_3.D$ adducts. These are isoelectronic with group 14 compounds.

7.1.1 Homoleptic compounds, AlR_3

Several methods are available for the preparation of triorganoaluminum compounds.
Trimethylaluminum is obtained industrially in the reaction of aluminum powder with methyl chloride and sodium metal in an autoclave:

https://doi.org/10.1515/9783110695274-007

$$Al + 3\ MeCl + 3\ Na \longrightarrow AlMe_3 + 3\ NaCl$$

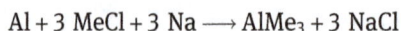

Other large-scale (industrial) preparations include direct synthesis from aluminum metal powder, olefins and hydrogen under pressure (50–200 bar) and heating (110–160 °C):

$$Al + 3/2\ H_2 + 3\ R\text{-}CH{=}CH_2 \longrightarrow Al(CH_2CH_2R)_3$$

For aromatic derivatives, the reaction of organomercury compounds with aluminum metal is the method of choice:

$$2Al + 3HgR_2 \longrightarrow 2AlR_3 + 3Hg$$

The olefins can add to aluminum hydride (AlH_3) or lithium alanate ($Li[AlH_4]$) to form R_2AlH, AlR_3 or $Li[AlR_4]$ (hydroalumination) depending upon the reaction conditions. A similar reaction of $Bu^i{}_2AlH$ with acetylenes gives a *cis*-isomer (Fig. 7.2).

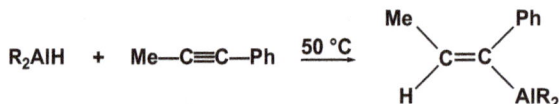

Fig. 7.2: Addition of aluminum hydrides to acetylenes.

The addition of olefins to aluminum trialkyls increases the chain length by insertion of the olefin and may finally result in the polymerization of the alkene:

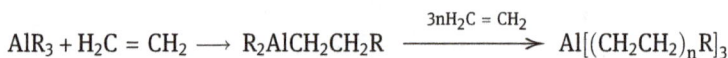

$$AlR_3 + H_2C = CH_2 \longrightarrow R_2AlCH_2CH_2R \xrightarrow{3nH_2C\,=\,CH_2} Al[(CH_2CH_2)_nR]_3$$

The phenyl group from $Al(C_6H_5)_3$ can be cleaved by the acidic hydrogen of acetylenes at moderate temperature (25–50 °C):

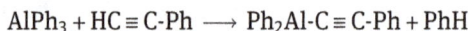

$$AlPh_3 + HC \equiv C\text{-}Ph \longrightarrow Ph_2Al\text{-}C \equiv C\text{-}Ph + PhH$$

Cyclopentadienyl derivatives of aluminum (also gallium and indium), $R_2MC_5H_5$, are prepared from the dialkylmetal chlorides and sodium cyclopentadienide:

$$R_2AlCl + NaC_5H_5 \longrightarrow R_2AlC_5H_5 + NaCl$$

When aluminum is a heteroatom in a ring, the dimerization is prevented. Such heterocycles can be formed by internal proton abstraction from biphenyl derivatives on heating (Fig. 7.3).

Trimethylaluminum and other alkyls with nonbulky substituents dimerize through dielectronic three-center bonds and the aryl derivatives by sharing the phenyl rings. Only AlR_3 compounds with R = *iso*-Pr, *tert*-Bu, C_6F_5, C_6Me_5 and $C_6H_3Me_3$-2,4,6 are planar monomeric molecules (sp^2 hybridization) due to steric reasons. Trimethylaluminum is monomeric only in the vapor phase, with a planar molecule based on sp^2 hybridization.

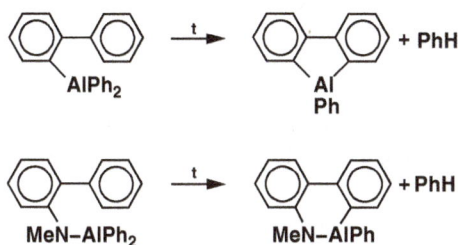

Fig. 7.3: Aluminum as a heteroatom in a ring.

Ethyl groups can also form tricenter bielectron bridges, resulting in the formation of triethylaluminum dimers $[Et_2Al(\mu\text{-}CH_2CH_3)]_2$. Dimethyl(pentafluorophenyl) aluminum is a dimer formed with bridging tricenter bielectronic methyl bridges.

Another type of compounds with three aluminum–carbon bonds are the methylene-bridged dialuminum derivatives $R_2Al\text{-}CH_2\text{-}AlR_2$, for example, with $R = CH(SiMe_3)_2$ or $C_6H_2Pr^i_3\text{-}2,4,6$ (Fig. 7.4).

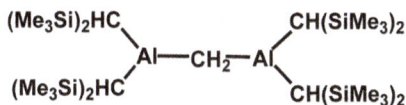

Fig. 7.4: Methylene-bridged dialuminum compound.

The triorganoaluminum derivatives are very sensitive to oxygen (sometimes pyrophoric) and to compounds with active hydrogen (water, alcohols, acids, amines and thiols). Controlled reactions with such reagents can have a preparative value for the synthesis of several classes of organolauminum compounds of general composition R_2AlX. Other reactions are also possible (Fig. 7.5).

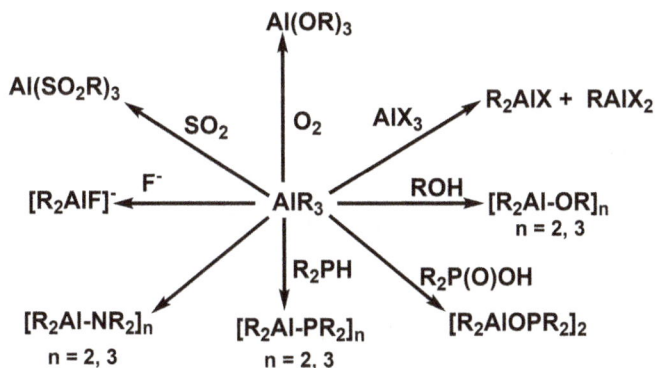

Fig. 7.5: Reactions of organoaluminum compounds.

7.1.2 Hypervalent species (anions), [AlR₄]⁻

Four-coordinated $[AlR_4]^-$ anions are formed when in the reactions of anhydrous aluminum halides with Grignard or organolithium reagents an excess of the later is used. Several tetrahedral anions $[AlR_4]^-$ with R = Me [1, 2], Et [3, 4] and Ph [5–7] are illustrations for this type.

The acetylenes react with lithium alanate (Li[AlH₄]) and form tetrasubstituted organoaluminum anions with hydrogen elimination:

$$4RC \equiv CH + LiAlH_4 \longrightarrow Li[Al(C \equiv CR)_4] + 4H_2$$

Metallation of benzene is possible with Na[AlEt₄] in the presence of sodium ethylate with formation of the tetraphenylalanate anion:

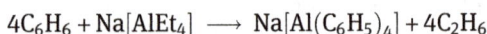

$$4C_6H_6 + Na[AlEt_4] \longrightarrow Na[Al(C_6H_5)_4] + 4C_2H_6$$

Cyclopentadiene, thiophene and furan react similarly on heating with M[AlH₄] to form anionic tetraorgano derivatives.

A number of organoaluminum compounds with four metal–carbon bonds are formed by addition of heterocyclic carbenes to the metal, as in the following adducts of trimethylaluminum [8, 9] and tris(pentafluorophenyl)-aluminum [10] with heterocyclic carbenes (Fig. 7.6).

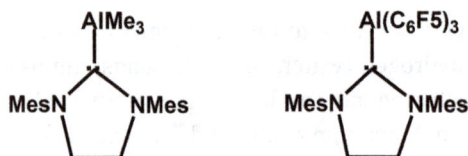

Fig. 7.6: Organoaluminum compounds with heterocyclic carbine donors.

7.1.3 Subvalent cations, [MR₂]⁺

The subvalent $[MR_2]^+$ cations are isoelectronic with group 12 species and are stabilized by coordination of two donor atoms to form $[MR_2.2D]^+$ cations, but $[TlR_2]^+$ which are isoelectronic with HgR₂ (a compound of group 12) are stable.

7.1.4 Triorganoaluminum donor adducts, AlR₃.D

Triorganoaluminum compounds are coordinatively unsaturated, and the tendency to occupy the fourth valence orbital of the metal results in the formation of AlR₃.D adducts with various donors. There are numerous such binary compounds, for example, $AlMe_3.NH_3$, $AlMe_3.NMe_3$, $AlMe_3.MeCN$, $AlMe_3.PPh_3$, $AlMe_3.THF$, $AlPh_3$. Et_2O, $AlPh_3.THF$, $AlPh_3.PMe_3$, $AlEt_3.E(CH_2SiMe_3)_3$ with $E = P$ or As, $Bu^t_3Al.PPr^i_3$, Al $(C_6F_5)_3.H_2O$, $Al(C_6F_5)_3.MeOH$, $Al(C_6F_5)_3.THF$, $Al(C_6F_5)_3.amine$ (amine = Me_2NH, Me $(PhCH_2)NH$, piperidine, $[Al(C_6F_5)_3X]^-$ (X = Cl, Br)) and many more, with tetrahe-drally coordinated aluminum (sp^3 hybridization).

Higher coordination numbers are common in inorganic coordination aluminum compounds, for example, five in $AlH_3.bipy$, six in $[AlF_6]^{3-}$ and $Al(acac)_3$, but are much rarer in organometallic compounds. Examples can be cited: the adducts of trimethylaluminum with 2,2-bipyridyl and with tetramethyltetrazene, in which the metal is five-coordinated (Fig. 7.7).

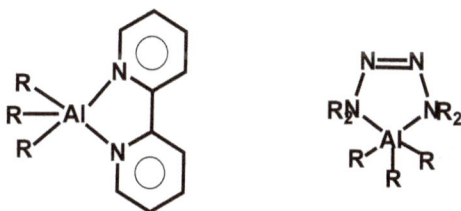

Fig. 7.7: Five-coordinated aluminum in nitrogen complexes.

The reaction of anhydrous aluminum trihalides with Grignard reagents produces adducts with diethyl ether, the solvent commonly used:

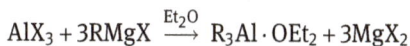

$$AlX_3 + 3RMgX \xrightarrow{Et_2O} R_3Al \cdot OEt_2 + 3MgX_2$$

The adducts are also formed in the reaction of mixtures of aluminum and magnesium with organic halides in diethyl ether, without previous preparation of a Grignard reagent:

$$2Al + 3Mg + 6RX + Et_2O \longrightarrow 2R_3Al \cdot OEt_2 + 3MgX_2$$

A compound with three aluminum–carbon bonds can be obtained by the addition of a heterocyclic carbene to a dimethylaluminum group, to form in diethyl ether an adduct with four-coordinated metal [11] (Fig. 7.8).

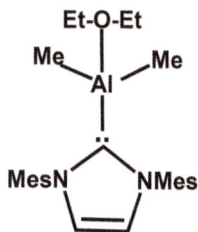

Fig. 7.8: An adduct with heterocyclic carbine.

Planar triorganoaluminum molecules can add two donor ligands to form complexes with five-coordinated metal centers. Examples are [Al(C₆F₅)₃.2MeCN] [12] and [Al(C₆F₅)₃{FAl(C₆F₅)₃}₂] [13] (Fig. 7.9).

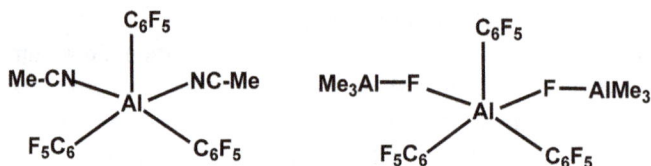

Fig. 7.9: Five-coordinated aluminum.

7.1.5 Inverse organoaluminum compounds

When a number of organoaluminum moieties are attached to a central carbon molecule, compounds described as inverse organometallic will be obtained. An illustrative selection is presented here.

One such compound is a tetrahedral molecule of tetrakis(dimethylaluminum) methane [C(AlMe₂)₂(AlMeCl₂)₂] [14] (Fig. 7.10).

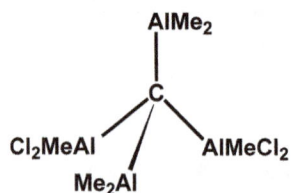

Fig. 7.10: Carbon-centered inverse organometallic compound of aluminum.

Other examples are the dihydro-9,10-anthrylene derivative [15] and the 1,2,4,5-tetrafluorophenylene compound (Fig. 7.11) [16].

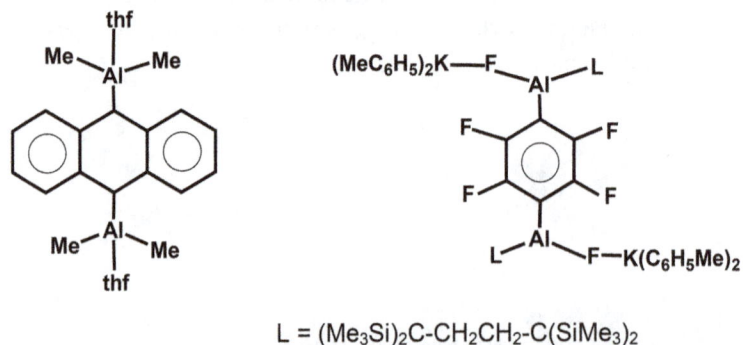

L = (Me₃Si)₂C-CH₂CH₂-C(SiMe₃)₂

Fig. 7.11: Inverse organoaluminum compounds with ring centers.

7.1.6 Organoaluminum inverse coordination complexes

The coordinative unsaturation of AlR_3 molecules is the source of some binuclear inverse coordination complexes, where an electron donor anion acts as a bridge between two metal atoms and functions as a coordination center in compounds of the general type $[R_3Al-X-AlR_3]^-$. The Al–X bonds are normal donor–acceptor bonds and examples can be cited with X = F^-, Cl^-, OH^-, PhO^-, PhS^-, N_3^-, NO_3^-, with R = alkyl, aryl or C_6F_5 [17, 18]. The complexes $[(\mu-F)(AlMe_3)_2]^-$ and $[(\mu-F)\{Al(C_6F_5)_3\}_2]^-$ are linear (Al–F–Al angle 180°) but $[(\mu-Cl)(AlMe_3)_2]^-$ [19] and $[(\mu-E)\{Al\{HC(SiMe_3)\}_2\}_2]^{2-}$ with E = S, Se, Te are bent (with Al–E–Al angles in the range 110–117°), unlike $[(\mu-O)\{HC(SiMe_3)_2\}_2]^{2-}$ which is linear (Fig. 7.12) [20].

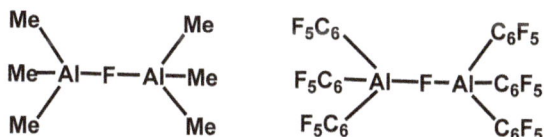

Fig. 7.12: Fluorine-centered organoaluminum inverse coordination complexes.

The triorganoaluminum molecules are excellent acceptors and with polytopic donor molecules, which can act as centroligands, they form a broad variety of inverse coordination complexes, that is, compounds in which a nonmetallic centroligand is connected to two or more acceptor organoaluminum moieties. Trimethylaluminum as electron pair acceptor forms a series of inverse coordination complexes by attachment to various ditopic nitrogen donors acting as centroligands, for example, pyrazine, 4,4′-bipyridine [21] and heterocycles like 1,3,5-trimethyl-1,3,5-triazinane and 1,4-diazabicyclooctane [22]. An inverse coordination complex with urotropine (hexamethylenetetramine) centroligand and three $AlMe_3$ addends has been described (Fig. 7.13). The sequential compounds with one and two $AlMe_3$ molecules were also obtained in the work cited [23]

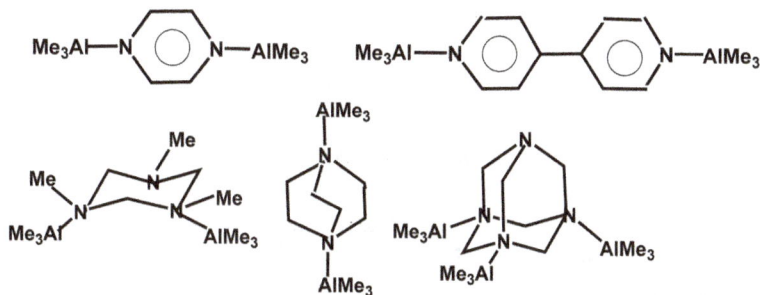

Fig. 7.13: Organoaluminum inverse coordination complexes.

Some spectacular compounds are the inverse coordination complexes with crown ethers or azacrowns as centroligands and trimethylaluminum addends (Fig. 7.14).

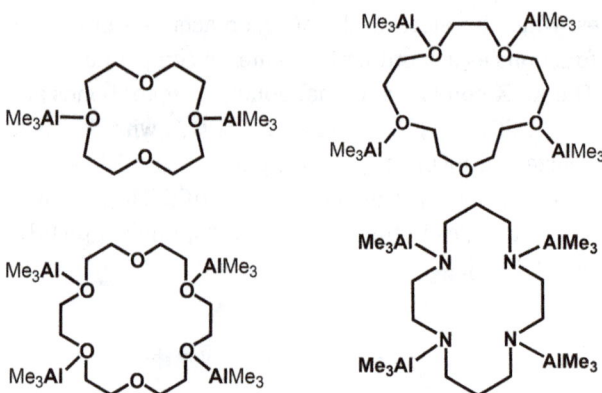

Fig. 7.14: Organoaluminum inverse coordination complexes with crown ethers as centroligands.

7.1.7 Organoaluminum R$_2$AlH hydrides

The alkylaluminum hydrides are associated supermolecules. In solution, the R$_2$AlH compounds are cyclic trimers formed through Al $\cdots$ H $\cdots$ Al electron-deficient bridges, while in the vapor phase they are dimers, for example, with R = Me (Fig. 7.15).

Fig. 7.15: Alkylaluminum hydrides.

7.1.8 Organoaluminum R$_2$AlX halides

The diorganoaluminum halides (R$_2$Al-X) are supramolecular oligomers, formed by association through donor–acceptor (coordinative) bonds, as cyclic dimers, trimers, tetramers and other oligomers with four-coordinated aluminum.

The diorganoaluminum chlorides ([R$_2$AlCl]$_2$) and sesquichlorides (R$_3$Al$_2$Cl$_3$) are cyclic dinuclear compounds, while the fluorides are cyclic tetramers, [R$_2$AlF]$_4$ (Fig. 7.16).

Alkylaluminum halides are formed in the reactions of aluminum trialkyls with anhydrous aluminum or zinc halides by substituent redistribution:

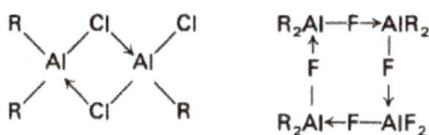

Fig. 7.16: Diorganoaluminum halides.

$$2AlR_3 + AlX_2 \longrightarrow 2R_2AlX$$

$$2AlR_3 + ZnX_2 \longrightarrow 2R_2AlX + ZnR_2$$

Equimolecular mixtures of R_2AlX and $RAlX$ are formed in the direct reaction of aluminum metal with alkyl halides, and these are associated as dimetallic sesquihalides:

$$3RX + 2Al \longrightarrow R_2AlX + RAlX_2 \longrightarrow R_3Al_2X_3$$

Dialuminum sesquichloride is also obtained from aluminum metal, aluminum trichloride, olefins and hydrogen under heating and pressure:

$$Al + AlCl_3 + 3C_2H_4 + 1,5H_2 \longrightarrow (C_2H_5)_3Al_2Cl_3$$

These reactions are applied for the industrial production of polymerization catalyst.

7.1.9 Organoaluminum functional derivatives (alkoxides, thiolates and amides)

The diorganoaluminum compounds (R_2Al-X, where X = OR, NHR, NR_2, PR_2, AsR_2, SR or SeR), prepared from trialkyl with alcohols, thiols, selenols and amines, are supramolecular oligomers containing four-coordinated aluminum, $[R_2Al$-$ER]_n$.

In diorganoaluminum compounds $[R_2Al$-$ER]_n$ the metal is four-coordinated. The association degree, that is, formation of dimers, trimers or tetramers, depends on the size of the functional organic groups. When in the OR', SR' or NR'_2 functional groups, R' is methyl, the trimers can be formed. With larger groups, dimers are more common.

Monoorganoaluminum groups with primary amines form tetrameric cubane structures $[RAl$-$NR']_4$, but hexamers $[RAl$-$NR']_6$, heptamers $[RAl$-$NR']_7$ and octamers $[RAl$-$NR'])_8$ have also been reported.

Diorganoaluminum compounds with a donor functional group substituent undergo self-assembly with the formation of supramolecular cyclic dimers $[Me_2Al$-CH_2 $NPr^i_2]_2$ and tetramers $[Bu^t_2Al$-$CN]_4$ (Fig. 7.17).

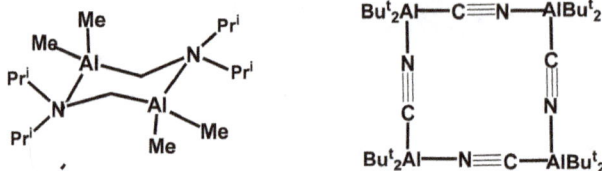

Fig. 7.17: Cyclic organoaluminum supermolecular oligomers.

7.1.10 Compounds with Al–Al bonds

Compounds with Al–Al bonds are rare but a few examples with bulky substituents can be cited, like $\{(Me_3Si)_2CH\}_2Al-Al\{CH(SiMe_3)_2\}_2$ and $(C_6H_2Pr^i_3)_2Al-Al(C_6H_3Pr^i_3)_2$ [24, 25] (Fig. 7.18).

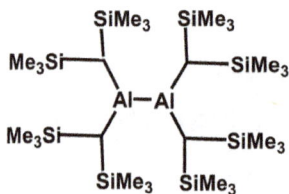

Fig. 7.18: Structure of a dialane.

The first univalent aluminum compound that is stable at room temperature is [AlCp*]$_4$, which contains π-bound C_5Me_5 [26, 27]. It was prepared by the treatment of [AlCl]$_x$ with MgCp*$_2$ or better by reductive dehalogenation of [{Cp*AlCl(μ-Cl)}$_2$] with potassium. Other neutral organoaluminum(I) compounds of this type, namely, [Al-C(SiMe_3)_3]$_4$ [28] and [Al-SiBu$^t_3$]$_4$ [29] have been reported. The former was prepared by the reaction of $(Me_3Si)_3CAlI_3$.THF with Na/K alloy.

7.2 Organogallium compounds

There is a similarity of composition and structure between organogallium and orga-noaluminum compounds but the chemical reactivities can be different.

7.2.1 Homoleptic derivatives, GaR$_3$

Triorganogallium derivatives can be obtained by halogen-alkyl exchange with Grignard reagents, organolithium, organozinc or organoaluminum compounds and from gallium metal with organomercury compounds. In basic solvents (e.g., diethyl ether), they are obtained as solvates:

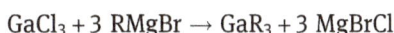

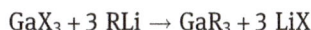

$$GaCl_3 + 3\ RMgBr \rightarrow GaR_3 + 3\ MgBrCl$$

$$GaX_3 + 3\ RLi \rightarrow GaR_3 + 3\ LiX$$

$$2\ GaX_3 + 3\ ZnR_2 \rightarrow 2\ GaR_3 + 3\ ZnX_2$$

$$GaX_3 + AlR_3 \rightarrow GaR_3 + AlX_3$$

$$2\ Ga + 3\ HgMe_2 \rightarrow 2\ GaMe_3 + 3\ Hg$$

Note: The reaction with dimethylmercury should be avoided because of the extreme toxicity of this compound and the hazards associated with its use!

Unlike their trialuminum analogues, the triorganogallium compounds are mono-meric, do not form bridged supramolecular dimers and are less reactive. The Lewis acidity decreases in the order Al > Ga > In, and this influences the comparative chemistry of these elements.

The cyclopentadienyl derivative, $GaMe_2C_5H_5$, forms supramolecular chains of Me_2Ga units bridged by C_5H_s rings in the solid state.

Triorganogallanes react with water, alcohols, amines, and other active hydrogen reagents to give the functional derivatives, R_2GaX with X = OH, OR, NR_2, SR, PR_2, AsR_2, etc., which are associated in the solid state with supramolecular arrays and in solution with cyclic dimeric, trimeric or tetrameric supermolecular oligomers.

The lower gallium trialkyls are pyrophoric, and the higher members of the fam-ily fume in air. The ethers, amines, phosphines and thioethers form adducts in which the coordination number of gallium has been increased to four. Their stabil-ity decreases in the order N > P > As. With active hydrogen compounds, including acetylenes, they eliminate a hydrocarbon and form functional derivatives, which are usually associated as cyclic or linear $(R_2Ga\text{-}ER')_n$, supramolecular structures.

A general overview of the typical reactions of triorganogallium derivatives is presented in Fig. 7.19.

7.2.2 Hypervalent anions

In the reaction with organolithium reagents, the 3:1 molar ratio between LiR and $GaCl_3$ must be observed, and an excess produces salts of tetrasubstituted anions $Li^+[GaR_4]^-$. Otherwise, a deficit of RLi gives disubstituted R_2GaI. Several tetraorga-nogallato anions are known, where R = CH_3, C_2H_5, CH_2SiMe_3, CF_2CF_3 and C_6F_5.

A unique compound with four gallium–carbon bonds is a supramolecular chain-like $C_5H_5GaEt_2$ [30] (Fig. 7.20).

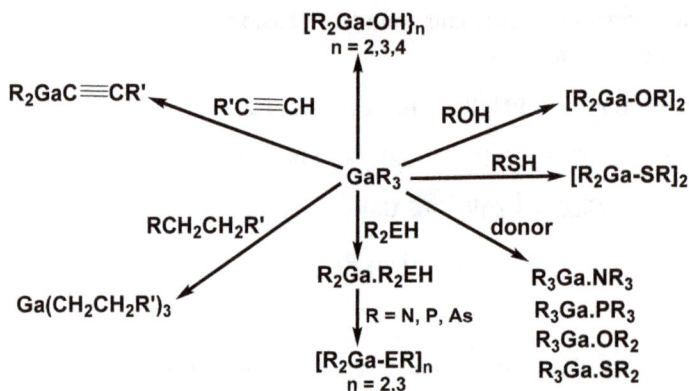

Fig. 7.19: Typical reactions of triorganogallium compounds.

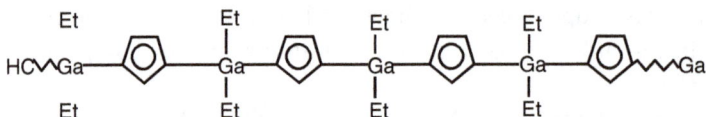

Fig. 7.20: Supramolecular chain of $C_5H_5GaEt_2$.

Other compounds with four gallium–carbon bonds are the adducts of heterocyclic carbenes to triorganogallium alkyls [31, 32] (Fig. 7.21).

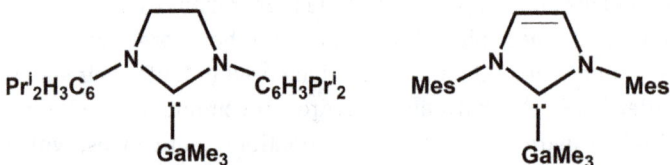

Fig. 7.21: Triorganogallium adducts of heterocyclic carbenes.

7.2.3 Subvalent cations

The subvalent diorganogallium cations are stabilized and known only as four-coordinated complexes with ammonia or ethylenediamine, for example, [Me$_2$Ga(NH$_3$)$_2$]$^+$ and [Me$_2$Ga(H$_2$NCH$_2$CH$_2$NH$_2$)]$^+$.

7.2.4 Inverse organogallium compounds

Inverse organometallic compounds of gallium are scarce. Examples are $(GaMe_2)_2$ C_6H_4-1,4 [33] and $(\mu_4$-C)$[Ga(CH(SiMe_3)_2)(PhPHO_2)]_4$ [34] (Fig. 7.22).

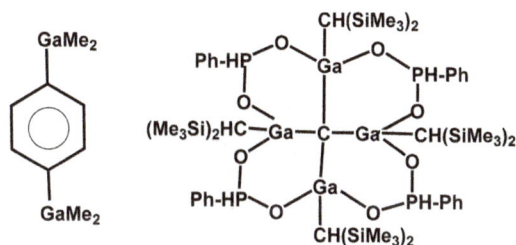

Fig. 7.22: Inverse organogallium compounds.

7.2.5 Diorganogallium halides, R_2GaX

Diorganogallium monohalides are prepared by cleavage of a Ga−R bond from GaR_3 compounds with hydrogen halides or halogens, by alkylation of gallium trihalides with organolithium reagents and by redistribution between GaR_3 and gallium trichloride.

The triorganogallium monohalides, except the fluorides, are dimeric in vapor phase and solution. The fluorides are supramolecular cyclic trimers or tetramers (Fig. 7.23).

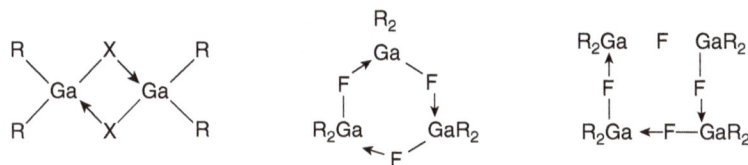

Fig. 7.23: Cyclic supermolecular oligomers of organogallium halides.

The monoorganogallium halides ($RGaX_2$) are less stable. They are prepared by substitution in triorganohalides with mild alkylating reagents (e.g., SiR_4, GeR_4, SnR_4 or ZnR_2), by redistribution between GaR_3 and GaX_3, by cleavage of Ga−R bonds in GaR_3 with anhydrous HCl, or by addition of $HGaCl_2$ to olefins.

A variety of hypervalent gallium anions, with tetrahedral geometry at the metal, $[R_3GaX]^-$ (with R = Me, Et; X = F, Br), $[R_2GaCl_2]^-$, $[RGaCl_3]^-$ and $[R_3Ga\text{-}X\text{-}GaR_3]^-$ are known, with lesser visibility.

7.2.6 Diorganogallium hydroxides, R$_2$GaOH

Dimethylgallium hydroxide displays an unprecedented diversity of structures, ranging from cyclic trimeric, tetrameric, hexameric supermolecular oligomers to a monodimensional helical polymolecular chain, depending on the solvent used for recrystallization [35] (Fig. 7.24).

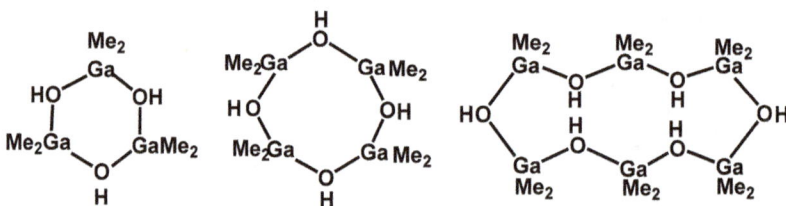

Fig. 7.24: Cyclic supramolecular oligomers of dimethylgallium hydroxide.

7.2.7 Diorganogallium functional derivatives

Among the diorganogallium functional derivatives are dimeric alkoxides [R$_2$Ga-OR′]$_2$, mercaptides [R$_2$Ga-SR′]$_2$, amides [R$_2$Ga-NR′$_2$]$_2$, phosphides, R$_2$Ga-PR′$_2$]$_2$ and arsenides [R$_2$Ga-AsR$_2$]$_2$. The compounds with small organic substituents, that is, methyl groups, are trimeric [Me$_2$GaPMe$_2$]$_3$ and [Me$_2$Ga-AsMe$_2$]$_3$ (Fig. 7.25).

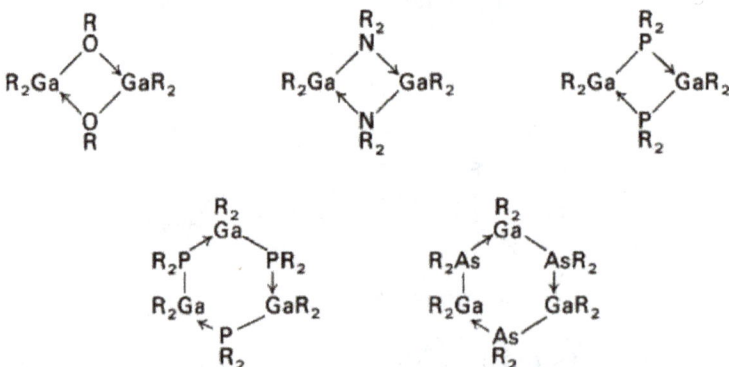

Fig. 7.25: More supramolecular organogallium cyclic oligomers.

7.2.8 Organogallium inverse coordination complexes

Two interesting inverse coordination complexes are formed by addition of trimethylgallium to nitrogen centroligands. Sequential addition of GaMe$_3$ molecules to hexamethylenetetramine (urotropine) affords the isolation of all members of the series $(CH_2)_6N_4 \cdot nGaMe_3$ with $n = 1$–4, the tetragallium compound being illustrated here. Another aesthetically attractive inverse coordination complex is formed by addition of triethylgallium to a hexaaza macrocycle [36] (Fig. 7.26).

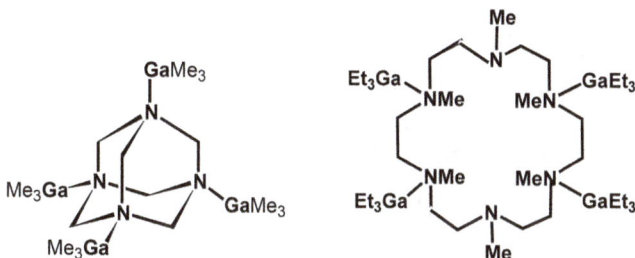

Fig. 7.26: Inverse coordination complexes of trialkylgallium.

7.3 Organoindium compounds

Organoindium chemistry is reminiscent of organoaluminum and organogallium chemistry.

7.3.1 Homoleptic compounds, InR$_3$

Trisubstituted derivatives, InR$_3$, are obtained by treatment of indium trihalides with organomagnesium, aluminum or lithium reagents and from the reaction of indium metal with organomercury compounds. The lower indium trialkyls are pyrophoric, are readily oxidized and react vigorously with water and active hydrogen compounds.

The structure of trimethylindium in the vapor phase is monomeric and planar, but in the solid state it is made of tetrameric units $(In_4(CH_3)_{12})$ formed through weak electron-deficient methyl bridges. Triphenylindium, on the other hand, is monomeric in the solid state, like triphenylgallium.

A monovalent organoindium compound η^5-cyclopentadienylindium, η^5-C$_5$H$_5$In, was prepared from indium trichloride and sodium cyclopentadienide. The solid state structure has an ionic character made of In$^+$ and C$_5$H$_5{}^-$ ions.

The triorganoindium compounds form tetracoordinated adducts, R$_3$In $\cdot$ D, with D = OR$_2$, SR$_2$, NR$_3$, etc.

7.3.2 Diorganoindium halides, R$_2$InX

Diorganoindium halides (R$_2$InX) are formed in reactions of indium trihalides with Grignard or organolithium reagents in appropriate ratios, or by cleavage of an In–R bond from InR$_3$ with halogens or haloforms. Organoindium dihalides, RlnX$_2$, are less well known. Phenyl derivatives can be prepared by the action of the halogens upon triphenylindium. Both mono- and dihalides, Ph$_n$InX$_{3-n}$ ($n = 1$, 2), are prepared by oxidative arylation of indium(I) halides with diphenylmercury to give Ph$_2$InX and PhInX$_2$.

The halides, Me$_2$InX and (C$_6$F$_5$)$_2$InX, form dimers in the vapor phase and in solution through halogen bridges with indium becoming four-coordinated. The phenyl derivatives (Ph$_2$InX) are associated in supramolecular chains through halogens but PhInI$_2$ is unexpectedly an ionic compound, [Ph$_2$In]$^+$[InI$_4$]$^-$.

7.3.3 Functional diorganoindium derivatives

Functional derivatives, associated as supramolecular cyclic oligomers (R$_2$In–X)$_n$, are known with X = OR ($n = 2$ and 3), NR$_2$ ($n = 2$), PR$_2$ ($n = 2$ and 3) and AsR$_2$ ($n = 2$ and 3). Dimer formation is general, but when R is small, cyclic trimers can also be formed (Fig. 7.27).

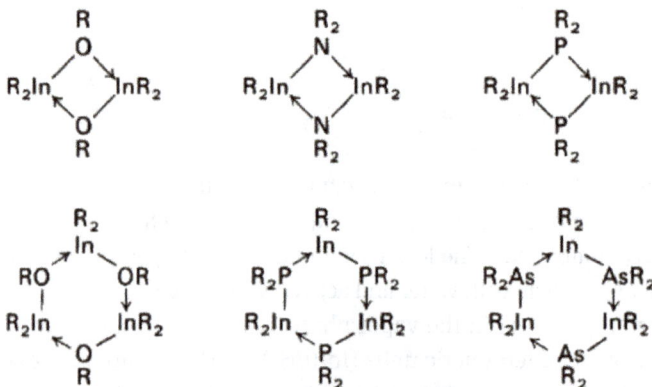

Fig. 7.27: Cyclic supramolecular organoindium oligomers.

7.4 Organothallium compounds

There are fundamental differences between organothallium chemistry and the organometallic chemistry of aluminum, gallium and indium, reflected in the unusual stability of the disubstituted derivatives, [R$_2$Tl]$^+$X$^-$. The cations [R$_2$Tl]$^+$ are isoelectronic with the diorganomercury compounds, R$_2$Hg, and possess linear structures. The trisubstituted derivatives (TlR$_3$), particularly the trialkyls, are relatively unstable.

7.4.1 Homoleptic derivatives, TlR$_3$

Trisubstituted derivatives can be prepared from thallium(III) chloride and Grignard reagents in tetrahydrofuran; in ether, only the disubstituted derivatives (R$_2$TlX) are formed. Thallium(I) iodide reacts with organolithium reagents in the presence of alkyl iodides in ether through the intermediacy of organothallium(I) compounds which disproportionate to organothallium(III) compounds and thallium metal:

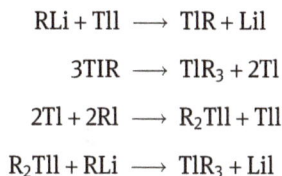

$$RLi + TlI \longrightarrow TlR + LiI$$

$$3TlR \longrightarrow TlR_3 + 2Tl$$

$$2Tl + 2RI \longrightarrow R_2TlI + TlI$$

$$R_2TlI + RLi \longrightarrow TlR_3 + LiI$$

Thallium metal has been formed in the synthesis of triphenylthallium from thallium(I)chloride and phenyllithium, confirming the intermediate formation of a monovalent thallium derivative:

$$3TlCl + 3PhLi \longrightarrow 3TlPh \longrightarrow TlPh_3 + 2Tl$$

Since R$_2$TlX compounds are readily obtained, their alkylation by organolithium reagents is convenient and can produce unsymmetrically substituted compounds, R$_2$TlR':

$$R_2TlX + R'Li \longrightarrow TlR'R_2 + LiX$$

Hypervalent tetraorganosubstituted thallium compounds are rare. Acetylene derivative anions $[Tl(C \equiv CR)_4]^-$ are obtained from TlCl$_3$.4NH$_3$ and sodium acetylides in liquid ammonia. Other tetrasubstituted derivatives are $[Tl(C_6F_5)_4]^-$ and $[Tl(C_6F_5)_2(C_6Cl_5)_2]^-$ anions:

$$(C_6X_5)_2TlBr + 2LiC_6X_5 \xrightarrow{\text{NBu}_4\text{Br}} [NBu_4]^+ [Tl(C_6X_5)_2]^-$$

$$X = F, Cl$$

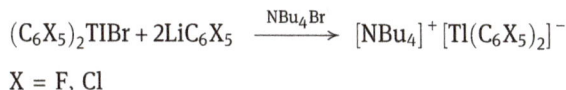

The trisubstituted TlR$_3$ derivatives with lower alkyl groups are pyrophoric; the others are also sensitive to oxygen, water and active hydrogen compounds. They pyrolyze more readily than their gallium and indium analogues, presumably *via* free radicals as suggested by the formation of biphenyl in the thermal decomposition of triphenylthallium.

The TlR$_3$ compounds show only weak acceptor properties, but Me$_3$Tl · NMe$_3$ and TlMe$_3$ · PMe$_3$ have been isolated. TlMe$_3$ does not coordinate arsines and forms only a very weak adduct with diethyl ether. However, Tl(C$_6$F$_5$)$_3$ forms a stable diethyl ether adduct.

7.4.2 Monovalent organothallium compounds, TlR

The only stable organic derivative of monovalent thallium is η^5-cyclopentadienyl thallium (η^5-C$_5$H$_5$Tl) obtained by treatment of thallium(I) sulfate with bis(cyclopentadienyl) mercury in alcoholic alkalies. The compound sublimes in vacuo and is monomeric in the vapor phase. In the molecule, the thallium ion is located above the C$_5$H$_5$ ring. In the solid state, the structure is supramolecular and consists of a chain-like array of alternating Tl$^+$ ions and C$_5$H$_5^-$ rings (Fig. 7.28).

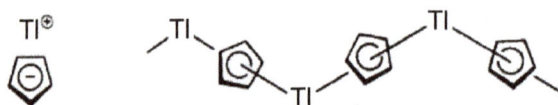

Fig. 7.28: Monomeric and supramolecular structure of cyclopentadienylthallium.

η^5-Cyclopentadienylthallium(I) is stable in air and readily transfers C$_5$H$_5$ groups to other metals. As such it is important as a reagent for the synthesis of transition metal cyclopentadienyls (metallocenes).

7.4.3 Diorganothallium halides, R$_2$TlX

The action of Grignard reagents on thallium trichloride stops at the disubstituted product, and yields are low because of the reducing effect of the organomagnesium compound upon trichloride. The bromides, R$_2$TlBr, are more readily obtained.

The reaction of thallium trichloride with organolithium and organomercury reagents can also be of use, and a surprising reaction of thallium trihalides with aryl-boronic acids produces diorganothallium halides:

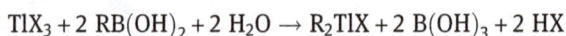

$$\text{TlX}_3 + 2\ \text{RB(OH)}_2 + 2\ \text{H}_2\text{O} \rightarrow \text{R}_2\text{TlX} + 2\ \text{B(OH)}_3 + 2\ \text{HX}$$

Some diorganothallium halides, for example, (C$_6$F$_5$)$_2$TlX are dimerized in solution via halogen bridges. The disubstituted halides (R$_2$TlX) are ionic in the solid state, are stable to 200–300 °C and are little soluble in organic solvents but dissolve readily in pyridine, owing to coordination. Thus, Me$_2$Tl–I and Me$_2$TlCl contain linear [Me–Tl–Me]$^+$ cations and halide anions.

7.4.4 Functional diorganothallium compounds, R$_2$TlX′

The functional derivatives (R$_2$TlX, where X = OMe, OEt, SMe, SeMe, NMe$_2$) obtained from halides with alkali-metal derivatives of alcohols, thiols or amines are dimers with cyclic structures (Fig. 7.29).

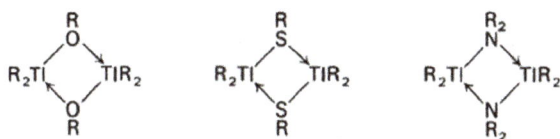

Fig. 7.29: Dimeric [R₂TlX]₂ supermolecules.

Dimethylthallium hydroxide, Me_2TlOH, is a strong base, which in water dissociates into $[Me_2Tl]^+$ and OH^- ions.

7.4.5 Functional monoorganothallium compounds, RTlX₂

An important reaction is the direct thallation reaction of aromatic compounds with thallium(III) carboxylates:

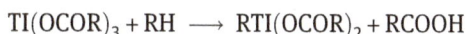

$$Tl(OCOR)_3 + RH \longrightarrow RTl(OCOR)_2 + RCOOH$$

Further reading

Anwar RA, Haque RS, Saleem Z, Iqbal MA Recent advances in synthesis of organometallic complexes of indium. Rev Inorg Chem 2020, 17, 107–51.

References

[1]   Wolf BM, Stuhl C, Maichle-Mössmer C, Anwander R Calcium tetraalkylaluminate and tetramethylgallate complexes supported by the bulky scorpionate ligand Tpt Bu,Me. Organometallics 2019, 38, 1614–21.
[2]   Nieland A, Mix A, Neumann B, Stammler H-G, Mitzel NW Cationic rare-earth-metal methyl complexes: A new preparative access exemplified for Y and Pr. Dalton Trans 2010, 39, 6753–60.
[3]   Michel O, Dietrich HM, Litlabø R, Törnroos KW, Maichle-Mössmer C, Anwander R. Tris (pyrazolyl)borate complexes of the alkaline-earth metals: alkylaluminate precursors and Schlenk-type rearrangements. Organometallics 2012, 31, 3119–27.
[4]   Sizov AI, Zvukova TM, Bulychev BM, Belsky VK Synthesis and properties of unsolvated bis (cyclopentadienyl)titanium alumohydride. Structure of {[(η5-$C_5H_5)_2$Ti(μ-H)]₂[(η⁵-C_5H_5)Ti(μ-H₂] Al_3(μ-H₄)(H)}₂·C_6H_6 a 12-nuclear titanium aluminum hydride complex with a short Al-Al bond length, and refined structure. J Organomet Chem 2000, 603, 167–73.
[5]   Pour N, Gofer Y, Major DT, Aurbach D Structural analysis of electrolyte solutions for rechargeable Mg batteries by stereoscopic means and DFT calculations. J Am Chem Soc 2011, 133, 6270–78.
[6]   Krieck S, Görls H, Westerhausen M. Synthesis and properties of calcium tetraorganylalanates with [Me₄₋ₙAlPhₙ]⁻ anions. Organometallics 2008, 27, 5052–57.

[7] Liu LL, Cao LL, Shao Y, Stephan DW Single electron delivery to Lewis pairs: an avenue to anions by small molecule activation. J Am Chem Soc 2017, 139, 10062–71.

[8] Wu MM, Gill AM, Yunpeng L, Falivene L, Yongxin L, Ganguly R, et al. Synthesis, structural studies and ligand influence on the stability of aryl-NHC stabilised trimethylaluminium complexes. Dalton Trans 2015, 44, 15166–74.

[9] Schmitt A-L, Schnee G, Welter R, Dagorne S Unusual reactivity in organoaluminium and NHC chemistry: Deprotonation of AlMe₃ by an NHC moiety involving the formation of a sterically bulky NHC-AlMe3 Lewis adduct. Chem Commun 2010, 46, 2480–82.

[10] Zhang Y, Miyake GM, Chen EY-X. Alane-based classical and frustrated Lewis pairs in polymer synthesis: rapid polymerization of MMA and naturally renewable methylene buthyrolactones into high-molecular-weight polymers. Angew Chem Int Ed 2010, 49, 10158–62.

[11] Schnee G, Bolley A, Gourlaouen C, Welter R, Dagorne S Synthesis and structural characterization of NHC-stabilized Al(III) and Ga(III) alkyl cations and use in the ring-opening polymerization of lactide. J Organomet Chem 2016, 820, 8–13.

[12] Shcherbina NA, Pomogaeva AV, Lisovenko AS, Kazakov IV, Gugin NY, Khoroshilova OV, et al. Structures and stability of complexes of E(C₆F₅)₃ (E = B, Al, Ga, In) with acetonitrile. Z Anorg Allg Chem 2020, 646, 873–81.

[13] Chen M-C, Roberts JAS, Seyam AM, Li L, Zuccaccia C, Stahl NG,, Marks TJ. Diversity in weakly coordinating anions. Mono- and polynuclear halo(perfluoroaryl)metalates as cocatalysts for stereospecific olefin polymerization: synthesis, structure, and reactivity. Organometallics 2006, 25, 2833–50.

[14] Wei P, Stephan DW Salts of the cation [(Cp*Cr)₄(μ-Cl)₃(μ-CH₂)₃AlMe]⁺ with the oxo- and methine-based aluminum anions [(Me₂Al)₂(μ-CH)(AlCl₂Me)₂]⁻ and [(Me₂Al)(μ₃-O)(AlCl₂Me) (AlMe₂Cl)]. Organometallics 2003, 22, 1992–94.

[15] Lehmkuhl H, Mehler K, Shakoor A, Krüger C, Tsay Y-H, Benn R, Rufinska A, Schroth G 9,10-Dihydro-9,10-anthrylen-Verbindungen des Aluminiums. Chem Ber 1985, 118, 4248–58.

[16] Kurumada S, Takamori S, Yamashita M An alkyl-substituted aluminium anion with strong basicity and nucleophilicity. Nature Chem 2020, 12, 36–39.

[17] Atwood JL, Newberry WR The interaction of aromatic hydrocarbons with organometallic compounds of the main group elements. J Organomet Chem 1974, 66, 15–21.

[18] Caputo CB, Hounjet LJ, Dobrovetsky R, Stephan DW Lewis acidity of organofluorophosphonium salts: hydrodefluorination by a saturated acceptor. Science 2013, 341, 1374–77.

[19] Wei P, Stephan DW Cationic and neutral phosphido-bridged pentamethylcyclopentadienyl –chromium dimers. Organometallics 2003, 22, 1712–17.

[20] Uhl W, Gerding R, Hahn I, Pohl S, Saak W, Reuter H The insertion of chalcogen atoms into Al-Al and Ga-Ga bonds. Polyhedron 1996, 15, 3987–92.

[21] Ogrin D, van Poppel LH, Barron AR Effects of solvent on the relative stability of mono and di-aluminium aryloxide complexes of bipyridines: Anomalous behavior of [(ᵗBu)₂Al(OPh)]₂(μ-4,4-bipy). Dalton Trans 2005, 1722–26.

[22] Bradford AM, Bradley DC, Hursthouse MB, Motevalli M Interactions of 1,4-diazabicyclo[2.2.2] octane with Group III metal trimethyls: Structures of Me₃M.N(C₂H₄)₃N.MMe₃ (M = aluminum, gallium). Organometallics 1992, 11, 111–15.

[23] Hill JB, Eng SJ, Pennington WT, Robinson GH Reaction of Me₃Ga and Me₃Al with the tertiary-tetraaza analog of adamantane, hexamethylenetetramine (N4-Ada). Syntheses and molecular structures of [(Me₃M)ₙN₄-Ada] (n = 1–4, M = Ga or Al). J Organomet Chem 1993, 445, 11–18.

[24] Pluta C, Pörschke K-R, Krüger C, Hildenbrand K. An Al . . . Al one-electronπ bond. Angew Chem Int Ed 1993, 32, 388–90.

[25] Wehmschulte RJ, Ruhlandt-Senge K, Olmstead MM, Hope H, Sturgeon BE, Power PP Reduction of a tetraaryldialane to generate aluminum-aluminum.pi-bonding. Inorg Chem 1993, 32, 2983–84.

[26] Dohmeier C, Robl C, Tacke M, Schnöckel H The tetrameric aluminum(I) compound [{Al(η^5-C$_5$Me$_5$)}$_4$]. Angew Chem Int Ed 1991, 30, 564–65

[27] Schulz S, Roesky HW, Koch HJ, Sheldrick GM, Stalke D, Kuhn A, Simple A Synthesis of [(Cp*Al)$_4$] and its conversion to the heterocubanes[(Cp*AlSe)$_4$] and [(Cp*AlTe)$_4$] (Cp* = η^5-C$_5$(CH$_3$)$_5$). Angew Chem Int Ed 1993, 32, 1729–31.

[28] Schnitter C, Roesky HW, Röpken C, Herbst-Irmer R, Schmidt H-G, Noltemeyer M. The behavior of [RAlX$_2 \cdot$ THF] compounds under reductive conditions: tetrakis[tris(trimethylsilyl) methylaluminum(I)] – a neutral aluminum(I) compound with σ-bound alkyl groups and a tetrahedral structure. Angew Chem Int Ed 1998, 37, 1952–55.

[29] Purath A, Dohmeier C, Ecker A, Schnöckel H, Amelunxen K, Passler T, Wiberg N. Synthesis and crystal structure of the tetraaluminatetrahedrane [Al$_4$(t-Bu)$_3$]$_4$, the second Al$_4$R$_4$ compound. Organometallics 1998, 17, 1894–96.

[30] Beachley OT, Rosenblum DB, Churchill MR, Lake CH, Krajkowski LM. Syntheses, characterization, crystal and molecular structures, and solution properties of Et$_2$Ga(C$_5$H$_5$) and EtGa(C$_5$H$_5$)$_2$. Organometallics 1995, 14, 4402–08.

[31] Wu MM, Gill AM, Yunpeng L, Yongxin L, Ganguly R, Falivene L, Garcia F. Aryl-NHC-group 13 trimethyl complexes: Structural, stability and bonding insights. Dalon Trans 2017, 46, 854–64.

[32] Horeglad P, Cybularczyk M, Trzaskowski B, Żukowska GZ, Dranka M, Zachara J Dialkylgallium alkoxides stabilized with N-heterocyclic carbenes: opportunities and limitations for the controlled and stereoselective polymerization of *rac*-lactide. Organometallics 2015, 34, 3480–96.

[33] Jutzi P, Izundu J, Sielemann H, Neumann B, Stammler H-G Bis- and Tris(dimethylgallyl) benzenes: synthesis, solid-state structures, and redistribution reactions. Organometallics 2009, 28, 2619–24.

[34] Uhl W, Voss M, Hepp A Reactions of the tetraalkyldigallium compound R$_2$Ga-GaR$_2$ [R = CH (SiMe$_3$)$_2$] with acidic reagents, Retention vs. cleavage of the Ga-Ga bond and formation of supramolecular aggregates via hydrogen bonding Z Anorg Allg Chem, 2011, 637, 1845–52.

[35] Kushwah N, Pal MK, Kumar M, Wadawale AP, Manna D, Ghanty TK, Jain VK. Structural diversity ranging from cyclic trimeric, tetrameric, hexameric to 1-D helix in dimethylgallium hydroxide. J Organomet Chem 2015, 781, 65–71.

[36] Coward KM, Jones AC, Steiner A, Bickley JF, Pemble ME, Boag NM, Rushworth SA, Smith LM Synthesis of ultra-high purity trialkylgallium MOVPE precursors. Crystal structures of triethylgallium and triisopropylgallium adducts with macrocyclic tertiary amines. J Mater Chem 2000, 10, 1875–80.

8 Organometallic compounds of group 14 metals

Of group 14 elements, only the organometallic compounds of tin and lead are treated here. Silicon and germanium are nonmetals and, contrary to current practice in the literature, their organic compounds will not be described as organometallics in this book.

The bonds of tin and lead to carbon bond have a pronounced covalent character. This bond is only moderately reactive, and the organometallic compounds of these elements (tin in particular) are rather stable to heat, oxygen and moisture and can be handled in open atmosphere.

Group 14 elements have in their valence shell two s-electrons and two p electrons which undergo a stable sp^3 hybridization to form four covalent bonds. However, the participation of d-orbitals is also possible and expands the covalency of the central atom to five, six and seven in sp^3d and sp^3d^2 hybridizations. The tendency to achieve higher coordination numbers is favored by electronegative substituents like fluorine which contract the diffuse d-orbitals so that they are able to participate in chemical bonding (Fig. 8.1).

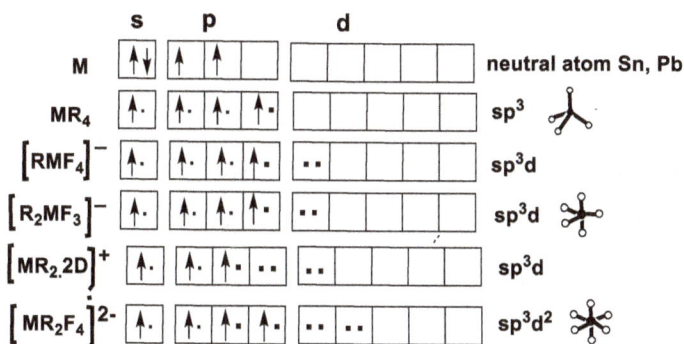

Fig. 8.1: The use of valence orbitals and electrons in the compounds of group 14.

8.1 Organotin compounds

Tetraorganotin compounds are four-coordinated monomers but the functional derivatives R_3SnX and R_2SnX_2 form dimeric or polymeric supramolecular associations in the solid state and coordinate Lewis bases.

The interest in organotins has been stimulated by their use as stabilizers for polyvinyl chloride, catalysts for polyurethane formation, as antioxidants and so on. Their biological activity stimulated their use as pesticides and promise as antitumor agents.

https://doi.org/10.1515/9783110695274-008

The tendency of tin to increase its coordination number beyond four gives rise to structural peculiarities which greatly limit the analogies between silicon, germanium and tin compounds.

8.1.1 Homoleptic species

Tetrasubstituted organotin derivatives (SnR_4) are readily prepared from Grignard reagents with tin tetrachloride for symmetrical derivatives, or with an organotin halide for unsymmetrical derivatives:

$$SnCl_4 + 4RMgX \quad \rightarrow \quad SnR_4 + 4MgClX$$

$$R_2SnCl_2 + 2R'MgX \quad \rightarrow \quad SnR_2R'_2 + 2MgClX$$

Organolithium reagents work even better, especially, for example, in the preparation of $Sn(C_6F_5)_4$.

Wurtz reactions, using tin tetrachloride, an organic chloride and sodium metal, are of only historical interest and seldom used because sodium metal reduces the tetrachloride to metallic tin:

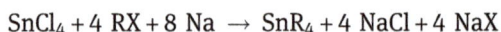

$$SnCl_4 + 4\ RX + 8\ Na \rightarrow SnR_4 + 4\ NaCl + 4\ NaX$$

Organoaluminum compounds can serve as alkylating agents and require the presence of complexing agents (tertiary amines, ethers and even sodium chloride) for the fixation of the aluminum chloride formed. This method also has a limited use:

$$3SnCl_4 + 4AlR_3 + 4NaCl \rightarrow 3SnR_4 + 4NaAlCl_4$$

Organometallic derivatives of sodium are employed for the formation of Sn-C≡C-R groups since sodium acetylides are readily available.

The tin–carbon bond is rather reactive and the tetraorganotins undergo a variety of specific reactions of cleavage, redistribution and insertion (Fig. 8.2).

A rather extensive series of compounds with four tin–carbon bonds are heterocycles with tin heteroatoms. Organotin heterocycles are prepared with Grignard and organolithium reagents, and also by hydride addition in some particular cases (Fig. 8.3).

Tristannacyclohexane has been synthesized in a three-step reaction, from organozinc reagents and magnesium coupling:

$$R_3SnX, R_2SnX_2$$

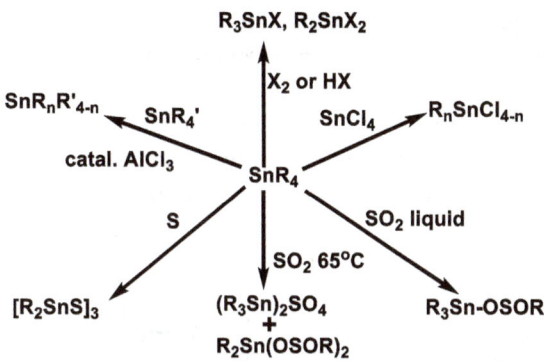

Fig. 8.2: Some general reactions of tetraorganotin compounds.

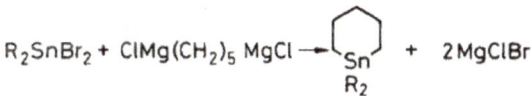

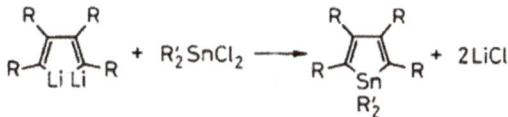

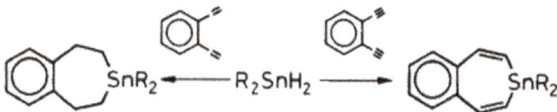

$$R_2SnBr_2 + ClMg(CH_2)_5 MgCl \longrightarrow \underset{R_2}{\overset{}{Sn}} + 2\,MgClBr$$

Fig. 8.3: Organotin heterocycles.

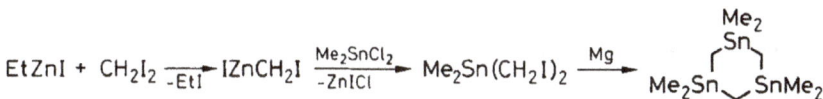

$$EtZnI + CH_2I_2 \xrightarrow[-EtI]{} IZnCH_2I \xrightarrow[-ZnICl]{Me_2SnCl_2} Me_2Sn(CH_2I)_2 \xrightarrow{Mg} \begin{array}{c} Me_2 \\ Sn \\ Me_2Sn \qquad SnMe_2 \end{array}$$

Fig. 8.4: Synthesis of tristannacyclohexane ring.

8.1.2 Hypervalent penta- and hexa-organotin compounds

By using d-orbitals, the coordination number of tin can be increased to five and six and by accepting one or two more electrons and pairing them into covalent bonds, some hypervalent tin anions with five or six tin–carbon bonds are formed.

A first example is the square pyramidal pentakis(pentafluoroethyl) anion ([Sn(C$_2$F$_5$)$_5$]$^-$) obtained from SnCl$_4$ with LiC$_2$F$_5$ in 1:5 molar ratio and from Sn(C$_2$F$_5$)$_4$ in a succession of reactions [1].

Another compound with five tin–carbon bonds is a spirobifluorene-methyl derivative [{SnMe(C$_6$H$_4$C$_6$H$_4$)$_2$}]$^-$, which was demonstrated crystallographically to have a distorted trigonal bipyramidal structure with the methyl group in equatorial position [2] (Fig. 8.5).

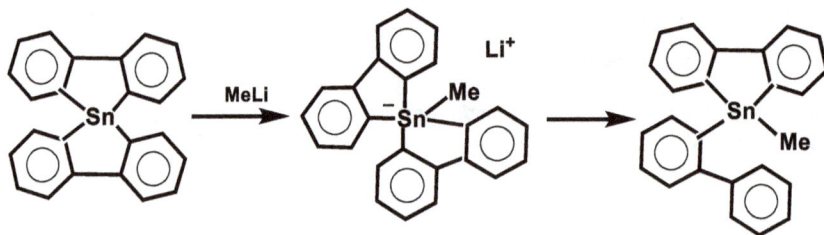

Fig. 8.5: Fluorene organotin compounds.

The hypervalent penta(2-furyl)tin anion ([Sn(2-C$_3$H$_3$O)$_5$]$^-$) and hexa(2-furyl)tin dianion ([Sn(2-C$_3$H$_3$O)$_6$]$^{2-}$) were prepared from 2-furyl-lithium with tetra(2-furyl)tin and with SnCl$_4$, respectively [3] (Fig. 8.6).

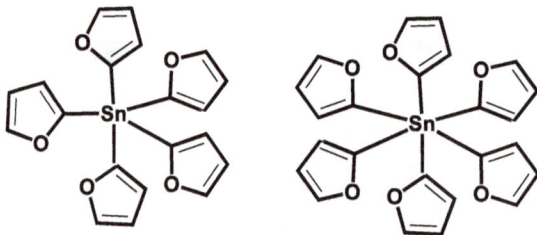

Fig. 8.6: Hypervalent organotin compounds.

Two hexakis(2-pyridyl)tin dianions ([Sn(C$_5$NR)$_6$]$^{2-}$ with R = Me and OBu^t) have been prepared as dilithium salts from Sn(2-Py^R)$_4$ with Li[2-Py^R], where Py^R is C$_5$H$_4$NR [4].

8.1.3 Diorganotin species

Compounds of the composition SnR$_2$ are apparently derivatives of divalent tin, but these compounds are in fact cyclic oligomers [SnR]$_n$ or ill-defined species containing SnR$_2$, SnR$_3$ or even SnR units, in polymeric irregular networks whose overall composition is SnR$_2$. When treated with elemental iodine, the Sn–Sn bonds in such materials are cleaved, and R$_3$SnI, R$_2$SnI$_2$ and RSnI$_3$ are identified as products.

Genuine organic derivatives of divalent tin have, however, now been prepared. These are the organotin(II) compounds such as dicyclopentadienyl (:Sn(η^5-C$_5$H$_5$)$_2$) and bis[bis(trimethylsilyl)methyl] (:Sn[CH(SiMe$_3$)$_2$]$_2$).

Dialkylstannylenes are formed as transient intermediates and can be trapped with suitable reagents. Thus, transition metal complexes of stannylenes can be generated, either directly from stable stannylenes or by indirect routes, but tin can be better described as tetravalent in these complexes.

Dicyclopentadienyltin(II) (:Sn(η^5-C$_5$H$_5$)$_2$, stannocene) is prepared from tin(II) chloride and sodium cyclopentadienide:

$$\text{SnCl}_2 + 2\,\text{Na}^+\text{C}_5\text{H}_5^- \quad \longrightarrow \quad :\text{Sn}\left(\text{C}_5\text{H}_5 - \eta^{\,5}\right)_2 + 2\text{NaCl}$$

This compound is a pentahapto complex (:Sn(η^5-C$_5$H$_5$)$_2$), both in the vapor phase and solid state with an angular structure. The mono-η^5-cyclopentadienyltin chloride is associated through weak chlorine bridges in the solid state. A pentamethylcyclopentadienyltin(II) cation, [(η^5-Me$_5$C$_5$)Sn:]$^+$, has also been obtained from :Sn(η^5-C$_5$Me$_5$)$_2$ and strong acids (Fig. 8.7).

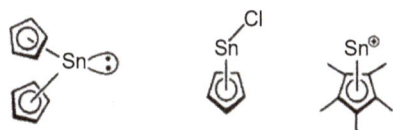

Fig. 8.7: Organotin cyclopentadienyls.

Bis[bis(trimethylsilyl)methyl]tin(II), :Sn[CH(SiMe$_3$)$_2$]$_2$, was prepared from tin(II) chloride and bis(trimethylsilyl)-methyllithium, or with tin(II) bis(trimethylsilyl)amide:

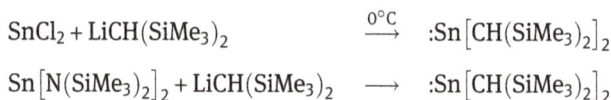

$$\text{SnCl}_2 + \text{LiCH}(\text{SiMe}_3)_2 \quad \xrightarrow{0°C} \quad :\text{Sn}\left[\text{CH}(\text{SiMe}_3)_2\right]_2$$

$$\text{Sn}\left[\text{N}(\text{SiMe}_3)_2\right]_2 + \text{LiCH}(\text{SiMe}_3)_2 \quad \longrightarrow \quad :\text{Sn}\left[\text{CH}(\text{SiMe}_3)_2\right]_2$$

The compound is monomeric in solution but dimeric in the solid state, with an uncommon Sn=Sn bent double bond and an Sn–Sn distance of 2.76 Å:

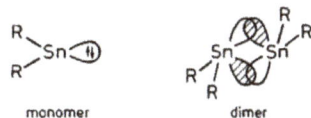

monomer   dimer   Fig. 8.8: Diorganotin monomer and dimer.

The compound :Sn[CH(SiMe$_3$)$_2$]$_2$ displays reactions typical for a diorganotin(II) compound (Fig. 8.9), that is, oxidative additions and complex formation with metal carbonyls to give, for example, the chromium complex [(Me$_3$Si)$_2$CH]$_2$Sn:Cr(CO)$_5$.

Dialkylstannylenes (:SnR$_2$) are formed in several reactions as transient intermediates and can be trapped by alkyl halides (oxidative addition). In the absence

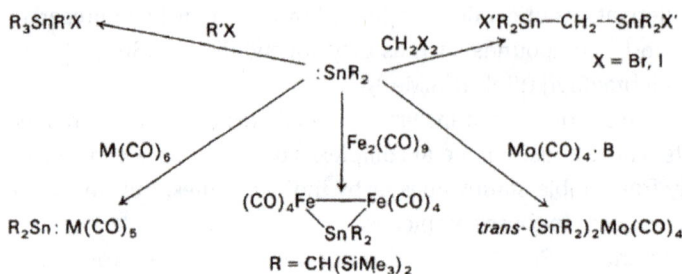

Fig. 8.9: Reactions of diorganotin compounds.

of a trapping agent, stannylenes polymerize to cyclopolystannanes. Stannylenes are formed by photolysis or pyrolysis of polystannanes, by thermolysis of tetraalkyl-distannanes, dihalodistannanes, or pentaalkyldistannanes, or by decomposition of stannylene–transition metal complexes.

Trisubstituted organotin free radicals, $\cdot SnR_3$, have a trigonal pyramidal structure and can be prepared by irradiation of $:Sn[CH(SiMe_3)_2]_2$. These free radicals are unusually stable and can be isolated (Fig. 8.10).

Fig. 8.10: Triorganotin free radical.

Hexaalkyldistannanes, $R_3Sn–SnR_3$, with bulky substituents dissociate reversibly at 180 °C (for R = 2,4,6-trimethylphenyl) or at 100 °C (for R = 2,4,6-triethylphenyl).

8.1.4 Tin π-complexes

Organotin compounds with aromatic groups π-bonded to the metal are known for tin. One such compound is $[Sn(\eta^6\text{-}C_6H_6)Cl_2][AlCl_4]$ with a supramolecular chain structure resulted from association through chlorine bridges [5]. The reaction of the molten salt $Sn[AlCl_4]_2$ with benzene yields a dinuclear compound with four $\eta^6\text{-}C_6H_6$-Sn moieties [6] (Fig. 8.11).

Another unusual compound is a complex in which a trigonally planar coordinated Sn(II) atom is encapsulated in the cavity of [2.2.2]paracyclophane [7] (Fig. 8.12).

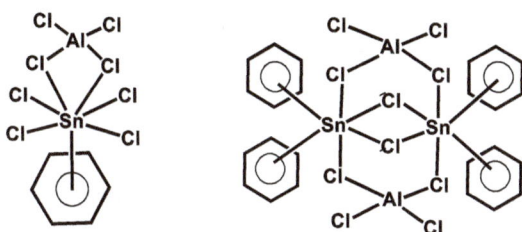

Fig. 8.11: Organotin-benzene π-complexes.

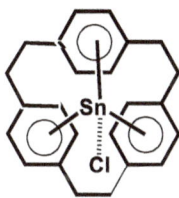

Fig. 8.12: A trigonal organotin π-complex.

8.1.5 Inverse organotin compounds

Organometallic compounds with organotin moieties as external or terminal groups attached to a carbon skeleton, that is, inverse organotin compounds, are quite numerous and varied. The simplest is tris(triphenyltin)methane ($HC(SnPh_3)_3$) [8]. Related compounds $HC(SnPhCl)_3$, $HC(SnPh_2I)_3$, $HC(SnPhI_2)_3$, $HCH(SnPhCl_2)_3$ and NC-C $(SnMe_3)_3$ have also been described [9]. A typical inverse organometallic compound is tetrakis(trimethyltin)methane ($C(SnMe_3)_4$) prepared from CCl_4 and Me_3SnCl with lithium metal in THF [1]. It has a tetrahedral molecular structure with carbon in the center [11] (Fig. 8.13).

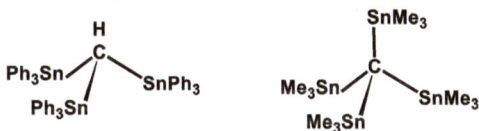

Fig. 8.13: Carbon-centered inverse organotin compounds.

A remarkable family of inverse organotin compounds is the series of composition $Ph_3Sn-(CH_2)_n-SnPh_3$ with $n = 1–8$ [12].

Other related compounds include 1,2-bis(trimethylstannyl)acetylene ($Me_3Sn-C\equiv C-SnMe_3$) and 1,4-bis(triphenyltin)butane-1,3-diyne [13] (Fig. 8.14).

More inverse organotin compounds have been described with aromatic centers, for example, a bis(triphenyltin) derivative of 4,4'-bis(methylene)-1,1'-biphenyl [14]

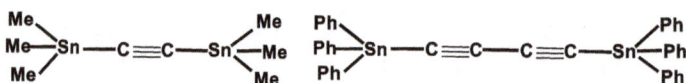

Fig. 8.14: Inverse organotin compounds derived from mono- and di-acetylene.

and two anthracene derivatives with two [15] and four [16] trimethyltin decorating groups (Fig. 8.15).

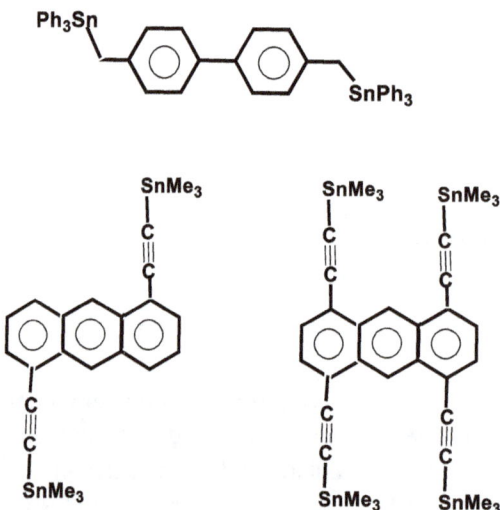

Fig. 8.15: More inverse organotin compounds.

A tetrakis(trimethyltin) derivative of cyclopentadiene 2,3,5,5-$C_5H_2(SnMe_3)_4$ [17] and a bis(triphenyltin)dipyrazolylmethane $H_2C(C_4N_2SnPh_3)_2$ [18] are also worth mentioning, to illustrate the diversity of inverse organotin compounds. The former is obtained from cyclopentadiene with excess of $Me_3Sn-NMe_2$ (Fig. 8.16).

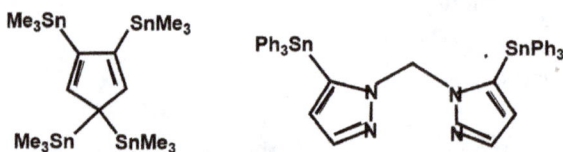

Fig. 8.16: Inverse organotin compounds with five-membered rings.

8.1.6 Organotin hydrides, R_nSnH_{4-n}

The organotin hydrides are tetrahedral monomers in the gas phase. They are highly reactive compounds and are used as intermediates in addition reactions and as reducing reagents in organic chemistry. Their preparation uses the reduction of organotin halides with lithium alanate or sodium borohydride:

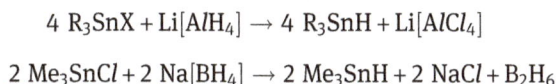

$$4\ R_3SnX + Li[AlH_4] \rightarrow 4\ R_3SnH + Li[AlCl_4]$$

$$2\ Me_3SnCl + 2\ Na[BH_4] \rightarrow 2\ Me_3SnH + 2\ NaCl + B_2H_6$$

8.1.7 Organotin halides, R_nSnX_{4-n}

There are several ways to prepare organotin halides, R_nSnX_{4-n}. On heating a tetraorganostannane with tin tetrahalide, a redistribution of substituents takes place with formation of organotin halides determined by the molar ratio of the reactants:

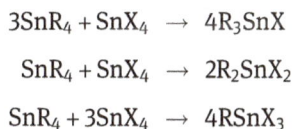

$$3SnR_4 + SnX_4 \rightarrow 4R_3SnX$$

$$SnR_4 + SnX_4 \rightarrow 2R_2SnX_2$$

$$SnR_4 + 3SnX_4 \rightarrow 4RSnX_3$$

In a direct synthesis, tin metal and an organic halide, RX, react at 60–180 °C (in the order of decreasing reactivity, X = I, Br, CI):

$$Sn + 2RX \xrightarrow{\text{catalyst}} R_2SnX_2$$

Benzyl chloride reacts with tin metal in boiling water to give tribenzyltin chloride $(PhCH_2)_3SnCl$, or in toluene to give dibenzyltin dichloride $(PhCH_2)_2SnCl_2$.

In the reaction of tin tetrachloride with aluminum alkyls, only three chlorine atoms are replaced with formation of trialkyltin chloride:

$$SnCl_4 + AlR_3 \rightarrow R_3SnCl + AlCl_3$$

Organotin bromides and iodides can also be synthesized by cleavage of organic groups with halogens:

$$SnR_4 \xrightarrow[-RX]{X_2} R_3SnX \xrightarrow[-RX]{X_2} R_2SnX_2$$

The structures of organotin halides have some peculiarities. In the vapor phase, the triorganotin halides (R_3SnX) are tetrahedral monomers, but in the solid state, they can be associated into supramolecular arrays. Thus, trimethyltin fluoride and trimethyltin chloride are chains in which the tin atoms are five-coordinated with trigonal bipyramidal geometry and the $SnMe_3$ triangles are bridged by halogens (Fig. 8.17). Triphenyltin chloride is, however, monomeric and tetrahedral in the solid state.

Diorganotin dihalides, R_2SnX_2, are tetrahedral monomers in the vapor phase, and the structure is preserved in the solid state as suggested by their low melting points but Me_2SnCl_2 and Et_2SnX_2 (X = CI, Br and I) show intermolecular Sn . . . X interactions.

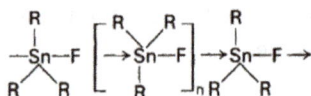

Fig. 8.17: Supramolecular self-assembly of triorganotin fluorides.

The fluorides are highly associated and consequently are high melting and insoluble as in Me_2SnF_2, which has a double fluorine-bridged supramolecular structure containing six-coordinated tin (Fig. 8.18).

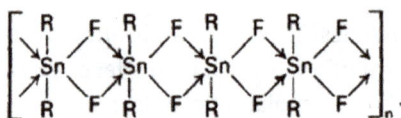

Fig. 8.18: Supramolecular self-assembly of diorganotin difluorides.

The organotin halides can further add halide ions to increase the coordination number to five, like in the hypervalent trigonal bipyramidal complex anions $[MeSnCl_4]^-$, $[Me_2SnCl_3]^-$, $[Ph_3SnCl_2]^-$ and $[Bu_3SnCl_2]^-$, and to six, as in the *trans*-octahedral anions $[Me_2SnF_4]^{2-}$, $[Me_2SnCl_4]^{2-}$ and $[Me_3SnCl_3]^{2-}$.

With donor molecules, the organotin halides form adducts which contain five- or six-coordinated tin and exhibit bipyramidal, trigonal or distorted-octahedral geometries.

Halogeno (chloro and bromo)-centered dinuclear inverse coordination complex anions are also known [19, 20] (Fig. 8.19).

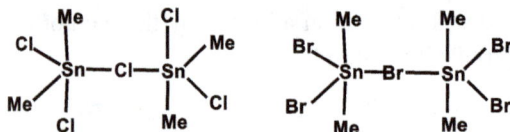

Fig. 8.19: Halogen-centered inverse coordination organotin complexes.

8.1.8 Organostannoxanes (organotin oxides)

Hexaorganotin distannoxanes ($R_3Sn-O-SnR_3$) are formed in the hydrolysis of triorganotin halides by the condensation of the hydroxide intermediates. Most distannoxanes are monomeric, some with linear Sn–O–Sn groups (e.g., with R = Bu^t, CH_2Ph, CH_2CH_4X-*ortho* with X = F, Cl, Br) and others with bent Sn–O–Sn groups (e.g., with R = Et, Ph, CH_2Tol and $C_6H_4CH_2NMe_2$) [21, 22].

Distannoxane dihalides, $XR_2SnOSnR_2X$, are obtained by heating a mixture of the oxide and halide:

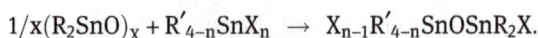

$$1/x(R_2SnO)_X + R'_{4-n}SnX_n \longrightarrow X_{n-1}R'_{4-n}SnOSnR_2X.$$

The dihalides, $XR_2SnOSnR_2X$, are dimeric in solution and contain four-membered Sn_2O_2 rings in the solid state with five-coordinated tin atoms. In the diorganotin oxide, $[R_2SnO]_x$ all the tin atoms are five-coordinated in polycyclic, supramolecular structures. The *tert*-butyl derivative, $[Bu^t_2SnO]_3$, contains a planar Sn_3O_3 ring (see Fig. 8.20).

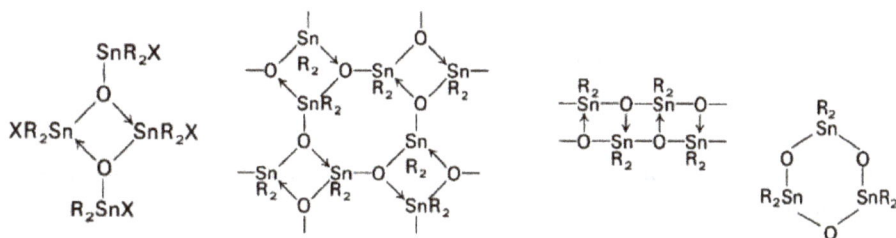

Fig. 8.20: Structures of organotin oxides.

Di-*tert*-butyltin oxide postulated to exist as the cyclic trimer, $[Bu^t_2SnO]_3$, in solution is the only other diorganotin oxide with tetracoordinate tin atoms. The cyclotristan-noxane ring is essentially planar [23].

8.1.9 Organotin hydroxides, $R_nSn(OH)_{4-n}$

Organotin hydroxides are obtained by alkaline hydrolysis of halides, but they are readily dehydrated:

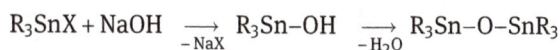

$$R_3SnX + NaOH \xrightarrow[-NaX]{} R_3Sn-OH \xrightarrow[-H_2O]{} R_3Sn-O-SnR_3$$

Organotin hydroxides, like their halide precursors, undergo supramolecular self-assembly with increased coordination number at tin. Trimethyltin hydroxide is associated in the solid state but Me_3SnOH is a cyclic dimer in solution [24] (Fig. 8.21).

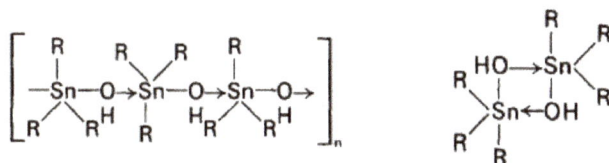

Fig. 8.21: Supramolecular structures of triorganotin hydroxides.

Organotin hydroxides are basic, reacting with organic acids to form organotin carboxylates and with dithiophosphoric acids to form dithiophosphates.

8.1.10 Organotin alkoxides ($R_nSn(OR')_{4-n}$) and related compounds

Organotin alkoxides, $R_nSn(OR')_{4-n}$, are obtained by treatment of organotin halides, distannoxanes and alcohols:

$$R_3SnX + R'OH + NR'_3 \rightarrow R_3Sn-OR' + HX \cdot NR''_3$$

$$R_3SnX + NaOR' \rightarrow R_3Sn-OR' + NaX$$

$$R_3SnOSnR_3 + 2R'OH \rightarrow 2R_3Sn-OR' + H_2O$$

The alkoxy derivatives insert molecules with double bonds forming a series of functional derivatives (Fig. 8.22).

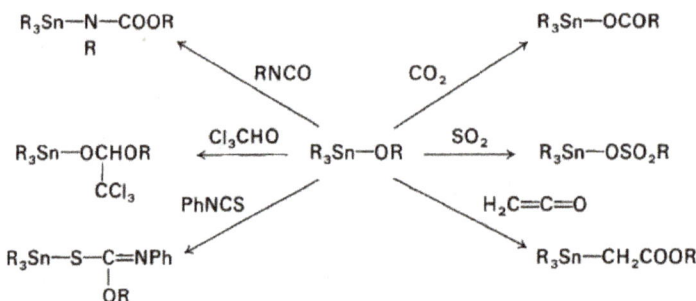

Fig. 8.22: Insertion reactions in triorganotin alkoxides.

In the solid state, $Me_3Sn-OMe$ is a supramolecular chain containing trigonal bipyramidal units of Me_3Sn groups connected by OMe bridges. Triethanolamine reacts with organotin triethoxide ($RSn(OEt)_3$) to form the so-called stannatranes which are penta-coordinated organotin compounds. Similar compounds are obtained from $(R_2SnO)_x$ with diethanolamine (Fig. 8.23).

Fig. 8.23: Penta-coordinated organotin compounds.

8.1.11 Organotin sulfides, selenides and tellurides

The Sn–S bond is stable to hydrolysis and the organotin sulfides can be synthesized even in water. The triorganotin sulfides are prepared from the corresponding chlorides and sodium or silver sulfide, or by treatment of oxides or hydroxides with hydrogen sulfide:

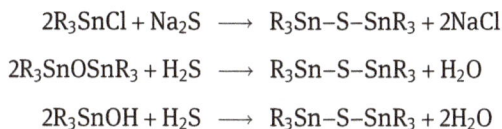

$$2R_3SnCl + Na_2S \longrightarrow R_3Sn-S-SnR_3 + 2NaCl$$

$$2R_3SnOSnR_3 + H_2S \longrightarrow R_3Sn-S-SnR_3 + H_2O$$

$$2R_3SnOH + H_2S \longrightarrow R_3Sn-S-SnR_3 + 2H_2O$$

Hexaorganotin disulfides, $R_3Sn-S-SnR_3$, with R = Ph [25] and R = Cyh [26] are examples. Selenium analogues are also known [27] (Fig. 8.24).

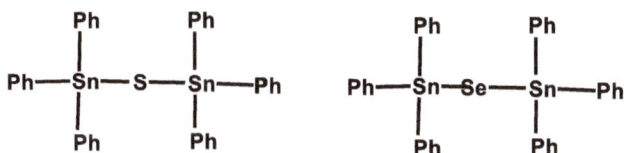

Fig. 8.24: Triphenyltin sulfide and selenide.

Diorganotin halides react with sodium sulfide, and diorganotin oxides react with hydrogen sulfide to form cyclic organotin sulfides (cyclostannathianes) (Fig. 8.25).

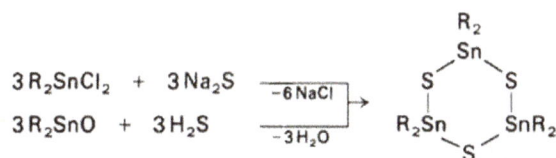

Fig. 8.25: Cyclic diorganotin sulfides.

The reaction of diorganotin dichloride with bulky *tert*-Bu groups with sodium sulfide, selenide or telluride yields four-membered, cyclic dimers (Fig. 8.26).

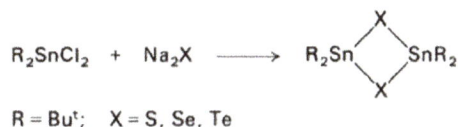

R = Bu^t; X = S, Se, Te

Fig. 8.26: Cyclic diorganotin dimers with a bulky substituent.

Dimethyltin dihydride reacts with elemental sulfur, selenium or tellurium to give the cyclic trimers $(Me_2SnX)_3$ (X = S, Se, Te).

   The reaction of organotin trihalides with sodium sulfide gives tricyclic tetramers $R_4Sn_4S_6$ (R = Me, *iso*-Pr, C(SiMe$_3$)$_3$, CH$_2$Ph, *para*-Tol, C$_6$F$_5$, C$_6$H$_2$Me-2.4.6, ferrocenyl) with adamantane-like structures. Selenium analogues $R_4Sn_4Se_6$ (R = Me, CH$_2$Ph, ferrocenyl) are also known (Fig. 8.27).

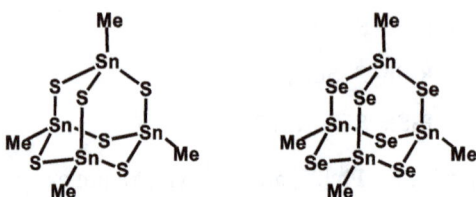

Fig. 8.27: Organotin adamantanes.

Tin-rich sulfur- and selenium-containing inorganic heterocycles have also been prepared (Fig. 8.28).

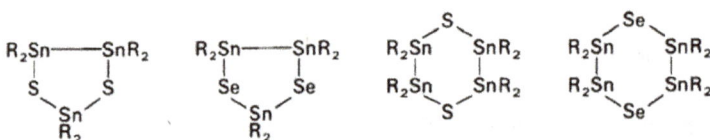

Fig. 8.28: Tin-rich inorganic heterocycles.

8.1.12 Organotin thiolates, $R_nSn(SR')_{4-n}$

Organotin thiolato derivatives are prepared by treatment of organotin halides, oxides, alkoxides, hydrides or amines with thiols or alkali metal mercaptides:

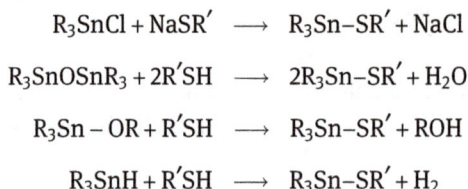

$$R_3SnCl + NaSR' \longrightarrow R_3Sn-SR' + NaCl$$

$$R_3SnOSnR_3 + 2R'SH \longrightarrow 2R_3Sn-SR' + H_2O$$

$$R_3Sn - OR + R'SH \longrightarrow R_3Sn-SR' + ROH$$

$$R_3SnH + R'SH \longrightarrow R_3Sn-SR' + H_2$$

8.1.13 Organotin amino-derivatives, $R_nSn(NR'R'')_{4-n}$

The organotin halides do not react with ammonia or primary or secondary amines to form substitution products, and only addition compounds are formed instead. Metallated amines are necessary to form compounds containing Sn–N bonds:

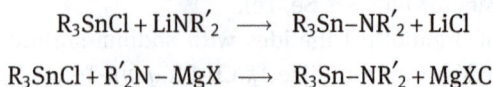

$$R_3SnCl + LiNR'_2 \longrightarrow R_3Sn-NR'_2 + LiCl$$

$$R_3SnCl + R'_2N - MgX \longrightarrow R_3Sn-NR'_2 + MgXCl$$

The Sn–N bond is sensitive to active hydrogen reagents such as water, alcohols and acids, which cleave the Sn–N bonds and yield substitution products.

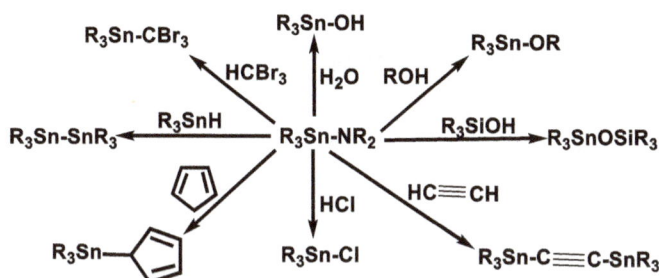

Fig. 8.29: Some reactions of triorganotin amino derivatives.

Small unsaturated molecules insert into Sn–N bonds. These reactions have a preparative value since the organotin group can be removed, leaving a purely organic compound.

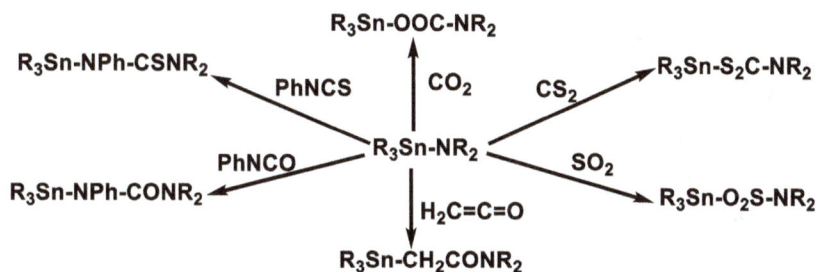

Fig. 8.30: Insertion reactions into Sn–N bonds.

8.1.14 Organostannazanes

Dinuclear stannazane ($R_3Sn-NMe-SnR_3$) groups can be obtained by the reaction of trimethyltin chloride and N-lithiated methylamine:

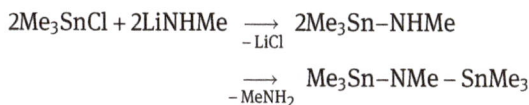

$$2Me_3SnCl + 2LiNHMe \xrightarrow[-LiCl]{} 2Me_3Sn-NHMe$$

$$\xrightarrow[-MeNH_2]{} Me_3Sn-NMe-SnMe_3$$

or by transamination of aminostannanes with primary amines:

$$2R_3Sn-NMe_2 + EtNH_2 \longrightarrow Me_3Sn-NEt-SnMe_3 + 2Me_2NH$$

Representatives of this class are $[Me_3Sn-NH_2-SnMe_3]^+$ [28] and $ClMe_2Sn-NBu^t-SnMe_2Cl$ [29].

Six-membered rings are obtained from diorganotin dichlorides with potassium amide in liquid ammonia and four-membered rings by successive reactions involving lithiations (Fig. 8.31).

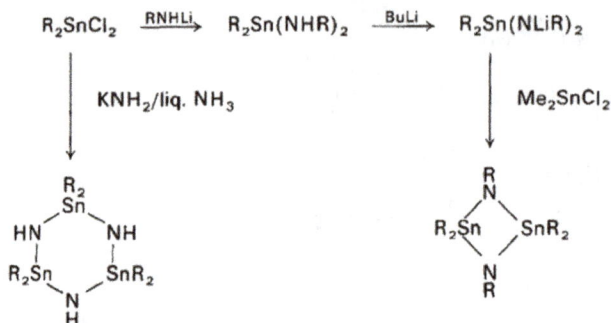

$$R_2SnCl_2 \xrightarrow{\ RNHLi\ } R_2Sn(NHR)_2 \xrightarrow{\ BuLi\ } R_2Sn(NLiR)_2$$

KNH$_2$/liq. NH$_3$ ⟍ ⟍ Me$_2$SnCl$_2$

Fig. 8.31: Organotin–nitrogen heterocycles.

A cubane ([Sn$_4$(μ_3-NSnMe$_3$)$_4$]$_4$) has been prepared from SnCl$_2$ with [LiN(SnMe$_3$)$_2$] [30] (Fig. 8.32).

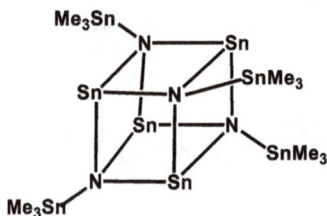

Fig. 8.32: The cubane structure of a tin–nitrogen compound.

8.1.15 Organotin inverse coordination complexes

Open, planar triangular inverse coordination complexes are known with oxygen, nitrogen, halogens and phosphorus as inverse coordination centers.

The planar oxo-centered compound, [(μ_3-O)(SnMe$_3$)$_3$]Cl, can be prepared in two ways:

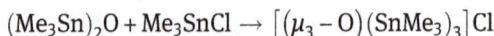

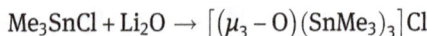

$$(Me_3Sn)_2O + Me_3SnCl \rightarrow \left[(\mu_3 - O)(SnMe_3)_3\right]Cl$$

$$Me_3SnCl + Li_2O \rightarrow \left[(\mu_3 - O)(SnMe_3)_3\right]Cl$$

and in solid state is associated through Sn–Cl→Sn bonds to form a supramolecular bidimensional structure [31].

Tristannylamines are prepared by the reaction of trialkyltin chlorides and lithium nitride [32] (Fig. 8.33).

A phosphorus-centered analogue is also known [33] (Fig. 8.34).

$$3R_3SnCl + Li_3N \longrightarrow \underset{\underset{SnR_3}{\overset{|}{N}}}{\overset{R_3Sn\diagdown\diagup SnR_3}{}} + 3\ LiCl$$

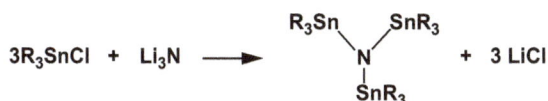

Fig. 8.33: Organotin inverse coordination complex with planar trigonal nitrogen coordination center.

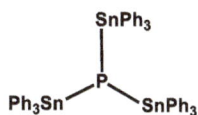

$$\underset{Ph_3Sn}{\overset{SnPh_3}{\underset{\diagup}{\overset{|}{P}}}}\diagdown SnPh_3$$

Fig. 8.34: Organotin inverse coordination complex with planar trigonal phosphorus coordination center.

Organotin inverse coordination complexes, as substituted ammonium and phosphonium derivatives [34] (Fig. 8.35), can be prepared by two methods:

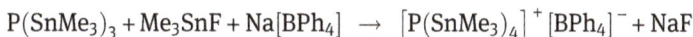

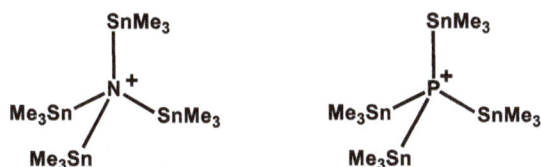

$$E(SnMe_3)_3 + M_3SnOTf \ \rightarrow \ \left[E(SnMe_3)_4\right]^+ [OTf]^- \quad E = N,\ P$$

$$P(SnMe_3)_3 + Me_3SnF + Na[BPh_4] \ \rightarrow \ \left[P(SnMe_3)_4\right]^+ [BPh_4]^- + NaF$$

$$\underset{Me_3Sn}{\overset{SnMe_3}{\underset{\diagup}{\overset{|}{N^+}}}}\diagdown SnMe_3 \qquad\qquad \underset{Me_3Sn}{\overset{SnMe_3}{\underset{\diagup}{\overset{|}{P^+}}}}\diagdown SnMe_3$$

$$Me_3Sn \qquad\qquad\qquad\qquad Me_3Sn$$

Fig. 8.35: Organotin inverse coordination complexes with tetrahedral nitrogen and phosphorus coordination centers.

Some interesting inverse coordination complexes of diorganotin moieties with oxygen, nitrogen, sulfur and halogens as coordination centers are triangular compounds that contain SnR_2 groups with five-coordinated tin forming an Sn_3X_3 ring; a nonmetal atom is embedded in the center and plays a structure-directing role. The oxygen- and nitrogen-centered rings in the compounds $[(\mu_3\text{-}O)\{But_2Sn)_3(\mu_2\text{-}OMe)_3]$ [35] and $[(\mu_3\text{-}N)(SnMe_2)_3(\mu_2\text{-}X)_3]$ with X = Cl, Br, I [36, 37] are planar (Fig. 8.36).

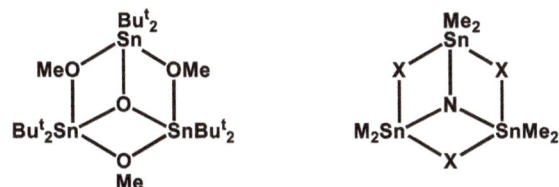

Fig. 8.36: Planar cyclic trinuclear inverse coordination complexes.

In the sulfur- [38] and chlorine [39]-centered complexes, the six-membered rings display chair conformation (Fig. 8.37).

Fig. 8.37: Chair-shaped cyclic trinuclear inverse coordination complexes.

Similar structures with a modified external ring have been described with carbonate [40], silanolate [41, 42] and phosphinate [43] external bridging linkers (Fig. 8.38).

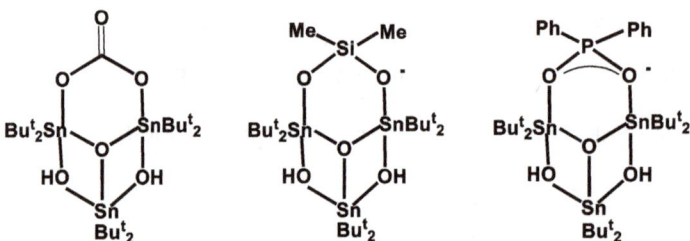

Fig. 8.38: More oxygen-centered organotin inverse coordination complexes.

Organotin derivatives of inorganic oxoanions are inverse coordination complexes, readily prepared due to the affinity of tin for oxygen groups. Thus, organotin sulfate [$(\mu_2\text{-}SO_4)\{SnMe_3(H_2O)\}_2$] [44], selenate [$(\mu_2\text{-}SeO_4)\{SnMe_3(H_2O)\}_2$] [45] and nitrate [$(\mu_3\text{-}NO_3)(SnPh_3Cl)_3$]$^-$ [46] are shown in Fig. 8.39.

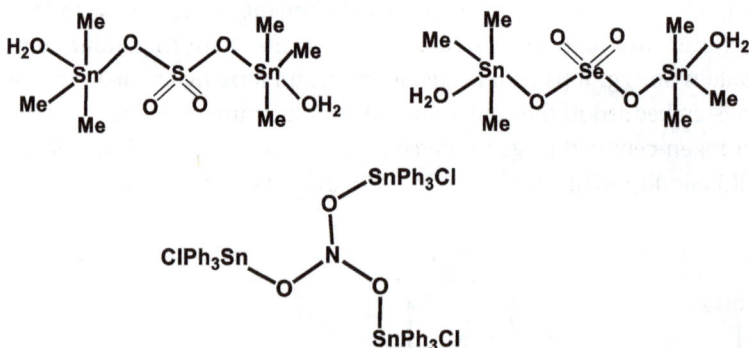

Fig. 8.39: Organotin inverse coordination complexes with inorganic anions as coordination centers.

A very large number of inverse coordination complexes are organotin caboxy-lates. Practically, any di- or polycarboxylic acid can form organotin inverse coordination complexes with a carboxylato centroligand. A selection is presented here.

Bimetallic carboxylates can be illustrated with the oxalato complex $[(\mu_2\text{-}C_2O_4)(SnPh_3)_2]$ [47] and with the succinato complex $[(\mu_2\text{-}OOCCH_2CH_2COO)(SnPh_3)_2]$ [48] (Fig. 8.40).

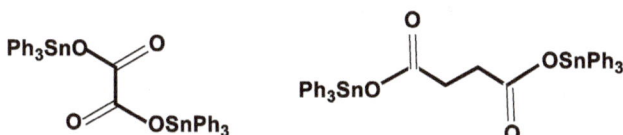

Fig. 8.40: Organotin inverse coordination complexes with dicarboxylato coordination centers.

The benzene-1,3,5-tricarboxylato anion forms a trinuclear organotin complex $[\{\mu_6\text{-}C_6H_3(COO)_3(SnPh_3)_3]$ [49, 50], and a tetrametallic complex $[(\mu_4\text{-}C_6H_2(COO)_4(SnPh_3)_4]$ has been prepared with benzene-1,2,4,5-tetracarboxylato anion [51] (Fig. 8.41).

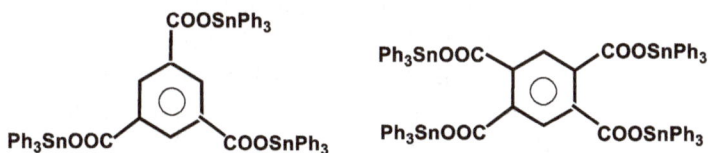

Fig. 8.41: Organotin inverse complexes with dicarboxylato coordination centers.

Obviously, carboxylic anions derived from heterocycles can form similar inverse coordination complexes, as shown with an organotin derivative of thiophene-2,5-dicarboxylate $[(\mu_2\text{-}H_2C_4S)\{COOSnPh_3)_2]$ [52] (Fig. 8.42).

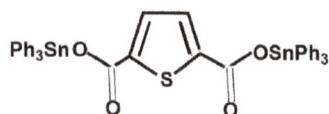

Fig. 8.42: Organotin inverse coordination complex with thiophenedicarboxylato coordination center.

Thiolato inverse coordination complexes represented by derivatives of thiols, thiocarboxylic anions and dithiocarbamates are readily formed. Just a few examples will illustrate this possibility.

Ethene-1,1,2,2-tetrathiolate forms a dinuclear organotin complex $[(\mu_4\text{-}C_2S_4)\{Sn(Me_3Si)_2CH_2CH_2SiMe_3)_2\}_2]$ [53], and another dinuclear complex can be illustrated with tetrathiafulvalene-2,3,6,7-tetrathiolate (Fig. 8.43) [54].

Thiolates derived from heterocycles can also play the role of centroligands in inverse coordination organotin complexes and derivatives of 1,3,5-triazine [(μ$_3$-C$_3$N$_3$)(CH$_2$SnR$_3$)$_3$] (R = Me, Ph) are examples [55] (Fig. 8.43).

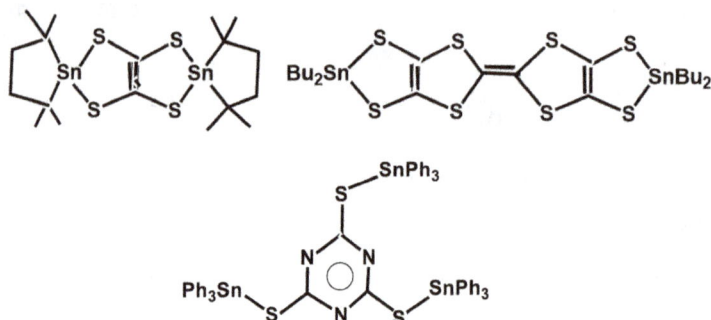

Fig. 8.43: Organotin inverse coordination complexes with thiolato coordination centers.

A numerous family of inverse coordination complexes include dithiocarbamates (Fig. 8.44). Thus, 1,4-bis(dithiocarbamato)-piperazine forms dinuclear complexes [(μ$_2$-S$_2$C-N(CH$_2$CH$_2$)$_2$N-CS$_2$)(SnPh$_3$)$_2$] [56] and [(μ$_2$-S$_2$C-N(CH$_2$CH$_2$)$_2$N-CS$_2$)(SnPh$_2$Cl)$_2$] [57]. A rare dithiocarbamate derived from diazepam forms a binuclear complex with two tricyclohexyltin moieties [58]. A trinuclear complex is formed by a triethyla-mino centroligand [(μ$_3$-N(CH$_2$CH$_2$NMe-CS$_2$)(SnClPh$_2$)$_3$] [59] (Fig. 8.44).

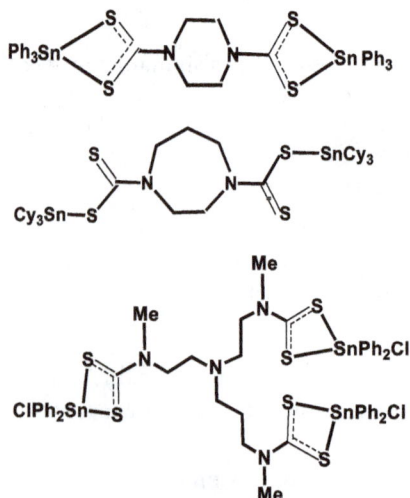

Fig. 8.44: Organotin inverse coordination complexes with dithiocarbamato coordination centers.

Nitrogen heterocycles as centroligands represent another family of organotin inverse coordination complexes. Examples are derivatives of pyrazine [(μ_2-$C_6H_4N_2$) ($SnCl_2Me_2$)$_2$] [60] and 4,4′-bipyridyl [(μ_2-bipy)($SnClPh_3$)$_2$] [61] (Fig. 8.45).

Fig. 8.45: Organotin inverse complexes with nitrogen heterocycles as coordination centers.

8.1.16 Organodi- and poly-stannanes

In spite of its clear metallic character, the tin atoms have a remarkable ability of concatenation, that is, formation of compounds with covalent Sn–Sn bond skeletons. Of course, this ability is far from that of the carbon, the top element of group 14, but it is possible to prepare well-defined linear polystannane chains of some length, also rings and cages. The Sn–Sn bonds are weak and reactive, but compounds with tin–tin bonds are stable toward oxygen and water.

Distannanes can be readily obtained in several ways:

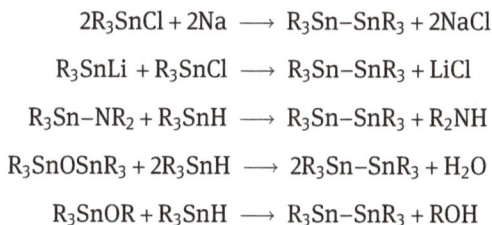

$$2R_3SnCl + 2Na \longrightarrow R_3Sn-SnR_3 + 2NaCl$$

$$R_3SnLi + R_3SnCl \longrightarrow R_3Sn-SnR_3 + LiCl$$

$$R_3Sn-NR_2 + R_3SnH \longrightarrow R_3Sn-SnR_3 + R_2NH$$

$$R_3SnOSnR_3 + 2R_3SnH \longrightarrow 2R_3Sn-SnR_3 + H_2O$$

$$R_3SnOR + R_3SnH \longrightarrow R_3Sn-SnR_3 + ROH$$

Excess alkali metal needs to be avoided since it cleaves the Sn–Sn bonds with formation of [$SnMe_3$]$^-$ anions:

$$Me_3SnBr + 2Na \xrightarrow{liq.NH_3} Me_3Sn-SnMe_3 \xrightarrow{2Na} 2Na^+ Me_3Sn^-$$

Numerous distannanes ($R_3Sn-SnR_3$, e.g., with R = Me, *iso*-Bu, Ph, CH_2Ph, *para*-Tol, Cyh, CF_2CF_3, C_6F_5) are known, and also $Me_2PhSi-(SnBu^t_2)_2-SiMe_2Ph$ [62].

Other compounds with Sn–Sn bonds are formed by connecting organotin heterocycles 2,2′,3,3′,4,4′,5,5′-octaphenyl-1H, 1′H-1,1′-bistannole [63] and 1,1′-di-t-butyl -2,2′,3,3′,4,4′,5,5′-octaethyl-1,1′-bistannole [64] and by introducing nitrogen functional substituents bis(2,6–bis(dimethylaminomethyl)phenyl-*C,N,N′*)-di-tin [65] (Fig. 8.46).

Distannenes (R_2SnSnR_2), believed to contain double bonds, have been known for many years [66]. The composition R_2SnSnR_2 suggests a structure with Sn=Sn double bonds, but the situation is different [67]. In the anion [Sn_2Ph_4]$^{2-}$, synthesized by the reaction of diphenyltin dichloride with Li in liquid NH_3 (and isolated as

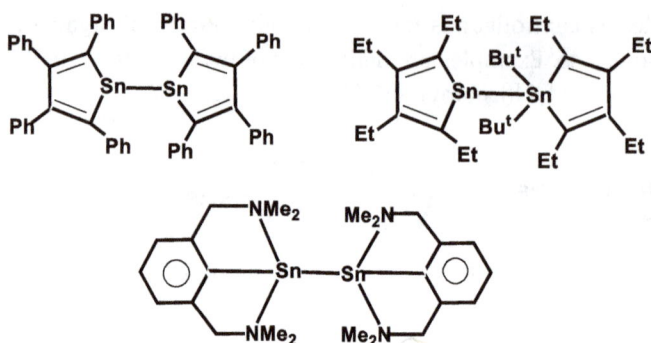

Fig. 8.46: Other distannane compounds.

Li(NH$_3$)$_4$] salt), the Sn–Sn interatomic distance is 2.905(3) Å, compared with the Sn–Sn distance in the nonionic Sn$_2$Ph$_6$ of 2.77 Å [68]. In the [K(18-crown-6)] salt of the dianion [Ph$_2$Sn–SnPh$_2$]$^{2-}$, the Sn–Sn bond length is 2.909(1) Å [69]. These long interatomic distances rule out a double bond character of the tin–tin bonds. In group 14, the heavier elements, tin and lead have a strong tendency to increase the lone-pair character versus the double bond character, with preference for structures with single Sn–Sn bonds and lone pairs at the metal (Fig. 8.47).

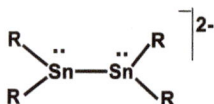

Fig. 8.47: The structure of distannene.

The Sn–Sn single bond character is demonstrated by the geometry of R$_2$SnSnR$_2$ in the solid state and by the SnSn interatomic distances, which are even longer than the Sn–Sn single bond lengths, due to repulsion of lone pairs. A result is that the R$_2$SnSnR$_2$ molecules dissociate in solution into pairs of heavy carbene analogues, :SnR$_2$.

An exception was, however, demonstrated with the compound (Bu$^t_2$MeSi)$_2$Sn = Sn(SiMeBu$^t_2$)$_2$, prepared by reacting SnCl$_2$.dioxane with Bu$^t_2$MeSiNa in THF. This is an authentic distannene and has a planar geometry, a shorter Sn=Sn bond, and is stable in both solid state and solution [70] (Fig. 8.48).

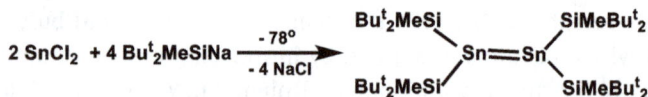

Fig. 8.48: The structure of an authentic distannene.

Distannynes (RSnSnR), formally tin analogues of alkynes, have been prepared with very bulky alkyl or aryl substituents. Unlike carbon, tin (and also lead) do not form authentic triple bonds. In compounds with tin (and lead), the triple bond has been transformed into a single bond and two nonbonded electron pairs.

The tin compound RSnSnR with R = C_6H_3-2,6-$(C_6H_3$-2,4,6-$C_6H_3Pr^i_2)_2$ has a *trans*-nonlinear structure, with a C–Sn–Sn bond angle of 125.1(2)°. Note that in the lead compound R–PbPbR with the same R, the C–Pb–Pb bond angle is 94.26(4)°. Another example is the distannyne RSnSnR, where R = 4-$Me_3SiC_6H_2(C_6H_3Pr^i_2$-2,6)$_2$, with Sn–Sn–C having 99.25(14)° and Sn–Sn length is 3.066 Å [71] (Fig. 8.49).

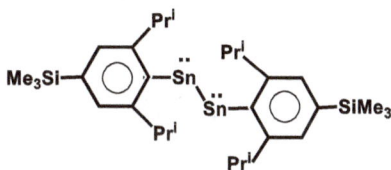

Fig. 8.49: The structure of a distannine.

Oligolinear polystannanes have been obtained with three to six tin atoms in the chain.

Tristannanes, X–SnR_2–SnR_2–SnR_2–X, are known with R = Me, X = $SiMe_3$; R = Bu^t, X = Cl; X = Bu^t; R = Ph, X = Bu^t, Ph, $Si(SiMe_3)_3$.

1,3-Diclorotristannane, $ClSnBu^t_2$-$SnBu^t_2$-$SnBu^t_2Cl$, was formed in a surprising reaction of $SnCl_4$ with Bu^tMgCl in hexane (45% yield), whereas the same reaction carried out in THF gave $ClSnBu^t_2Cl$ and in toluene gave $ClBu^t_2SnSnBu^t_2Cl$ [72].

Tristannane (Ph_3Sn–$SnPh_2$–$SnPh_3$) was obtained in a reaction of $M(SnPh_3)_2$ with Ph_3Sn–$SnPh_3$ (M = Ca,Sr) [12].

Several tetrastannanes X-SnR_2-SnR_2-SnR_2-SnR_2-X with R = Bu^t, X = Br, I, SPh^t, $SiMe_2Ph$, and also Ph_3Sn-$SnBu^t_2$-$SnBu^t_2$-$SnPh_3$ have been described. A preparation is available by reduction of organotin dihalides with sodium in liquid ammonia, in the presence of monohalides. The four polystannanes $Ph_3Sn(Bu^t_2Sn)_nSnPh_3$, synthesized from $I(SnBu^t_2)_nI$ with $Li[SnPh_3]$ are all-*trans*-configured (Fig. 8.50).

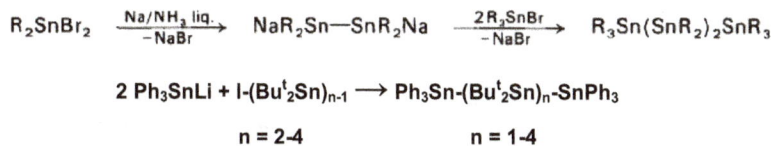

$$R_2SnBr_2 \xrightarrow[-NaBr]{Na/NH_3 \text{ liq.}} NaR_2Sn-SnR_2Na \xrightarrow[-NaBr]{2R_3SnBr} R_3Sn(SnR_2)_2SnR_3$$

$$2\ Ph_3SnLi + I\text{-}(Bu^t_2Sn)_{n-1} \longrightarrow Ph_3Sn\text{-}(Bu^t_2Sn)_n\text{-}SnPh_3$$

$$n = 2\text{-}4 \qquad\qquad n = 1\text{-}4$$

Fig. 8.50: Preparation of chain-like polystannanes.

The halogen-terminated tetrastannane chain $X(Bu^t_2Sn)_4X$ (X = Br, I) has been synthesized by controlled cleavage of the related cyclotetrastannane, $(Bu^t_2Sn)_4$, with iodine in toluene. Similarly prepared was $PhS(SnPh_2)_4SPh$ by the reaction of cyclotetrastannane, $(Bu^t_2Sn)_4$, with diphenyldisulfide (PhSSPh) [73].

Silylated tetrastannane (PhMe$_2$Si(SnBu$^t_2$)$_4$SiMePh) was obtained along with the tristannane (PhMe$_2$Si(SnBu$^t_2$)$_3$SiMe$_2$Ph) in a reaction of Cl(Bu$^t_2$Sn)$_3$Cl with PhMe$_2$-SiLi [74].

Pentastannanes and hexastannanes are zigzag chains of compositions Ph$_3$Sn (SnBu$^t_2$)$_3$SnPh$_3$ and Ph$_3$Sn(SnBu$^t_2$)$_3$SnPh$_3$ and have been described quite early [75] (Fig. 8.51).

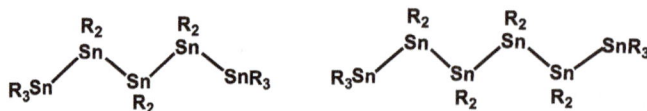

Fig. 8.51: Chain-like penta- and hexastannanes.

More recently, anionic [Ph$_2$SnSnPh$_2$]$^{2-}$, {Sn$_4$Ph$_4$]$^{4-}$ and hexatin [Sn$_6$Ph$_{12}$]$^{2-}$ species (Fig. 8.52) have been obtained in a reaction of Ph$_2$SnCl$_2$ with the Zintl-phase K$_4$Sn$_9$ [76].

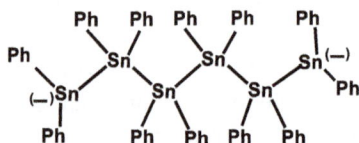

Fig. 8.52: A dianionic chain-like hexastannane.

Branched tetrastannanes have been prepared by condensing organotin trihydrides with trimethylstannylamine:

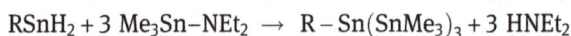

$$RSnH_2 + 3\ Me_3Sn-NEt_2 \rightarrow R-Sn(SnMe_3)_3 + 3\ HNEt_2$$

Branched anions of the type [Sn(SnPh$_3$)$_3$]$^{2-}$ (Fig. 8.53) are formed by redistribution of phenyl groups in an unusual reaction [12]:

$$M(SnPh_3)_2 + 6Ph_3Sn-SnPh_3 \rightarrow M[Sn(SnPh_3)_3]_2 + 6\ SnPh_4, \quad M = Cs,\ Sr$$

A branched-chain pentastannane (R = Ph) (Fig. 8.53) was synthesized and by a succession of reactions and structurally characterized by x-ray diffraction [12, 77, 78]:

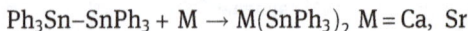

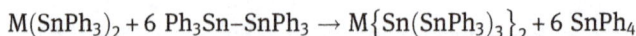

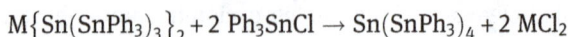

$$Ph_3Sn-SnPh_3 + M \rightarrow M(SnPh_3)_2\ M = Ca,\ Sr$$

$$M(SnPh_3)_2 + 6\ Ph_3Sn-SnPh_3 \rightarrow M\{Sn(SnPh_3)_3\}_2 + 6\ SnPh_4$$

$$M\{Sn(SnPh_3)_3\}_2 + 2\ Ph_3SnCl \rightarrow Sn(SnPh_3)_4 + 2\ MCl_2$$

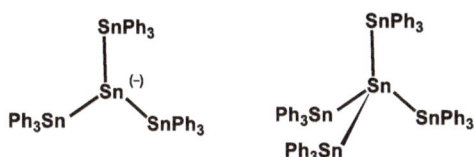

Fig. 8.53: Branched-chain polystannanes.

Cyclic polystannanes are remarkable tin analogues of cycloalkanes. The compounds of the composition SnR_2 were first believed to be derivatives of divalent tin, but many of these compounds are in fact cyclic polymers, $[SnR_2]_n$.

The decomposition of organotin hydrides in the presence of amine or alcohol catalysts produces cyclopolystannanes and hydrogen (Fig. 8.54).

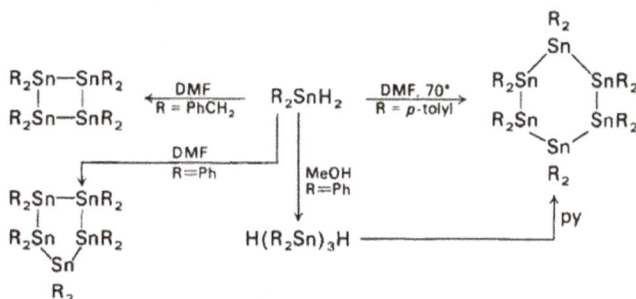

Fig. 8.54: Formation of some cyclopolystannanes.

Cyclopolystannanes result from the reaction between diorganotin amides and hydrides:

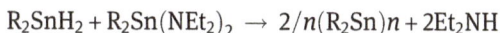

$$R_2SnH_2 + R_2Sn(NEt_2)_2 \rightarrow 2/n(R_2Sn)n + 2Et_2NH$$

At present time, cyclostannanes with three, four, five and six tin atoms in the ring have been reported.

Three-membered ring compounds, $(R_2Sn)_3$, are known with bulky ligands $R = C_6$-H_3Et_2-2,6 [79] and $C_6H_2Pr^i$-2,4,6 [80] (Fig. 8.53). The Sn_3 ring was formed in the reaction of Ar_2SnCl_2 with lithium naphthalenide.

Four-membered ring cyclic compounds $(R_2Sn)_4$ with $R = Ph$, *tert*-butyl, *tert*-amyl [81], $C_6H_3Mes_2$-2,6 [82], CH_2SiMe_3 [83] and Sn_4R_7X with $X = Br$ [163], Me [84] have been reported (Fig. 8.55). In the reaction of $Bu^t_2SnCl_2$ with Bu^tMgCl the main product was the cyclotetrastannane $[SnBu^t_2]_4$; the chain-like tristannane $(Bu^t_3Sn–SnBu^t_2–SnBu^t_3)$ and distannane $(Bu^t_3Sn–SnBu^t_3)$ were also isolated from the reaction product [85]. Several cyclohexastannanes $(R_2Sn)_6$ with $R = Me$ [86], Ph [87], CH_2Ph [88], CF_2CF_3 [89]

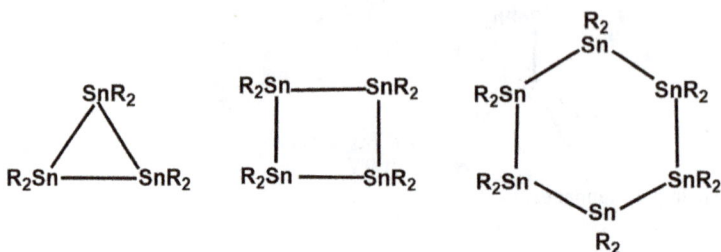

Fig. 8.55: Three-, four- and six-membered ring cyclopolystannanes.

containing the Sn_6 ring of chair conformation (Fig. 8.55) have been prepared by reductive reactions of diorganoditin halides.

Apparently, there is no structurally characterized monocyclic pentamer $[(R_2Sn]_5$.

Some interesting bicyclic stannanes with propelane structures and $R = C_6H_3Et_2$-2,6 [66], $C_6H_3(OPr^i)_2$–2,6 [91] and $C_6H_3Mes_2$-2.6 [92] are known (Fig. 8.56).

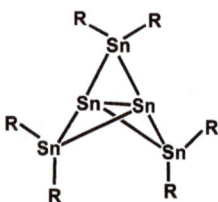

Fig. 8.56: Bicyclic pentastannanes with propelane structure.

A bicyclic hexastannane, $Sn(SnPh_2SnPh_2)_3Sn$ [93], and a bicyclic anionic octastannane, $(Sn_2Ar_2)(SnBu)(SnAr)(Sn_2Ar_2)_2$ with $Ar = C_6H_3Et_2$-2,6 [94], contain two fused rings (Fig. 8.57).

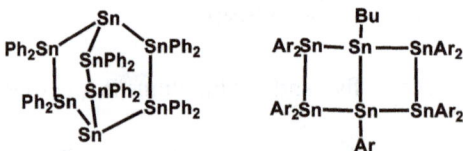

Fig. 8.57: Fused ring bicyclic polystannanes.

Polycyclic stannanes with cubane structure $[Sn_8(C_6H_3Et_2$-2,6$)_8]$ [95] and $[Sn_8(2,6$-$MesC_6H_3)_4]$ [96] and pentagonal prismatic structure $[SnC_6H_3Et_2$-2,6$]_{10}$ and $[SnC_6H_3\{CH(SiMe_3)_2\}]_{10}$ have been obtained by reductive elimination of hydrogen from appropriate monoaryltin trihydrides. The pentagonal prismatic cage, $Sn_{10}(C_6H_3Et_2)_{10}$, was formed in the pyrolysis of the cyclotristannane, $[Sn(C_6H^3Et_2$-2,6$)_2]_3$, along with the octamer [97] (Fig. 8.58).

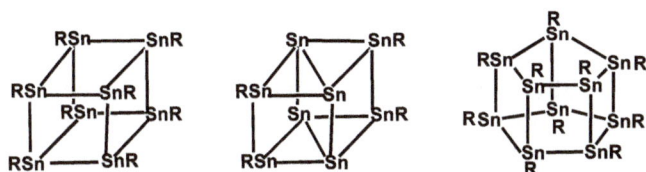

Fig. 8.58: Organotin [RSn]$_n$ cages with $n = 8$ and 10.

A unique heptatin cluster with pentagonal bipyramidal geometry is also worth mentioning here. It contains in the apical positions two bulky $C_6H_3Pr^i_2$-2,6 organic groups attached to tin [98] (Fig. 8.59).

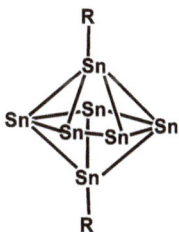

Fig. 8.59: Pentagonal bipyramidal heptatin cluster.

Further reading

Rabiee N, Safarkhani M, Amini MM Investigating the structural chemistry of organotin (IV) compounds: Recent advances. Rev Inorg Chem 2018, 38, 13–45.

Caseri W Initial organotin chemistry. J Organomet Chem 2014, 751, 20–24.

Chandrasekhar V, Nagendran S, Baskar V Organotin assemblies containing Sn-O bonds. Coord Chem Rev 2002, 235, 1–52.

Tiekink ERT Structural chemistry of organotin carboxylates: A review of the crystallographic literature. Appl Organomet Chem 1991, 5, 1–23.

Sita LR Heavy-metal organic chemistry: Building with tin. Acc Chem Res 1994, 27, 191–97.

Sita LR Structure/property relationships of polystannanes. Adv Organomet Chem 1995, 38, 189–243.

Power PP Bonding and reactivity of heavier Group 14 element alkyne analogues. Organometallics 2007, 26, 4362–72.

Sekiguchi A, Sakurai H Cage and cluster compounds of silicon, germanium, and tin. Adv Organomet Chem 1995, 37, 1–38.

Sekiguchi A, Lee V Heavy cyclopropenes of Si, Ge, and Sn. A new challenge in the chemistry of Group 14 elements. Chem Rev 2003, 103, 1429–47.

Schrenk C, Schnepf A Metalloid Sn clusters: Properties and the novel synthesis via a disproportionation reaction of a monohalide. Rev Inorg Chem 2014, 34, 93–118.

8.2 Organolead compounds

Because of the lower metal–carbon bond strength, organolead derivatives decompose at moderate temperatures (100–200 °C), are slowly oxidized in air and are somewhat light sensitive. The Pb–C bond is stable to moisture, although those containing Pb–C$_6$F$_5$ groups are more readily hydrolyzed.

Lead forms the same types of compounds as tin, namely, tetrasubstituted derivatives PbR$_4$ and R$_n$PbX$_{4-n}$, where X = H, halogen, OH, OR, NRR', SR, etc., addition compounds with bases in which the lead atom has a coordination number greater than four and oligomers and polymers with Pb–Pb, Pb–O–Pb, Pb–S–Pb or Pb–NPb backbones. However, the rich diversity of organotin compounds is not matched by lead.

8.2.1 Homoleptic compounds, PbR$_4$

The past industrial interest in tetraethyllead stimulated the intensive investigation of synthetic methods for tetrasubstituted derivatives.

Grignard and organolithium reagents are of most utility for laboratory purposes. Since lead(IV) chloride is unstable, lead(II) chloride is used as a starting material. Organolead(II) derivatives, PbR$_2$, are intermediates which disproportionate to either PbR$_4$ or R$_3$Pb–PbR$_3$ as final products. The overall reactions are:

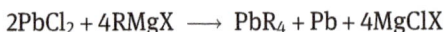

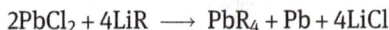

$$2PbCl_2 + 4RMgX \longrightarrow PbR_4 + Pb + 4MgClX$$

$$3PbCl_2 + 6RMgX \longrightarrow Pb_2R_6 + Pb + 3MgX_2 + 3MgCl_2$$

$$2PbCl_2 + 4LiR \longrightarrow PbR_4 + Pb + 4LiCl$$

The deposition of lead metal is avoided if an alkyl iodide is added. The organic group R must be the same as that in the Grignard or organolithium reagent (M = MgX or Li):

$$2PbCl_2 + 6MR + 2RI \rightarrow 2PbR_4 + 2MI + 4MCl$$

The use of lead(IV) salts, Pb(OCOCH$_3$)$_4$ or K$_2$[PbCl$_6$], offers no advantage over PbCl$_2$; however, these are used in the synthesis of acetylene derivatives:

$$K_2[PbCl_6] + 4LiC \equiv CR \rightarrow Pb(C \equiv CR)_4 + 2KCl + 4LiCl$$

Lead(IV) acetate is employed in the synthesis of tetraethyl- and tetramethyllead:

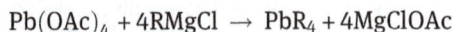

$$Pb(OAc)_4 + 4RMgCl \rightarrow PbR_4 + 4MgClOAc$$

The large-scale industrial availability of triethylaluminum stimulated its use in the synthesis of tetraethyllead:

$$6PbX_2 + 4AlEt_3 \rightarrow 3PbEt_4 + 3Pb + 4AlX_3$$

Ethyl iodide in alcoholic alkalies is electrolyzed with lead cathodes to prepare tetraethyllead. The complex $Na[Et_3Al\text{-}F\text{-}AlEt_3]$, melting at 35 °C, is used as an electrolyte with lead anodes:

$$4Na[Et_3Al\text{--}F\text{--}AlEt_3] + 3Pb \xrightarrow{e^-} 3PbEt_4 + 4NaEt_3AlF$$

The NALCO industrial process produces tetraethyl- and tetramethyllead by the electrolysis of a Grignard reagent and an alkyl halide with lead anodes:

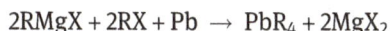

$$2RMgX + 2RX + Pb \rightarrow PbR_4 + 2MgX_2$$

The chemical properties of tetraorganolead derivatives reflect the weak Pb–C bond, and Pb–C bond cleavage reactions are used in the synthesis of organolead halides. Tetraorganolead compounds can also transfer organic groups to metals and nonmetals by distribution and act as alkylating agents (Fig. 8.60).

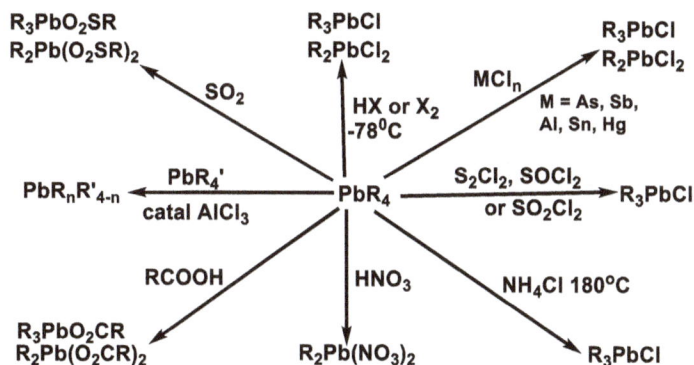

Fig. 8.60: Typical reactions of tetraorganolead compounds.

8.2.2 Heterocycles with lead heteroatoms

Saturated lead heterocycles are prepared by Grignard reagents, and organolithium reagents are used for the synthesis of tricyclic aromatic compounds (Fig. 8.61).

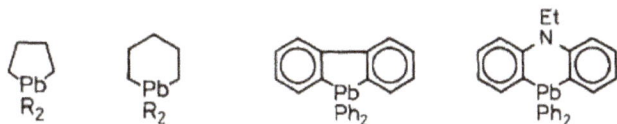

Fig. 8.61: Organolead heterocyclic compounds.

Spirocyclic compounds have been prepared with the aid of Grignard reagents (Fig. 8.62).

$$3 \; Ph_2Pb \, (OCOCH_3)_2 \quad + \quad 3 \; H_2S \quad \longrightarrow$$

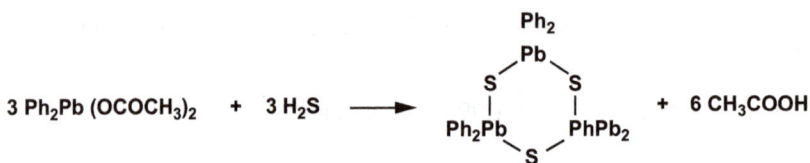

$$+ \quad 6 \; CH_3COOH$$

Fig. 8.62: Organolead spirocyclic compound.

8.2.3 Subvalent species

Only the bis-cyclopentadienyllead(II) derivatives, $:Pb(\eta^5\text{-}C_5R_5)_2$, and the purple $:Pb$ $[CH(SiMe_3)_2]_2$ are established monomeric derivatives of divalent lead.

Bis(η^5-cyclopentadienyl)lead(II) has an angular sandwich structure in the vapor phase but a polymeric structure in the solid state (Fig. 8.63).

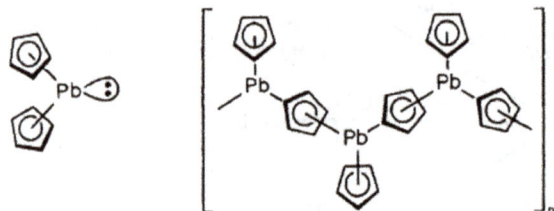

Fig. 8.63: Monomeric and supramolecular $Pb(\eta^5\text{-}C_5H_5)_2$.

Cleavage of $:Pb(\eta^5\text{-}C_5H_5)_2$ with acids leads to mono-cyclopentadienyl derivatives, C_5H_5PbX.

The bis(trimethylsilyl)methyl derivative $:Pb[CH(SiMe_3)_2]_2$ is prepared from $PbCl_2$ and $LiCH(SiMe_3)_2$ like the tin analogue and also forms metal carbonyl complexes.

8.2.4 Inverse organolead compounds

A typical inverse organometallic compound is the tetrahedral $[(\mu_4\text{-}C)(PbBrPh_2)_4]$ prepared from Ph_3PbLi and CCl_4 in THF [99]. Similar compounds are $Ph_3Pb\text{-}CCl_2\text{-}PbPh_3$ and $Ph_3Pb\text{-}CH_2\text{-}PbPh_3$ [100] (Fig. 8.64).

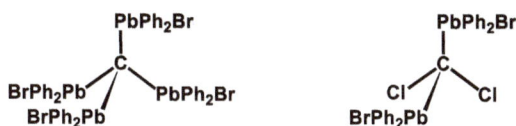

Fig. 8.64: Inverse organolead compounds.

8.2.5 Organolead halides, R_nPbX_{4-n}

Derivatives with $n = 3$ and 2 are prepared from tetrasubstituted compounds by Pb–C bond cleavage. Cleavage with halogens or with the halides of other elements can be stopped at the R_3PbX or R_2PbX stages.

Dilead derivatives, $R_3Pb–PbR_3$, can be cleaved to give triorganolead halides, R_3PbX.

Monoalkyllead triiodides ($RPbI_3$) are reaction products of lead(II) iodide with RI in the presence of an $SbMe_3$ catalyst.

Solid organolead halides contain lead in higher than four coordination, as in diphenyllead dichloride which is a chlorine double-bridged supramolecular array. Triphenyllead chloride and bromide consist of halogen-bridged chains made up of trigonal bipyramidal units (Fig. 8.65).

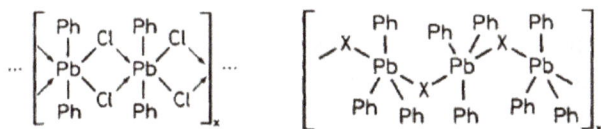

Fig. 8.65: Supramolecular self-assembly of organolead halides.

Hypervalent anionic species containing five-coordinated lead atoms of the type $[R_3PbX_2]^-$ and $[R_2PbX_3]^-$ such as $[NMe_4]^+[Ph_3PbCl_2]^-$ are known.

The cationic species $[Ph_4Pb_2I_2]^{2+}$ and $[Ph_4Pb_2I_3]^+$ are formed in methanolic solutions of Ph_3PbX and Ph_2PbX_2 in addition to mononuclear species. The $[Me_3Pb]^+$ cations are formed in methylene chloride solutions of $[Me_3Pb]^+[MeAlCl_3]^-$ obtained from Me_3PbCl and $MeAlCl_2$.

8.2.6 Organolead hydroxides, $R_nPb(OH)_{4-n}$

The hydroxides are obtained by the hydrolysis of the corresponding halide in alcoholic alkali solutions or by wet silver oxide. These are ionic compounds which form weakly basic aqueous solutions and react with organic and inorganic acids to form the corresponding salts.

Hydroxylated dimethyllead species are formed in aqueous sodium perchlorate solutions (Fig. 8.66).

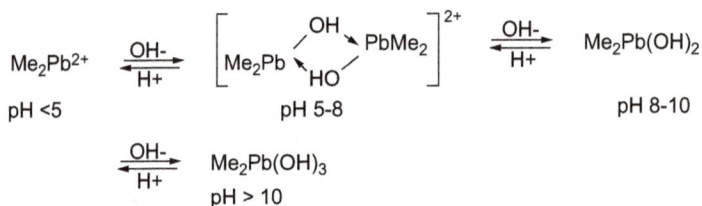

$$\text{Me}_2\text{Pb}^{2+} \quad \xrightleftharpoons[\text{H+}]{\text{OH-}} \quad \left[\text{Me}_2\text{Pb} \overset{\displaystyle\text{OH}}{\underset{\displaystyle\text{HO}}{\diagup\hspace{-0.5em}\diagdown}} \text{PbMe}_2 \right]^{2+} \quad \xrightleftharpoons[\text{H+}]{\text{OH-}} \quad \text{Me}_2\text{Pb(OH)}_2$$

pH <5   pH 5-8   pH 8-10

$$\xrightleftharpoons[\text{H+}]{\text{OH-}} \quad \text{Me}_2\text{Pb(OH)}_3$$

pH > 10

Fig. 8.66: Organolead hydroxo compounds.

Solid triphenyllead hydroxide consists of associated supramolecular zigzag chains.

8.2.7 Organolead oxides

Diplumboxanes or triorganolead oxides, $R_3Pb-O-PbR_3$, are obtained by hydrolysis of triorganolead halides, followed by condensation of the intermediate hydroxides:

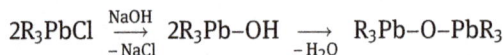

$$2R_3PbCl \xrightarrow[-\text{NaCl}]{\text{NaOH}} 2R_3Pb-OH \xrightarrow[-\text{H}_2\text{O}]{} R_3Pb-O-PbR_3$$

The diorganolead oxides, RPbO, are amorphous, insoluble and infusible, suggesting a polymeric structure. The plumbonic acid, $RPbO_2H$, are also polymeric.

8.2.8 Organolead alkoxides

Triorganolead oxides react with alcohols to form alkoxides:

$$R_3Pb-O-PbR_3 + R'OH \longrightarrow R_3Pb-OR' + R_3Pb-OH$$

but a better procedure starts with the halides:

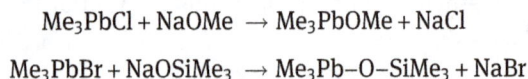

$$Me_3PbCl + NaOMe \rightarrow Me_3PbOMe + NaCl$$

$$Me_3PbBr + NaOSiMe_3 \rightarrow Me_3Pb-O-SiMe_3 + NaBr$$

The alkoxides undergo insertion reactions (Fig. 8.67).

$R_3Pb-S-C-OR$ $\xleftarrow{CS_2}$ R_3Pb-OR $\xrightarrow{CO_2}$ $R_3Pb-O-C-OR$

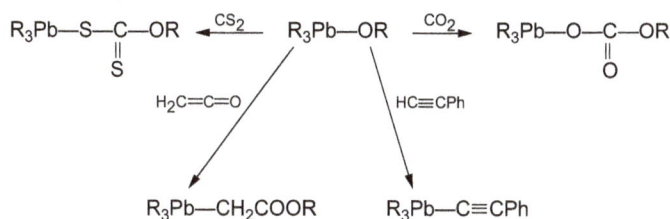

Fig. 8.67: Some reactions of organolead alkoxides.

8.2.9 Organolead carboxylates, $R_nPb(OCOR')_{4-n}$

The only well-defined monoorganolead compounds belong to this class. Along with $RPb(OCOR')_3$, di- and triorgano-substituted lead derivatives, $R_2Pb(OCOR')_2$ and $R_3Pb(OCOR)$, are obtained by the cleavage of tetrasubstituted compounds with carboxylic acids:

$$PbR_4 + R'COOH \rightarrow R_3PbOCOR' + RH$$

These reactions are less vigorous than with halogens and, therefore, easier controlled.

Organolead carboxylates are often preferred over the halides as starting materials, owing to their ready availability through this synthesis.

Organolead dicarboxylates and mercury(II) carboxylates react to produce monoorganolead derivatives:

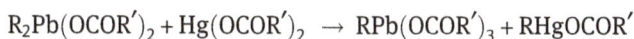

$$R_2Pb(OCOR')_2 + Hg(OCOR')_2 \rightarrow RPb(OCOR')_3 + RHgOCOR'$$

The tetracarboxylates can also be used with diorganomercury derivatives:

$$Pb(OCOR')_4 + HgR_2 \rightarrow RPb(OCOR')_3 + RHgOCOR'$$

Solid $Me_3PbOCOCH_3$ consists of associated supramolecular chains of planar, trigonal bipyramidal Me_3Pb groups bridged by acetato fragments.

8.2.10 Organolead sulfide

Triorganolead sulfides are obtained by the reaction of halides with sodium sulfide:

$$2R_3PbX + Na_2S \longrightarrow R_3Pb-S-PbR_3 + 2NaX$$

The diorganolead sulfides are cyclic trimers. The phenyl derivative is prepared from diphenyllead diacetate and hydrogen sulfide (Fig. 8.68).

$$3Ph_2Pb(OCOCH_3)_2 + 3H_2S \longrightarrow$$

$$+ 6CH_3COOH$$

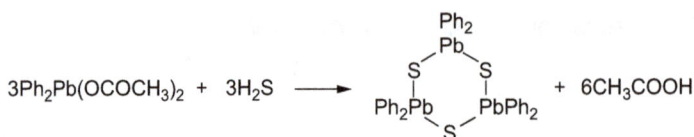

Fig. 8.68: Cyclic organolead sulfide.

Heating $Ph_3Pb-PbPh_3$ with sulfur in benzene produces $Ph_3Pb-S-PbPh_3$ and $[Ph_2PbS]_3$. Carbon disulfide serves as a sulfur source in the conversion of triphenyl-lead hydroxide to the corresponding sulfide.

8.2.11 Organolead thiolates $R_nPb(SR')_{4-n}$

The lead–sulfur bond is stable toward water. Mercapto derivatives are prepared from organolead chlorides, hydroxides or thiolates:

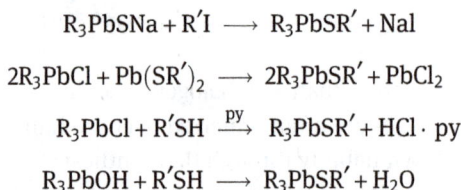

$$R_3PbSNa + R'I \longrightarrow R_3PbSR' + NaI$$

$$2R_3PbCl + Pb(SR')_2 \longrightarrow 2R_3PbSR' + PbCl_2$$

$$R_3PbCl + R'SH \xrightarrow{py} R_3PbSR' + HCl \cdot py$$

$$R_3PbOH + R'SH \longrightarrow R_3PbSR' + H_2O$$

Diorganolead derivatives, $R_2Pb(SR)_2$, are prepared similarly.

8.2.12 Organolead compounds with Pb–Pb bonds

Diplumbanes are obtained by the reaction of lead(II) chloride and Grignard reagents:

$$3PbCl_2 + 6RMgX \longrightarrow 6[PbR_2] \longrightarrow R_3Pb-PbR_3 + Pb$$

or by reduction of triorganolead halides with sodium in liquid ammonia.

The mechanism of the formation of diplumbanes via Grignard reactions is given by the following sequence:

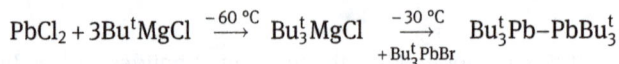

$$PbCl_2 + 3Bu^tMgCl \xrightarrow{-60\ ^\circ C} Bu_3^tMgCl \xrightarrow[+ Bu_3^tPbBr]{-30\ ^\circ C} Bu_3^tPb-PbBu_3^t$$

Simultaneous oxidation and hydrolysis with hydrogen peroxide and ice of Ph_3PbLi yields the branched pentaplumbane (Fig. 8.69).

The PbR_2 species may be cyclic polyplumbanes but their structures have not been established.

$$\text{Ph}_3\text{Pb}\text{—}\overset{\overset{\displaystyle \text{PbPh}_3}{|}}{\underset{\underset{\displaystyle \text{PbPh}_3}{|}}{\text{Pb}}}\text{—PbPh}_3$$

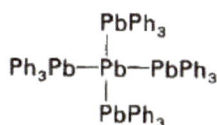

Fig. 8.69: A branched tetraplumbane.

References

[1]   Wiesemann M, Klösener J, Neumann B, Stammler HG, Hoge B. On pentakis(pentafluoroethyl) stannate, [Sn(C$_2$F$_5$)$_5$]$^-$, and the gas-free generation of pentafluoroethyllithium, LiC$_2$F$_5$. Chem-Eur J 2018, 24, 1838–43.

[2]   Saito M, Imaizumi S, Tajima T, Ishimura K, Nagase S Synthesis and structure of pentaorganostannate having five carbon substituents. J Am Chem Soc 2007, 129, 10974–75.

[3]   Gebauer I, Gräsing D, Matysik J, Zahn S, Zeckert K Penta- and hexaorganostannate(iv) complexes based on O-heterocyclic ligands. Dalton Trans 2017, 46, 8279–85.

[4]   Schrader I, Zeckert K, Zahn S. Dilithium Hexaorganostannate(IV) Compounds. Angew Chem Int Ed 2014, 53, 13698–700.

[5]   Rodesiler PF, Amma EL, Auel T Metal ion-aromatic complexes. XXII. Preparation, structure, and stereochemistry of tin (II) in π-benzenetin di(tetrachloroaluminate)-benzene. J Am Chem Soc 1975, 97, 7405–10.

[6]   Schmidbaur H, Probst T, Huber B, Steigelmann O, Müller G Arene complexes of p-block elements: [(η^6-C$_6$H$_6$)$_2$SnCl(AlCl$_4$)]$_2$. The first bis(arene) coordination compound of a Group 14 element. Organometallics 1989, 1567–69.

[7]   Probst T, Steigelmann O, Riede J, Schmidbaur H. GeII and SnII complexes of [2.2.2] paracyclophane with threefold internal η^6 coordination. Angew Chem Int Ed 1990, 29, 1397–98.

[8]   Wendji AS, Lutter M, Stratmann LM, Jurkschat K. Syntheses, structures, and complexation studies of tris(organostannyl)methane derivatives. Chemistry Open 2016, 5, 554–65.

[9]   Hillwig R, Harms K, Dehnicke K Die Kristallstruktur von Tris(trimethylstannyl) acetonitril. Crystal structure of tris(trimethylstannyl)-acetonitrile. Z Naturforsch B 1997, 52, 145–48.

[10] Dimmel DR, Wilkie CA, Lamothe PJ The mass spectra of (trimethylstannyl)-methanes: Occurrence of tin-carbon double bonds. Org Mass Spectrom 1975, 10, 18–25.

[11] Klinkhammer K, Kühner S, Regelmann B, Weidlein J. Die Kristall- und molekülstruktur von tetrakis(trimethylstannyl) methan. J Organomet Chem 1995, 496, 241–43.

[12] Englich U, Ruhlandt-Senge K, Uhlig F Novel triphenyltin substituted derivatives of heavier alkaline earth metals. J Organomet Chem 2000, 613, 139–47.

[13] Wrackmeyer B, Khan E, Badshah A, Molla E, Thoma P, Tok OL, Milius W, Kempe R, Senker J. Tetra(alkynyl)silanes, a 3,6-disila-triyne, a 3,6,9-trisila-tetrayne, a 1,3,4,6-tetrasiladiyne, and bis(trimethylstannyl)ethyne. Molecular Structures and Solid-state NMR Studies Z Naturforsch B 2010, 65, 119–27.

[14] Rojas-León I, Hernández-Cruz MG, Vargas-Olvera EC, Höpfl H, Alnasr H, Jurkschat K Dinuclear organotin compounds carrying naphthylene- and biphenylene-spacer groups. J Organomet Chem 2020, 920, 121344.

[15] Lamm J-H, Glatthor J, Weddeling J-H, Mix A, Chmiel J, Neumann B, Stammler H-G, Mitzel NW Polyalkynylanthracenes – Syntheses, structures and their behaviour towards UV irradiation. Org Biomol Chem 2014, 12, 7355–65.

[16] Niermeier P, Teichmann L, Neumann B, Stammler H-G, Mitzel NW 1,4,5,8-Tetraethynylanthracene – synthesis, UV/Vis absorption spectroscopy and its application as building block for tetradentate acceptor molecules. Eur J Org Chem 2018, 2018, 6780–86.

[17] Lenze N, Neumann B, Stammler H-G, Jutzi P Tetrakis(trimethylstannyl)cyclopentadiene. Improved synthesis, ^{1}H-, ^{13}C- and ^{119}Sn-NMR data and crystal structure. J Organomet Chem 2000, 608, 86–88.

[18] Hill MS, Mahon MF, McGinley JMG, Molloy KC The synthesis and characterisation of C-triorganometallated (metal=Sn, Si) bis-(thienyl)- and bis-(pyrazolyl)alkanes, including the crystal structure of [(Ph$_3$Sn)C$_3$N$_2$]$_2$CH$_2$. Polyhedron 2001, 20, 1995–2002.

[19] Yamin MB, Boshaala MAA, Hamid OA, Hoong-Kun Fun IAR, O Bin S. Crystal structure of ferrocenium μ-chloro-bis[dichlorodimethylstannate(IV)], C$_{14}$H$_{22}$Cl$_5$FeSn$_2$. Z Kristallogr. 1998, 213, 515–16.

[20] Razak IA, Fun HK, Yamin BM, Boshaala AMA, Chinnakali K. Ferrocenium μ-bromo-bis [dibromodimethylstannate(IV)]. Acta Crystallogr Sect C Cryst Struct Commun 1998, 54, 912–14.

[21] Zhang F-X, Kuang D-Z, Feng Y-L, Wang J-Q, Xu Z-F Synthesis, crystal structures, and quantum chemistry of the μ-oxygen-bis[tri(o-chlorobenzyl)tin] and μ-oxygen-bis[tri(o-fluorobenzyl)tin]. Wuji Huaxue Xuebao 2002, 18, 1057–62.

[22] Padělková Z, Weidlich T, Kolářová L, Eisner A, Císařová I, Zevaco TA, Ruzicka A. Products of hydrolysis of C,N-chelated triorganotin(IV) chlorides and use of products as catalysts in transesterification reactions. J Organomet Chem 2007, 692, 5633–45.

[23] Puff H, Schuh W, Sievers R, Zimmer R A Diorganotin oxide with a planar tin-chalcogen six-membered ring. Angew Chem Int Ed 1981, 20, 591–591.

[24] Dey S, Schönleber A, Mondal S, Ali SI, van Smaalen S. Role of steric hindrance in the crystal packing of Z′ = 4 superstructure of trimethyltin hydroxide. Cryst Growth Des 2018, 18, 1394–400.

[25] D'yachenko OA, Zolotoi AB, Atovmyan LO, Mirskov RG, Voronkov MG Crystal and molecular structure of hexaphenyldistannathiane [(C$_6$H$_5$)$_3$Sn]$_2$S. Dokl Akad Nauk SSSR (Russ Proc Nat Acad Sci USSR) 1977, 237, 863–66.

[26] Zheng X-F, Yin L, Liu X-C, Tian L-J μ-Sulfido-bis[tricyclohexyltin(IV)]. Acta Crystallogr Sect E Struct Reports Online 2007, 63, m1744–m1744.

[27] Krebs B, Jacobsen H-J Kristall- und molekülstruktur von hexaphenyldistannylselenid (C$_6$H$_5$)$_3$SnSeSn(C$_6$H$_5$)$_3$. J Organomet Chem 1979, 178, 301–08.

[28] Hillwig R, Harms K, Dehnicke K, Müller U Elementorganisch substituierte Ammonium-Salze. Die Kristallstrukturen von [HN(SnMe$_3$)$_3$]I, [H$_2$N(SnMe$_3$)$_2$][SnMe$_3$Cl$_2$] und [N(AsMe$_3$)$_2$]Br. Z Anorg Allg Chem 1997, 623, 676–82.

[29] Hausen HD, Kuhnle R, Weidlein J Zur Struktur des Diazadimethylstannetidinderivats [(CH$_3$)$_2$Sn(NC(CH$_3$)$_3$)$_2$GaCH$_3$]$_2$. Z Naturforsch B 1995, 50, 1419–23.

[30] Eichler JF, Just O, Rees WS, Synthesis and characterization of imidocubanes with exocube GeIV and SnIV substituents: [M(μ$_3$-NGeMe$_3$)]$_4$ (M = Sn, Ge, Pb); [Sn(μ$_3$-NSnMe$_3$)]$_4$. Inorg Chem 2006, 45, 6706–12.

[31] Räke B, Müller P, Roesky HW, Usón I. Synthesis and structural characterization of graphite-like [(Me$_3$Sn)$_3$O]Cl. Angew Chem Int Ed 1999, 38, 2050–52.

[32] Appel A, Kober C, Neumann C, Nöth H, Schmidt M, Storch W. Synthesis and structure of tris (trialkylstannyl)- and tris(dialkylhalostannyl)amines; stabilization of the Sn$_3$N skeleton by intramolecular Sn–X–Sn bridges. Chem Ber 1996, 129, 175–89.

[33] Cummins CC, Huang C, Miller TJ, Reintinger MW, Stauber JM, Tannou I, Tofan D, Toubaei A, Velian A, Wu G The stannylphosphide anion reagent sodium bis(triphenylstannyl) phosphide:

Synthesis, structural characterization, and reactions with indium, tin, and gold electrophiles. Inorg Chem 2014, 53, 3678–87.

[34] Driess M, Monsé C, Merz K, van Wüllen C. Perstannylated ammonium and phosphonium ions: organometallic onium ions that are also base-stabilized stannylium ions. Angew Chem 2000, 39, 3684–86.

[35] Ballivet-Tkatchenko D, Burgat R, Plasseraud L, Richard P The ionic tin (IV) complex tri-μ_2-methoxy-μ_3-oxo-tris[di-tert-butyltin(IV)] tri-μ_2-methoxy-bis[tert-butyldimethoxystannate(IV)]. Acta Crystallogr Sect E Struct Reports Online 2004, 60, m830–2.

[36] Kober C, Kroner J, Storch W. Tris(chlorodimethylstannyl)amine, a molecule with a planar Sn_3N skeleton stabilized by intramolecular Sn-Cl-Sn bridges. Angew Chem Int Ed 1993, 32, 1608–10.

[37] Singh N, Bhattacharya S. Synthesis and structural studies of an electron deficient hexatitanium cluster dication stabilized by novel tritin anions. Dalton Trans 2011, 40, 2707–10.

[38] Dornsiepen E, Dehnen S. Behavior of organotin sulfide clusters towards zinc compounds. Eur J Inorg Chem 2019, 2019, 4306–12.

[39] Singh N, Bhattacharya S Synthesis and structural studies of an electron deficient hexatitanium cluster dication stabilized by novel tritin anions. Dalton Trans 2011, 40, 2707–10.

[40] Reuter H, Wilberts H On the structural diversity anions coordinate to the butterfly-shaped $[(R_2Sn)_3O(OH)_2]^{2+}$ cations and vice versa. Can J Chem 2014, 92, 496–507.

[41] Cervantes-Lee F, Sharma HK, Haiduc I, Pannell KH A unique self-assembled tricyclic stannasiloxane containing a planar Sn_3SiO_5 fused 6.4.4 tricyclic ring system. J Chem Soc Dalton Trans 1998, 1–2.

[42] Beckmann J, Jurkschat K, Schürmann M, Suter D, Willem R. Synthesis and molecular structure of a tricyclic stannasiloxane containing a novel $SiSn_3O_3F_2$ structural motif. Organometallics 2002, 21, 3819–22.

[43] Beckmann J, Dakternieks D, Duthie A, Jurkschat K, Mehring M, Mitchell C, Schurmann M. The isoelectronic replacement of E = P^+ and Si in the trinuclear organotin–oxo clusters [Ph_2E $(OSn^tBu_2)_2O\cdot^tBu_2Sn(OH)_2$]. Eur J Inorg Chem 2003, 2003, 4356–60.

[44] Molloy KC, Quill K, Cunningham D, McArdle P, Higgins T. A reinvestigation of the structures of organotin sulphates and chromates, including the crystal and molecular structure of bis (trimethyltin) sulphate dihydrate. J Chem Soc Dalton Trans 1989, 267–73.

[45] Diop CAK, Toure A, Diop A, Bassene S, Sidibe M, Diop L, Mahon MF, Molloy KC, Russo U. J Soc Ouest-Afr Chim 2007, 23, 49.

[46] Diop T, Diop L, Michaud F, Ardisson JD $Et_4N[NO_3(SnClPh_3)_2(SnPh_3NO_3)]$: A trinuclear organostannate complex and related derivatives. Main Gr Met Chem 2013, 36, 83–88.

[47] Diop L, Mahieu B, Mahon MF, Molloy KC, Okio KYA Bis(triphenyltin) oxalate. Appl Organomet Chem 2003, 17, 881–82.

[48] Ng SW. Bis(triphenyltin) succinate and its complex with dimethyl sulfoxide and ethanol, and its complex with hexamethylphosphoramide. Acta Crystallogr Sect C Cryst Struct Commun 1998, 54, 745–50.

[49] Tian L-J, Yu F-Y, Sun Y-X, Ding Y-J. ($\mu3$-1,3,5-Benzenetricarboxylato)-tris[triphenyltin(IV)] trichloromethane solvate. Acta Crystallogr Sect E Struct Reports Online 2006, 62, m1203–4.

[50] Ma C, Han Y, Zhang R, Wang D. Self-assembled triorganotin(IV) moieties with 1,3,5-benzenetricarboxylic acid: syntheses and crystal structures of monomeric, helical, and network triorganotin(IV) complexes. Eur J Inorg Chem 2005, 2005, 3024–33.

[51] Xiao X, Sui C, Han L, Liu J, Feng B Self-assembly of triorganotin(IV) moiety with 1,2,4,5-benzenetetracarboxylic acid: Syntheses, characterizations, and influence of solvent on the molecular structure (II). Heteroat Chem 2017, 28, e21356.

[52] Zhao L, Liang J, Yue G, Deng X, He Y. Bis(triphenylstannyl) thiophene-2,5-dicarboxylate. Acta Crystallogr Sect E Struct Reports Online 2009, 65, m722–m722.

[53] Yan C, Xu Z, Xiao XQ, Li Z, Lu Q, Lai G, Kira M Reactions of an isolable dialkylstannylene with carbon disulfide and related heterocumulenes. Organometallics 2016, 35, 1323–28.

[54] Xie J, Boyn J-N, Filatov AS, McNeece AJ, Mazziotti DA, Anderson JS Redox, transmetallation, and stacking properties of tetrathiafulvalene-2,3,6,7-tetrathiolate bridged tin, nickel, and palladium compounds. Chem Sci 2020, 11, 1066–78.

[55] Haiduc I, Mahon MF, Molloy KC, Venter MM Synthesis and spectral characterisation of organotin (IV) 1,3,5-triazine-2,4,6-trithiolato complexes, including the crystal structures of 1,3,5-(R$_3$Sn)$_3$C$_3$N$_3$S$_3$ (R = Me, Ph). J Organomet Chem 2001, 627, 6–12.

[56] Poplaukhin P, Tiekink ERT (μ$_2$-Pyridinealdazine-κ^4 N N': N''N''')bis[bis(N,N-di-n-propyldithiocarbamato-κ2 S,S ')cadmium(II)]. Acta Crystallogr Sect E Struct Reports Online 2008, 64, m1176–m1176.

[57] Fuentes-Martínez JP, Toledo-Martínez I, Román-Bravo P, García Y Garcia P, Godoy-Alcántar C, López-Cardoso M, Morales-Rojas H. Diorganotin(IV) dithiocarbamate complexes as chromogenic sensors of anion binding. Polyhedron 2009, 28, 3953–66.

[58] Cotero-Villegas AM, Pérez-Redondo MDC, López-Cardoso M, Toscano A, Cea-Olivares R. Organotin(IV) azepane dithiocarbamates: Synthesis and characterization of the first organotin (IV) complexes with seven-membered cyclic dithiocarbamates. Phosphorus Sulfur Silicon Relat Elem 2020, 195, 498–506.

[59] Reyes-Martínez R, García PGY, López-Cardoso M, Höpfl H, Tlahuext H Self-assembly of diphenyltin(iv) and tris-dithiocarbamate ligands to racemic trinuclear cavitands and capsules. Dalton Trans 2008, 6624–27.

[60] Cunningham D, McArdle P, McManus J, Higgins T, Molloy K. An X-ray crystallographic investigation of the structures of pyrazine adducts of diphenyltin dichloride and dimethyltin dichloride. J Chem Soc Dalton Trans 1988, 2621–27.

[61] Ma C, Zhang J, Zhang R Syntheses, characterizations, and crystal structures of organotin(IV) chloride complexes with 4,4′-bipyridine. Heteroat Chem 2004, 15, 338–46.

[62] Fischer R, Schollmeier T, Schürmann M, Uhlig F Syntheses of novel silyl substituted distannanes. Appl Organomet Chem 2005, 19, 523–29.

[63] Haga R, Saito M, Yoshioka M Reversible redox behavior between stannole dianion and bistannole-1,2- dianion. J Am Chem Soc 2006, 128, 4934–35.

[64] Kuwabara T, Saito M 1,1′-Di-tert-butyl-2,2′,3,3′,4,4′,5,5′-octaethyl-1,1′-bistannole. Acta Crystallogr Sect E Struct Reports Online 2011, 67, m949–m949.

[65] Jambor R, Kašná B, Kirschner KN, Schürmann M, Jurkschat K [{2,6-(Me$_2$NCH$_2$)$_2$C$_6$H$_3$}Sn]$_2$: An iIntramolecularly coordinated diorganodistannyne. Angew Chem Int Ed 2008, 47, 1650–53.

[66] Goldberg DE, Hitchcock PB, Lappert MF, Thomas KM, Thorne AJ, Fjeldberg TT, Haaland A, Schilling BER. Subvalent group 4B metal alkyls and amides. Part 9. Germanium and tin alkene analogues, the dimetallenes M$_2$R$_4$ [M = Ge or Sn, R = CH(SiMe$_3$)$_2$]: X-ray structures, molecular orbital calculations for M$_2$H$_4$, and trends in the series M$_2$R′$_4$ [M = C, S. J Chem Soc, Dalton Trans 1986, 2387–94.

[67] Power PP π-Bonding and the lone pair effect in multiple bonds between heavier main group elements. Chem Rev 1999, 99, 3463–504.

[68] Scotti N, Zachwieja U, Jacobs H Tetraammin-Lithium-Kationen zur Stabilisierung phenylsubstituierter Zintl-Anionen: Die Verbindung [Li(NH$_3$)$_4$]$_2$[Sn$_2$Ph$_4$]. Z Anorg Allg Chem 1997, 623, 1503–05.

[69] Fischer RC, Pu L, Fettinger JC, Brynda MA, Power PP. Very large changes in bond length and bond angle in a heavy group 14 element alkyne analogue by modification of a remote ligand substituent. J Am Chem Soc 2006, 128, 11366–67.

[70] Lee VY, Fukawa T, Nakamoto M, Sekiguchi A, Tumanskii BL, Karni M, Apeloig Y (^{t}Bu$_2$MeSi)$_2$SnSn(SiMe^{t}Bu$_2$)$_2$: A distannene with a >SnSn< double bond that is stable both in the solid state and in solution. J Am Chem Soc 2006, 128, 11643–51.

[71] Peng Y, Fischer RC, Merrill WA, Fischer J, Pu L, Ellis BD, Fettinger JC, Herber RH, Power PP. Substituent effects in ditetrel alkyne analogues: Multiple *vs.* single bonded isomers. Chem Sci 2010, 1, 461–68.

[72] Sharma HK, Miramontes A, Metta-Magaña AJ, Pannell KH Reaction between ^{t}BuMgCl and SnCl$_4$, Illustrating the solvent-dependent predominant formation of Cl(^{t}Bu$_2$Sn)$_n$Cl (THF, n = 1; Toluene, n = 2; hexane, n = 3) and the subsequent wavelength-selective photochemical transformation of n = 3→2→1. Organometallics 2011, 30, 4501–04.

[73] Puff H, Breuer B, Gehrke-Brinkmann G, Kind P, Reuter H, Schuh W, Wald W, Weidenbrück G Bindungsabstände zwischen organylsubstituierten Zinnatomen. J Organomet Chem 1989, 363, 265–80.

[74] Sharma HK, Metta-Magaña A, Pannell KH 1,3-Dichloro-1,1,2,2,3,3-hexa-tert-butyltristannane: Thermolysis involving trapping of stannylene, ^{t}Bu$_2$Sn: And reactivity with various nucleophiles. Inorg Chim Acta 2018, 475, 3–7.

[75] Adams S, Dräger M. Polystannanes Ph$_3$Sn-(^{t}Bu$_2$Sn)$_n$-SnPh$_3$ (n = 1–4): A Route to Molecular Metal Angew Chem Int Ed 1987, 26, 1255–56.

[76] Wiesler K, Suchentrunk C, Korber N Syntheses and crystal structures of ammoniates with the phenyl-substituted polytin anions Sn$_2$Ph, *cyclo*-Sn$_4$Ph, and Sn$_6$Ph. Helv Chim Acta 2006, 89, 1158.

[77] Gilman H, Cartledge FK Tetrakis(triphenylstannyl)tin. J Organomet Chem 1966, 5, 48–56.

[78] Neumann WP, Schwarz A. A new route to dialkylstannanediyls (dialkylstannylenes) and their preparative application. Angew Chem Int Ed 1975, 14, 812–812.

[79] Cardin CJ, Cardin DJ, Constantine SP, Drew MGB, Rashid H, Convery MA, Fenske D. Syntheses and crystal structures of [Sn{2-[(Me$_3$Si)$_2$C]C$_5$H$_4$N}R] [R = C$_6$H$_2$Pr$^i_3$-2,4,6 or CH(PPh$_2$)$_2$], two novel heteroleptic tin (II) compounds derived from [Sn-{2-[(Me$_3$Si)$_2$C]C$_5$H$_4$N}Cl], and for [{Sn (C$_6$H$_2$Pr$^i_3$-2,4,6)$_2$}$_3$], a structural redetermination. J Chem Soc Dalton Trans 1998, 2749–56.

[80] Masamune S, Sita LR, Williams DJ. Cyclotristannoxane (R$_2$SnO)$_3$ and cyclotristannane (R$_2$Sn)$_3$ systems. Synthesis and crystal structures. J Am Chem Soc 1983, 105, 630–31.

[81] Puff H, Bach C, Schuh W, Zimmer R. Bindungsabstände zwischen organylsubstituierten zinnatomen. J Organomet Chem 1986, 312, 313–22.

[82] Handford RC, Wheeler TA, Don T. Structure and bonding in a diamond-shaped tin cluster possessing a *cyclo*-Sn$_4$ core. Chem Eur J 2020, 26, 6126–29.

[83] Belsky VK, Zemlyansky NN, Kolosova ND, Borisova IV. Synthesis and structure of tetrakis[bis (trimethylsilylmethyl)tin]. J Organomet Chem 1981, 215, 41–48.

[84] Cardin CJ, Cardin DJ, Convery MA, Devereux MM, Kelly NB. Synthesis, structure and reactivity of the novel tetratin species bromoheptakis(2,6-diethylphenyl)-cyclotetrastannabutane. J Organomet Chem 1991, 414, C9–11.

[85] Lechner M-L, Fürpaß K, Sykora J, Fischer RC, Albering J, Uhlig F. Functionalized tetrastannacyclobutanes, J Organomet Chem 2009, 694, 4209–15.

[86] Puff H, Breuer B, Gehrke-Brinkmann G, Kind P, Reuter H, Schuh W, Wald W, Weidenbrück G. Bindungsabstände zwischen organyl substituierten Zinnatomen: III. Offenkettige Verbindungen. J Organomet Chem 1989, 363, 265–80.

[87] Preut H, Mitchell TN. 1,1,2,2,3,3,4,4,5,5,6,6-Dodecamethyl-1,2,4,5-tetrastannacyclohexane. Acta Crystallogr Sect C Cryst Struct Commun 1991, 47, 951–53.

[88]   Dräger M, Mathiasch B, Ross L, Ross M. Kristallstruktur, Schwingungs- und NMR-Spektren von Dodecaphenylcyclohexastannan $(Ph_2Sn)_6$. Z Anorg Allg Chem 1983, 506, 99–109.

[89]   Puff H, Bach C, Reuter H, Schuh W Bindungsabstände zwischen organylsubstituierten zinnatomen. J Organomet Chem 1984, 277, 17–28.

[90]   Klösener J, Wiesemann M, Niemann M, Neumann B, Stammler HG, Höge B. Synthesis and reactivity of donor-stabilized bis(pentafluoroethyl)stannylene $[Sn(C_2F_5)_2(D)_n]$ (D = THF, DMAP, PMe_3, $[Sn(C_2F_5)_3]^-$). Chem-Eur J 2018, 24, 4412–22.

[91]   Drost C, Hildebrand M, Lönnecke P. Synthesis and crystal structure of a novel distannylstannanediyl and a rare pentastannapropellane. Main Group Met Chem 2002, 25, 93–98.

[92]   Erickson JD, Fettinger JC, Power PP. Reaction of a germylene, stannylene, or plumbylene with trimethylaluminum and trimethylgallium: insertion into Al–C or Ga–C Bonds, a reversible metal–carbon insertion equilibrium, and a new route to diplumbenes. Inorg Chem 2015, 54, 1940–48.

[93]   Sita LR, Bickerstaff RD. Isolation and molecular structure of the first bicyclo[2.2.0] hexastannane. J Am Chem Soc 1989, 111, 3769–70.

[94]   Steller BG, Fischer RC, Flock M, Hill MS, Liptrot DJ, McMullin CL, Rajabi NA, Kathrin T, Wilson ASS Reductive dehydrocoupling of diphenyltin dihydride with $LiAlH_4$: Selective synthesis and structures of the first bicyclo[2.2.1]heptastannane-1,4-diide and bicyclo[2.2.2]octastannane-1,4-diide. Chem Commun. 2020, 56, 336–39

[95]   Sita LR, Kinoshita I. Octakis(2,6-diethylphenyl) octastannacubane. Organometallics 1990, 9, 2865–67.

[96]   Eichler BE, Power PP Synthesis and characterization of $[Sn_8(2,6-Mes_2C_6H_3)_4]$ (Mes = 2,4,6-Me $_3C_6H_2$): A main group metal cluster with a unique structure. Angew Chem Int Ed. 2001, 40, 796

[97]   Sita LR, Kinoshita I. Decakis(2,6-diethylphenyl)-decastanna[5]prismane: Characterization and molecular structure. J Am Chem Soc 1991, 113, 1856–57.

[98]   Rivard E, Steiner J, Fettinger JC, Giuliani JR, Augustine MP, Power PP Convergent syntheses of $[Sn_7\{C_6H_3-2,6-(C_6H_3-2,6-{}^iPr_2)_2\}_2]$: A cluster with a rare pentagonal bipyramidal motif. Chem Commun 2007, 4919–21.

[99]   Kroon J, Hulscher JB, Peerdeman AF. The molecular structure of tetrakis (diphenylbromoplumbyl)methane in the crystal. J Organomet Chem 1970, 23, 477–85.

[100] Willemsens LC, van der Kerk G. Investigations on organolead compounds VIII. Reactions of (triorganoplumbyl)metal reagents with polychloromethanes; tetrakis(triphenylplumbyl) methane and related compounds. J Organomet Chem 1970, 23, 471–75.

9 Organometallic compounds of group 15 metals

9.1 Organoantimony compounds

Antimony forms a very broad diversity of organosubstituted compounds. Homoleptic species include triorganostibines (tertiary stibines), $Sb^{III}R_3$, tetrasubstituted cations (stibonium ions), $[Sb^VR_4]^+$, and pentasubstituted compounds (stiboranes), Sb^VR_5. Hexasubstituted anions, $[Sb^VR_6]^-$, are also known. Heteroleptic compounds comprise organoantimony hydrides, halides, oxides and hydroxocompounds (stibinic and stibonic acids), sulfur, nitrogen and other heteroatom compounds, compounds with Sb–Sb bonds and metal complexes with stibine ligands.

9.1.1 Homoleptic compounds, SbR₃

The molecular structure is trigonal pyramidal with stereochemically active lone pair. Crystallographically confirmed molecular structures are available for several compounds, including $SbBu^t_3$ [1], $Sb(CH_2Ph)_3$ [2], $Sb(C_6F_5)_3$ [3], $Sb\{(C_6H_3(CF_3)_2\text{-}2,4\}_3$ [4] and $SbMes_2Ph$ [5].

Grignard alkylation is frequently employed for synthesis of tertiary stibines, but the lower alkyl derivatives can be difficult to separate from the solvent:

$$SbCl_3 + 3\,RMgCl \longrightarrow SbR_3 + 3\,MgCl_2$$

Organoaluminum compounds yield triorganostibines from antimony(III) oxide:

$$Sb_2O_3 + 2\,AlR_3 \longrightarrow 2\,SbR_3 + Al_2O_3$$

Reduction of triorganodihalogenostibines (R_3SbX_2) with zinc, lithium borohydride or hydrazine hydrate also gives triorganostibines:

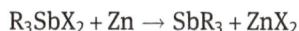

$$R_3SbX_2 + Zn \rightarrow SbR_3 + ZnX_2$$

This reaction is used to purify stibines prepared with Grignard reagents by conversion to R_3SbX_2 which is isolated from the ether and then reduced.

The direct synthesis from elemental antimony and organic halides was used for the preparation of tris(perfluoromethyl)stibine, $Sb(CF_3)_3$, from CF_3I at 165–170 °C, and tris(pentafluorophenyl)-antimony, $Sb(C_6F_5)_3$, from $(C_6F_5)_2TlBr$.

Trialkylstibines are air sensitive; the lower alkyls are pyrophoric. The aromatic derivatives are air stable.

With halogens, the triorganostibines undergo oxidative-addition reactions to form the dihalides, R_3SbX_2, and with alkyl halides to produce stibonium halides $[R_3SbR']^+X^-$. Mercury(II) chloride cleaves triarylstibines to form R_2SbCl and $RHgCl$,

https://doi.org/10.1515/9783110695274-009

but $CuCl_2$, $TlCl_3$, $FeCl_3$ and the phosphorus, arsenic and antimony tri- or pentahalides oxidize them to R_3SbCl_2.

9.1.2 Organoantimony heterocycles

Dimagnesium compounds give heterocyclic stibines which contain three Sb–C bonds (Fig. 9.1).

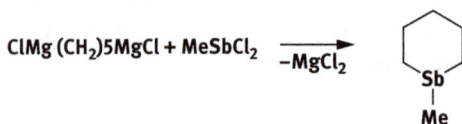

$ClMg(CH_2)5MgCl + MeSbCl_2 \xrightarrow[-MgCl_2]{}$

Fig. 9.1: Six-membered heterocyclic organoantimony compound.

Heterocyclic stiboles were prepared with organolithium reagents (R = Ph) (Fig. 9.2).

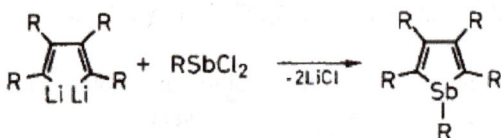

$+ RSbCl_2 \xrightarrow[-2LiCl]{}$

Fig. 9.2: Five-membered heterocyclic organoantimony compound.

A reactive dicoordinated organoantimony stibabenzene is obtained by dehydrohalogenation of a heterocyclic chloride (Fig. 9.3).

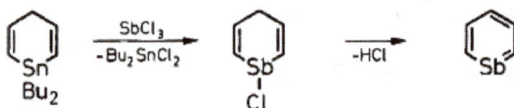

$\xrightarrow[-Bu_2SnCl_2]{SbCl_3}$ $\xrightarrow{-HCl}$

Fig. 9.3: Stibabenzene ring.

9.1.3 Pentavalent SbR₅ compounds

This class of pentavalent organoantimony compounds comprises $SbPh_5$ [6], $Sb(C_6F_5)_5$ [7] and $Sb(C_6H_4CF_3\text{-}4)_5$ [8]. Solid pentaphenylantimony has an unusual square pyramidal trigonal geometry unlike the bipyramidal geometry of $AsPh_5$ and other pentacoordinated compounds. Solid pentamethylantimony and Sb(*para*-$C_6H_4CH_3)_5$, however, have trigonal bipyramidal structures. The differing molecular geometries of SbR$_5$ compounds are surprising.

The aryl-pentasubstituted derivatives are stable. They can be obtained by reactions of arylantimony(V) halides with organolithium reagents, organozinc and Grignard reagents:

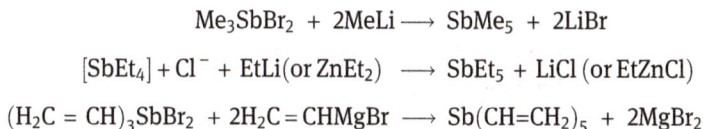

$$Me_3SbBr_2 + 2MeLi \longrightarrow SbMe_5 + 2LiBr$$

$$[SbEt_4] + Cl^- + EtLi(or\ ZnEt_2) \longrightarrow SbEt_5 + LiCl\ (or\ EtZnCl)$$

$$(H_2C = CH)_3SbBr_2 + 2H_2C = CHMgBr \longrightarrow Sb(CH{=}CH_2)_5 + 2MgBr_2$$

Mixed derivatives can be obtained with reagents containing different organic groups:

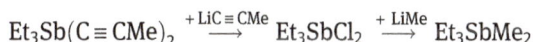

$$Et_3Sb(C \equiv CMe)_2 \xrightarrow{+\ LiC \equiv CMe} Et_3SbCl_2 \xrightarrow{+\ LiMe} Et_3SbMe_2$$

Pentaphenylantimony is obtained by the reaction of $[SbPh_4]^+Br^-$ or antimony pentachloride with phenyllithium.

9.1.4 Subvalent stibonium [Sb^{V}R$_4$]$^+$ cations

Subvalent tetrahedral antimony(V) cations $[SbMe_4]^+$ [9], $[SbPh_4]^+$ [10] and $Sb(C_6F_5)_4]^+$ [11] have been structurally characterized.

The tetraorganostibonium salts are prepared by the quaternization of tertiary stibines with alkyl halides. Aromatic stibines require trimethyloxonium tetrafluoroborate in liquid sulfur dioxide:

$$SbPh_3 + Me_3O^+BF_4{}^- \rightarrow Ph_3SbMe^+ + BF_4{}^- + Me_2O$$

If the excess of halogen is avoided, the pentaorgano derivatives can be cleaved to stibonium halides with halogens, but this reaction is not suited as a preparative procedure:

$$SbR_5 + X_2 \rightarrow SbR_4^+ X^- + RX$$

Grignard arylation of triphenylantimony dichloride also yields a stibonium salt:

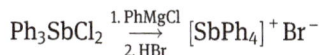

$$Ph_3SbCl_2 \xrightarrow[2.\ HBr]{1.\ PhMgCl} [SbPh_4]^+Br^-$$

Tetraphenylstibonium bromide is obtained from a Friedel–Crafts reaction:

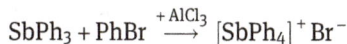

$$SbPh_3 + PhBr \xrightarrow{+\ AlCl_3} [SbPh_4]^+Br^-$$

Five-coordinated, oxygen-containing insoluble R$_4$Sb-OH derivatives are formed from tetraalkylstibonium halides and moist silver oxide.

The reduction of stibonium halides with Li[AlH$_4$] produces triorganostibines:

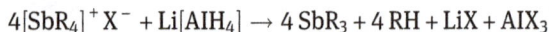

$$4[SbR_4]^+X^- + Li[AlH_4] \rightarrow 4\,SbR_3 + 4\,RH + LiX + AlX_3$$

9.1.5 Hypervalent Sb^V anions: hexasubstituted anions, [SbR$_6$]$^-$

The highest degree of organosubstitution is achieved in anions produced from pentaphenylantimony and phenyllithium:

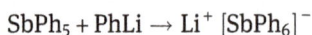

$$SbPh_5 + PhLi \rightarrow Li^+\,[SbPh_6]^-$$

9.1.6 Inverse organoantimony compounds

A few inverse organoantimony compounds are shown here: Ph$_2$Sb–CH$_2$–SbPh$_2$ [12], Ph$_2$Sb(CH$_2$)$_3$SbPh$_2$ [13] and (PhC $\equiv$ C)$_2$Sb(CH$_2$)$_3$Sb(C $\equiv$ CPh)$_2$ [14] (Fig. 9.4).

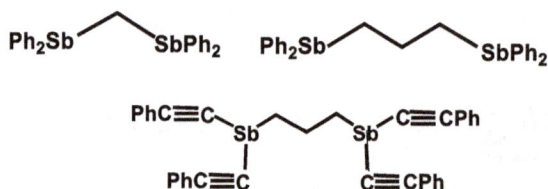

Fig. 9.4: Inverse organoantimony compounds.

Aromatic derivatives are also known [15] (Fig. 9.5).

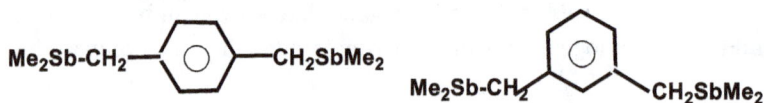

Fig. 9.5: Aromatic inverse organoantimony compounds.

9.1.7 Organoantimony(III) halides

Diorganoantimony halides(R$_2$SbX) are molecules with a trigonal pyramidal molecular geometry. The dimethylantimony iodide (Me$_2$SbI) is associated in solid state with formation of a supramolecular zigzag chain with Sb $\cdots$ I 366.7(1) Å, Sb . . . I 171.87(4)° and Sb–I . . . Sb 116.83(3)° [16].

The trigonal pyramidal molecular geometry was established crystallographically for Ph$_2$SbCl [17] and [C$_6$H$_2$(CF$_3$)$_3$]$_2$SbCl [18].

Among $RSbX_2$ molecules, the trigonal pyramidal geometry has been established crystallographically for Bu^tSbCl_2 [19], $PhSbX_2$ (X = Cl, Br, I) [20] and [{$(Me_3Si)_2CH$}] $SbCl_2$ [21]. The iodide $MeSbI_2$ is associated into chains with Sb–I . . . Sb–I long and short interatomic distances.

The syntheses of these compounds are based upon redistribution reactions:

$$SbPh_3 + SbCl_3 \rightarrow PhSbCl_2$$

$$3\,PhSbCl_2 + 2\,PBr_3 \rightarrow 3\,PhSbBr_2 + 2\,PCl_3$$

$$PhSbCl_2 + 2\,NaI \rightarrow PhSbI_2 + 2\,NaCl$$

Organoantimony(III) halides are seldom made by the Grignard synthesis but with *tert*-butylmagnesium chloride, the diorgano derivative, R_2SbCl, is obtained from antimony(III) chloride.

To achieve partial substitution in SbX_3, a reagent of lower reactivity is required; organolead and organotin compounds are suitable for the chlorides, and organosilicon compounds are satisfactory for the fluorides:

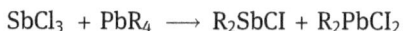

$$SbCl_3 + PbR_4 \longrightarrow R_2SbCl + R_2PbCl_2$$

$$SbF_3 + 2[PhSiF_5]^{2-} \longrightarrow Ph_2SbF + 2[SiF_6]^{2-}$$

Pyrolysis of triorganodihalides R_3SbX_2, the redistribution between inorganic trihalides and tertiary stibines, and the reduction of stibonic and stibinic acids with sulfur dioxide and hydroiodic acid in hydrochloric medium or with tin(II) chloride are used in the synthesis of organoantimony(III) halides:

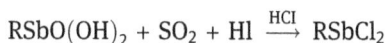

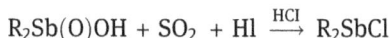

$$R_3SbX_2 \xrightarrow{\Delta} R_2SbX + RX$$

$$SbX_3 + SbR_3 \longrightarrow R_2SbX + RSbX_2$$

$$RSbO(OH)_2 + SO_2 + HI \xrightarrow{HCl} RSbCl_2$$

$$R_2Sb(O)OH + SO_2 + HI \xrightarrow{HCl} R_2SbCl$$

The direct synthesis from alkyl halides and metallic antimony can be used for the methyl derivatives, Me_2SbCl and $MeSbCl_2$.

Convenient preparations of phenylantimony(III) chlorides are the redistribution reactions:

$$SbPh_3 + SbCl_3 \rightarrow PhSbCl_2$$

$$3\,PhSbCl_2 + 2\,PBr_3 \rightarrow 3\,PhSbBr_2 + 2\,PCl_3$$

$$PhSbCl_2 + 2\,NaI \rightarrow PhSbI_2 + 2\,NaCl_3$$

The iodides can be prepared by double exchange with sodium iodide:

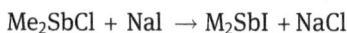

$$Me_2SbCl + NaI \rightarrow M_2SbI + NaCl$$

Other related Sb^{III} compounds are $[PhSbCl_4]^{2-}$ with square pyramidal molecular geometry and $[PhSb^{III}X_3]^-$ (X = Cl, Br) [22]with a psi-trigonal bipyramidal geometry with the phenyl group in equatorial position [23].

The chlorides can be prepared by addition of chlorine to chlorostibines:

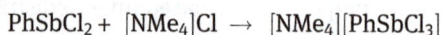

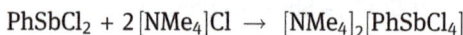

$$PhSbCl_2 + [NMe_4]Cl \longrightarrow [NMe_4][PhSbCl_3]$$

$$PhSbCl_2 + 2[NMe_4]Cl \longrightarrow [NMe_4]_2[PhSbCl_4]$$

An interesting case is the unique compound $[SbMe_4]_2[MeSbI_4]$ [24] made of a tetrahedral $[SbMe_4]^+$ cation and the $[MeSbI_4]^{2-}$ anion (the latter with a square pyramidal geometry with apical methyl) (Fig. 9.6).

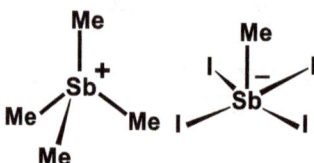

Fig. 9.6: Mixed anion–cation organoantimony compound.

Compounds of the $[R_2Sb^{III}X_2]$ series with known structures are $[Ph_2Sb^{III}Cl_2]$, $[Ph_2Sb^{III}Br_2]^-$ [25], $[Ph_2Sb^{III}Br_2]^-$ [26] and $[Ph_2SbI_2]^-$ [27] with psi-trigonal bipyramidal geometry and the phenyl groups in equatorial position:

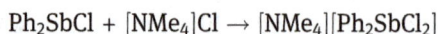

$$Ph_2SbCl + [NMe_4]Cl \rightarrow [NMe_4][Ph_2SbCl_2]$$

Tetraorganoantimony(V) halides, R_4SbX. The halides Me_4SbF, Ph_4SbCl [28], $SbPh_4Br$ [29, 30] are molecular compounds with tetragonal pyramidal geometry with the halogen in axial position. In the solid state, Me_3SbF is associated into supramolecular chains with Sb–F–Sb bridges and six-coordinated antimony.

Tetramethylantimony fluoride is prepared from pentamethylantimony and KHF_2 or HF:

$$SbMe_5 + KHF_2(or HF) \longrightarrow Me_4Sb-F$$

Related pseudohalides are prepared by similar reactions (R = Me, Ph; X = N_3, CN, SCN):

$$SbMe_5 + HX \longrightarrow Me_4Sb-X$$

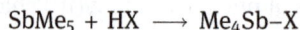

Triorganoantimony(V) dihalides, R_3SbX_2. There are several compounds of this class whose molecular structures have been established by X-ray diffraction. These include Me_3SbF_2 [31], Mes_3SbF_2 [32], $(PhCH_2)_3SbBr_2$ [33], Cyh_3SbBr_2 [34], Ph_3SbI_2 [35] and $(Me_3SiCH_2)_3SbI_2$ [36]. In all, the halogens occupy the axial positions in a trigonal bipyramidal geometry (Fig. 9.7).

Aromatic derivatives, R_3SbX_2, precipitate on treatment of aromatic stibines with halogens. Triphenylantimony difluoride is formed in the reaction of Ph_3SbO with

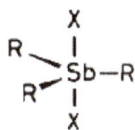

Fig. 9.7: Trigonal bipyramidal geometry of R_3SbX_2 compounds.

SF_4 or in the fluorination of $SbPh_3$ with XeF_2. The reaction of antimony pentachloride with diphenylmercury yields Ph_3SbCl_2 and PhHgCl.

The dihalides decompose above their melting points with formation of R_2SbX. Hydrolysis yields $R_3Sb(OH)X$ and $R_3Sb(OH)$. With tertiary phosphines, they form tetracoordinated anionic species by transfer of organic groups to phosphorus:

$$Me_3SbBr_2 + PR_3 \longrightarrow [R_3PMe]^+ [Me_2SbBr_2]^-$$

Diorganoantimony(V) trihalides, R_2SbX_3. The compounds Ph_2SbCl_2Br, $Ph_2SbClBr_2$ and Ph_2SbBr_3 [37] are trigonal bipyramidal molecules with the phenyl groups in equatorial positions and two axial halogens, associated into supramolecular chains with halogen bridges.

Dimethylantimony trichloride is prepared by chlorination of Me_2SbCl and bis(chlorovinyl)antimony trichloride by addition of acetylene to antimony pentachloride.

Other alkyl derivatives are obtained by the chlorination of distibines, $R_2Sb\text{-}SbR_2$. An ionic dimer of Me_2SbCl_3, namely $[SbMe_4]^+[SbCl_6]^-$, is obtained from $SbCl_3$ and Me_2InCl.

Solid Ph_2SbCl_3 is a supramolecular dimer with chloride bridges of unequal lengths (Fig. 9.8), while Ph_2SbBr_3, $Ph_2SbClBr_2$ and Ph_2SbCl_2Br are monomeric with trigonal bipyramidal structures.

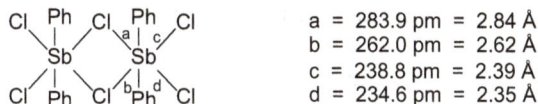

a = 283.9 pm = 2.84 Å
b = 262.0 pm = 2.62 Å
c = 238.8 pm = 2.39 Å
d = 234.6 pm = 2.35 Å

Fig. 9.8: The structure of supramolecular dimeric Ph_2SbCl_3.

Aromatic derivatives $ArSbCl_4$ are obtained from aryldiazonium salts with antimony chlorides ($SbCl_3$ or $SbCl_5$), from stibonic acids and hydrochloric acid, by the reaction of phenylhydrazine hydrochloride with antimony pentachloride in the presence of copper(II) chloride and oxygen, and by chlorination of diarylchlorostibines. The fluoride Ph_2SbF_3 is obtained by fluorination of Ph_2SbF with xenon difluoride and by treatment of $Ph_2Sb(O)OH$ with SF_4.

The tetrahalides react with alkylammonium salts to form arylpentachloroantimonates, $[RNH_3]^+[RSbCl_5]^-$. These anions have been established in octahedral $[PhSb^VCl_5]^-$ [38] and $[PhSbBr_5]^-$ [39].

The tendency of antimony to increase its coordination number from five to six results in addition of halide and pseudohalide ions:

$$Ph_2SbCl_3 + [NMe_4]^+ X^- \longrightarrow [NMe_4]^+ [Ph_2SbCl_3X]^-$$

$$X = Cl, Br, N_3, NCS$$

Ionic organoantimony(V) halides. A cation $[Mes_3SbCl]^+$ [40] is known, having a tetrahedral structure with the chlorine in apical position.

Anions of the type $[R_2Sb^VX_4]^2$ are represented by $[Ph_2Sb^VCl_4]^{2-}$ [41] which is an octahedral complex with four chlorine atoms in equatorial positions and the two phenyl groups in axial positions.

9.1.8 Oxygen-containing organoantimony compounds

Organoantimony oxides and hydroxides. The organoantimony oxides RSbO and $R_2Sb-O-SbR_2$ are anhydrides of stibonous $RSb(OH)_2$ and stibinous R_2Sb-OH, respectively. The acids cannot be isolated and their anhydrides are obtained by alkaline hydrolysis of organoantimony(III) halides, or by in situ reduction of arylstibonic acids with sulfur dioxide followed by alkaline hydrolysis. Distiboxanes are also formed by thermal disproportionation of organoantimony(III) oxides:

$$4\,RSbO \longrightarrow (R_2Sb)_2O + Sb_2O_3$$

or by cleavage of triarylstibines with acids, followed by treatment with alkalies.

The insoluble monoorganoantimony(III) oxides are polymeric. The soluble, low melting distiboxanes, $R_2Sb-O-SbR_2$, are monomeric in the solid state.

Treatment of distiboxanes with carboxylic acids leads to diorganoantimony(III) carboxylates, $R_2Sb-OCOR'$.

Stibine oxides, R_3SbO, and triorganoantimony(V) hydroxides, $R_3Sb(OH)_2$, are interconvertible. Trialkylantimony hydroxides are formed in the hydrolysis of trichlorides, but the trifluoromethyl derivative $(CF_3)_3SbCl_2$ on hydrolysis yields the ionic compound $[H_3O]^+[(CF_3)_3SbCl_2(OH)]^-$, which can be converted with wet silver oxide into $Ag^+[(CF_3)_3Sb(OH)_3]^-$.

Trimethylantimony(V) hydroxide, $Me_3Sb(OH)_2$, is dehydrated in vacuo to form trimethylstibine oxide, Me_3SbO.

Triarylantimony(V) hydroxides are obtained by hydrolysis of dihalides in alkaline medium, or by oxidation of triarylstibines with hydrogen peroxide in acetone or with HgO in ether. They are dehydrated to form stibine oxides:

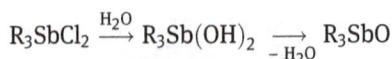

$$R_3SbCl_2 \xrightarrow{H_2O} R_3Sb(OH)_2 \xrightarrow[-H_2O]{} R_3SbO$$

The oxides, R_3SbO, are polymeric.

A dihydroxo compound $(CH_3)_3Sb(OH)_2 \cdot 7H_2O$ has been prepared from $(CH_3)_3SbCl_2$ and NaOH and has a trigonal bipyramidal molecular geometry with the OH groups in axial positions [42].

Triarylantimony(V) hydroxides, which exist as $[R_3Sb(OH)_3]^-$ anions in aqueous solution, are strong bases and precipitate metal hydroxides in reactions with their salts.

Stibonic acids, $RSbO(OH)_2$. Aliphatic derivatives are unknown but stable derivatives are prepared from aryldiazonium salts with antimony halides or by the precipitation of the $[Ar-N_2]^+[SbCl_4]^-$ salt from a hydrochloric solution of the diazonium salt on treatment with antimony(III) chloride, followed by alkaline treatment (producing nitrogen evolution) and re-acidification to give the stibonic acid. The hydrolysis of arylantimony tetrachlorides, $RSbCl_4$, is also used.

The stibonic acids may be polymeric, but a hydrated six-coordinated structure $[RSb(OH)_5]^-[H^+]$ is possible, analogous to the inorganic anion, $[Sb(OH)_6]^-$.

The aromatic stibonic acids form organoantimony(V) tetrachlorides, $RSbCl_4$, with concentrated hydrochloric acid and are used for the purification of acids. With sulfur dioxide and hydrogen iodide in hydrochloric acid solution, the acids are reduced to aryldichlorostibines.

Stibinic acids, $R_2Sb(O)H$. Dimethylstibinic acid is prepared by the hydrolysis of dimethylantimony trichloride, Me_2SbCl_3, or by the wet oxidation of tetramethyl-distibine, $Me_2Sb-SbMe_2$.

Aromatic derivatives are formed as a mixture with stibonic acids in the reaction of arylhydrazines with antimony trichloride in the presence of copper(I) chloride, or from aryldiazonium salts with monoorgano-antimony(III) compounds:

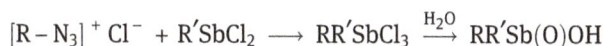

$$[R-N_3]^+ Cl^- + R'SbCl_2 \longrightarrow RR'SbCl_3 \xrightarrow{H_2O} RR'Sb(O)OH$$

or by oxidation of triarylstibines with hydrogen peroxide in alkaline medium. Pure compounds are obtained by hydrolysis of diarylantimony(V) trichlorides.

Heterocyclic stibinic acids are formed by cyclodehydration of stibonic acids (Fig. 9.9).

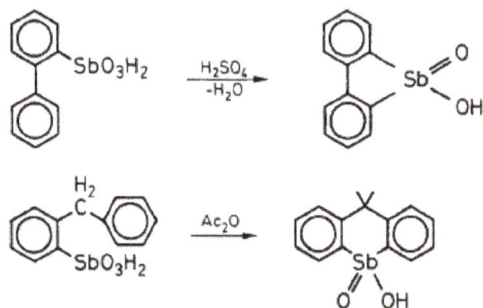

Fig. 9.9: Heterocyclic stibinic acids from aromatic stibonic acids.

Stibinous-acid esters, RSb(OR')$_2$ are prepared from organoantimony dihalides and sodium alkoxides:

$$RSbCl_2 + 2NaOR' \longrightarrow RSb(OR')_2 + 2NaCl$$

or directly from dihalides and alcohols (diols) in the presence of bases (Fig. 9.10).

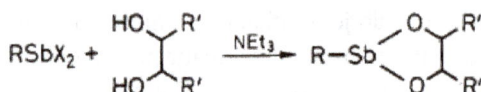

Fig. 9.10: Stibinous acid esters.

Stibinous-acid esters, R$_2$Sb-OR' are obtained from dialkylantimony halides and sodium alkoxides:

$$R_2Sb - Cl + NaOR' \longrightarrow R_2Sb-OR' + NaCl$$

Dialkoxyantimony trialkoxides, R$_n$Sb(OR')$_{5-n}$ can be prepared from the corresponding trichlorides and sodium alkoxides at low temperature:

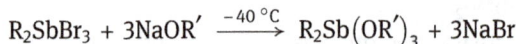

$$R_2SbBr_3 + 3NaOR' \xrightarrow{-40\,°C} R_2Sb(OR')_3 + 3NaBr$$

The Me$_2$Sb(OMe)$_3$ derivative is a supramolecular dimer (Fig. 9.11).

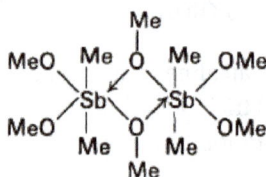

Fig. 9.11: Dimeric Me$_2$Sb(OMe)$_3$ supermolecule.

Tetraorganoantimony alkoxides R$_4$Sb-OR' can be made by cleavage of pentasubstituted derivatives with alcohols or phenols.

9.1.9 Sulfur-containing organoantimony compounds

Stibine sulfides, R$_3$SbS, are prepared by oxidative addition of sulfur to tertiary stibines, by treatment of trialkylantimony hydroxides with hydrogen sulfide, or by reaction of triorganoantimony dichlorides with hydrogen sulfide in ammonia-alcoholic solutions (Fig. 9.12).

$$R_3Sb \ + \ S \ \xrightarrow{\quad} \ R_3SbS \ \xleftarrow{\quad} \ R_3Sb(OH)_2 \ + \ H_2S$$

$$\uparrow$$

$$R_3SbX_2 \ + \ H_2S$$

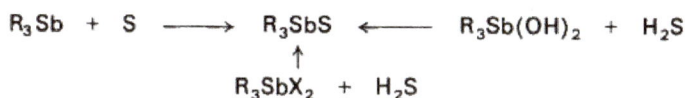

Fig. 9.12: Formation of R_3SbS compounds.

Organoantimony(III) sulfides, RSbS and $R_2Sb–S–SbR_2$, are prepared by the reaction of chloroorganostibines or the corresponding oxides with hydrogen sulfide.

Arylantimony sulfides are obtained from the oxides with carbon disulfide in the presence of ammonia, or with dithiocarbamates. The RSbS compounds are polymeric.

Organoantimony(lll) dithiolates, $RSb(SR')_2$, are diesters of thioacids, obtained from monoorganoantimony(III) oxides or dihalides with thiols. Heterocyclic esters are prepared from dithiols (Fig. 9.13).

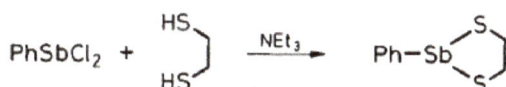

Fig. 9.13: Synthesis of an organoantimony–sulfur heterocycle.

Diorganoantimony thiolates, R_2Sb-SR', are formed in reactions of halides or oxides with sulfur reagents:

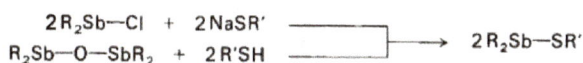

$$\left. \begin{array}{l} 2\,R_2Sb\!-\!Cl \ + \ 2\,NaSR' \\ R_2Sb\!-\!O\!-\!SbR_2 \ + \ 2\,R'SH \end{array} \right\} \longrightarrow 2\,R_2Sb\!-\!SR'$$

or by cleavage of triphenylstibine:

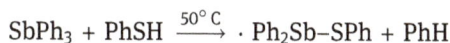

$$SbPh_3 + PhSH \xrightarrow{50^\circ C} \cdot Ph_2Sb\!-\!SPh + PhH$$

Triorganoantimony dithiolates $R_3Sb(SR')_2$ result from reactions of triorganoantimony dihalides with thiols:

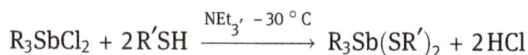

$$R_3SbCl_2 + 2\,R'SH \xrightarrow{NEt_3, \ -30\,^\circ C} R_3Sb(SR')_2 + 2\,HCl$$

Tetraalkylantimony thiolates, R_4Sb-SR' are obtained by cleavage of pentaalkyls with thiols:

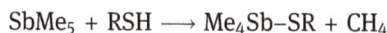

$$SbMe_5 + RSH \longrightarrow Me_4Sb\!-\!SR + CH_4$$

9.1.10 Nitrogen-containing organoantimony compounds

Organodiaminostibines $RSb(NR_2)_2$ and diorganoaminostibines $R_2Sb\text{-}NR_2$ are prepared from the corresponding halides and lithiated amines:

$$RSbX_2 + 2LiNR'_2 \longrightarrow RSb(NR'_2) + 2LiX$$

$$R_2SbX + LiNR'_2 \longrightarrow R_2Sb-NR'_2 + LiX$$

Four-membered Sb_2N_2 rings like $[Bu^t_2Sb-NBu^t]_2$ (Fig. 9.14) [43] are formed with organo-antimony(III) moieties. Using Bu^t_2SbCl as reagent, a *tert*-butyl-substituted stibinoamine $Bu^t_2SbN(H)Bu^t$, an isopropyl-substituted interpnictogen $Bu_2Sb^tN(HiPr^i$, with LiNHR and a primary stibinoamine $Bu^t_2SbNH_2$ are obtained. Condensation of $Bu^t_2SbNH_2$ leads to a distibazane compound $(Bu^t_2Sb)_2NH$ with elimination of ammonia [44].

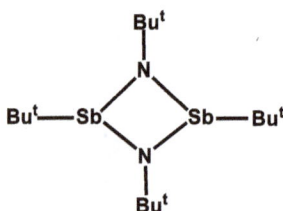

Fig. 9.14: A cyclic Sb_2N_2 ring compound.

The reaction of triorganoantimony dihalides with ammonia gives amino-stibonium salts $[R_3Sb-NH_2]^+X^-$.

Triarylstibine imines, $R_3Sb = NR'$, are obtained from tertiary stibines and the sodium salt of *N*-bromacetamide or *N*-chlorosulfonamides and by treatment of triarylantimony dichlorides with sodium amide [45]. The dimers of R_3SbNR' are cyclic compounds with Sb_2N_2 rings, for example, $[ClPh_2Sb = NCH_2Ph]_2$ [46] (Fig. 9.15).

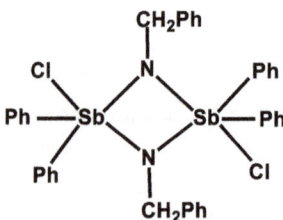

Fig. 9.15: Dimeric $[Ph_2ClSb-NCH_2Ph]_2$.

9.1.11 Organoantimony compounds with Sb–Sb bonds

Distibanes. Tetraorganodistibanes, $R_2Sb-SbR_2$, are formed by the action of organic free radicals upon metallic antimony mirrors, by reduction of dialkylantimony bromide with sodium in liquid ammonia or with magnesium in THF, or by the reaction of alkyl halides with antimony in the presence of sodium or lithium (also in liquid ammonia).

A number of tetraorganodistibanes, $R_2Sb–SbR_2$ with R = Et, Ph, Mes, $CH(SiMe_3)_2$, have been reported [47]. Aromatic derivatives are formed in the reduction of diaryliodostibines with sodium hypophosphite and tetraphenyldistibane $Ph_2Sb–SbPh_2$ [48] is a typical example.

Methyldistibane compounds display an unprecedented diversity. Thus, neutral derivatives with four organic groups, $Me_2Sb^{III}–Sb^{III}Me_2$ [49, 50], monocationic compounds with five organic groups $[Me_2Sb^{III}–Sb^{V}Me_3]^+$ [51, 52] and dicationic compounds with six organic groups $[Me_3Sb^{V}–Sb^{V}Me_3]^{2+}$ [53] have been described (Figure 9.16).

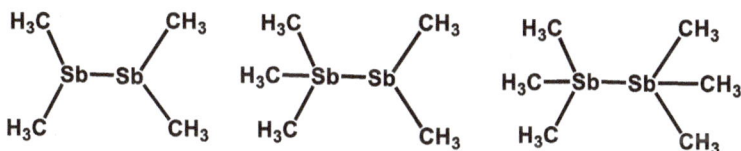

Fig. 9.16: The variety of methyl distibane derivatives.

Tetraorganodistibanes are cleaved by halogens and hydrogen halides. On heating above 200 °C, they disproportionate to form tertiary stibines and metallic antimony. The connection of two antimony heterocycles through Sb–Sb bonds leads to another type of distibanes [54] (Fig. 9.17).

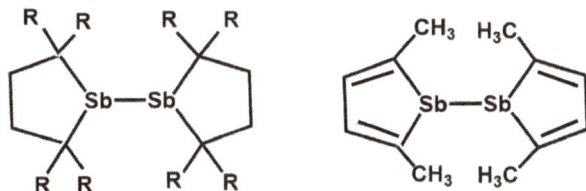

Fig. 9.17: Antimony heterocycles connected as distibane compounds.

Distibenes, that is, compounds with Sb=Sb double bonds [55–57], are formed with bulky organic substituents which prevent the oligomer or polymer formation of RSb moieties (Fig. 9.18).

Longer polyantimony chains are a rarity, and only a tristibane anion $[Bu^t_2Sb–Sb–SbBu^t_2]^-$ has been structurally characterized (as potassium salt) [58] (Fig. 9.19).

Cyclic stibanes. The insoluble $[RSb]_n$ compounds are cyclic oligomers or linear polymers. The solid tetramer, $[Bu'Sb]_4$ (obtained from Bu_2SbLi and iodine or from Bu_2SbCl and magnesium in THF), and the hexamer, $[PhSb]_6$ (prepared from phenylstibine $PhSbH_2$), are cyclic stibanes:

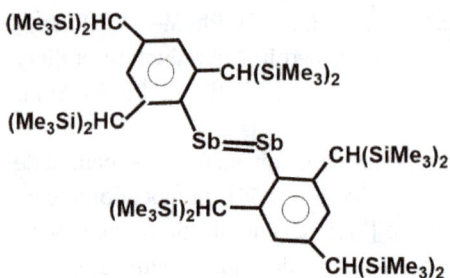

Fig. 9.18: Structure of a distibene.

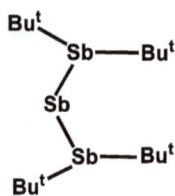

Fig. 9.19: Structure of a tristibane chain.

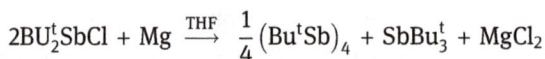

$$2BU_2^t SbCl + Mg \xrightarrow{THF} \frac{1}{4}(Bu^t Sb)_4 + SbBu_3^t + MgCl_2$$

Other preparations of [RSb]$_n$ derivatives include the reduction of stibonic acids with sodium dithionite or hypophosphite, the decomposition of arylantimony hydrides or the condensation of hydrides with aryldichlorostibines. The presence of halogen in the product is interpreted in terms of a polymeric structure with halogen terminal groups Cl(SbR)$_n$Cl but no crystal structure analysis for this type of compound is available.

A cyclic trimer with trimethylsilyl substituents [Sb(CH(SiMe₃)₂]₃ [59] seems to be the sole representative of this family (Fig. 9.20).

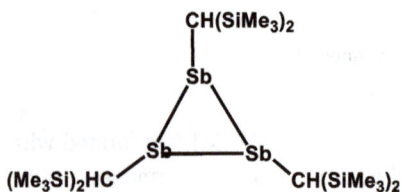

Fig. 9.20: Structure of a cyclic tristibane.

Several four-membered cyclostibanes are known with various organic substituents, including [SbBu^t]$_4$ [60], [Sb(HC(SiMe₃)₂)]$_4$ [61] and others [62, 63] (Fig. 9.21).

Five-membered [Sb(CH₂Bu^t)]$_5$ [64] and six-membered [SbPh]$_6$ [65] are rarities but well-documented cyclostibanes (Fig. 9.22).

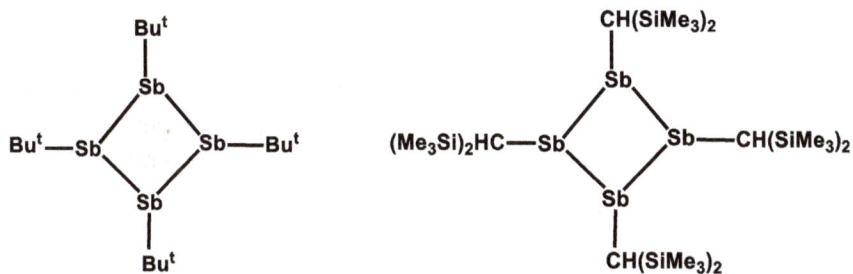

Fig. 9.21: Four-membered cyclostibanes.

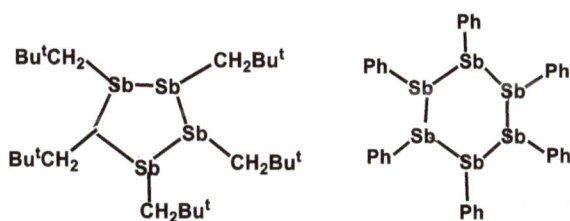

Fig. 9.22: Five- and six-membered cyclostibanes.

Bi- and tricyclic stibanes are mentioned: a pentaantimony bicyclic Sb$_5${C$_6$H$_3${CH$_2$-NMe$_2$-2,6}$_2$}$_3$ compound [66], a bicyclic Sb$_4$–Sb$_4$ compound (R = CH(SiMe$_3$)$_2$) [67] and a tricyclic octastibane, Sb$_8${HC(SiMe$_3$)$_2$}$_4$ [68] (Fig. 9.23).

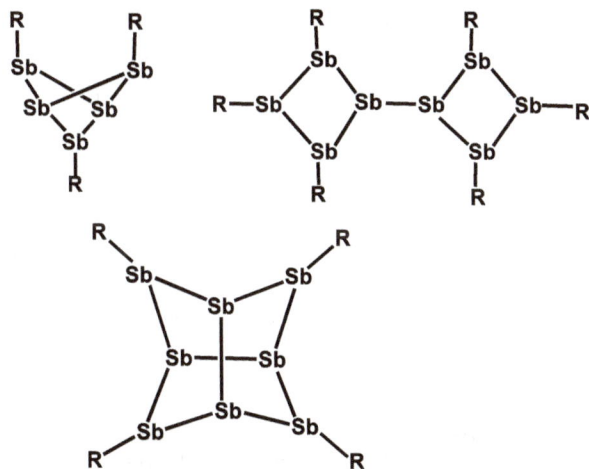

Fig. 9.23: Bi- and tricyclic polystibanes.

9.1.12 Organoantimony(III) compounds as donor ligands

Having a lone pair, the triorganostibines can act as donors to form a large number of metal complexes. Only a short list is mentioned, to illustrate the versatility of triorganostibines as ligands, illustrated with just a few chemical diagrams (Fig. 9.24):

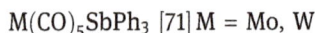

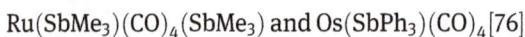

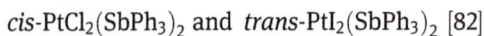

$Cr(CO)_5SbPh_3$ [69]

$M(CO)_5SbMe_3$, M = Cr, W [70]

$M(CO)_5SbPh_3$ [71] M = Mo, W

$[MnCO)_5SbPh_3]^+$ and $ReCl(CO)_3SbPh_3$ [72]

$Fe(CO)_4(SbBu^t_3)$ [73]

$Fe(CO)_4(SbPh_3)$ [74]

$Ru(CO)_4(SbPh_3)$ [75]

$Ru(SbMe_3)(CO)_4(SbMe_3)$ and $Os(SbPh_3)(CO)_4$ [76]

$CoI_3(SbPh_3)_2$ [77]

mer-$RhCl_3(SbPh_3)_3$ [78]

trans-$PdCl_2(SbPr^i_3)_2$ [79]

cis-$PtBr_2(SbPh_3)_2$ [80]

$Pt(SbMe_3)_4$ [81]

cis-$PtCl_2(SbPh_3)_2$ and *trans*-$PtI_2(SbPh_3)_2$ [82]

$[Cu(SbPh_3)_4]^+$ and $[Ag(SbPh_3)_4]^+$ [83]

$[Ag(SbPh_3)_4]^+$ [84, 85]

$Au(SbMe_3)_2$ and $Au[Sb(Ph_2Mes)_3]$ [86]

$[Au(PPh_3)_4]^+$ [87]

$AlI_3(SbPri_3)$ $GaBr_3(SbEt_3)$ $InCl_3(SbEt_3)$ $InCl_3(SbPr^i_3)$ [88]

$GaBut_3(SbMe_3)$ [89]

Complexes with distibane ligands can also be prepared and examples are (μ-Me_2Sb–$SbMe_2$)[$Cr_2(CO)_5$]$_2$ [90], [(μ-Me_2Sb–$SbMe_2$)($GaBu^t$)$_2$], [(μ-Et_2Sb–$SbEt_2$)(MBu^t_3)$_2$] [91] and [(μ-Ph_2Sb–$SbPh_2$){$W(CO)_5$}$_2$] [92].

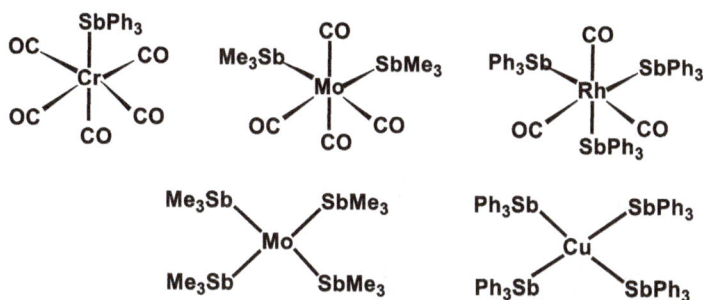

Fig. 9.24: Examples of triorganostibane ligand complexes.

Rare tristibane donor complexes [{μ$_2$-Sb{HC(SiMe$_3$)$_2$}$_3$Fe(CO)$_4$] [93] and a branched tetrastibane complex [(μ$_3$-Sb){SbAr)$_3$(μ$_2$-TiCp$_2$)$_3$] (Ar = C$_6$H$_4$CH$_2$NMe$_2$) [94] are also known (Fig. 9.25).

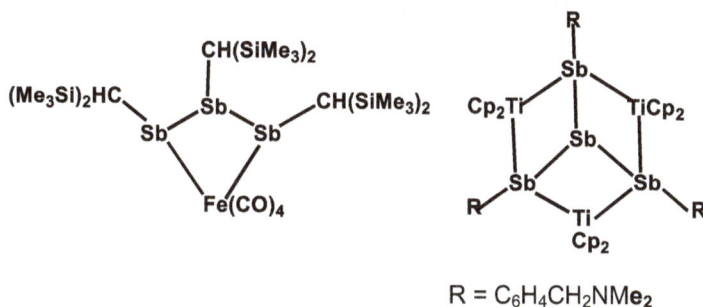

R = C$_6$H$_4$CH$_2$NMe$_2$

Fig. 9.25: Tri- and tetrastibane ligands.

9.1.13 Organoantimony inverse coordination complexes

This category of compounds provides various examples in which a single nonmetallic central atom is surrounded by a number of organoantimony moieties. The cyclic poly-stibanes can also serve as centroligands in inverse coordination complexes. Examples are known with a cyclotristibane cyclo-[Sb{CH(SiMe$_3$)$_2$}$_3${W(CO)$_5$}] [95], with a cyclote-trastibane, cyclo-[(SbBu)$_4${W(CO)$_5$}$_2$], and with a cyclo-pentastibane, cyclo-[Sb(CH$_2$-SiMe$_3$)$_5${W(CO)$_5$}] (Fig. 9.26).

The oxo-centered cation [(μ$_3$-O)(SbMe$_2$)$_3$]$^+$ [96], molecular azo-centered [(μ$_3$-N)(SbR$_2$)$_3$] (R = Me, Ph) [97] and iodo-centered (μ$_4$-I)(SbI$_2$Ph)$_4$]$^-$ anion [98] also illustrate this category (Fig. 9.27).

Organoantimony derivatives of inorganic oxoacids represent a second family of inverse coordination complexes and examples include nitrato [(μ$_2$-NO$_3$)(SbPh$_4$)$_3$]$^-$,

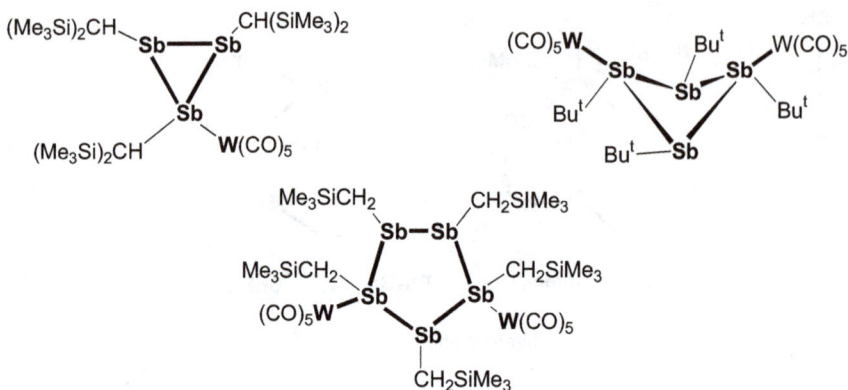

Fig. 9.26: Organocyclostibanes as centroligands in inverse coordination complexes.

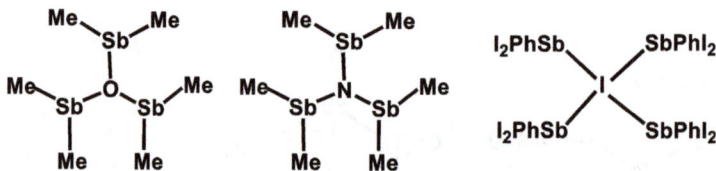

Fig. 9.27: Organoantimony inverse coordination complexes.

sulfato [(μ$_2$-SO$_4$)(SbPh$_4$)$_2$] [99] and carbonato [(μ$_2$-CO$_3$)(SbMe$_4$)$_2$] [100] and [(μ$_3$-CO$_3$) (SbPh$_4$)$_3$] [101] compounds (Fig. 9.28).

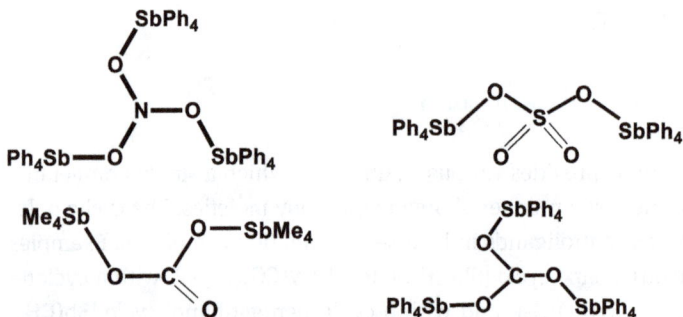

Fig. 9.28: Organoantimony inverse coordination complexes with inorganic centroligands.

Carboxylato anions are versatile centroligands and form numerous organoantimony inverse coordination complexes. A selection of illustrative examples include oxalato [μ$_2$-C$_2$O$_4$)(SbPh$_2$)$_2$] [102], succinato [(μ$_2$-OOC–CH$_2$CH$_2$COO)(SbPh$_4$)] [103], phthalato [(μ$_2$-C$_6$H$_4$(COOSbPh$_4$-1,2)$_2$] [104] and pyridine dicarboxylato [(μ$_2$-C$_6$H$_3$N(COOSbPh$_4$-2,6)$_2$] [105] compounds (Fig. 9.29).

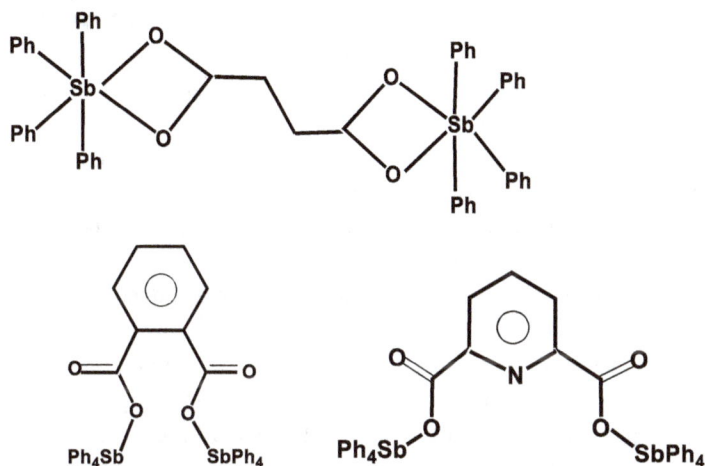

Fig. 9.29: Organoantimony inverse coordination complexes with carboxylato centro-ligands.

Further reading

Breunig HJ, Rösler R Organoantimony compounds with element-element bonds. Coord Chem Rev 1997, 163, 33–53.

Copolovici DM, Bojan RV, Raţ CI, Silvestru C. New chiral organoantimony(III) compounds containing intramolecular N→Sb interactions – Solution behaviour and solid state structures. Dalton Transactions 39, 6410–18.

Said MA, Kumara-Swamy KC, Poojary DM, Clearfield A, Veith M, Huch V Dinuclear and tetranuclear cages of oxodiphenylantimony phosphinates: Synthesis and structures. Inorg Chem 1996, 35, 3235

9.2 Organobismuth compounds

Bismuth has more metallic character than its higher congeners in group 15. The bismuth–carbon bond is less thermally stable and is readily cleaved. Thus, organobismuth hydrides decompose at –50 °C, as do the compounds containing Bi–Bi bonds, and the hydrolysis products of organobismuth halides are better described.

Nitrogen- and sulfur-containing organobismuth compounds are little known.

9.2.1 Homoleptic BiR₃ compounds

Bismuth metal or bismuth tribromide reacts with diorganomercury compounds. Grignard or organolithium compounds are effective, and organosodium compounds are used for the preparation of acetylenic derivatives. Organoaluminum reagents react with bismuth(III) oxide to form triorganosubstituted compounds. Other methods include the electrochemical synthesis of BiR_3 derivatives by electrolysis of the salt $Na^+[R_3Al\text{-}Et]^-$ with a bismuth anode, and the decomposition of the aryldiazonium salts, $[ArN_2]^+[BiCl_4]^-$, in the presence of metallic copper.

Solid triphenylbismuth is trigonal pyramidal with different angles of rotation of the phenyl groups about the Bi–C bonds.

The trialkylsubstituted bismuth derivatives are pyrophoric in air, decompose on distillation, and are readily cleaved by halogens. The aromatic derivatives are more stable in air. Strong acids cleave all organobismuth derivatives to inorganic bismuth (III) compounds, but weak acids lead to partial cleavage of aromatic groups to R_2BiX and $RBiX_2$ derivatives. The triorganobismuth compounds (bismuthines) are weaker donors than tertiary arsines and stibines, but some transition metal complexes of triphenylbismuth are known.

9.2.2 Pentaorgano-substituted derivatives, BiR₅

Pentaphenylbismuth is prepared from phenyllithium with triphenylbismuth dichloride:

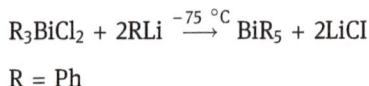

$$R_3BiCl_2 + 2RLi \xrightarrow{-75\ °C} BiR_5 + 2LiCI$$

$$R = Ph$$

The product decomposes at ca. 100 °C and is converted to tetraphenylbismuthonium salts with bromine in CCl_4 or HCl in ether. With excess of phenyllithium, it forms a hypervalent hexaphenyl anion $[BiPh_6]^-$.

9.2.3 Subvalent bismuth cations, [Biⱽ R₄]⁺

Aromatic bismuthonium derivatives can be prepared from pentaphenylbismuth by bromine cleavage at −78 °C:

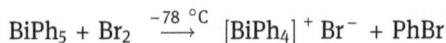

$$BiPh_5 + Br_2 \xrightarrow{-78\ °C} [BiPh_4]^+ Br^- + PhBr$$

Tetraphenylbismuthonium chloride is prepared by cleavage of pentaphenylbismuth with hydrogen chloride and can be converted to a more stable tetraphenylborate:

$$\text{BiPh}_5 + \xrightarrow[\substack{-78\ °C \\ -\text{phH}}]{\text{HCl}} [\text{BiPh}_4]^+ \text{Cl}^- \xrightarrow{\text{NaBPh}_4} [\text{BiPh}_4]^+ [\text{BiPh}_4]^-$$

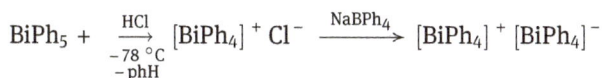

9.2.4 Organobismuth halides

Organobismuth(III) mono- and dihalides, R_nBiX_{3-n}, have been obtained by re-distribution between trisubstituted derivatives and bismuth(III) halides:

$$2\text{BiR}_3 + \text{BiX}_3 \longrightarrow 3\text{R}_2\text{BiX} \quad (X = \text{Cl, Br, I})$$

$$\text{BiR}_3 + 2\text{BiX}_3 \longrightarrow 3\text{RBiX}_2$$

or by arylation of bismuth(III) bromide with tetraphenyllead. Organolead reagents are also used for the synthesis of vinylbismuth dichloride:

$$\text{Pb}(\text{CH=CH}_2)_4 + 2\text{BiCl}_3 \xrightarrow{\text{CCl}_4} 2\text{H}_2\text{C=CH}-\text{BiCl}_2 + (\text{H}_2\text{C=CH})_2\ \text{PbCl}_2$$

and an organotin heterocycle is employed to prepare a halide which is the precursor of the unstable bismabenzene (Fig. 9.30).

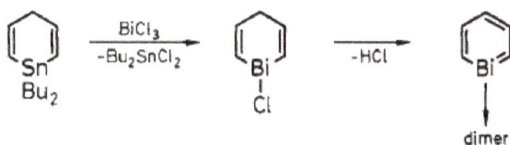

Fig. 9.30: Formation of the bismabenzene ring.

Organobismuth halides are also obtained by cleavage of trisubstituted derivatives with hydrogen halides, iodine chloride, phosphorus or arsenic trichlorides, acyl chlorides, mercury(II) and thallium(III)chlorides.

The reactive organobismuth(III) halides are sensitive to moisture and alcohols and pyrophoric in air. Diarylbismuth halides are biologically active, showing strong sternutatory action, and are toxic.

The hydrolysis of dihalides leads to oxides, RBiO, but dimethylbismuth bromide forms a hydroxide, $R_2\text{Bi}-\text{OH}$.

The halides react with sodium ethoxide and thiols to give ethoxy derivatives, $R_2\text{Bi}-\text{OEt}$, and thiolates, $R_2\text{Bi}-\text{SR}'$, respectively.

The anions $[\text{Ph}_2\text{BiX}_2]$" are formed by addition of halide or pseudohalide anions ($X = \text{Cl, Br, CN, SCN, N}_3$) to a diphenylbismuth halide.

Triorganobismuth(V) dihalides. The $R_3\text{BiX}_2$ compounds are prepared by addition of halogens – in stoichiometric amounts and under controlled conditions to avoid Bi–C bond cleavage – to trisubstituted derivatives:

$$BiR_3 + X_2 \xrightarrow[CHCl_3]{0\,°C} R_3BiX_2$$

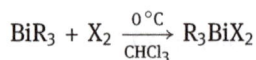

Sulfuryl chloride, sulfur monochloride, thionyl chloride and iodine trichloride are also used.

Solid triphenylbismuth dichloride is trigonal bipyramidal with organic groups in equatorial position and the halogens in axial position. Conductivity measurements in acetonitrile show no dissociation as $[R_3BiX]^+X^-$.

The thermal stability of the triorganobismuth dichlorides decreases in the order F > CI > Br > I, the iodides decomposing as low as −60 °C.

The dihalides are reduced with hydrazine to triorganosubstituted compounds.

9.2.5 Oxygen-containing organobismuth compounds

The triarylbismuth hydroxyhalides, $R_3Bi(OH)X$, are formed by treatment of dihalides with aqueous ammonia, and dihydroxides, $R_3Bi(OH)_2$, by treatment with wet silver oxide.

The solid, oxygen-containing salt, $[Ph_3Bi-O-BiPh_3]^{2+}[ClO_4]_2$, contains four-coordinated bismuth.

References

[1]   Heimann S, Kuczkowski A, Bläser D, Wölper C, Haack R, Jansen G, Schulz S. Syntheses and solid-state structures of $Et_2SbTeEt$ and $Et_2BiTeEt$. Eur J Inorg Chem 2014, 2014, 4858–64.

[2]   Becker G, Mundt O, Sachs M, Breunig HJ, Lork E, Probst J, Silvestru A. Untersuchungen am Chlordiphenyl- und Tribenzylstiban sowie am Tribenzyldibromstiboran - Molekülstrukturen und Isotypie. Z Anorg Allg Chem 2001, 627, 699–714.

[3]   Yang M, Tofan D, Chen CH, Jack KM, Gabbaï FP Digging the sigma-hole of organoantimony Lewis acids by oxidation. Angew Chem Int Ed 2018, 57, 13868–72.

[4]   Cornet SM, Dillon KB, Goeta AE The identification of some new antimony(III) compounds containing fluoroxyl ligands by ^{19}F solution-state NMR spectroscopy: Crystal and molecular structure of (Ar' = 2,6-$(CF_3)_2C_6H_3$; Ar" = 2,4-$(CF_3)_2C_6H_3$). Inorg Chim Acta 2005, 358, 844–48.

[5]   Bojan VR, Fernández EJ, Laguna A, López-de-luzuriaga JM, Monge M, Olmos ME, Puelles RC, Silvestru C Study of the coordination abilities of stibine ligands to gold(I). Inorg Chem 2010, 49, 5530–41.

[6]   Beauchamp AL, Bennett MJ, Cotton FA A reinvestigation of the crystal and molecular structure of pentaphenylantimony. J Am Chem Soc 1968, 90, 6675–80.

[7]   García-Monforte MA, Alonso PJ, Ara I, Menjón B, Romero P. Solid-state and solution structure of a hypervalent AX_5 compound: $Sb(C_6F_5)_5$. Angew Chem Int Ed. 2012, 51, 2754

[8]   Schröder G, Okinaka T, Mimura Y, Watanabe M, Matsuzaki T, Hasuoka A, Yamamoto Y, Matsukawa S, Akiba K. Syntheses, crystal and solution structures, ligand-exchange, and ligand-coupling reactions of mixed pentaarylantimony compounds. Chem-Eur J 2007, 13, 2517–29.

[9]   Pressprich MR, Bond MR, Willett RD Structures of [(CH$_3$)$_4$P]$_2$CoCl$_4$, [(CH$_3$)$_4$Sb]$_2$CuCl$_4$ and
       [(CH$_3$)$_4$Sb]$_2$ZnBr$_4$ and a correlation of pseudo-antifluorite [(CH$_3$)$_4$P]$_2$MX$_4$ salts. J Phys Chem
       Solids 2002, 63, 79–88.

[10]   Sharutin VV, Sharutina OK, Efremov AN μ$_2$-Oxo-bis[(2,5-dinitrophenoxo)triarylantimony]:
       syntheses, structures, and reactions with pentaarylantimony. Russ J Coord Chem 2020, 46,
       42–52.

[11]   Yang M, Pati N, Bélanger-Chabot G, Hirai M, Gabbaï FP Influence of the catalyst structure in
       the cycloaddition of isocyanates to oxiranes promoted by tetraarylstibonium cations. Dalton
       Trans 2018, 47, 11843–50.

[12]   Hill AM, Levason W, Webster M, Albers I Complexes of distibinomethane ligands. Iron,
       cobalt, nickel, and manganese carbonyl complexes. Organometallics 1997, 16, 5641–47.

[13]   Holmes NJ, Levason W, Webster M Ruthenium-(II) and -(III) complexes of a ditertiary stibine
       ligand. The effect of co-ordination on stibine ligand geometry. J Chem Soc Dalton Trans 1998,
       3457–62.

[14]   Paver MA, Joy JS, Coles SJ, Hursthouse MB, Davies J Organo-functionalised arsine and stibine
       organometallics; syntheses and structural characterisations of 1,3-[(PhC≡C)$_2$Sb]$_2$(CH$_2$)$_3$, As
       (C≡CPh)$_3$, R$_2$AsCH$_2$AsR$_2$ [R = Me$_3$SiC≡C-, (Me$_3$Si)$_2$N- and 2-SPy] with π-stacking in the latter.
       Polyhedron 2003, 22, 211–16.

[15]   Levason W, Matthews ML, Reid G, Webster M Synthesis and properties of new ditertiary
       stibines based upon o-, m- or p-xylyl and m- or p-phenylene backbones and their complexes
       with tungsten, iron and nickel carbonyls. Dalton Trans 2004, 51.

[16]   Breunig HJ, Althaus H, Rösler R, Lork E Neue Synthesen von Me$_2$SbX (X = Cl, I) und
       Kristallstrukturen von Me$_2$SbI und [(Me$_3$Si)$_2$CH]$_2$SbCl. Z Anorg Allg Chem 2000, 626,
       1137–40.

[17]   Becker G, Mundt O, Sachs M, Breunig HJ, Lork E, Probst J, et al. Element-element bonds. X.
       Studies of chloro(diphenyl)stibane, tribenzylstibane and tribenzyldibromostiborane -
       Molecular structures and isotypism. Z Anorg Allg Chem 2001, 627, 699–714.

[18]   Burford N, Macdonald CLB, LeBlanc DJ, Cameron TS Synthesis and characterization of bis
       (2,4,6-tris(trifluoromethyl)phenyl) derivatives of arsenic and antimony: x-ray crystal
       structures of As(R$_F$)$_2$Cl, Sb(R$_F$)$_2$Cl, and Sb(R$_F$)$_2$OSO$_2$CF$_3$. Organometallics 2000, 19, 152–55.

[19]   Mourad Y, Atmani A, Mugnier Y, Breunig HJ, Ebert KH Structure and electroreduction of
       Sb^{t}BuCl$_2$. Reactivity and electrochemical behaviour of (Sb^{t}Bu)$_4$. J Organomet Chem 1994,
       476, 47–54.

[20]   Mundt O, Becker G, Stadelmann H, Thurn H. Element-Element-Bindungen. VII.
       Intermolekulare Wechselwirkungen bei Dihalogen(phenyl)stibanen. Z Anorg Allg Chem 1992,
       617, 59–71.

[21]   Mohammed MA, Ebert KH, Breunig HJ Intermolecular interactions of RSbCl$_2$ and formation of
       cyclic organoantimony(III) chalcogenides, (RSbE)$_n$ [R = (Me$_3$Si)$_2$CH, E = S, Se, Te]. Z
       Naturforsch B 1996, 51, 149–52.

[22]   James SC, Norman NC, Orpen AG Pyridine adducts of arylbismuth(III) halides. J Chem Soc
       Dalton Trans 1999, 2837–43.

[23]   Hall M, Sowerby DB Phenylchloroantimon(III)ates; their preparations, and the crystal
       structures of Me$_4$N[PhSbCl$_3$], [Hpy]$_2$[PhSbCl$_4$], and Me$_4$N[Ph$_2$SbCl$_2$]. J Organomet Chem 1988,
       347, 59–70.

[24]   Breunig HJ, Ebert KH, Gülec S, Dräger M, Sowerby DB, Begley MJ, Behrens U.Strukturen von
       [Me$_4$Sb]$_2$[MeSbI$_4$]MeSbI$_2$, und [Me$_4$Sb]I. Darstellung von Me$_3$Sb·MeSbI$_2$ und Farbwechsel bei
       Me$_4$Sb$_2$·Me$_2$SbBr. J Organomet Chem 1992, 427, 39–48.

[25]   Sheldrick WS, Martin C Darstellung und Kristallstrukturen von Chloro- und
       Bromophenylantimonaten(III) [Ph$_2$SbX$_2$]$^-$ und [Ph$_2$Sb$_2$X$_7$]$^{3-}$ (X = Cl, Br). Preparation and crystal

structures of chloro- and bromophenylantimonates(III) [Ph$_2$SbX$_2$]⁻ and [Ph$_2$Sb$_2$X$_7$]$_3$. (X = Cl, Br). Z Naturforsch B 1992, 47, 919–24.

[26] Sharma P, Rosas N, Toscano A, Hernandez S, Shankar R, Cabrera A Crystal structures of Bu$_4$N [PhSbBr$_2$Cl] and Bu$_4$N[Ph$_2$SbBr$_2$]. Main Group Met Chem 1996, 19, 21–27.

[27] Sheldrick WS, Martin C. Preparation and crystal structures of phenyliodoantimonates(III). Structural correlation for haloantimonates(III). Z Naturforsch B 1991, 46, 639–46.

[28] Sharutin VV, Sharutina OK, Gerasimenko AV Phenylation of antimony(V) organic compounds with pentaphenylantimony. The structure of tetraphenylantimony chloride. Russ J Coord Chem 2003, 29, 89–92.

[29] Ferguson G, Glidewell C, Lloyd D, Metcalfe S Effect of the counter-ion on the structures of tetraphenylantimony(v)-stibonium compounds: Crystal and molecular structures of tetraphenylantimony(v) bromide, perchlorate, and tetraphenylborate. J Chem Soc Perkin Trans 2 1988, 731–35.

[30] Knop O, Vincent BR, Cameron TS Pentacoordination and ionic character: Crystal structure of Ph$_4$SbBr. Can J Chem 1989, 67, 63–70.

[31] Schwarz W, Gudek HJ Kristall- und Molekulstrukturen von Tetramethylantimon-fluorid (CH$_3$)$_4$SbF und Trimethylantimondifluorid (CH$_3$)$_3$SbF$_2$. Z Anorg Allg Chem. 1978, 444, 105–11.

[32] Yang M, Gabbaï FP Synthesis and properties of triarylhalostibonium cations. Inorg Chem 2017, 56, 8644–50.

[33] Becker G, Mundt O, Sachs M, Breunig HJ, Lork E, Probst J, Silvestru A Untersuchungen am Chlordiphenyl- und Tribenzylstiban sowie am Tribenzyldibromstiboran - Molekülstrukturen und Isotypie. Z Anorg Allg Chem. 2001, 627, 699–714

[34] Khosa MK, Mazhar M, Ali S, Molloy K, Dasgir S, Shaheen F. Synthesis, Spectroscopic (FT-IR, ^{1}H, ^{13}C, mass spectrometry), and biological investigation of five-coordinated germanium-substituted tricyclohexyl antimony dipropionates: crystal structure of tricyclohexylantimony dibromide. Turk J Chem 2006, 30, 731–43

[35] MacDonald DJ, Jennings MC, Preuss KE. A new polymorph of dichloridotriphenylantimony. Acta Crystallogr Sect C Cryst Struct Commun 2010, 66, m137–m140.

[36] Hendershot DG, Pazik JC, George C, Berry AD Synthesis and characterization of neopentyl- and [(trimethylsilyl)methyl]antimony compounds. Molecular structures of (Me$_3$CCH$_2$)$_3$Sb, (Me$_3$CCH$_2$)$_3$SbI$_2$, (Me$_3$SiCH$_2$)$_3$Sb, and (Me$_3$SiCH$_2$)$_3$SbI$_2$. Organometallics 1992, 11, 2163–68.

[37] Bone SP, Sowerby DB Crystal structures of diphenylantimony(V) tribromide, dibromide chloride, and bromide dichloride. J Chem Soc Dalton Trans 1979, 718–22.

[38] Zaitseva EG, Medvedev SV, Aslanov LA Zh Strukt Khim. Russ J Struct Chem 1990, 31, 110–11.

[39] Zaitseva EG, Medvedev SV, Aslanov LA Crystal and molecular structures of cesium phenylpentachloroantimonate Cs[PhSbCl$_5$], potassium phenylpentabromoantimonate K [PhSbBr$_5$], and cesium hexachloroantimonate Cs[SbCl$_6$]. J Struct Chem 1990, 31, 92–97

[40] Yang M, Gabbai FP. Synthesis and properties of triarylhalostibonium cations. Inorg Chem. 2017, 56, 8644–50.

[41] Pop A, Silvestru A, Juárez-Pérez EJ, Arca M, Lippolis V, Silvestru C. Organoselenium(ii) halides containing the pincer 2,6-(Me$_2$NCH$_2$)$_2$C$_6$H$_3$ ligand – An experimental and theoretical investigation. Dalton Trans 2014, 43, 2221–33.

[42] Lang G, Klinkhammer KW, Recker C, Schmidt A. Dihydroxytrimethylstiboran - einige Eigenschaften und Struktur. Z Anorg Allg Chem 1998, 624, 689–93.

[43] Briand GG, Chivers T, Parvez M The relative stabilities of PhE(NH-t-Bu)$_2$ and PhE(μ-N-t-Bu)$_2$EPh (E = As, Sb, and Bi): X-ray structures of {Li$_2$[PhAs(N-t-Bu)$_2$]}$_2$ and PhE(μ-N-t-Bu)$_2$EPh (E = Sb, Bi). Can J Chem 2003, 81, 169–74.

[44] Jockisch A, Schmidbaur H The molecular structure of tris(dimethylstibino)amine. Z Naturforsch B 1998, 53, 1386–88.

[45] Ringler B, Von Hänisch C, Novel stibano amines: synthesis and reactivity towards group 13 element organics, Z Anorg Allg Chem 2016, 462, 294–98.

[46] Copsey MC, Gallon SB, Grocott SK, Jeffery JC, Russell CA, Slattery JM Synthetic and structural studies of cyclodistib(V)azanes. Inorg Chem 2005, 44, 5495–500.

[47] (a) Lorenz IP, Rudolph S, Piotrowski H, Polborn K. Reactions of $K_2[Fe(CO)_3(PPh_3)]$: Reductive Sb-Sb coupling with Ph_2SbCl to form *trans*-$[Fe(CO)_3(PPh_3)(Sb_2Ph_4)]$ and salt metathesis with Me_3SbCl_2 to yield *trans*-$[Fe(CO)_3(PPh_3)(SbMe_3)]$. Eur J Inorg Chem 2005, 82–85; (b) Kuczkowski A, Heimann S, Weber A, Schulz S, Bläser D, Wölper C. Structural characterization of Et_4Sb_2 and Et_4Bi_2. Organometallics 2011, 30, 4730–4735; (c) Schulz S, Heimann S, Kuczkowski A, Bläser D, Wölper, C. The origin of thermochromic behavior in distibines: Still an open question. Organometallics, 2013, 32, 3391–3394; (d) Becker G, Freudenblum H, Witthauer C. Trimethylsilylverbindungen der Vb-Elemente. VI. Synthese, Molekul- und Kristallstruktur des Tetrakis(trimethylsilyl)distibans im Vergleich mit Tetraphenyldistiban. Z Anorg Allg Chem 1982, 492, 37–51; (e) Breunig HJ, Moldovan O, Nema M, Rosenthaler U, Raţ CI, Varga R. Formation of [Cp2TiSbMe2]2, [Cp2TiSb(SiMe3)2]2 and [Cp2TiCl]2·2Mes4Sb2. J Organomet Chem 2011, 696, 523–526; (f) Cowley AH, Nunn CM, Westmoreland DL. Structure of tetramesityldistibane. Acta Crystallogr Section C Cryst Struct Commun 1990, 46(5), 774–776; (g) Balázs L, Breunig HJ, Silvestru C, Varga R. Synthesis and Crystal Structure of *meso*-R(Ph)Sb-Sb(Ph)R [R = $(Me_3Si)_2CH$]. Z Naturforsch 2005, 60b,1321–1323.

[48] (a) Breunig HJ, Pawlik J. Synthesen und Kristallstrukturuntersuchungen von Pentacarbonylwolfram-Komplexen mit Tetra-tert.-butylcyclotetrastiban und Tetraphenyldistiban als Liganden. Z Anorg Allg Chem 1995, 621, 817–22; (b) Becker G, Freudenblum H, Witthauer C.Trimethylsilylverbindungen der Vb-Elemente. VI. Synthese, Molekül- und Kristallstruktur des Tetrakis(trimethylsilyl)distibans im Vergleich mit Tetraphenyldistiban. Z Anorg Allg Chem 1982, 492, 37–51.

[49] Mundt O, Riffel H, Becker G, Element—Element Bonds, III. Intermolecular Sb···Sb Interactions in Crystalline Tetramethyldistibane. Z Naturforsch B 1984, 39, 317–22

[50] Ashe AJ, Ludwig EG, Oleksyszyn J, Huffman JC The structure of tetramethyldistibine. Organometallics 1984, 3, 337–38.

[51] Althaus H, Breunig HJ, Lork E Crystal structure of $[Me_3Sb\text{-}SbMe_2]_2[(MeSbBr_3)_2]$, a trimethylstibine adduct of the dimethylstibenium ion or a stibinostibonium salt? Chem Commun 1999, 1971–72.

[52] Hering C, Lehmann M, Schulz A, Villinger A. Chlorine/methyl exchange reactions in silylated aminostibanes: a new route to stibinostibonium cations. Inorg Chem 2012, 51, 8212–24.

[53] Minkwitz R, Hirsch C Reaktionen von Trimethylstiban in den supersauren Systemen XF/MF5 (X=H, D; M=As, Sb), Kristallstrukturen von $(CH_3)_3SbD^+SbF_6^-$ Und $(CH_3)_3SbSb(CH_3)_3{}^{2+}(SbF_6)$ SO_2. Z Anorg Allg Chem 1999, 625, 1674–82.

[54] Ishida S, Hirakawa F, Furukawa K, Yoza K, Iwamoto T. Persistent antimony- and bismuth-centered radicals in solution. Angew Chem Int Ed 2014, 53, 11172–76.

[55] Sasamori T, Arai Y, Takeda N, Okazaki R, Furukawa Y, Kimura M, Nagase S, Tokitoh N. Structures and properties of kinetically stabilized distibenes and dibismuthenes, novel doubly bonded systems between heavier group 15 elements. Bull Chem Soc Jpn 2002, 75, 661–75.

[56] Tokitoh N, Arai Y, Sasamori T, Okazaki R, Nagase S, Uekusa H, Ohashi Y. A Unique Crystalline-State Reaction of an Overcrowded Distibene with Molecular Oxygen: The First Example of a Single Crystal to a Single Crystal Reaction with an External Reagent. J Am Chem Soc 1998, 120, 433–34.

[57] Majhi PK, Ikeda H, Sasamori T, Tsurugi T, Mashima KN, Tokitoh K. Inorganic-salt-free reduction in main-group chemistry: synthesis of a dibismuthene and a distibene. Organometallics 2017, 36, 1224–26.

[58] Althaus H, Breunig HJ, Probst J, Rösler R, Lork E The crystal structure of a trinuclear anion, [(t-Bu$_2$Sb)$_2$Sb]$^-$ formed by reaction of cyclo-(t-Bu$_4$Sb$_4$) with potassium, and the four-membered rings, cyclo-(t-Bu$_4$E$_n$Sb$_{4-n}$) (E = P, n = 0-3; E = As, n = 0-2). J Organomet Chem 1999, 585, 285–89.

[59] Breunig HJ, Rösler R, Lork E A cyclotristibane: Synthesis and crystal structure of cyclo-[(Me$_3$Si)$_2$CH]$_3$Sb$_3$. Organometallics 1998, 17, 5594–95.

[60] Mundt O, Becker G, Wessely H-J, Breunig HJ, Kischkel H. Element-Element-Bindungen. I. Synthese und Struktur des Tetra(tert-butyl)tetrarsetans und des Tetra(tert-butyl) tetrastibetans. Z Anorg Allg Chem 1982, 486, 70–89.

[61] Ates M, Breunig HJ, Ebert K, Gülec S, Kaller R, Dräger M. Syntheses and structures of isopropyl-and (bis(trimethylsilyl)methyl)-antimony rings and catena-tri-and catena-tetrastibanes by reaction of organoantimony rings with distibanes. Organometallics 1992, 11, 145–50.

[62] Opris LM, Silvestru A, Silvestru C, Breunig HJ, Lork E. Syntheses and chemistry of hypervalent cyclo-R$_4$Sb$_4$, cyclo-(RSbE)$_n$ [R = 2-(Me$_2$NCH$_2$)C$_6$H$_4$, E = O, S] and precursors. Dalton Trans 2004, 3575–85.

[63] Kekia OM, Jones RL, Rheingold AL An extraordinarily distorted cyclo-tetrastibine. Crystallographic structure of a new polymorph of cyclo-(σ-C$_5$Me$_5$Sb$_4$). Organometallics 1996, 15, 4104–06.

[64] Balázs G, Balázs L, Breunig HJ, Lork E Synthesis, chemistry, and structures of neopentyl and (trimethylsilyl) methyl antimony and bismuth oligomers. Organometallics 2003, 22, 2919–24.

[65] Breunig HJ, Ebert KH, Gülec S, Probst J. Syntheses, Structures, and equilibria of o-, m-, p-tolyl- and phenylantimony rings. Chem Ber 1995, 128, 599–603.

[66] Dostál L, Jambor R, Růžička A, Holeček J Syntheses and structures of Ar$_3$Sb$_5$ and Ar$_4$Sb$_4$ compounds (Ar = C$_6$H$_3$-2,6-(CH$_2$NMe$_2$)$_2$. Organometallics 2008, 27, 2169–71.

[67] Balázs G, Breunig HJ, Lork E, Mason S Neutron diffraction crystallography of meso-R(H)Sb-Sb (H)R and reactions of R(H)Sb-Sb(H)R and RSbH$_2$ (R = (Me3Si)2CH) leading to tungsten carbonyl complexes, methylstibanes, and antimony homocycles. Organometallics 2003, 22, 576–85.

[68] Breunig HJ, Rösler R, Lork E. Sb$_8$R$_4$, R = (Me$_3$Si)$_2$CH - A polycyclic organostibane. Angew Chem Int Ed 1997, 36, 2237–38.

[69] Carty AJ, Taylor NJ, Coleman AW, Lappert MF The chromium–heavy group 5 donor bond: A comparison of structural changes within the series [Cr(CO)$_5$(XPh$_3$)] (X = P, As, Sb, or Bi) via their X-ray crystal structures. J Chem Soc, Chem Commun 1979, 639–40.

[70] Breunig HJ, Borrmann T, Lork E, Moldovan O, Raţ CI, Wagner RP Syntheses, crystal structures and DFT studies of [Me$_3$EM(CO)$_5$] (E = Sb, Bi; M = Cr, W), cis-[(Me$_3$Sb)$_2$Mo(CO)$_4$], and [^{t}Bu$_3$BiFe (CO)$_4$]. J Organomet Chem 2009, 694, 427–32.

[71] Aroney MJ, Buys IE, Davies MS, Hambley TW Crystal structures of [W(CO)$_5$(PPh$_3$)], [M(CO)$_5$ (AsPh$_3$)] and [M(CO)$_5$(SbPh$_3$)] (M = Mo or W): A comparative study of structure and bonding in [M(CO)$_5$(EPh$_3$)] complexes (E = P, As or Sb; M = Cr, Mo or W). J Chem Soc Dalton Trans 1994, 2827–34.

[72] Holmes NJ, Levason W, Webster M Triphenylstibine substituted manganese and rhenium carbonyls: Synthesis and multinuclear NMR spectroscopic studies. X-ray crystal structures of ax-[Mn$_2$(CO)$_9$(SbPh$_3$)], [Mn(CO)$_5$(SbPh$_3$)][CF$_3$SO$_3$] and fac-[Re(CO)$_3$Cl(SbPh$_3$)$_2$]. J Organomet Chem 1998, 568, 213–23.

[73]   Rheingold AL, Fountain ME Tetracarbonyl(tri-tert-butylstibine)iron, [Fe(CO)$_4$(Sb(C$_4$H$_9$)$_3$)]. Acta Crystallogr Sect C Cryst Struct Commun 1985, 41, 1162–64.

[74]   Bryan RF, Schmidt WC Metal–metal bonding in co-ordination complexes. Part XII. Crystal structure of tetracarbonyl-(triphenylstibine)iron. J Chem Soc Dalton Trans 1974, 2337–40.

[75]   Forbes EJ, Jones DL, Paxton K, Hamor TA Synthesis and crystal structure of tetracarbonyl (triphenylstibine)ruthenium. J Chem Soc Dalton Trans 1979, 879–82.

[76]   Martin LR, Einstein FWB, Pomeroy RK Axial-equatorial isomerism in the complexes M(CO)$_4$(L) (M = Fe, Ru, Os; L = Group 15 ligand). Crystal structures of ax-Ru(CO)$_4$(AsPh$_3$), ax-Ru(CO)$_4$ (SbMe$_3$), and eq-Os(CO)$_4$(SbPh$_3$). Inorg Chem 1985, 24, 2777–85.

[77]   Godfrey SM, McAuliffe CA, Pritchard RG. Extreme symbiosis: The facile one-step synthesis of the paramagnetic cobalt(III) complex of triphenylantimony, CoI$_3$(SbPh$_3$)$_2$, from the reaction of triphenylantimonydiiodine with unactivated coarse grain cobalt metal powder. J Chem Soc Chem Comm. 1994, 45–46.

[78]   Cini R, Tamasi G, Defazio S, Corsini M, Berrettini F, Cavaglioni A Unusual hetero-atomic RhSCNSb(Rh) co-ordination ring: Synthesis and X-ray structure of [Rh(N1,S2-2-thiopyrimidinato)2(N1(Sb),S2(Rh)-2-thiopyrimidinato){Sb(C$_6$H$_5$)$_3$}] and long time sought structure of mer-[RhCl$_3${Sb(C$_6$H$_5$)$_3$}$_3$]. Polyhedron 2006, 25, 834–42.

[79]   Phadnis PP, Jain VK, Varghese B Preparation and characterization of tris(iso-propyl)stibine complexes of palladium and platinum. Appl Organomet Chem 2002, 16, 61–64.

[80]   Sharma P, Cabrera A, Sharma M, Alvarez C, Arias JL, Gomez RM, Hernandez S Trans Influence of triphenylstibine: Crystal structures of cis-[PtBr$_2$(SbPh$_3$)$_2$], trans-[PtBr(Ph)(SbPh$_3$)$_2$], [NMe$_4$] [PtBr$_3$(SbPh$_3$)], and cis-[PtBr$_2$(SbPh$_3$)(PPh$_3$)]. Z Anorg Allg Chem 2000, 626, 2330–34

[81]   Miyamoto TK Synthesis of water-soluble platinum(II) complexes stabilized with trimethylstibane. Stibane transfer in aqueous solution. Chem Lett 1994, 23, 2031–32.

[82]   Wendt OF, Scodinu A, Elding LI Trans influence of triphenylstibine. Crystal and molecular structures of cis-[PtCl$_2$(SbPh$_3$)$_2$] and trans-[PtI$_2$(SbPh$_3$)$_2$]. Inorg Chim Acta 1998, 277, 237–41.

[83]   Bowmaker GA, Effendy HRD, Kildea JD, De Silva EN, Skelton BW, White AH Lewis-base adducts of group 11 metal(I) compounds. LXIV: Syntheses, spectroscopy structures of some 1:4 adducts of copper (I) and silver(I) perchlorates with triphenylarsine and triphenylstibine. Aust J Chem 1997, 50, 539–52.

[84]   Kuprat M, Schulz A, Thomas M, Villinger A Synthesis and characterization of a stable non-cyclic bis(amino)arsenium cation. Can J Chem 2018, 96, 502–12.

[85]   Cingolani A, Effendy, Hanna JV, Pellei M, Pettinari C, Santini C, Skelton BW, White A. H Crystal structures and vibrational and solution and solid-state (CPMAS) NMR spectroscopic studies in triphenyl phosphine, arsine, and stibine silver(I) bromate systems, (R$_3$E)$_x$AgBrO$_3$ (E = P, As, Sb; x = 1–4). Inorg Chem 2003, 42, 4938–48.

[86]   Bojan VR, Fernandez EJ, Laguna A, Lopez-de-luzuriaga JM, Monge M, Olmos ME, Puelles RC, Silvestru C. Inorg Chem 2010. 49, 5530

[87]   Li Y-Z, Ganguly R, Hong KY, Li Y, Tessensohn ME, Webster R, Tessensohn ME, Webster R, Leong WK Stibine-protected Au$_{13}$ nanoclusters: Syntheses, properties and facile conversion to GSH-protected Au$_{25}$ nanocluster. Chemical Science 2018, 9, 8723–30

[88]   Greenacre VK, Levason W, Reid G Trialkylstibine complexes of boron, aluminum, gallium, and indium trihalides: synthesis, properties, and bonding. Organometallics 2018, 37, 2123–35

[89]   Schulz S, Kuczkowski A, Nieger M. Lewis acid–base adduct Me$_3$Sb-Ga(t-Bu)$_3$. J Chem Crystallogr 2010, 40, 1163–66.

[90]   Breunig HJ, Geshner I, Geshner ME, Lork E Syntheses and coordination chemistry of di-, tri-, and tetrastibanes, R$_2$Sb(SbR′)$_n$SbR$_2$ (n = 0, 1, 2). J Organomet Chem 2003, 677, 15–20.

[91]   Kuczkowski A, Schulz S, Nieger M, Saarenketo P Reactivity of distibanes toward trialkylalanes and -gallanes: syntheses and X-ray structures of bisadducts and heterocycles. Organometallics 2001, 20, 2000–06.

[92]   Breunig HJ, Pawlik J Synthesen und Kristallstrukturuntersuchungen von Pentacarbonylwolfram-Komplexen mit Tetra-*tert.*-butylcyclotetrastiban und Tetraphenyldistiban als Liganden. Z Anorg Allg Chem 1995, 621, 817–22.

[93]   Balász G, Breunig HJ, Lork E, .Syntheses and crystal structures of [μ-(Me$_3$SiCH$_2$Sb)$_5$-Sb1,Sb3-{W(CO)$_5$}$_2$] and [{(Me$_3$Si)$_2$CHSb}$_3$Fe(CO)$_4$] - Two cyclic complexes with antimony ligands. Z Anorg Allg Chem 2001, 627, 1855–58.

[94]   Breunig HJ, Lork E, Moldovan O, Raţ CI, Rosenthal U, Silvestru C Polynuclear titanocene complexes with antimony ligands: [(Cp$_2$Ti)$_2$(SbR$_2$)$_2$] (R = Et), [(Cp$_2$Ti)$_3$(SbR)$_3$Sb] [R = 2-(Me$_2$NCH$_2$)C$_6$H$_4$] and [(Cp$_2$Ti)$_5$(SbR)$_2$Sb$_7$] (R = Me$_3$SiCH). Dalton Trans 2009, 5065–67.

[95]   Balázs G, Breunig HJ, Lork E Reaktionen von Cyclostibanen (RSb)$_n$ [R = (Me$_3$Si)$_2$CH, n = 3; Me$_3$CCH$_2$, n = 4, 5] mit den Übergangsmetallcarbonyl-Komplexen [W(CO)5(thf)], [Cp$_x$Mn(CO)$_2$(thf)], [Cp$_x$Cr(CO)$_3$], und [Co$_2$(CO)$_8$]; Cpx = MeC$_5$H$_4$. Z Anorg Allg Chem 2003, 629, 1937–42.

[96]   Breunig HJ, Mohammed MA, Ebert KH Preparation and structure of [(Me$_2$Sb)$_3$O]Br, a planar stibino oxonium salt. Polyhedron 1994, 13, 2471–72.

[97]   (a) Jockisch A, Schmidbaur H, The Molecular Structure of Tris(dimethylstibino)amin) Z Naturforsch B. 1998, 53, 1386–88; (b) Balázs L, Breuning HJ, Kruger T, Lork E, Formation and Structure of Tris(diphenylstibino)amine. Z Naturforsch. B, 2001, 56, 1325–1327

[98]   Von Seyerl J, Scheidsteger O, Berke H, Huttner G [(PhSbI$_2$)$_4$I]⁻, ein quadratisch planarer komplex von iodid. J Organomet Chem 1986, 311, 85–89.

[99]   Sharutin VV, Sharutina OK, Pakusina AP, Platonova TP, Gerasimenko AV, Bukvetskii BV, Synthesis and structure of tetra- and triphenylantimony organosulfonates. Russ J Coord Chem 2004, 30, 13–22.

[100] Lang G, Klinkhammer KW, Recker C, Schmidt A. Dihydroxytrimethylstiboran - einige Eigenschaften und Struktur. Z Anorg Allg Chem 1998, 624, 689–93.

[101] Sharutin VV, Sharutina OK, Efremov AN Arylantimony Derivatives of Three-Coordinated Carbon. Russ J Inorg Chem 2020, 65, 45–51.

[102] Millington PL, Sowerby DB Phenylantimony(V) oxalates: Isolation and crystal structures of [SbPh$_4$][SbPh$_2$(ox)$_2$], [SbPh$_3$(OMe)]$_2$ox and (SbPh$_4$)$_2$ox. J Chem Soc Dalton Trans 1992, 1199–204.

[103] Sharutin VV, Sharutina OK, Gubanova YO, Andreev PV, Somov NV, Synthesis and structure of [(μ$_4$-succinato)- hexadecaphenyltetraantimony] triiodide solvate with benzene [(Ph$_4$Sb)$_2$O$_2$CCH$_2$CH$_2$CO$_2$(Ph$_4$Sb)$_2$][I$_3$]$_2$·4PhH. Koord Khim (Russ Coord Chem) 2014, 40, 559.

[104] Sharutin VV, Sharutina OK, Senchurin VS, Platonova IP, Nasonova NV, Pakusina AP, Gerasimenko AV, Sergienko SS. Koord Khim (Russ Coord Chem) 2001, 27, 710.

[105] Gubanova YO, Sharutin VV, Sharutina OK, Petrova KY Specific features of the reactions of pentaphenylantimony with polyfunctional heterocyclic carboxylic acids. Russ J Gen Chem 2020, 90, 1664–69.

10 Organometallic compounds of group 12 metals

10.1 Organozinc compounds

Organozinc compounds played an important role as intermediates in synthetic organic chemistry before the discovery of Grignard reagents. The first organozinc compounds were prepared by Frankland in 1849, and they are among the first organometallic compounds known. Interest in organozinc compounds was recently revived, and new applications as intermediates in organic synthesis have brought them again into use.

Disubstituted compounds, ZnR_2, and monosubstituted functional derivatives, RZnX (X = halogen, OR, NR_2, SR, etc.), are all well known, and hypervalent three-coordinated $[ZnR_3]^-$ and four-coordinated $[ZnR_4]^{2-}$ derivatives are also described.

10.1.1 Homoleptic compounds, ZnR_2

Diorganozinc compounds can be prepared from zinc metal, or better zinc–copper alloy, and alkyl iodides:

$$2Zn + 2Rl \rightarrow 2RZnl \rightarrow ZnR_2 + Znl_2$$

Zinc halides react with Grignard reagents, organolithium or aluminum compounds to form diorganozinc compounds (Fig. 10.1)

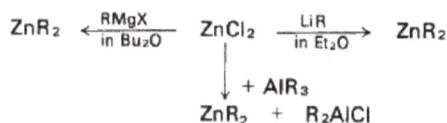

Fig. 10.1: Preparations of diorganozinc compounds.

For aryl derivatives, the reaction of organomercury compounds with zinc metal in refluxing xylene is a convenient procedure:

$$HgR_2 + Zn \rightarrow ZnR_2 + Hg$$

Compounds with R = Me, Et [1], C_6F_5 [2] and $C_6H_3(CF_3)_2$-2,4,6 [3] can serve as examples for their linear structure.

The cyclopentadienyl zinc compound $(\eta^5\text{-}C_5H_5)Zn\text{-}CH_3$ (prepared from $Zn(CH_3)I$ and NaC_5H_5) is monomeric in vapor phase but in the solid state, it displays a unique supramolecular chain structure in which the η^5-cyclopentadienyl groups alternate with zinc atoms.

https://doi.org/10.1515/9783110695274-010

The pentamethylcyclopentadienyl compounds (η^5-C_5Me_5)Zn-R with R = Me, Et, Ph, Mes (prepared by comproportionation of Zn(η^5-C_5Me_5)$_2$ with ZnR$_2$) are monomeric half sandwich complexes [4].

The diorganozinc compounds (ZnR$_2$) are very sensitive to oxygen; the lower alkyls are spontaneously flammable and the higher alkyls fume in air. Unlike organomagnesium compounds, the organozinc compounds do not react with carbon dioxide; hence, this gas served in earlier years of organometallic chemistry as a protective atmosphere.

10.1.2 Hypervalent anions [ZnR$_3$]$^-$ and [ZnR$_4$]$^{2-}$

Alkali metal organometallics react with disubstituted zinc derivatives to form hypervalent anions with three and four organic groups attached to the metal. Thus, diphenylzinc reacts with phenyllithium to form Li$^+$[ZnPh$_3$]$^-$ and Li$_3$[Zn$_2$Ph$_7$]$^{3-}$ of unknown structure. Similarly, dimethylzinc forms the compounds Li$_2$[ZnMe$_4 \cdot$ OEt$_2$] and Li$_2$[ZnMe$_4$] with methyllithium in ether. In both Li$_2$[ZnMe$_4$] and Li$_2$[Zn(C$\equiv$CH)$_4$], a tetrahedral arrangement of organic groups about the metal was found by X-ray diffraction.

Triorganozinc anions [ZnR$_3$]$^-$ with R = Et [5], *tert*-Bu [6], Ph [7], Mes, C$_6$F$_5$ [8], CH$_2$SiMe$_3$ [9], C$\equiv$CPh [10] with various counter ions have been structurally characterized by X-ray diffraction as trigonal planar compounds. Tetraorganozinc anions [ZnR$_4$]$^{2-}$ with R = Me [11] and C$\equiv$CPh [12] display tetrahedral geometry.

10.1.3 Diorganozinc donor adducts, R$_2$Zn.D

The zinc atom can make use of its available p-orbitals, by accepting lone electron pairs from donor molecules to form weak complexes. Simple ethers form relatively unstable complexes, but cyclic ethers (THF, dioxane, etc.) and chelating diethers such as dimethoxyethane increase the stability (Fig. 10.2).

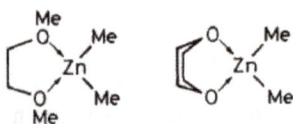

Fig. 10.2: Diether complexes of ZnMe$_2$.

With dioxane, a supramolecular chain is formed by dimesitylzinc [13], which also forms an adduct with 4,4'-bipyridyl [14] (Fig. 10.3).

The diamines form similar complexes, for example, ZnPh$_2$.NMe$_3$, Zn(C$_6$F$_5$)$_2$.2Py [15], as well as chelates with bidentate amines [16, 17] (Fig. 10.4).

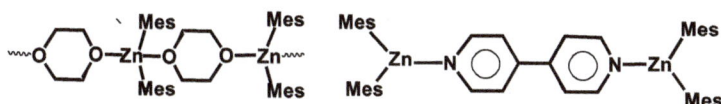

Fig. 10.3: Adducts of dimesitylzinc.

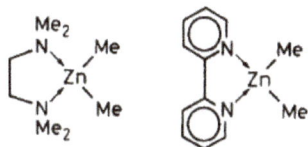

Fig. 10.4: Diamine adducts of dimethylzinc.

A unique inverse coordination complex is formed by dimethylzinc with an aza-crown as exodonor centroligand [18] (Fig. 10.5).

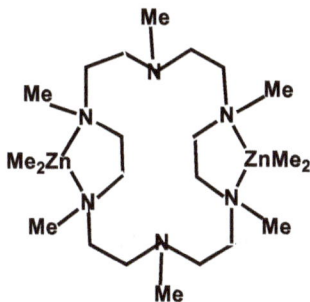

Fig. 10.5: Azacrown inverse coordination complex of dimethylzinc.

The complexes with phosphines, arsines and sulfides are unstable. Electron withdrawing groups at zinc increase the acceptor power of the metal and its tendency to form complexes. Thus, stable tertiary phosphine complexes such as $Zn(C_6F_5).2PPh_3$ can be obtained [19].

10.1.4 Organozinc hydrides, RZnH

Simple organozinc hydrides (RZnH) are difficult to obtain. Reducing diorganozincs with lithium alanate gives RZnH in THF, but conversion to RZn_2H_3 occurs:

$$ZnR_2 + LiAlH_4 \xrightarrow{\text{THF}} RZnH \xrightarrow{20\,^\circ C} RZn_2H_3, \ R = Me, Ph$$

Diorganozinc derivatives, ZnR_2, react with zinc hydride in the presence of pyridine to form a cyclic trimer (Fig. 10.6).

$$ZnR_2 + ZnH_2 + py \xrightarrow{THF} (RZnH \cdot py)_3$$

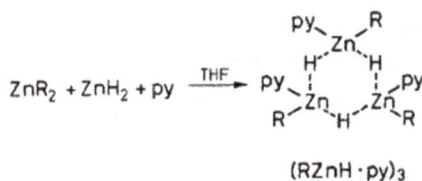

Fig. 10.6: Trimeric organozinc hydride.

10.1.5 Organozinc halides, RZnX

The formation of organozinc analogues of Grignard reagents by a reaction of zinc metal and alkyl halides can be achieved only in strongly polar solvents, such as dimethylformamide, diglyme or dimethylsulfoxide:

$$Zn + RX \rightarrow RZnX \, (X = Br)$$

For the synthesis of ketones from acyl halides, compounds with a lower reactivity than Grignard reagents are often useful. The latter are transformed into organozinc halides and are used without isolation:

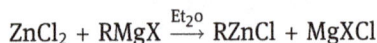

$$ZnCl_2 + RMgX \xrightarrow{Et_2O} RZnCl + MgXCl$$

Organozinc halides can also be prepared by the reaction of disubstituted derivatives and zinc halides. The equilibrium:

$$ZnR_2 + ZnX_2 \rightleftharpoons 2RZnX$$

is strongly shifted to the right.

The structure of organozinc halides in solution probably involves both coordinative solvation (in suitable solvents) and intermolecular association (in noncoordinating solvents). The RZnX derivatives are polymerized in the solid state. For example, ethylzinc iodide is a supramolecular structure consisting of macromolecular chains involving Zn_3I_3 rings with the metal four-coordinated and the iodine atoms three-coordinated.

10.1.6 Organozinc alkoxides

The disubstituted organozinc compounds are sensitive to the action of reagents containing active hydrogen. Water and alcohols cleave only one of the organic groups, forming, respectively, bridged dimers $[RZn-OH]_2$, and tetramers $[RZn-OR']_4$ with cubane structures. These tetramers are cleaved by pyridine to form cyclic dimers (Fig. 10.7). When the alcohol contains a bulky organic group, cyclic dimers and trimers containing three-coordinated zinc are formed, such as [EtZn–OCPh$_3$]$_2$, [EtZn–OCHPh$_2$]$_3$, [EtZn–OC$_6$F$_5$]$_2$ and [EtZn–OC$_6$Cl$_5$]$_2$.

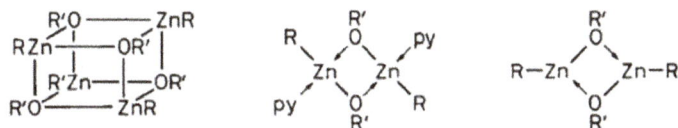

Fig. 10.7: Organozinc alkoxides.

Disubstituted ZnR$_2$ compounds and thiols give insoluble polymers, [RZn–SR']$_n$ (R = Me) or oligomers (n = 5, 6 or 8), containing unusual Zn$_n$S$_n$ polyhedral cages.

10.1.7 Organozinc amides

Primary and secondary amines can cleave Zn–R bonds to form amino derivatives associated through coordination into supermolecules, as the cyclic dimer [(MeZn–NPh$_2$].

The compound obtained from dimethylzinc and triphenylphosphinimine is a tetramer [(MeZn–N = PPh$_3$)$_4$ with a cubane structure (Fig. 10.8).

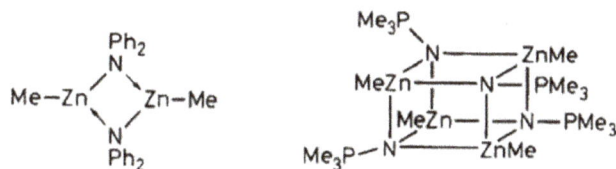

Fig. 10.8: Organozinc amino derivatives.

Amino derivatives react with carbon dioxide, phenylisocyanate and phenylisothiocyanate to form addition compounds by insertion of the reagent molecule into the Zn–N bond (Fig. 10.9).

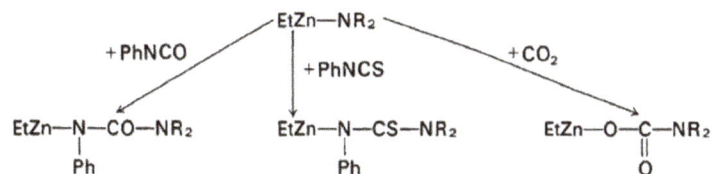

Fig. 10.9: Insertion reactions of organozinc amino derivatives.

These products are associated supermolecules. Thus, EtZn–NPhC(O)R (where R = OMe, NPh₂) are cyclic trimers and the compound with R = Me is a tetramer. With pyridine, the trimers are degraded to dimers (Fig. 10.10).

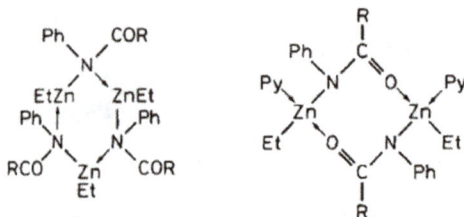

Fig. 10.10: Cyclic supramolecular oligomers.

10.1.8 Compounds with Zn–Zn bonds

Such type of compounds are rare, but a few examples can be cited: bis(cyclopenta-denyl)dizinc (η^5-C₅Me₅)Zn-Zn(η^5-C₅Me₅) [20] and a *cyclo*-trizinc compound with the same η^5-C₅Me₅ units [21] (Fig. 10.11).

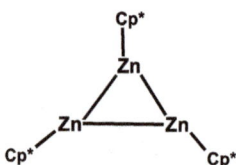

Fig. 10.11: Cyclic trizinc compound.

10.1.9 Inverse coordination organozinc complexes

A curiosity in organozinc chemistry is represented by the oxo-centered inverse coordination complexes formed by diethylzinc with sodium oxide [(μ_5-O)Na₂(ZnEt₂)₃] [22], and heavier metal oxides [(μ_6-O)M₂(ZnEt₂)₃] (M = K, Rb) [23]. In these complexes, the zinc atoms occupy equatorial positions, and the alkali metal atoms are in axial positions (Fig. 10.12).

10.2 Organocadmium compounds

Organocadmium compounds are less thoroughly investigated than those of zinc and magnesium; however, they are sometimes used as reagents in organic syntheses, for example, for the preparation of ketones from acyl halides. Unlike those of magnesium and zinc, organocadmium reagents do not react with the carbonyl group of the ketone formed.

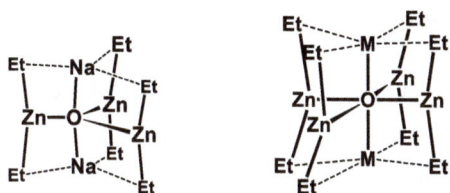

Fig. 10.12: Inverse coordination complexes with diethylzinc ligands and alkali metal oxides as coordination centers.

Organo-disubstituted derivatives, CdR$_2$ and CdRR′, and monosubstituted derivatives, RCdX (X = halogen, OR, SR, etc.), are known. The coordinative unsaturation of cadmium makes possible the formation of adducts with donors, the association of the functional derivatives RCdX, and the formation of hypervalent anionic species with three and four Cd–R bonds of the type [CdR$_3$]$^-$ and [CdR$_4$]$^{2-}$.

10.2.1 Homoleptic compounds, CdR$_2$

Diorganocadmium (CdR$_2$) compounds, structurally investigated with X-ray diffractometry, with R = Me [24], Ph [25] and C$_6$F$_5$ [26], display a linear molecular geometry. The bis(alkylcyclopentadienyl)cadmium has a particular structure with antiparallel orientation of the η^1-five-membered rings [27] (Fig. 10.13).

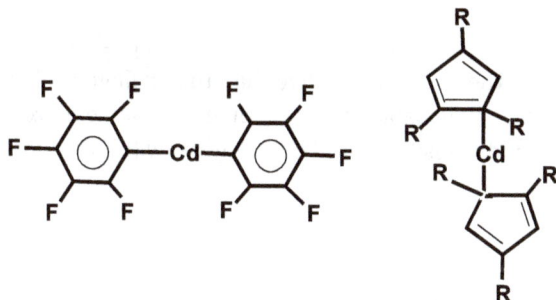

Fig. 10.13: Two unusual organozinc compounds.

There are several methods of preparing CdR$_2$ compounds.

The reaction of free cadmium (unlike zinc and magnesium) metal with organomercury derivatives cannot be used, since the organocadmium compounds are difficult to separate from the equilibrium mixture formed:

$$Cd + HgR_2 \rightleftharpoons CdR_2 + Hg$$

Divinylcadmium has been obtained by an exchange reaction between divinylmercury and dimethylcadmium.

The CdR_2 compounds are prepared from organomagnesium or lithium reagents and anhydrous cadmium halides:

$$CdX_2 + 2RMgX \longrightarrow CdR_2 + 2MgX_2$$

$$CdX_2 + 2RLi \longrightarrow CdR_2 + 2LiX$$

Adding hexamethylphosphortriamide (HMPA), which precipitates the magnesium halide as a complex, facilitates the isolation of the organocadmium derivative.

Organolithium reagents are particularly suitable for the synthesis of aromatic compounds of cadmium, for example, phenyl and pentafluorophenyl derivatives.

Bis(pentafluorophenyl)- and bis(pentachlorophenyl)cadmium can be prepared by thermal decarboxylation of organic salts:

$$Cd(OCOC_6F_5)_2 \rightarrow Cd(C_6F_5)_2 + 2CO_2$$

and allyl derivatives by exchange between dimethylcadmium and the appropriate boron compounds:

$$3CdMe_2 + 2B(CH_2CR = CHR')_3 \rightarrow 3Cd(CH_2CR = CHR')_2 + 2BMe_3$$

Thallium can also transfer organic groups to cadmium, as in the following preparation of bis(pentafluorophenyl)cadmium:

$$(C_6F_5)_2TlBr + Cd \xrightarrow{160\ °C} Cd(C_6F_5)_2 + TlBr$$

Dialkylcadmium derivatives are distillable, monomeric liquids, decomposing thermally at temperatures above 150 °C. They are less reactive than the analogous zinc compounds and are not spontaneously flammable in air, although they are oxidized to peroxides, $Cd(OOR)_2$. The Cd–C bond is easily cleaved by halogens, with formation of cadmium halides.

10.2.2 Hypervalent anions, $[CdR_3]^-$

The tendency to increase the coordination number and to reduce the coordinative unsaturation of CdR_2 compounds is reflected in the formation of anions containing three and four organic groups attached to cadmium. Thus, diphenylcadmium reacts with phenyllithium to form an unstable salt:

$$CdPh_2 + LiPh \rightarrow Li^+ [CdPh_3]^-$$

and the tetrahedral $[Cd(C{\equiv}CR)_4]^{2-}$ anion has been prepared:

$$Cd(SCN)_2 + 4KC \equiv CR + Ba(SCN)_2 \rightarrow Ba[Cd(C \equiv CR)_4] + 4KSCN$$

10.2.3 Diorganocadmium donor adducts, CdR$_2$.D

The direct synthesis with alkyl iodides and cadmium metal in HMPA yields a complex:

$$2RI + 2Cd \xrightarrow{HMPA} CdR_2 \cdot 2HMPA + CdI_2$$

The acidity of organocadmium compounds is much less than their organozinc analogues. Thus, the 2,2′-bipyridyl complex, [CdMe$_2$,bipy] [28], is unstable and the dioxane complex, [CdMe$_2$.dioxane]$_x$, is dissociated in solution (Fig. 10.14).

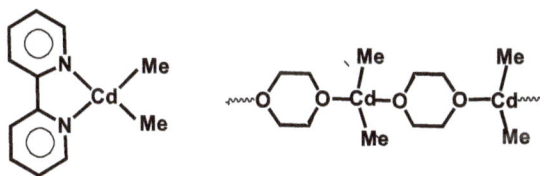

Fig. 10.14: Two adducts of dimethylcadmium.

10.2.4 Organocadmium halides, RCdX

Organocadmium halides, RCdX, have only been recently isolated pure, although they have long been used in ethereal solutions for preparative purposes as reagents in organic chemistry. Adding anhydrous cadmium chloride to Grignard reagents in ether yields CdR$_2$ and RCdX, depending upon the reagent ratio. The RCdX derivatives do not play a role comparable to that of Grignard reagents, since it is more convenient to use solutions of CdR$_2$ compounds in organic preparations. They are, however, important for the preparation of the asymmetric derivatives, RCdR′, through reactions between RCdX and R′MgX compounds.

Organocadmium halides are obtained by redistribution reactions of cadmium dialkyls with dihalides:

$$CdR_2 + CdX_2 \rightarrow 2RCdX$$

and by the reaction of Grignard reagents with cadmium halides:

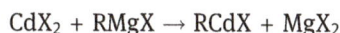

$$CdX_2 + RMgX \rightarrow RCdX + MgX_2$$

The halides, RCdX, are infusible crystalline solids, decomposing at ca. 100 °C, monomeric in dimethylsulfoxide solution.

The trimethylsilyl derivatives [CdX{C(SiMe$_2$Ph)}] with X = Cl [29] and Br [30] are self-assembled supramolecular dimers (Fig. 10.15).

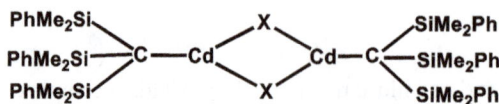

Fig. 10.15: Dimeric organocadmium halides.

10.2.5 Organocadmium functional compounds, RCdX′

Cadmium, like zinc, can form associated functional derivatives, RCdX′. Dimethyl-cadmium reacts with alcohols to form dimers, [(MeCd–OR]$_2$ (R = $tert$-Bu), and tet-ramers, [MeCd–OR]$_4$, and with thiols to form insoluble polymers, [MeCd-SR]$_n$, or low-molecular-weight oligomers ($n = 4$ when R = $tert$-Bu and $n = 6$ when R = iso-Pr). The tetramers have cubane structures (Fig. 10.16).

The phosphide derivatives, for example, MeCd-PBu$^t_2$, are cyclic trimers [31]. Other examples are [C$_6$F$_5$Cd-OH]$_4$ [32] and [Bu^{t}Cd-OBu^t]$_4$ [33].

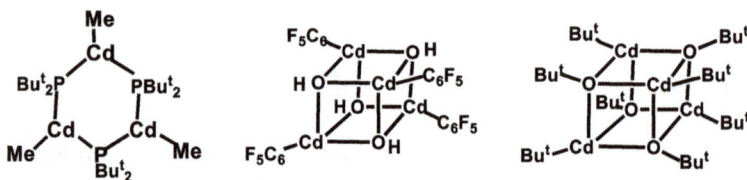

Fig. 10.16: Self-assembled organocadmium compounds.

10.3 Organomercury compounds

The first organomercury compound was obtained in 1853 by E. Frankland by the action of methyl iodide on mercury metal under sunlight irradiation. A large num-ber of organomercury compounds were synthesized for pharmacological purposes, but their role in chemotherapy has now been completely superseded. The use of organomercurial fungicides is also on the decline, owing to their high toxicity. Inter-est in the biological activity of organomercury compounds, particularly methylmer-cury species, has been focused in recent years, after the world famous poisoning accident in Japan, known as Minamata disease, in which biomethylation of inorganic mercury salts was implicated.

The great synthetic utility of organomercury compounds is based upon the abil-ity of mercury to transfer organic groups to other metals and nonmetals.

The major types of organomercury compounds include HgR$_2$, RHgX and their ad-dition compounds, but the acceptor ability of mercury in its organic derivatives is only moderate.

10.3.1 Homoleptic compounds, HgR$_2$

The monomeric diorganomercury compounds, HgR$_2$, are linear, reflecting sp hybridization.

One of the most versatile laboratory procedures for the disubstituted derivatives uses Grignard reagents, for example, for the synthesis of bis(pentafluorophenyl) mercury, Hg(C$_6$F$_5$)$_2$:

$$HgX_2 + 2RMgX \rightarrow HgR_2 + 2MgX_2$$

Organolithium reagents can be used equally successfully, and the use of organoaluminum compounds facilitated by sodium chloride is also possible:

$$3HgCl_2 + 2AlR_3 + 2NaCl \rightarrow 3HgR_2 + 2NaAlCl_4$$

Cyclopentadienyl derivatives of mercury are prepared by treating cyclopentadienyl sodium or thallium(I) with mercury(II) chloride.

Treating red mercury(II) oxide with triethylboron in aqueous alkali yields diethylmercury.

The original method of Frankland, based upon the reaction of alkyl and aryl halides or sulfates with sodium amalgam, is now seldom used:

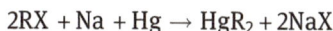

$$2RX + Na + Hg \rightarrow HgR_2 + 2NaX$$

Thermal decarboxylation of some mercury(II) carboxylates with elimination of carbon dioxide can be used for compounds containing rather electronegative organic groups, for example, bis(pentafluorophenyl)mercury:

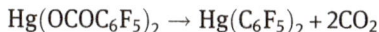

$$Hg(OCOC_6F_5)_2 \rightarrow Hg(C_6F_5)_2 + 2CO_2$$

Acetylenic derivatives can be mercurated directly with inorganic complexes:

$$2R-C\equiv CH + K_2HgI_4 + 2KOH \longrightarrow Hg(C\equiv CH)_2 + 4KI + 2H_2O$$

Disubstituted organomercury compounds are also prepared by reduction of organomercury halides, RHgX, with sodium, copper, alkali metal stannites and hydrazine hydrate, or by their disproportionation in reactions with tertiary phosphines, alkali metal iodides and other reagents:

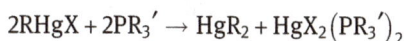

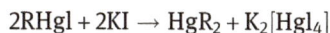

$$2RHgX + 2Na \rightarrow HgR_2 + Hg + 2NaX$$

$$2RHgX + 2PR_3' \rightarrow HgR_2 + HgX_2(PR_3')_2$$

$$2RHgI + 2KI \rightarrow HgR_2 + K_2[HgI_4]$$

The reduction with sodium iodide in ethanol or acetone is a standard procedure.

Unsymmetrically substituted diorganomercury derivatives, RHgR', can be prepared by Grignard reactions (RHgX + R'MgX) or by redistribution of differently substituted symmetrical compounds:

$$HgR_2 + HgR'_2 \rightleftharpoons 2R-Hg-R$$

The distribution is statistical when R and R′ are similar and nonstatistical when R and R′ are different.

Dialkylmercury derivatives are very toxic, volatile liquids exhibiting moderate thermal stability. The analogous aromatic derivatives are stable solids; some are light sensitive. The chemical reactivity of the diorganomercury compounds is much lower than with their zinc and cadmium analogues, for example, in not being sensitive to moisture and air.

10.3.2 Organomercury heterocycles

Some organomercury compounds were formulated with nonlinear C–Hg–C groupings. One such example is *ortho*-phenylene mercury prepared from *ortho*-dibromobenzene and sodium amalgam. First, a dimeric structure was assigned, and then an early X-ray investigation suggested a hexameric structure in which all mercury atoms would be coplanar and the C–Hg–C bonds are collinear. A more accurate redetermination showed that this compound is trimeric and has a trinuclear structure (Fig. 10.17).

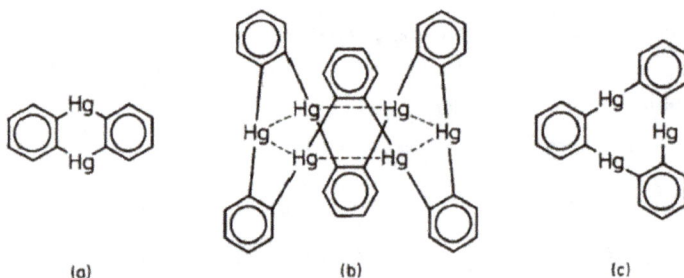

Fig. 10.17: Organomercury heterocyclic structures.

The trimeric structure was also found in the perfluorophenylene similarity, [Hg (*ortho*-C$_6$F$_4$)]$_3$ [34, 35], and numerous derivatives of this compound are known.

The compound prepared from *ortho*-,*ortho*′-dilithiobiphenyl and mercury(II) chloride is not a heterocyclic monomer with bent C–Hg–C bonds as suggested earlier, but a tetramer (Fig. 10.18).

10.3.3 Diorganomercury donor adducts, HgR$_2$.D

The tendency of mercury to increase its coordination number is only moderate; sp hybridization is more stable, for example, disubstituted derivatives do not form

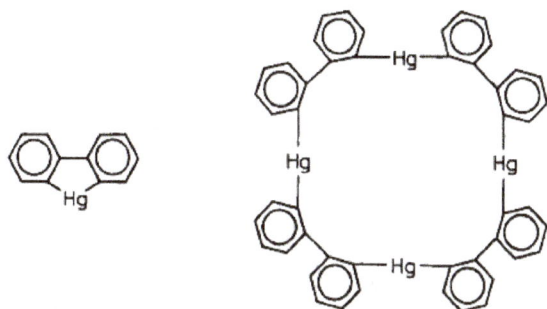

Fig. 10.18: A tetrameric organomercury heterocyclic structure.

complexes with tertiary amines and phosphines. Perfluoro- and perchlorophenyl derivatives are exceptions with enhanced acceptor ability, and bis(pentafluoro-phenyl) mercury forms adducts with bipyridyl and tetraphenyldiphosphinoethane (diphos), while diphenylmercury does not (Fig. 10.19).

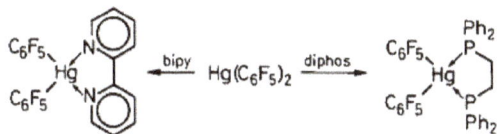

Fig. 10.19: Adducts of $Hg(C_6F_5)_2$.

The trifluoromethyl derivative, $Hg(CF_3)_2$, behaves similarly. Four-coordinated mercury has also been reported in the salts of the complex anion $[Hg(CF_3)_2I_2]^{2-}$.

10.3.4 Organomercury halides

Monosubstituted organomercury derivatives are readily obtained. Alkyl iodides react with mercury metal after photochemical initiation to form organomercury halides. The method can be applied to the synthesis of perfluoroalkyl derivatives:

$$RI + Hg \rightarrow RHgI$$

By appropriate adjustment of the reagent ratio and reaction conditions, some methods cited for disubstitution can also be used for the preparation of monosubstituted compounds, for example, the reaction of mercury(II) chloride with Grignard reagents:

$$HgCl_2 + RMgCl \rightarrow RHgCl + MgCl_2$$

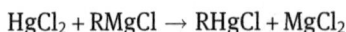

or the reaction of sodium amalgam with alkyl halides or sulfates:

$$3RX + 2HgNa \rightarrow RHgX + HgR_2 + 2NaX$$

Organomercury halides are formed by redistribution of diorgano derivatives with mercury(II) halides:

$$HgR_2 + HgX_2 \longrightarrow 2RHgX$$

A specific synthesis of aromatic derivatives involves the reduction of diazonium and iodonium salts (R = aryl) with elimination of nitrogen or iodobenzene:

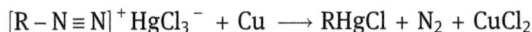

$$[R-N \equiv N]^+ \, Cl^- + Hg \longrightarrow RHgCl + N_2$$

$$[R-N \equiv N]^+ HgCl_3{}^- + Cu \longrightarrow RHgCl + N_2 + CuCl_2$$

$$[R-1-R]^+ Cl + Hg \longrightarrow RHgCl + Rl$$

The analysis of arylboronic acids is based upon the quantitative transfer of aromatic groups to mercury:

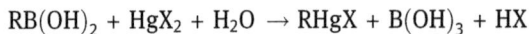

$$RB(OH)_2 + HgX_2 + H_2O \rightarrow RHgX + B(OH)_3 + HX$$

The action of sulfinic acids on mercury(II) chloride was popular in the past:

$$RSO_2H + HgCl_2 \rightarrow RHgCl + SO_2 + HCl$$

The RHgX compounds are also formed in reactions of mercury(II) halides with organic derivatives of tin, lead, antimony, bismuth, cadmium, thallium and other metals. The transfer of organic groups from these metals to mercury proceeds easily.

Monosubstituted organomercury compounds, RHgX, are crystalline solids, sometimes vacuum sublimable. They are water-soluble when $X = F$, NO_3, $\frac{1}{2} SO_4$, ClO_4, etc., with formation of RHg^+ cations.

Mercuration is an important synthesis for monosubstituted, mainly aromatic organomercury derivatives. In this long-known reaction, aromatic compounds act on mercury(II) acetate or perchlorate. Aromatic amines, phenols and ethers are all more reactive than unsubstituted benzene, while nitro- and halogeno-substituted benzenes react very slowly:

$$RH + HgX_2 \rightarrow RHgX + HX$$

Mercury salts undergo addition to olefins, in the presence of alcohols (oxymercuration):

$$H_2C = CH_2 + HgX_2 + ROH \rightarrow ROCH_2CH_2HgX + HX$$

or amines (aminomercuration):

$$H_2C = CH_2 + HgX_2 + HNR_2 \rightarrow R_2NCH_2CH_2HgX + HX$$

These two reactions have been widely applied in organic synthesis.

10.3.5 Organomercury hydroxides, RHgOH

The older literature contains references to organomercury hydroxides, RHgOH (pre-pared by the reaction of halides with silver oxide or potassium hydroxide in alco-hol). These are actually mixtures of tris(organomercuryl)-oxonium hydroxides, [O $(HgR)_3]^+OH^-$, and bis(organomercury) oxides, RHg–O–HgR.

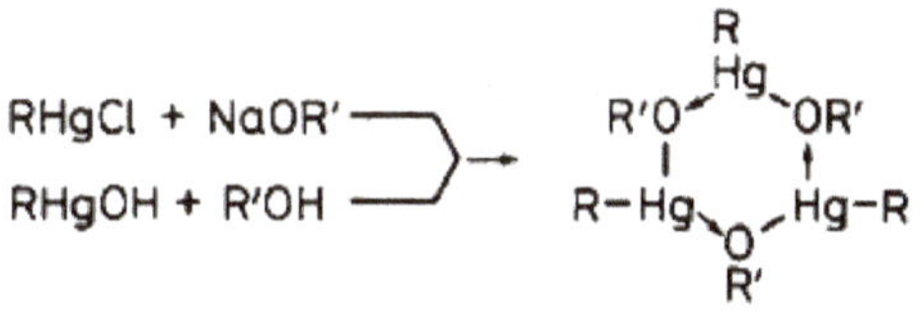

RHgCl + NaOR' —⟩
RHgOH + R'OH —⟩ →

$$
\begin{array}{c}
R \\
Hg \\
R'O^{\nearrow} \quad {}^{\searrow}OR' \\
| \qquad \uparrow \\
R-Hg_{\searrow}{}_{O}{}_{\swarrow}Hg-R \\
R'
\end{array}
$$

Fig. 10.20: Supramolecular trimeric organomercury alkoxides.

10.3.6 Organomercury alkoxides, RHg-OR'

The trimeric organomercury alkoxides [RHg–OR']$_3$ have a cyclic structure (Fig. 10.20). The trimethylsiloxy derivative, [MeHg–OSiMe$_3$]$_4$, is, however, monomeric in solu-tion, unlike zinc and cadmium analogues, but in the solid state is tetrameric, with a cubane-type structure.

10.3.7 Organomercury sulfides

The reaction of methylmercury bromide with sodium sulfide produces methylmer-cury sulfide, which can be converted to a tris(methylmercuryl)sulfonium salt (R = Me) (Fig. 10.21).

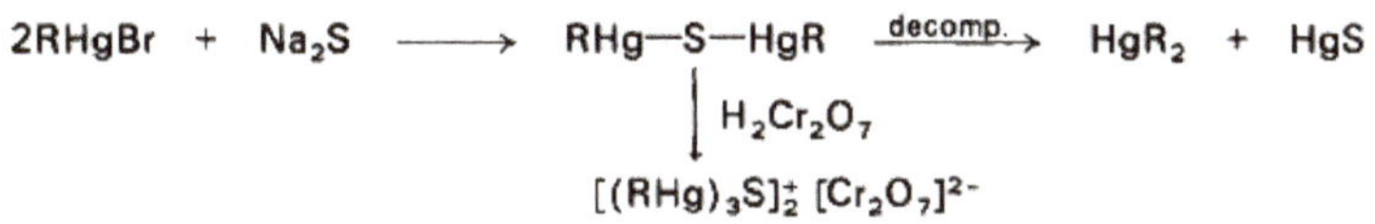

$$2RHgBr + Na_2S \longrightarrow RHg-S-HgR \xrightarrow{decomp.} HgR_2 + HgS$$
$$\Big\downarrow H_2Cr_2O_7$$
$$[(RHg)_3S]_2^+ \, [Cr_2O_7]^{2-}$$

Fig. 10.21: Formation of organomercury sulfides.

The arylmercury dithiophosphates, RHg–SP(S)(OR')$_2$, represent another class of the little investigated mercury–sulfur derivatives.

10.3.8 Inverse organomercury compounds

Some inverse organomercury compounds are formed in mercuration reactions of aromatic compounds. The mercuration of benzene with mercury(II) acetate proceeds in acetic acid in an autoclave, and thiophene can be easily mercurated with an aqueous solution of mercury(II) acetate. The reaction is used to eliminate thiophene from benzene, since the dimercurated product is insoluble. Furan forms a tetramercury derivative (Ac = $OCOCH_3$) (Fig. 10.22).

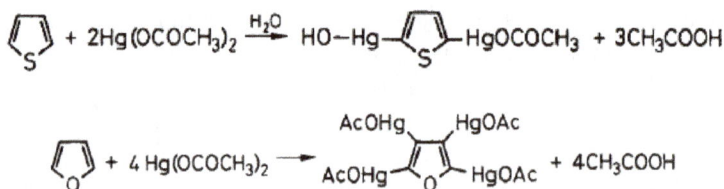

Fig. 10.22: Inverse organometallic compounds derived from thiophene and furan.

Other inverse organomercury compounds are 1,3,5-tris(chloromercury)benzene [19], 1,2-bis(chloromercury)tetrafluorobenzene [36, 37] and tetrakis(nitrato-mercury)methane (Fig. 10.23) [38].

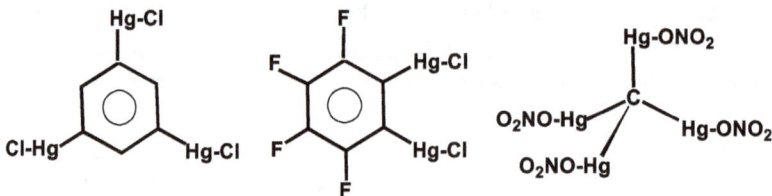

Fig. 10.23: More inverse organomercury compounds.

10.3.9 Organomercury inverse coordination complexes

Halogeno-centered inverse coordination complexes, with nearly planar molecular geometry, are known with bis(pentafluorophenyl) moieties connected to the central halogen (Fig. 10.24) [39].

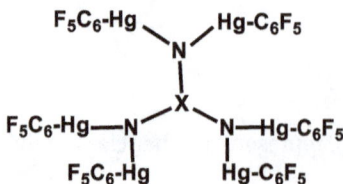

Fig. 10.24: Organomercury inverse coordination complexes with halogeno coordination centers.

A planar molecule of 1,3,5-triazine centroligand decorated with methylthiolato mercury moieties is another example of inverse coordination complex [40] (Fig. 10.25).

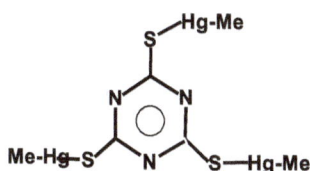

Fig. 10.25: Organomercury inverse coordination complex derived with 1,3,5-triazine trithiolato centroligand.

10.3.10 Compounds with Hg–Hg bonds

Surprisingly, very few structurally characterized representatives are known. Examples of RHg–HgR compounds with bulky aromatic substituents [41, 42] can be cited (Fig. 10.26).

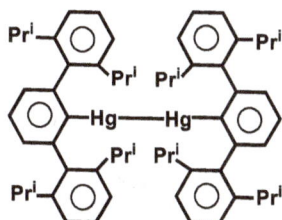

Fig. 10.26: A dimercury compound with bulky substituents.

References

[1]   Bacsa J, Hanke F, Hindley S, Odedra R, Darling GR, Jones AC, Steiner IA. The Solid-State Structures of Dimethylzinc and Diethylzinc. Angew Chem Int Ed 2011, 50, 11685–87.
[2]   Sun Y, Piers WE, Parvez M. The solid-state structure of bis(pentafluorophenyl)zinc. Can J Chem 1998, 76, 513–17.
[3]   Brooker S, Bertel N, Stalke D, Noltemeyer M, Roesky HW, Sheldrick GM, Edelmann FT. Main-group chemistry of the 2,4,6-tris(trifluoromethyl)phenyl substituent: X-ray crystal structures of [2,4,6-(CF$_3$)$_3$C$_6$H$_2$]$_2$Zn, [2,4,6-(CF$_3$)$_3$C$_6$H$_2$]$_2$Cd(MeCN) and [2,4,6-(CF$_3$)$_3$C$_6$H$_2$]$_2$Hg. Organometallics 1992, 11, 192, 195.
[4]   Resa I, Álvarez E, Carmona E. Synthesis and structure of half-sandwich zincocenes. Z Anorg Allg Chem 2007, 633, 1827–31.
[5]   Dell'Aera M, Perna FM, Vitale P, Altomare A, Palmieri A, Maddock LCH, Bole LJ, Kennedy AR, Hevia E, Capriati V. Boosting conjugate addition to nitroolefins using lithium tetraorganozincates: Synthetic strategies and structural insights. Chem-Eur J 2020, 26, 8742–48.
[6]   Boss SR, Coles MP, Haigh R, Hitchcock PB, Snaith R, Wheatley AEH. Ligand and metal effects on the formation of main-group polyhedral clusters. Angew Chem Int Ed 2003, 42, 5593–96.

[7]   Wallach C, Mayer K, Henneberger T, Klein W, Fässler TF. Intermediates and products of the reaction of Zn(ii) organyls with tetrel element Zintl ions: Cluster extension versus complexation. Dalton Trans 2020, 49, 6191–98.

[8]   Garratt S, Guerrero A, Hughes DL, Bochmann M. Arylzinc complexes as new initiator systems for the production of isobutene copolymers with high isoprene content. Angew Chem Int Ed 2004, 43, 2166–69.

[9]   Westerhäusen M, Gückel C, Habereder T, Vogt M, Warchhold M, Nöth H. Synthesis of strontium and barium bis{tris[(trimethylsilyl)methyl]zincates} via the transmetallation of bis[(trimethylsilyl)methyl]zinc. Organometallics 2001, 20, 893–99.

[10] Putzer MA, Neumuller B, Dehnicke K. Synthese und Kristallstrukturen der Zinkate [Na(12-Krone-4)$_2$][Zn{N(SiMe$_3$)$_2$}$_3$] und [Na(12-Krone-4)$_2$]$_2$[Zn(C≡C-Ph)$_3$(THF)][Zn(C≡C-Ph)$_3$]. Z Anorg Allg Chem 1997, 623, 539–44.

[11] Weiss E, Wolfrum R. Die Kristallstruktur des Lithiumtetramethylzinkats. Chem Ber 1968, 101, 35–40.

[12] Gao J, Li S, Zhao Y, Wu B, Yang X-J. Reactions of α-diimine-stabilized Zn–Zn-bonded compounds with phenylacetylene. Organometallics 2012, 31, 2978–85.

[13] Ashraf S, Jones AC, Bacsa J, Steiner A, Chalker PR, Beahan P, Hindley S, Odedra R, Williams PA, Heys PN. MOCVD of vertically aligned ZnO nanowires using bidentate ether adducts of dimethylzinc. Chem Vap Depos 2011, 17, 45–53.

[14] Irwin M, Krämer T, McGrady JE, Goicoechea JM. On the structural and electronic properties of [Zn$_2$(4,4′-bipyridine)(mes)]n– (n = 0–2), a homologous series of bimetallic complexes bridged by neutral, anionic, and dianionic 4,4′-bipyridine. Inorg Chem 2011, 50, 5006–14.

[15] Lennartson A, Håkansson M. Diphenyldipyridinezinc(II): Partial spontaneous resolution of an organometallic reagent. Acta Crystallogr Sect C Cryst Struct Commun 2009, 65, m205–7.

[16] Clegg W, García-Álvarez J, García-Álvarez P, Graham DV, Harrington RW, Hevia E, Kennedy AR, Mulvey RE, Russo L. Synthesis, structural authentication, and structurally defined metallation reactions of lithium and sodium DA-zincate bases (DA = diisopropylamide) with Phenylacetylene. Organometallics 2008, 27, 2654–63.

[17] Chen GJ, Zeng SX, Lee CH, Chang YL, Chang CJ, Ding S, Chen H-Y, Wu K-H, Chang I-J. Synthesis of zinc complexes bearing pyridine derivatives and their application of ε-caprolactone and L-Lactide polymerization. Polymer 2020, 194, 122374.

[18] Coward KM, Jones AC, Steiner A, Bickley JF, Smith LM, Pemble ME. Dimethylzinc adducts with macrocyclic amines. J Chem Soc Dalton Trans 2000, 3480–82.

[19] Messinis AM, Luckham SLJ, Wells PP, Gianolio D, Gibson EK, O'Brien HM, Sparkes HA, Davis SA, Callison J, Elorriaga D, Hernandez-Fajardo O, Bedford B. The highly surprising behaviour of diphosphine ligands in iron-catalysed Negishi cross-coupling. Nature Catal 2019, 2, 123–33.

[20] Resa I, Carmona E, Gutierrez-Puebla E, Monge A. Decamethyldizincocene, a stable compound of Zn(I) with a Zn-Zn bond. Science 2004, 305, 1136–38.

[21] Freitag K, Gemel C, Jerabek P, Oppel IM, Seidel RW, Frenking G, Banh H, Dilchert K, Fischer RA. The σ-aromatic clusters [Zn$_3$]$^+$ and [Zn$_2$Cu]: embryonic brass. Angew Chem Int Ed 2015, 54, 4370–74.

[22] Miller LZ, Shatruk M, McQuade DT. Alkali metal oxides trapped by diethylzinc. Chem Commun 2014, 50, 8937–40.

[23] Hedström A, Lennartson A. Assembly of ethylzincate compounds into supramolecular structures. J Organomet Chem 2011, 696, 2269–73.

[24] Hanke F, Hindley S, Jones AC, Steiner A. The solid state structures of the high and low temperature phases of dimethylcadmium. Chem Commun 2016, 52, 10144–46.

[25] Braun U, Böck B, Nöth H, Schwab I, Schwartz M, Weber S, Wietelmann U. Reactions of group 13 and 14 hydrides and group 1, 2, 13 and 14 organyl compounds with (tert-Butylimino) (2,2,6,6-tetramethylpiperidino)borane. Eur J Inorg Chem 2004, 3612–28.

[26] Strasdeit H, Büsching I, Duhme AK, Pohl S. Structure of the two-coordinate cadmium complex bis(pentafluorophenyl)-cadmium(II), [Cd(C$_6$F$_5$)$_2$]. Acta Crystallogr Sect C Cryst Struct Commun 1993, 49, 576–78.

[27] Bentz D, Wolmershäuser G, Sitzmann H. Bis(alkylcyclopentadienyl)-cadmium complexes and alkylcyclopentadienyl(mesityl)cadmium: Synthesis and structure. Organometallics 2006, 25, 3175–78.

[28] Almond MJ, Beer MP, Drew MGB, Rice DA. The adduct formed between dimethylcadmium and 2,2′-bipyridyl, 2,2′-bipyridyldimethylcadmium(II): A crystallographic and spectroscopic study. Organometallics 1991, 10, 2072–76.

[29] Al-Juaid SS, Eaborn C, Habtemariam A, Hitchcock PB, Smith JD, Tavakkoli K, Wietelmann U. The preparation and crystal structures of the compounds (Ph$_2$MeSi)$_3$CMCl (M = Zn, Cd, or Hg). J Organomet Chem 1993, 462, 45–55.

[30] Al-Juaid SS, Buttrus NH, Eaborn C, Hitchcock PB, Smith JD, Tavakkoli K. The preparation and crystal structures of the sterically hindered lithium alkylchloroacadmate [Li(thf)$_4$][Li(thf)$_2$ (μ-Cl)$_4${CdC(SiMe$_3$)$_3$}$_2$].thf (thf = tetrahydrofuran), the alkylcadmium halides [{Cd[C(SiMe$_3$)$_3$] Cl}$_4$] and [{Cd[C(SiMe$_2$Ph)$_3$]$_2$. J Chem Soc, Chem Commun 1988, 1389–91.

[31] Benac BL, Cowley AH, Jones RA, Nunn CM, Wright TC. Potential precursors to electronic materials: Three coordinate cadmium in [MeCd(μ-tert-Bu$_2$P)]$_3$, the first cadmium diorganophosphide. J Am Chem Soc 1989, 111, 4986–88.

[32] Weidenbruch M, Herrndorf M, Schäfer A, Pohl S, Saak W. Pentafluorphenylverbindungen des Zinks und Cadmiums: Bildung und Strukturen von (C$_6$F$_5$)$_2$Zn(thf)$_2$ und von tetramerem C$_6$F$_5$CdOH. J Organomet Chem 1989, 361, 139–45.

[33] Nöth H, Thomann M. Metal tetrahydroborates and tetrahydroborato metalates. 20. Reactions of cadmium tetrahydroborate with alcoholates and phenolates - X-ray structure of [MeCd(OCMe$_3$)]$_4$. Chem Ber 1995, 128, 923–27.

[34] Shubina ES, Tikhonova IA, Bakhmutova EV, Dolgushin FM, Antipin MY, Bakhmutov VI, Sivaev IB, Teplitskaya LN, Chizhevsky IT, Pisareva IV, Bregadze VI, Epstein LM, Shur VB. Crown compounds for anions: sandwich and half-sandwich complexes of cyclic trimeric perfluoro-o-phenylenemercury with polyhedral closo-[B$_{10}$H$_{10}$]$^{2-}$ and closo-[B$_{12}$H$_{12}$]$^{2-}$ Anions. Chem-Eur J 2001, 7, 3783–90.

[35] Haneline MR, Gabbaï FP. Polymorphism of trimeric perfluoro-ortho-phenylene mercury, [Hg(o-C$_6$F$_4$)]$_3$. Z Naturforsch B 2004, 59, 1483–87.

[36] Beckwith JD, Tschinkl M, Picot A, Tsunoda M, Bachman R, Gabbaï FP. Interaction of the bifunctional Lewis acid 1,2-Bis(chloromercurio)-tetrafluorobenzene with aldehydes, nitriles, and epoxides. Organometallics 2001, 20, 3169–74.

[37] Yakovenko AA, Gallegos JH, Antipin MY, Masunov A, Timofeeva TV. Crystal morphology as an evidence of supramolecular organization in adducts of 1,2-bis(chloromercurio) tetrafluorobenzene with organic esters. Cryst Growth Des 2011, 11, 3964–78.

[38] Grdenić D, Korpar-Čolig B, Matković-Čalogović D. Synthesis and crystal structures of tetrakis(nitratomercurio) methane monohydrate and bis(sulphatomercurio)bis(aquamercurio) methane. J Organomet Chem 1996, 522, 297–302.

[39] Naumann D, Schulz F. Strukturen von neuen Bis(pentafluorophenyl)-halogenomercuraten [{Hg(C$_6$F$_5$)$_2$}$_3$(μ-X)]- (X = Cl, Br, I). Z Anorg Allg Chem 2005, 631, 715–18.

[40] Cecconi F, Ghilardi CA, Midollini S, Orlandini A. Organomercury derivatives of the 2,4,6-trimercaptotriazine (H$_3$TMT). X-ray crystal structure of (HgMe)$_3$(TMT). J Organomet Chem 2002, 645, 101–04.

[41] Zhu Z, Brynda M, Wright RJ, Fischer RC, Merrill WA, Rivard E, Fettinger JC, Olmstead MM, Power PP. Synthesis and characterization of the homologous M-M bonded series Ar'MMAr' (M = Zn, Cd, or Hg; Ar' = C_6H_3-2,6-(C_6H_3-2,6-Pr^i_2)$_2$) and related arylmetal halides and hydride species. J Am Chem Soc 2007, 129, 10847–57.

[42] Casas JS, Castellano EE, García-Tasende MS, Sánchez A, Sordo J, Vázquez-López EM, Zukerman-Schpector J. Deprotonation reactions of 2-thiouracil with [2-(pyridin-2-yl)phenyl] mercury (II) acetate. Structural and spectroscopic effects. J Chem Soc, Dalton Trans 1996, 1973–78.

Part III: Organometallic compounds of transition metals

General

The transition metals are elements with partly filled d or f orbitals, either as atoms or in the zero, positive or negative oxidation states. In the third period, the 3d-level is being populated starting with scandium ($3d^1$) and ending with copper ($3d^{10}$). The elements from scandium to nickel ($3d^9$) are thus transition metals. Copper is also considered a transition metal because it is d^9 in some derivatives. The general chemical behavior of copper justifies its inclusion among transition metals.

The electronic structure of transition metals is shown in Tab. 10.1. In the series from Sc to Cu ($Z = 21–29$) and from Y to Ag ($Z = 39–47$), the 3d- and 4d-levels are being occupied stepwise. In the lanthanide family ($Z = 57–71$), the 4f-level is being occupied by 14 electrons, followed by the filling of the 5d-level in the series from Hf to Au ($Z = 72–79$). These metals are grouped in three transition series, corresponding to the 3d, 4d and 5d levels; the lanthanides (4f-level) and the actinides (5f-level) are inner-transition metals.

The formation of the organometallic derivatives of the transition elements is dominated by the tendency of the metals to achieve a noble gas configuration by the full occupation of (n-1)d, ns and np electron shells. This is achieved by accepting additional electrons from the ligands. As a result the formation of MR, or MR_mX_n type compounds where R is a σ-bonded organic group is not typical, because the formation of n covalent bonds (n = the valence of the metal) does not fill the (n-1)d levels. The formation of compounds with organic ligands able to donate enough π-electrons to complete a noble gas configuration is required.

In Chapter 2 the bonds between transition metals and electron-donating organic ligands were discussed. The valence electrons are shared by the metal and the ligand, and back donation from occupied d-orbitals of the metal into vacant orbitals (usually antibonding) of the ligand plays an important role.

In principle, any unsaturated or aromatic organic molecule or radical can act as a π-ligand. A potentially planar network of sp^2-hybridized carbon atoms possessing unhybridized p_z-orbitals with π-electrons can bond to a single transition metal atom. This condition is satisfied by a number of molecules with the planar skeletons (Fig. 11.1). Most of these are known to form transition metal complexes.

Molecules with large dimensions or with branched structures (Fig. 11.2) can bond either partially to a single metal atom, or entirely to a set of two or more atoms of transition metal atoms, usually connected by metal-metal bonds. These units are rather diversified but less used in the formation of π-complexes.

The simple monoolefins form only a single bond, while the polyolefins can use all or just a part of their p_z-orbitals and π-electrons to form bonds. The number of electrons accepted from the ligand depends upon the number required to achieve the next higher noble gas configuration (see the "effective atomic number" or the "18 electron rule" in Section 2.6).

https://doi.org/10.1515/9783110695274-011

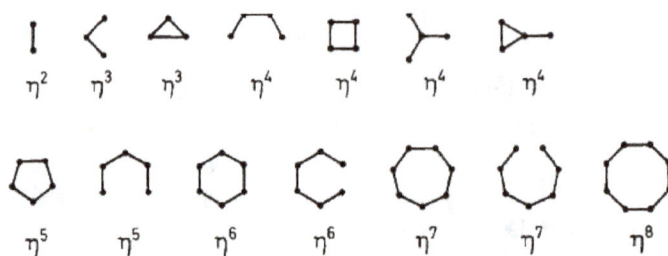

Fig. 11.1: Schematic representation of polycarbon planar molecules capable of π-bonding.

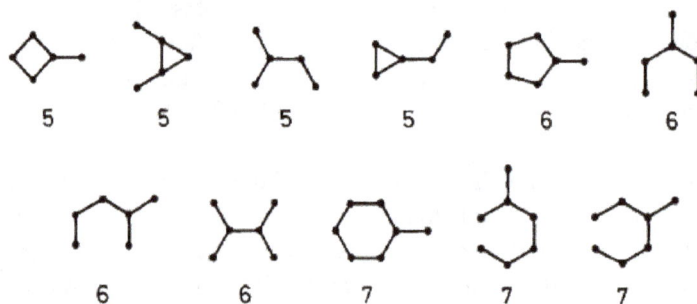

Fig. 11.2: Schematic representation of branched planar molecules capable of π-bonding.

The structure and chemical behavior of transition-metal organometallic compounds is determined largely by the ligand, but it is useful to compare the properties of different metal derivatives of the same ligand. It is thus more convenient to classify the transition-metal derivatives according to the nature of the ligand, or more exactly, according to the number of electrons contributed by the ligand to attaining the noble gas configuration of the central atom. In the π-complexes known, this number varies from two to eight. The ligands are classified as follows:

a) **Two-electron ligands:** carbon monoxide (:CO), carbon monosulfide (:CS), carbon monoselenide (:CSe), organic isocyanides (:C = N-R), carbenes (:CR$_2$), cyanide (:CN⁻), monoolefins or isolated double bonds.

b) **Three-electron ligands:** η^3-allyl (η^3-C$_3$H$_5$), carbyne (:C-R), cyclopropenyl (η^3-C$_3$R$_3$) groups.

c) **Four-electron ligands:** cyclobutadiene, butadiene, cyclopentadiene (as η^4-C$_5$H$_6$), hexadiene-1,3, other molecules containing a butadiene fragment, and the trimethylenemethyl radical.

d) **Five-electron ligands:** cyclopentadienyl (η^5-C$_5$H$_5$), cyclohexadienyl, etc.

e) **Six-electron ligands:** benzene and other aromatic molecules, borazine (inorganic benzene) B$_3$N$_3$H$_6$.

f) **Seven-electron ligands:** tropyllium cation (η^7-C$_7$H$_7$).

g) **Eight-electron ligands:** cyclooctatetraene (η^8-C$_8$H$_8$).

The formation of a σ-metal-carbon bond, metal-metal bond or any other single covalent bond (for example, M-X, where X = halogen, OR, OH, SR, NRR', etc.) contributes a single electron to the metal. Therefore, σ-alkyl and σ-aryl groups, as well as other groups attached to the metal through a single covalent bond, are considered in computing the electron balance as one-electron ligands.

11 Organometallic compounds with two electron ligands

11.1 Metal carbonyls

The metal carbonyls are compounds containing various combinations of transition metal atoms with carbon monoxide molecules. In these compounds, carbon monoxide is attached through a metal–carbon bond to a transition metal atom in a low oxidation state (usually zero or ± 1).

This class includes binary mononuclear compounds $M(CO)_x$, polynuclear compounds, of the general formula $M_x(CO)_y$ and heterobimetallic carbonyls (containing two different metals) of the type $M_xM'_y(CO)_z$. Anionic and cationic species are also formed when some CO ligands are replaced by electron pairs.

The binary metal carbonyls which can be isolated as stable compounds are listed in Tab. 11.1. Other compositions are known, which have been identified by low-temperature matrix isolation.

Tab. 11.1: Stable binary metal carbonyls.

Group 5	Group 6	Group 7	Group 8	Group 9	Group 10
$V(CO)_5$	$Cr(CO)_6$	$Mn_2(CO)_{10}$	$Fe(CO)_5$	$Co_2(CO)_8$	$Ni(CO)_4$
	$Mo(CO)_6$	$Mn_4(CO)_{16}$	$Fe_2(CO)_9$	$Co_4(CO)_{12}$	
	$W(CO)_6$	$Tc_2(CO)_{10}$	$Fe_3(CO)_{12}$	$Co_6(CO)_{16}$	
		$Re_2(CO)_{10}$	$Ru(CO)_5$	$Rh_2(CO)_8$	
			$Ru_2(CO)_9$	$Rh_4(CO)_{12}$	
			$Ru_3(CO)_{12}$	$Rh_6(CO)_{16}$	
			$Os(CO)_5$	$Ir_2(CO)_8$	
			$Os_2(CO)_9$	$Ir_4(CO)_{12}$	
			$Os_2(CO)_9$		
			$Os_4(CO)_{13}$		
			$Os_5(CO)_{16}$		
			$Os_6(CO)_{183}$		
			$Os_7(CO)_{21}$		
			$Os_8(CO)_{23}$		

https://doi.org/10.1515/9783110695274-012

There are numerous heterobimetallic metal carbonyls, illustrated in Tab. 11.2.

Tab. 11.2: Heterobimetallic metal carbonyls.

Binuclear	$MnRe(CO)_{10}$	$MnCo(CO)_9$	$ReCo(CO)_9$	$CoRh(CO)_7$
Trinuclear	$Mn_2Fe(CO)_{14}$	$Mn_2Ru(CO)_{14}$	$FeRu_2(CO)_{12}$	
	$MnReFe(CO)_{14}$	$Mn_2Os(CO)_{14}$	$Fe_2Ru(CO)_{12}$	
	$Re_2Fe(CO)_{14}$	$Re_2Os(CO)_{14}$	$Ru_2Os(CO)_{12}$	
			$RuOs_2(CO)_{12}$	
Tetranuclear	$Co_2Rh_2(CO)_{12}$	$Co_2Rh_2(CO)_{13}$		
	$Co_3Rh(CO)_{12}$			
	$Rh_3Ir(CO)_{12}$			
Hexanuclear	$Re_2Fe_4(CO)_{24}$	$Co_2Rh_4(CO)_{16}$		

11.1.1 The structure of metal carbonyls

The molecular structures of most metal carbonyls have been established by X-ray diffraction and investigated by spectroscopic techniques.

All metal carbonyls obey the 18-electron rule, with each CO molecule contributing a pair of electrons to the effective atomic number. The metals of odd atomic number cannot form neutral, mononuclear metal carbonyls; these metals form dinuclear or polynuclear carbonyls containing metal–metal bonds, or metal–carbonyl anions. The only exception is vanadium, which forms paramagnetic $V(CO)_6$, a compound with only 17 electrons in the valence shell of vanadium. The $[V(CO)_6]^-$ anion is preferred and diamagnetic.

Carbon monoxide acts as a ligand in several different ways. The common ones are as follows:

a) as terminal group, M–CO, with each carbon monoxide molecule attached to a single metal atom;

b) as symmetrical bimetallic bridge, with a carbon monoxide molecule connecting two metal atoms;

c) as symmetrical trimetallic bridge, centered above a triangular face of a polyhedral cluster (Fig. 11.3).

Some unsymmetrical modes involve the participation of the π-system of the CO unit in bonding (Fig. 11.4).

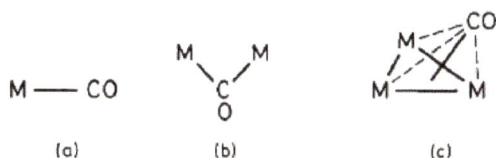

Fig. 11.3: Terminal bonding and symmetrical bridging of CO.

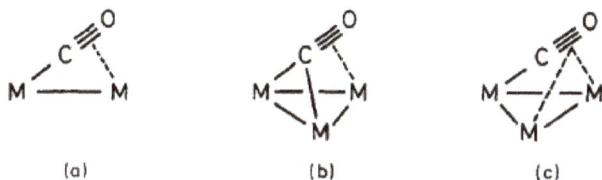

Fig. 11.4: Unsymmetrical bonding of CO.

These seldom occur in binary metal carbonyls, but were identified in metal carbonyl anions or in substituted derivatives. Unsymmetrical M–CO $\cdots$ M bridges observed in $Fe_3(CO)_{12}$ or $Fe_4(CO)_{12}$ are sometimes described as "semibridging."

Head-to-tail bridging of carbon monoxide, M-C$\equiv$O$\rightarrow$M, in which the coordinated CO molecule can further act as a donor through its oxygen, not observed in binary metal carbonyls, may occur when a strong acceptor of oxygen is available, as illustrated with organoaluminum compounds (Fig. 11.5).

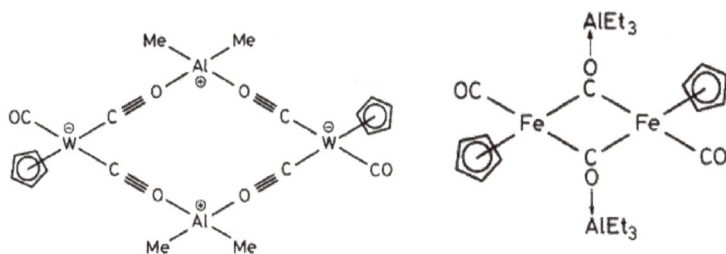

Fig. 11.5: Head-to-tail bridging of carbon monoxide.

The molecular geometries of mononuclear metal carbonyls are as expected for a metal atom surrounded by n carbonyl groups: $Ni(CO)_4$ tetrahedral, $Fe(CO)_5$ trigonal bipyramidal and $Cr(CO)_6$ octahedral (Fig. 11.6).

The structures of binuclear metal carbonyls such as $Mn_2(CO)_{10}$, $Fe_2(CO)_9$ and $Co_2(CO)_8$ were less logically predictable. The manganese compound $Mn_2(CO)_{10}$ is made up of two $Mn(CO)_5$ groups joined by a metal–metal bond oriented in the solid

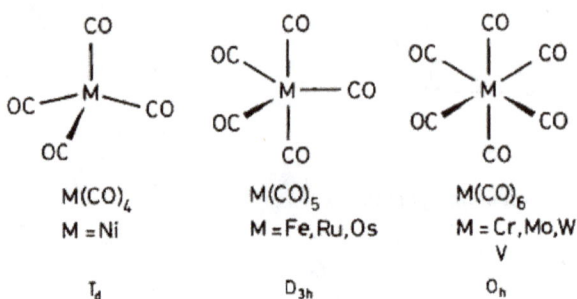

Fig. 11.6: Molecular geometries of mononuclear metal carbonyls.

with staggered CO groups. The binuclear carbonyls of iron and cobalt contain carbonyl bridges in addition to metal–metal bonds (Fig. 11.7).

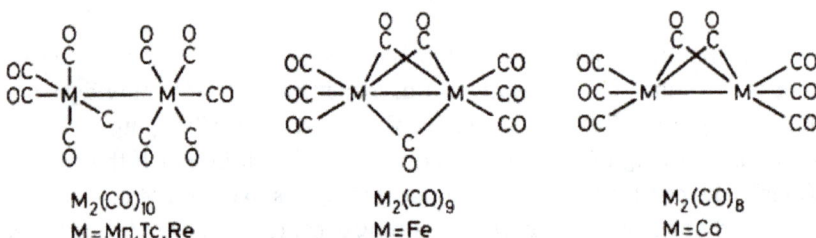

Fig. 11.7: Binuclear metal carbonyls.

The trinuclear carbonyls, $Fe_3(CO)_{12}$, $Os_3(CO)_{12}$ and $Ru_3(CO)_{12}$, contain metal–metal bonded triangles. The structures are, however, different with and without bridges, despite similar compositions. The structure of $Fe_3(CO)_{12}$ can be deduced from that of Fe_2 $(CO)_9$ by replacing a CO bridge in the latter with a $Fe(CO)_4$ bridge. The CO bridges in $Fe_3(CO)_{12}$ are unsymmetrical. The ruthenium and osmium compounds are isostructural (Fig. 11.8).

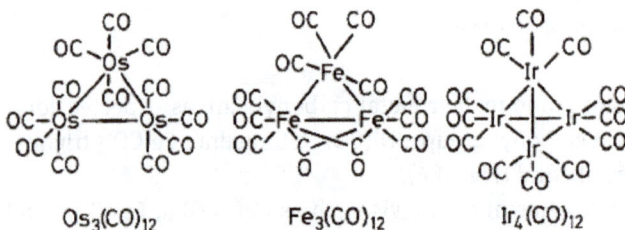

Fig. 11.8: Trinuclear metal carbonyls.

The structures of the tetranuclear carbonyls, $Co_4(CO)_{12}$ and $Rh_4(CO)_{12}$, are built upon a tetrahedral cluster of metal atoms. The triangle at the base of the pyramid has three CO bridges on the edges. The iridium compound, $Ir_4(CO)_{12}$, also has a tetrahedral structure, but without CO bridges. The hexanuclear carbonyl, $Rh_6(CO)_{16}$, contains an octahedral cluster of rhodium atoms and four trimetallic bridges, in addition to two terminal CO groups at each rhodium atom (Fig. 11.9).

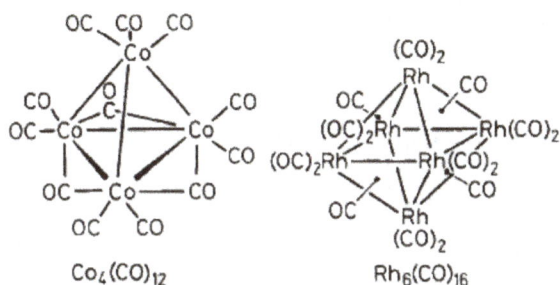

Fig. 11.9: Tetranuclear and hexanuclear metal carbonyls.

The polynuclear carbonyls, $Os_5(CO)_{16}$, $Os_6(CO)_{18}$ and $Os_7(CO)_{21}$, are bridge-free and contain only terminal-carbonyl ligands, attached to the polymetallic clusters.

Some unusual heterobimetallic cadmium–iron carbonyls with cyclic structures and the CO ligand at the iron sites are also known, namely $[CdFe(CO)_4]_4$ tetramer and the $[(bipy)CdFe(CO)_4]_3$ trimer (bipy = 2,2′-bipyridine) (Fig. 11.10).

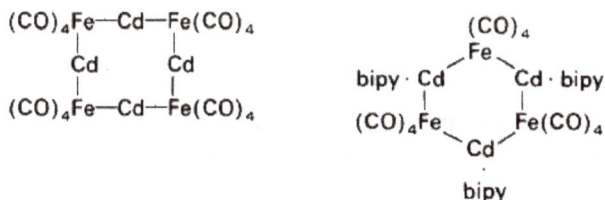

Fig. 11.10: Heterobimetallic cadmium–iron carbonyls.

The analogues $MFe(CO)_4$ (M = Zn, Hg, Pb) are also known. Some are polymers with chain structures.

11.1.2 Preparation of metal carbonyls

In general, the preparation of metal carbonyls requires high pressures of carbon monoxide, but some normal pressure syntheses have also been developed. The most important metal carbonyls are now available commercially.

Direct reactions of metals with CO are used for the synthesis of iron and nickel carbonyls.

The reductive carbonylation of metal oxides, salts, etc. is the most common procedure for the preparation of metal carbonyls. In metal carbonyls the oxidation state of the metal is zero, therefore their synthesis from metal oxides, sulfides, salts or other compounds, requires a reducing agent. This can be an active metal (sodium, magnesium, aluminum, zinc), an aluminum or zinc alkyl and hydrogen. Since the reactions are usually carried out under carbon monoxide pressure, this can also serve as a reducing agent. Examples are the syntheses of chromium, tungsten and rhenium carbonyls:

$$CrCl_3 + Al + 6\,CO \rightarrow Cr(CO)_6 + AlCl_3$$

$$WCl_6 + 6\,CO + 2\,Al(C_2H_5)_3 \rightarrow W(CO)_6 + 2\,AlCl_3 + 3C_4H_{10}$$

Photolytic and thermal reactions of mononuclear carbonyls are used for the synthesis of polynuclear carbonyls. Thus, UV photolysis of $Fe(CO)_5$ is used for the preparation of $Fe_2(CO)_9$ and thermolysis of $Os_3(CO)_{12}$ leads to osmium carbonyl clusters of higher nuclearity like $Os_4(CO)_{13}$ and $Os_6(CO)_{18}$:

$$2\,Fe(CO)_5 \rightarrow Fe_2(CO)_9 + CO$$

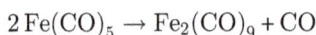

Double exchange (salt metathesis) of metal carbonylate salts with metal carbonyl halides are applied in the preparation of heterobimetallic metal carbonyls:

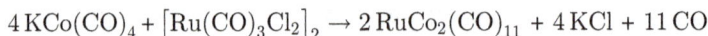

$$4\,KCo(CO)_4 + \left[Ru(CO)_3Cl_2\right]_2 \rightarrow 2\,RuCo_2(CO)_{11} + 4\,KCl + 11\,CO$$

11.1.3 Metal–carbonyl anions

The metal–carbonyl anions are isoelectronic with neutral metal carbonyls and formally result by replacing a carbon monoxide ligand by an electron pair, thereby satisfying the 18-electron rule.

Metal carbonyl anions are prepared by reacting dinuclear metal carbonyls with alkali metals or their amalgams:

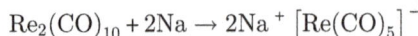

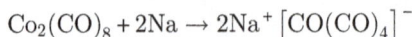

$$Mn_2(CO)_{10} + 2Na \rightarrow 2Na^+ \left[Mn(CO)_5\right]^-$$

$$Re_2(CO)_{10} + 2Na \rightarrow 2Na^+ \left[Re(CO)_5\right]^-$$

$$Co_2(CO)_8 + 2Na \rightarrow 2Na^+ \left[Co(CO)_4\right]^-$$

The reduction of iron pentacarbonyl by sodium amalgam in liquid ammonia produces a mononuclear anion:

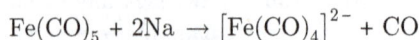

$$Fe(CO)_5 + 2Na \rightarrow \left[Fe(CO)_4\right]^{2-} + CO$$

Iron carbonyls react with alcoholic alkalis or organic bases to form anions conserving the cluster size:

$$Fe(CO)_5 + 4OH^- \rightarrow Fe(CO)_4^{2-} + 2H_2O + CO_3^{2-} \ Fe_2(CO)_9 + 4OH^-$$
$$\rightarrow Fe_2(CO)_8^{2-} + 2H_2O + CO_3^{2-} \ Fe_3(CO)_{12} + 4OH^-$$
$$\rightarrow Fe_3(CO)_{11}^{2-} + 2H_2O + CO_3^{2-}$$

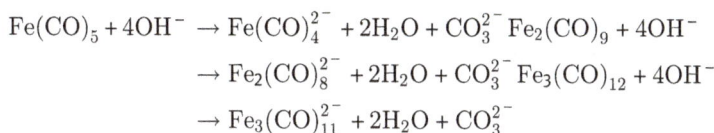

Reduction of iron pentacarbonyl by sodium amalgam in THF forms a dinuclear anion:

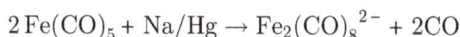

$$2\,Fe(CO)_5 + Na/Hg \rightarrow Fe_2(CO)_8{}^{2-} + 2CO$$

The reduction of nickel tetracarbonyl with sodium in liquid ammonia and with lithium amalgam in THF yields bi- and trinuclear species:

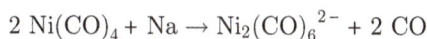

$$2\,Ni(CO)_4 + Na \rightarrow Ni_2(CO)_6{}^{2-} + 2\,CO$$

$$3\,Ni(CO)_4 + Li/Hg \rightarrow Ni_3(CO)_8{}^{2-} + 4\,CO$$

Group 16 metal carbonyls are reduced with sodium amalgam or in liquid ammonia, to form binuclear anions:

$$2\,M(CO)_6 + Na/Hg \rightarrow M_3(CO)_{10}{}^{2-} + 2\,CO, \ M = Cr, Mo, W$$

Heterobimetallic metal–carbonyl anions can be prepared by the reaction of neutral metal carbonyls with anionic carbonyls in photochemical condensation reactions:

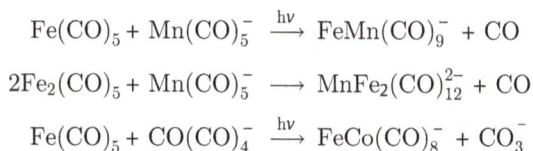

$$Fe(CO)_5 + Mn(CO)_5^- \xrightarrow{h\nu} FeMn(CO)_9^- + CO$$

$$2Fe_2(CO)_5 + Mn(CO)_5^- \longrightarrow MnFe_2(CO)_{12}^{2-} + CO$$

$$Fe(CO)_5 + CO(CO)_4^- \xrightarrow{h\nu} FeCo(CO)_8^- + CO_3^-$$

11.1.4 Metal–carbonyl cations

Binary metal–carbonyl cations are rare and include binary compounds $[Mn(CO)_6]^+$, $[Tc(CO)_6]^+$, $[Re(CO)_6]^+$, $[Cu(CO)_n]^+$ (with $n = 1$, 3 and 4) and $[Ag(CO)_2]^+$.

The cations are prepared from metal carbonyl halides with Lewis acids:

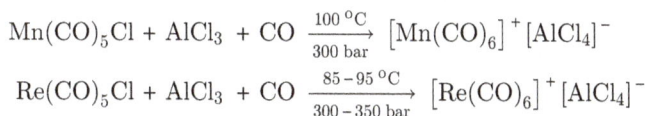

$$Mn(CO)_5Cl + AlCl_3 + CO \xrightarrow[300\ \text{bar}]{100\ ^\circ C} [Mn(CO)_6]^+[AlCl_4]^-$$

$$Re(CO)_5Cl + AlCl_3 + CO \xrightarrow[300-350\ \text{bar}]{85-95\ ^\circ C} [Re(CO)_6]^+[AlCl_4]^-$$

Copper–carbonyl cations are formed when Cu_2O in $HFSO_3$ or CF_3SO_3H absorbs CO to form $[Cu(CO)_4]^+$; addition of sulfuric acid yields unstable $[Cu(CO)_3]^+$ in equilibrium with $[Cu(CO)]^+$. Under similar conditions, silver oxide forms only $[Ag(CO)_2]^+$.

The copper–carbonyl cation $[Cu(CO)]^+$ has been obtained by the reaction of Cu $[AsF_6]$ with carbon monoxide.

More common are triphenylphosphine-substituted metal carbonyl cations and hydride derivatives.

11.1.5 Metal carbonyl halides

There are numerous compounds containing CO and halogen atoms coordinated to a metal atom. Interestingly, some metals which do not form stable binary carbonyls (e.g., gold, copper and palladium) are able to form metal carbonyl halides.

Metal carbonyl halides can be prepared by reactions of metal carbonyls with halogens. A reaction of choice is the cleavage of metal–metal bonds in polynuclear metal carbonyls with halogens, but direct substitution of CO is also possible. Paramagnetic $Cr(CO)_5I$ is obtained by the oxidation of the $[Cr_2(CO)_{10}]^{2-}$ anion with iodine and $Mn_2(CO)_{10}$ is also cleaved by halogens:

$$Mn_2(CO)_{10} + X_2 \longrightarrow 2Mn(CO)_5X \xrightarrow[-CO]{\Delta} [Mn(CO)_4X]_2.$$

Iron carbonyls, $Fe(CO)_5$ and $Fe_3(CO)_{12}$, react with iodine to form the compounds $Fe(CO)_4I_2$, $Fe_2(CO)_8I$ and $Fe(CO)_4I$.

In some cases, the reactions of noble metal halides with CO produce metal carbonyl halides. Thus, carbonylation of platinum(II) chloride forms $Pt(CO)_2Cl_2$, which on heating dimerizes to $[Pt(CO)Cl]_2$. A gold derivative, $Au(CO)Cl$, is formed from $AuCl_3$ and CO in $SOCl_2$. Most metal carbonyl halides are mononuclear, but some are associated through halogen bridging, for example, $[(CO)_4MnCl]_2$ and $[(CO)_3RuF_2]_4$ (Fig. 11.11).

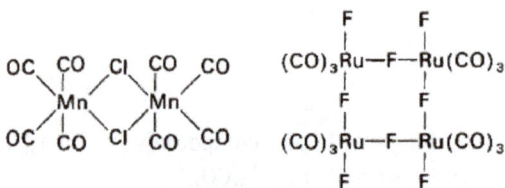

Fig. 11.11: Metal carbonyl halides associated through halogen bridging.

Carbonyl halide anions of the type $[M(CO)_5X]^-$, where $M = Cr$, Mo, W and $X =$ halogen, are obtained by substitution of carbon monoxide in $M(CO)_6$ with halide anions:

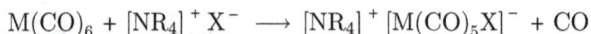

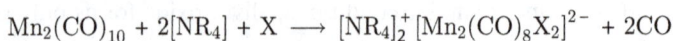

$$M(CO)_6 + [NR_4]^+X^- \longrightarrow [NR_4]^+[M(CO)_5X]^- + CO$$

$$M = Cr, Mo, W; \ X = Cl, Br, l$$

$$Mn_2(CO)_{10} + 2[NR_4] + X \longrightarrow [NR_4]_2^+[Mn_2(CO)_8X_2]^{2-} + 2CO$$

It seems that this reaction can be more general and can be applied to anions other than halides, for example, for a difluorodithiophosphinate:

$$Cr(CO)_6 + [PS_2F_2]^- \rightarrow [Cr(CO)_4PS_2F_2]^- + 2CO$$

The procedure was also applied to form $[M(CO)_5(NO_3)]^{1-}$ (M = Cr, W), $[M(CO)_5$ $(RCOO)]^-$ (M = Cr, Mo, W) and $[M(CO)_5SH]^{1-}$ (M = Mo, W) but surprisingly it was little used with other anions.

11.2 Metal thiocarbonyls

Carbon monosulfide can replace carbon monoxide in metal carbonyls to form thiocarbonyl complexes.

Only a single binary compound, $Ni(CS)_4$, prepared by co-condensation of nickel atoms with carbon monosulfide has been reported. All other derivatives contain carbon monoxide, cyclopentadienyl groups, phosphines, phosphites or a combination of these, as in the carbonyl-thiocarbonyls, $M(CO)_5(CS)$ (M = Cr, Mo, W), $Fe(CO)_4(CS)$ and the cyclopentadienyl derivatives $(\eta^5\text{-}C_5H_5)Mn(CO)_{3-n}(CS)_n$ ($n = 1$–3) and $\eta^5\text{-}C_5H_5Co(CS)_2$. An osmium compound containing a phosphine, $Os(CS)Cl_2(CO)_2(PPh_3)_2$, is known [1].

The thiocarbonyl group acts as a bridging ligand in $[(\eta^5\text{-}C_5H_5)Fe(CO)(CS)]_2$ and $[(\eta^5\text{-}C_5H_5)Mn(CS)(NO)]_2$ (Fig. 11.12).

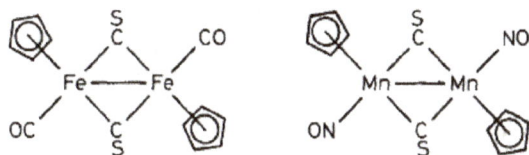

Fig. 11.12: Metal thiocarbonyl dimers with CS bridging.

Bridging thiocarbonyl is also found in a heterobimetallic manganese–iron compound $[Mn(CO)_4Fe(\eta^5\text{-}C_5H_5)(NO)(\mu_2\text{-}CS)_2(NO)]$ [2] and in a dinuclear ruthenium compound $(RuCp^*)(\mu_2\text{-}CS)(\mu_2\text{-}NPh)$ [3].

A trimetallic thiocarbonyl bridge is present in $(\eta^5\text{-}C_5H_5)_3Co_3(\mu_3\text{-}CS)_2$ and $(\eta^5\text{-}C_5H_5)_3Ni_3(\mu_3\text{-}CS)(CO)$ [4] (Fig. 11.13).

Theoretical calculations suggest that CS is both a better σ-donor and π-acceptor than CO, and this is confirmed by the selective replacement of CO rather than CS by phosphines. Also, M–CS bonds are shorter than M–CO bonds, indicating a higher degree of metal–ligand double bonding.

Carbon monosulfide could be a versatile ligand, but the syntheses of its complexes are not very attractive due to the reagents used. Metal–carbonyl anions, for example,

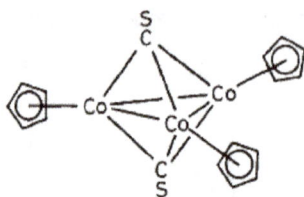

Fig. 11.13: Trimetallic thiocarbonyl bridging.

$[M_2(CO)_{10}]^{2-}$ (M = Cr, W), react with thiophosgene to produce $M(CO)_5(CS)$, and [Fe $(CO)_4]^{2-}$ gives $Fe(CO)_4(CS)$. The CS ligand can also be introduced by carbon disulfide:

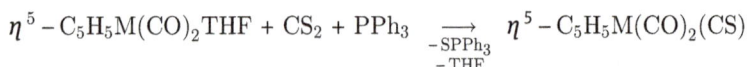

$$\eta^5 - C_5H_5M(CO)_2THF + CS_2 + PPh_3 \xrightarrow[-THF]{-SPPh_3} \eta^5 - C_5H_5M(CO)_2(CS)$$

M = Mn, Re

11.3 Metal selenocarbonyls

The third chalcogenide, CSe, can also form transition metal complexes, as in $(\eta^6\text{-}C_6H_5COOMe)Cr(CO)_2(CSe)$ [5], $RuCl_2(CO)(SCe)(PPh_3)_2$ [6] and $Cr(CO)_5(CSe)$. The CSe molecule is an even stronger π-acceptor ligand than either CS or CO, and forms shorter metal–carbon bonds in $(\eta^6\text{-}C_6H_5COOMe)Cr(CO)_2(CSe)$.

Other selenocarbonyl complexes include a carbene ligand compound, $RuCl_2$ $(CSe)(PCyh_3)\{C_3NMes_2\}$ [7], and dimethylpyrazolylborato complex anion, $[W(CO)_2$ $(CSe)\{HB(N_2C_3Me_2)\}]^-$. Bridging selenocarbonyl is found in a W-C = Se-M (with M = Cu, Au) complexes with the dimethylpyrazolylborato ligand $M(CO)_2\{HB(N_2C_3Me_2\}(\mu_2\text{-}CSe)$ $Cu(PPh_3)_2$ and $M(CO)_2\{HB(N_2C_3Me_2\}(\mu_2\text{-}CSe)Au(PPh_3)$ [8], and $Mo(CO)_2\{HB(N_2C_3Me_2\}$ $(\mu_2\text{-}CSe)RuCp*(CO)_2$ [9].

11.4 Metal–isocyanide complexes

Organic isocyanides:C = N-R are formally analogous to carbon monoxide and form some similar transition metal complexes. In many respects, the isocyanide metal complexes are analogous to metal carbonyls, for example, $M(CNR)_6$ (M = Cr, Mo, W), $Fe(CNR)_5$ or $Ni(CNR)_4$. However, there are important differences:

a)   The tendency to form metal–isocyanide cations is stronger.
b)   No metal isocyanide anions are known.
c)   Few polynuclear, metal–isocyanide complexes are known (only for nickel, while this metal does not form neutral polynuclear carbonyls).
d)   Copper, silver and gold, which do not form stable metal carbonyls, coordinate up to four isocyanide molecules.
e)   Isocyanide complex species are known, which have no carbonyl analogues.

Mixed compounds containing carbonyls, halides and cyclopentadienyls are known, as listed:

– *Metal-carbonyl isocyanides*

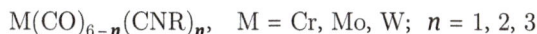

$$M(CO)_{6-n}(CNR)_n, \quad M = Cr, Mo, W; \; n = 1, 2, 3$$

$$Fe(CO)_{5-n}(CNR)_n, \quad n = 1, 2$$

$$Ni(CO)_{4-n}(CNR)_n, \quad n = 1, 2, 3$$

– *Metal-isocyanide halides*

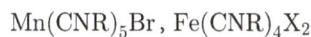

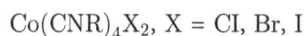

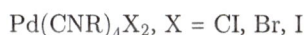

$$Mn(CNR)_5Br, \; Fe(CNR)_4X_2$$

$$Co(CNR)_4X_2, \; X = CI, Br, I$$

$$Pd(CNR)_4X_2, \; X = CI, Br, I$$

– *Metal-isocyanide phosphine complexes*

$$Fe(CNR)_3\{PMe_3\}_2, \; R = Ph, NCC_6H_3Me_2-2,6; \; CNCH_2Bu^t \,[10]$$

$$Re(CNMe)_4(PMePh_2)_2 \,[11]$$

– *Cyclopentadienylmetal isocyanides:*

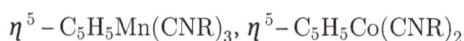

$$TiCp^*_2(CNMes)_2 \,[12]$$

$$\eta^5 - C_5Me_5W(CNBut)_2 \,[13]$$

$$\eta^5 - C_5H_5Mn(CNR)_3, \; \eta^5 - C_5H_5Co(CNR)_2$$

$$\eta^5 - C_5H_5Fe(CNR)_2X, \; X = CI, Br, I$$

$$\eta^5 - [C_5H_5Fe(CNR)_2]_2$$

$$\eta^5 - [C_5H_5Ni(CNR)]_2$$

Like carbon monoxide, isocyanide can coordinate as bridging groups (Fig. 11.14).

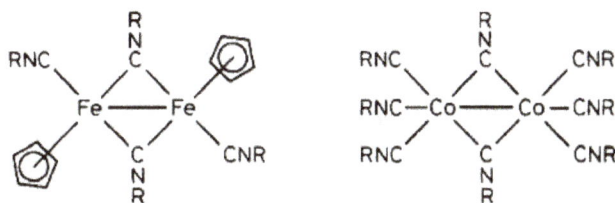

Fig. 11.14: Isocyanide bridging in dinuclear compounds.

The structures of mononuclear isocyanides are analogous to those of the metal carbonyls, for example, $Fe(CNBu')_5$, $Ru(CNBu')_5$, $Co_2(CNBu')_8$ and $Cr(CNPh)_6$.

The isocyanides are formed by replacement of carbon monoxide in metal carbonyls and their derivatives, or by direct reaction of a metal salt with an isocyanide.

Complexes with zero-oxidation state of the metal can be obtained from an anhydrous metal halide, the isocyanide and sodium amalgam.

An alternative route is the alkylation of metal-cyano complexes, for example, ferricyanide, with ethyl iodide or dimethyl sulfate.

The reduction of mononuclear isocyanide cations can lead to dinuclear compounds:

$$[CO(CNR)_5]^+ PF_6^- \xrightarrow{K/Hg\,in\,THF} Co_2(CNR)_8,\ R = Bu^t$$

11.5 Metal–carbene complexes and related compounds

Divalent-carbon compounds (carbenes):CR_2 were for many years postulated as intermediates in organic reactions. Such species can be stabilized by coordination to a transition metal through an sp^2-hybridized orbital with an electron pair available for donation and a vacant dz-orbital available for back-donation. An important event occurred after the isolation of the first stable heterocyclic carbenes, which allowed the synthesis of metal carbene complexes by direct reactions.

The carbene ligand is a monohapto, two-electron donor; related complexes are known in which the ligand is a cumulated polyene attached to a metal through a terminal carbon atom (Fig. 11.15).

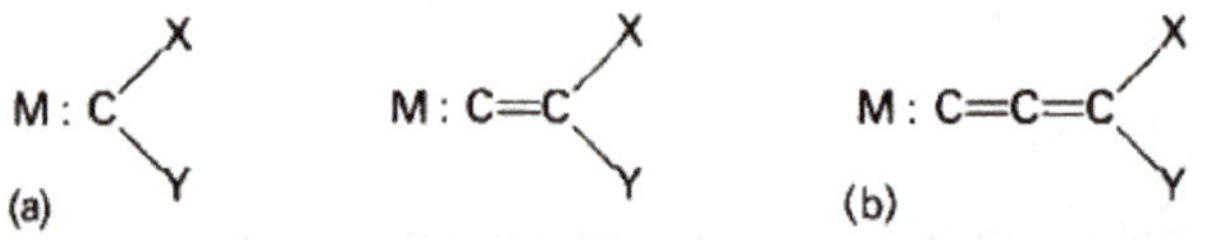

Fig. 11.15: Carbene ligand complexes.

In (a), X and Y can be identical (H, R, OR, NR_2, SR) or different. In (b), frequently X = Y = CN (dicyanovinylidene complexes).

The first metal–carbene complex was prepared by treatment of tungsten hexacarbonyl with organolithium reagents, followed by protonation and then reaction with diazomethane, or alkylation with trialkyloxonium tetrafluoroborate (Fig. 11.16). This reaction has been extended to Cr, Mo, Mn, Tc, Re, Fe and Ni carbonyls.

A less-common carbene is formed in the reaction of a 1,1-dichlorocyclopropene with sodium carbonyl dichromate to give complex of diphenylcyclopropenylidene [14] (Fig. 11.17). A ruthenium complex with a cyclopropylidene has also been described [15].

$$W(CO)_6 + LiR \longrightarrow (CO)_5W:C\begin{smallmatrix}O^-Li^+\\R\end{smallmatrix} \xrightarrow{H^+} (CO)_5W:C\begin{smallmatrix}OH\\R\end{smallmatrix}$$

R = Me, Ph

$$\xrightarrow{Me_3O^+BF_4^-} (CO)_5W:C\begin{smallmatrix}OCH_3\\R\end{smallmatrix}$$

(with CH_2N_2 step leading to OCH_3 product)

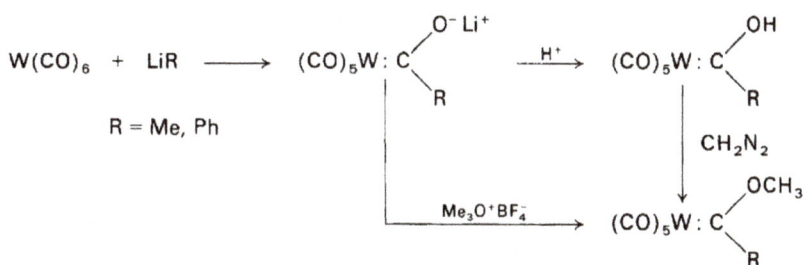

Fig. 11.16: Formation of metal–carbene complexes.

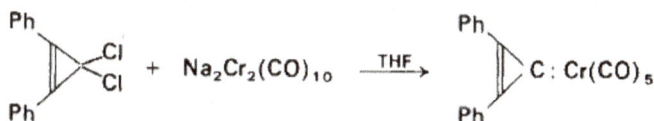

$$\text{(Ph, Ph cyclopropene)CCl}_2 + Na_2Cr_2(CO)_{10} \xrightarrow{THF} \text{(Ph, Ph cyclopropene)C:Cr(CO)}_5$$

Fig. 11.17: A rare carbene complex.

The addition of alcohols to isocyanide complexes also yields carbene complexes (Fig. 11.18).

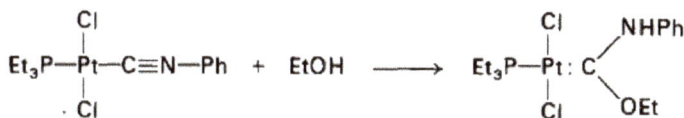

$$Et_3P-Pt(Cl)(Cl)-C\equiv N-Ph + EtOH \longrightarrow Et_3P-Pt(Cl)(Cl):C\begin{smallmatrix}NHPh\\OEt\end{smallmatrix}$$

Fig. 11.18: Conversion of isocyanide into a carbene ligand.

and has been extended to the synthesis of complexes containing four carbene ligands attached to the same metal atom:

$$[M(:CNMe)_4]^{2+} + 4MeNH_2 \rightarrow \{M[:C(NHMe)_2]_4\}^{2+}$$

Coordinated carbon disulfide can be converted into a carbene ligand by electrophilic attack with alkyl halides:

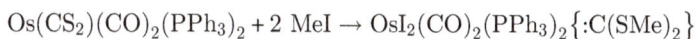

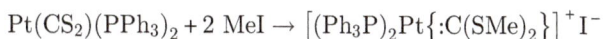

$$Os(CS_2)(CO)_2(PPh_3)_2 + 2\ MeI \rightarrow OsI_2(CO)_2(PPh_3)_2\{:C(SMe)_2\}$$

$$Pt(CS_2)(PPh_3)_2 + 2\ MeI \rightarrow [(Ph_3P)_2Pt\{:C(SMe)_2\}]^+I^-$$

Diphenylcarbene complexes are known for chromium [Cr(CO)$_5$(:CPh$_2$)] [16] and rhodium [Rh(η^5-C$_5$H$_5$)(:CPh$_2$)] [17] Fig. 11.19).

The first difluorocarbene complex resulted by fluorine abstraction from a σ-trifluoromethyl derivative:

$(CO)_5Cr : C\begin{smallmatrix}Ph\\Ph\end{smallmatrix}$ $Cp_2Rh : C\begin{smallmatrix}Ph\\Ph\end{smallmatrix}$

Fig. 11.19: Diphenylcarbene complexes.

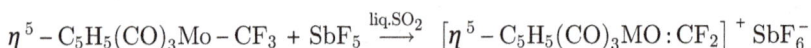

$$\eta^5 - C_5H_5(CO)_3Mo - CF_3 + SbF_5 \xrightarrow{\text{liq.SO}_2} \left[\eta^5 - C_5H_5(CO)_3MO : CF_2\right]^+ SbF_6^-$$

The trigonal planar carbene ligands are better σ-donors and weaker π-acceptors than carbon monoxide.

Carbene complexes in which at least one of the carbon substituents is an atom with electron pairs (X = OR, SR, NR$_2$) are stabilized by (p–p) π-interaction (Fig. 11.20).

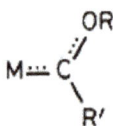

$M\cdots C\begin{smallmatrix}OR\\R'\end{smallmatrix}$

Fig. 11.20: Stabilized carbene complexes.

Compounds with vinylidene and allylidene ligands, closely related to carbene complexes, are illustrated in Fig. 11.21.

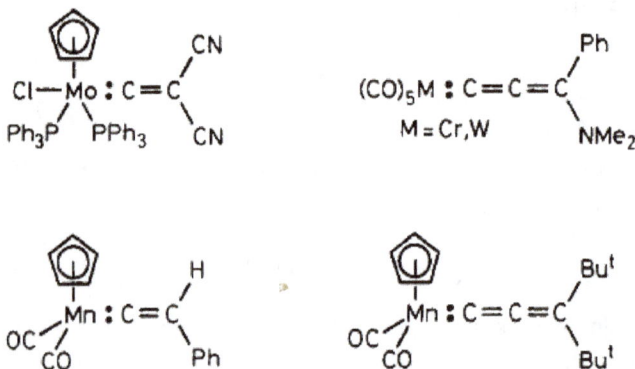

Cl—Mo : C=C (Ph$_3$P, PPh$_3$) with CN, CN

$(CO)_5M : C=C=C$ with Ph, NMe$_2$ M = Cr,W

Mn : C=C (OC, CO) with H, Ph

Mn : C=C=C (OC, CO) with Bu^t, Bu^t

Fig. 11.21: Vinylidene and allylidene complexes.

Addition, substitution, rearrangement, and other reactions can occur in the carbene ligand without cleavage from the metal, for example, the conversion of a phenylmethoxycarbene into a diphenylcarbene ligand (Fig. 11.22).

Two general features of metal carbenes should be noted: a) only few compounds with more than one carbene ligand are known, and b) the carbene complexes are usually neutral (seldom cationic or anionic).

An important development in the chemistry of carbene complexes was the isolation of stable carbene derived from nitrogen heterocycles, that is, imidazolin-2-

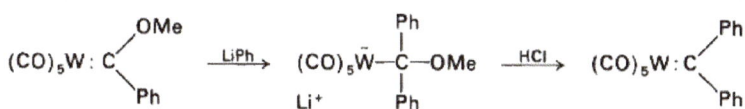

Fig. 11.22: Reaction of coordinated carbene.

ylidenes. This lead to the synthesis of numerous main group [18] and transition metal complexes, including lanthanides [19] and uranium [20–22] by various methods, in which the stable carbene was used as a ligand (Fig. 11.23), or was generated in the process of complex formation [23, 24] (Fig. 11.24).

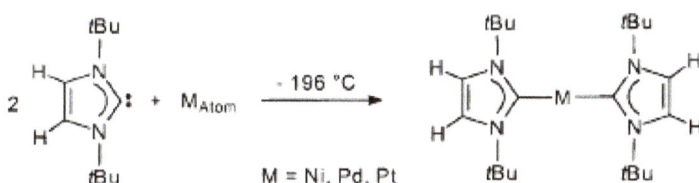

Fig. 11.23: Metal complexes with a nitrogen heterocyclic carbene ligand.

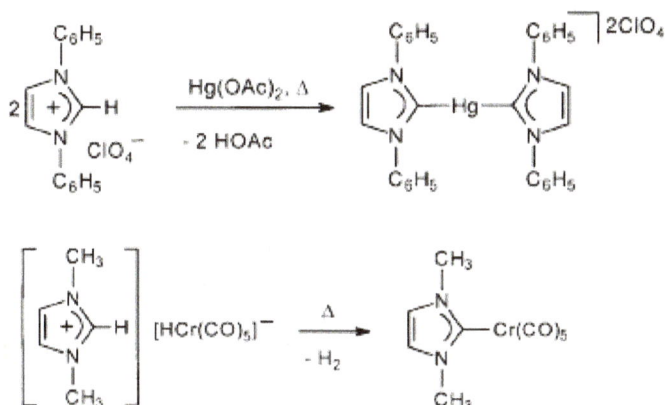

Fig. 11.24: Generation of carbene ligands in the coordination process.

A convenient method for the synthesis of carbene complexes is the cleavage of electron-rich olefins [25] (Fig. 11.25).

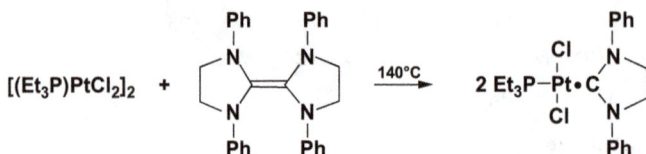

Fig. 11.25: Formation of carbene complex by cleavage of an electron-rich olefin.

11.6 Olefinic complexes

Transition metals are able to form complexes with olefins in which a C = C double bond contributes two electrons to the metal.

The first compound between a transition metal and an olefin, a platinum complex of ethylene, $K[PtCl_3(C_2H_4)]$, was obtained in 1827 by Zeise. In 1938, olefinic complexes of palladium with olefins were also obtained. The nature of these platinum and palladium complexes remained obscure for a long time. It was not until 1951 that the first satisfactory explanation of the metal–olefin bond was made by Dewar, and in 1953, the Chatt-Duncanson model was suggested. In Zeise's salt, $K[PtCl_3(C_2H_4)]$, the olefin is situated perpendicular to the plane formed by the central platinum atom and the chlorine ligands (Fig. 11.26).

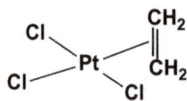

Fig. 11.26: The anion of Zeise's salt, $K[PtCl_3(C_2H_4)]$.

The mono- and polyolefins form complexes with almost all transition metals. Their stability varies with the nature of the metal and olefin and is strongly influenced by both the metal and the olefinic carbon atom substituents.

The monoolefins occupy a single coordinative site, as monodentate ligands. In polyolefins with nonconjugated (isolated) double bonds, each C = C bond acts as an independent donor, and two situations can arise (Fig. 11.27):

Fig. 11.27: Coordination modes of mono- and diolefins.

- the two isolated C = C bonds of the polyolefin act as a bidentate ligand;
- isolated C = C groups bridge different metal atoms (doubly monodentate ligand).

Examples of tridentate or tetradentate olefins (i.e., with three or four isolated double bonds attached to the same metal atom) are known.

Olefin complexes can be prepared using one of the following procedures:
- addition of the olefin to a metal salt (usually halide):

$$MX_n + \| \rightarrow \| \rightarrow MX_n$$

- substitution of carbon monoxide or other ligands (L = : CNR, PR_3, etc.):

$$ML_n + \| \rightarrow \| \rightarrow ML_{n-1}$$

- reduction of a metal cation in the presence of the olefin and an additional ligand (usually a phosphine):

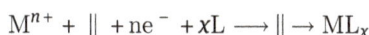

$$M^{n+} + \| + ne^- + xL \longrightarrow \| \rightarrow ML_x$$

- by the gas-phase reaction of the olefin with metal-atom vapors:

$$M^\circ + \| \longrightarrow M \leftarrow \|$$

- hydride abstraction from σ-alkyl derivatives of the metal by the triphenylmethyl cation (Fig. 11.28).

$$M{-}C_2H_5 \; + \; Ph_3C^+ \quad \longrightarrow \quad {\overset{H_2C}{\underset{H_2C}{\|}}}{\mapsto}M^+ \; + \; Ph_3CH$$

Fig. 11.28: Formation of olefin complexes by hydride abstraction.

11.6.1 Monoolefin complexes

Binary complexes containing only monoolefins attached to the metal atom include the ethylene complexes, $M(C_2H_4)_n$ with M = Co ($n = 1$ and 2), M = Ni ($n = 1$, 2 and 3), M = Pd ($n = 1$, 2 and 3), M = Cu ($n = 1$, 2 and 3), M = Au ($n = 1$) or $Ni(CF_2 = CF_2)_n$ ($n = 1$, 2 and 3), prepared by the metal vapor synthesis technique, with the complexes isolated in a low-temperature matrix.

Cyclic monoolefins coordinate as monoolefins and form various complexes, for example, cyclooctene coordinated to copper [26] and gold [27], cyclooctene coordinated to tungsten [28], platinum [29], copper [30], silver [31] and cyclocta-1,5-diene coordinated to gold [32].

Heteroleptic olefin complexes. The olefin is often accompanied by additional ligands, forming heteroleptic olefin complexes, as in olefin–metal carbonyl complexes, cyclopentadienylmetal carbonyl olefin complexes or metal–halide olefin complexes (Fig. 11.29).

Examples can be cited with monometallic titanium [Ti(η^5-C_5Me_5)$_2$(η^2-C_2H_4)] [33], niobium and tantalum [(η^2-C_2H_4)M{OSiBu$^t_3$)$_3$}] (M = Nb, Ta) [34] compounds and

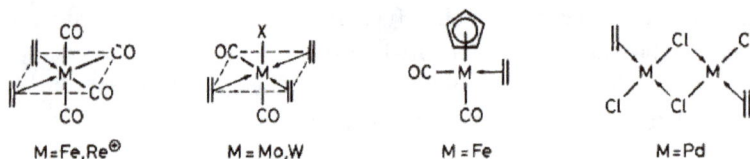

Fig. 11.29: Various heteroleptic olefin complexes.

bimetallic zirconium $[(\mu\text{-}C_2H_4)(\{ZrRCp_2)_2]$ (R = Me, Et) [35], hafnium $[(\mu\text{-}C_2H_4)\{Hf\{Br_3(PEt_3)_2\}_2]$ [36] compounds (Fig. 11.30).

Get e-alerts.

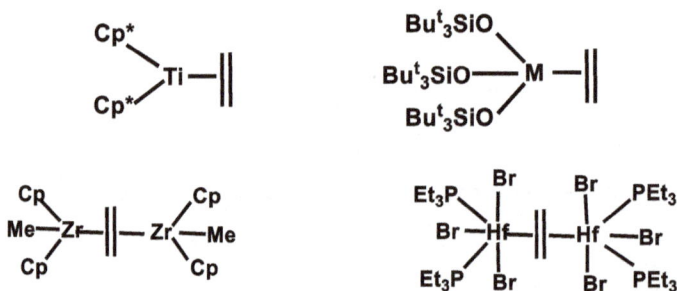

Fig. 11.30: Monoolefin complexes.

Ethylene–carbonyl derivatives of manganese are formed by the reaction of ethylene and manganese–pentacarbonyl chloride, or by hydride abstraction from the σ-ethyl derivative (Fig. 11.31).

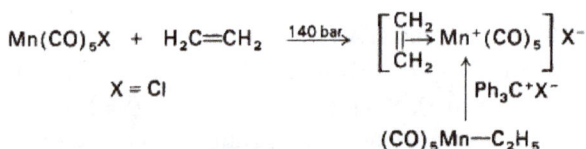

Fig. 11.31: Formation of a manganese–olefin complex.

More compounds of this class with other transition metals are described, namely, with rhenium $[(\eta^2\text{-}C_2H_4)Re(CO)(NO)(PPr^i_3)_2]$ [37],

$$\text{iron } [(\eta^2 - C_2H_4)Fe(R_2PCH_2CH_2PR_2)] \text{ (R = Et, Ph) [38, 39],}$$

$$\text{ruthenium } [(\eta^2 - C_2F_4)Ru(CO)_2(PPh_3)_2] \text{ [40],}$$

$$\text{osmium } [(\eta^2 - C_2H_4)Os(CO)_4] \text{ [41],}$$

cobalt $[\{\eta^2 - MeOOC - HC = CH - COOMe\}Co(PMe_3)_3]$ [42],

rhodium $[(\eta^2 - C_2H_4)RhF(PPr^i_3)_2]$ [43]

iridium trans $- [(\eta^2 - C_2H_4)IrF(acac)(PPr^i_3)_2]$ [44]

Nickel tetracarbonyl reacts with the activated olefins, acrylonitrile, acrolein, fumaronitrile, to form Ni(olefin)$_2$. The reduction of nickel(II) acetylacetonate in the presence of ethylene and triphenylphosphine yields a diphosphine–ethylene complex [45] (Fig. 11.32).

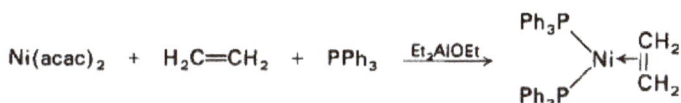

Fig. 11.32: Formation of a mixed nickel olefin–phosphine complex.

Other ethylene–nickel complexes are $[(\eta^2\text{-}C_2H_4)Ni(PPr^i_3)_2]$ [46],

$[(\eta^2 - C_2H_4)Ni(Bu^t_2PCHCH_2PBu^t_2)]$ [47] and $[(\eta^2 - C_2H_4)Ni\{MeC(CH_2PPh_2)_3\}]$ [48].

The starting material of choice for palladium π-olefin complexes is the benzonitrile derivative, $(PhCN)_2PdCl_2$ (Kharasch reagent), in which the weakly bonded nitrile is replaced by olefins. Palladium also forms anionic $[Pd(olefin)Cl_3]^-$ and neutral $[Pd(olefin)_2Cl_2]$ and $[Pd(olefin)(PR_3)_2]$ complexes. Some more recent palladium–olefin complexes are $[Pd(C_2H)(C_6F_5)_3]$ [49], $[Pd(C_2F_4)(PPh_3)_2]$ [50] and $[Pd\{C_2(NC)_2C = C(CN)_2\}(PPh_3)_2]$ [51] (Fig. 11.33).

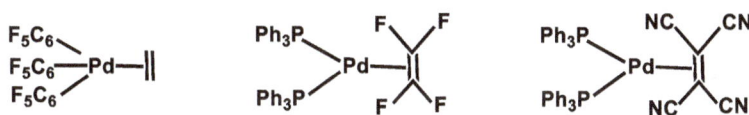

Fig. 11.33: Various palladium–olefin complexes.

The platinum π-olefin complexes, $M[PtX_3(olefin)]$ (X = Cl [52, 53], Br [54] $[PtX_2(olefin)]_2$ and trans-$[PtX_2(olefin)_2]$ were prepared long before the understanding of the nature of metal–olefin bond. The platinum–ethylene triphenylphosphine complex $[(\mu\text{-}C_2H_4)PtCl_2(PPh_3)]$ [55] has also been prepared.

A rare gold complex with mixed olefin–pyrazolylborato ligands, $[(\mu\text{-}C_2H_4)Au[(BPz^{2R}_3)_3]$ R = Pr^i [56], has been reported [56] (Fig. 11.34).

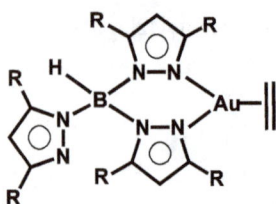

Fig. 11.34: An olefin–gold complex.

11.6.2 Bis(olefin) complexes

Numerous complexes with two olefin molecules have been described. As examples are cited, some molybdenum *trans*-[Mo(η^2-C$_2$H$_2$)$_2$(PMe$_3$)$_4$] [57], tungsten *trans*-[W(η^2-C$_2$H$_2$)$_2$(PMe$_3$)$_4$] [58] and *trans*-[W(η^2-C$_2$H$_2$)$_2$(CO)$_4$] [59] complexes (Fig. 11.35).

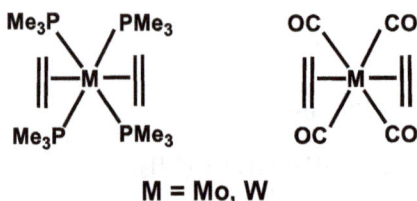

M = Mo, W

Fig. 11.35: Molybdenum and tungsten bis(ethylene) complexes.

A manganese complex is formed by ligand transfer with ethylene, replacing η^5-C$_5$H$_5$ and biphenyl ligands in a surprising reaction [60] (Fig. 11.36).

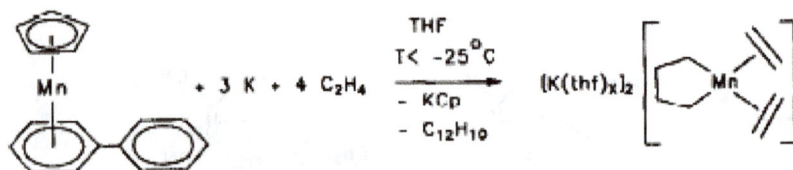

Fig. 11.36: Unusual formation of a bis(ethylene)manganese complex.

Rhenium–pentacarbonyl chloride reacts with ethylene (250 bar) to form a disubstituted derivative, [Re(η^2-C$_2$H$_4$)$_2$(CO)$_4$]$^+$. An unstable technetium cation, [Tc(η^2-C$_2$H$_4$)$_2$(CO)$_4$]$^+$, is also known:

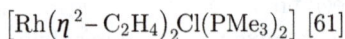

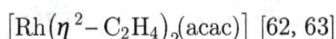

$$\left[\text{Rh}\left(\eta^2- \text{C}_2\text{H}_4\right)_2 \text{Cl}(\text{PMe}_3)_2\right] \ [61]$$

$$\left[\text{Rh}\left(\eta^2- \text{C}_2\text{H}_4\right)_2(\text{acac})\right] \ [62, 63]$$

$$\left[\text{Ir}\left(\eta^2- \text{C}_2\text{H}_4\right)_2(\text{acac})\right] \ [64]$$

$$[Ni(\eta^2 - C_2H_4)_2(PCy_3)] \ [65]$$

$$cis - [Pt(\eta^2 - C_2H_4)_2Cl_2] \ [66]$$

$$[Ag(\eta^2 - C_2H_4)_2\{N(SO_2CF_3)_2\}] \ [67]$$

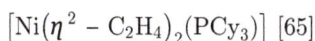

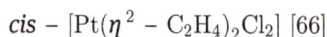

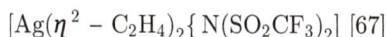

More bis(olefin) complexes are known (Fig. 11.37).

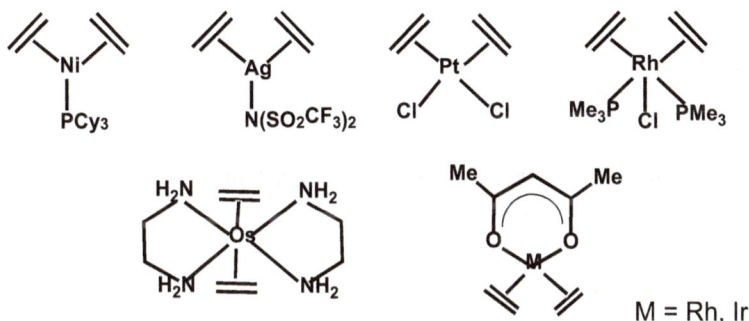

Fig. 11.37: More bis(olefin) complexes.

Palladium(II) chloride reacts with liquid olefins, for example, ethylene under pressure, to form the dimers $[Pd(C_2H_2)Cl(\mu_2\text{-}Cl)]_2$ [68]. Another dinuclear compound is the platinum complex or Zeise's dimer $[Pt(C_2H_2)Cl(\mu_2\text{-}Cl)]_2$ (Fig. 11.38).

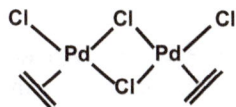

Fig. 11.38: Chlorine-bridged palladium–diethylene complex.

Rhodium salts catalyze the oligomerization and hydrogenation of olefins, via intermediate formation of complexes. The dimers, $[RhCl(C_2H_4)_2]_2$, containing four-coordinated rhodium (with a square planar geometry) and chlorine bridges have been prepared from $RhCl_3 \cdot H_2O$ and olefins (Fig. 11.39).

Fig. 11.39: Chlorine-bridged rhodium–tetraethylene complex.

11.6.3 Tris(olefin) and tetrakis(olefin) complexes

Three olefin molecules can coordinate to the same metal atom in some cases, and triolefin complexes are known with rhodium [69], iridium and platinum [70, 71] (Fig. 11.40), copper [72, 73], silver [74] and gold [75].

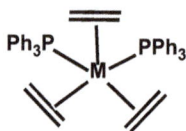

M = Rh, Ir, Pt   Fig. 11.40: Tris(olefin) complexes.

Cationic tris(olefin) gold(I) complexes are formed with ethylene $[Au(\eta^2\text{-}C_2H_4)_3]^+$ [76] and cyclooctene $[Au(\eta^2\text{-}C_8H_{14})]^+$ [77] (Fig. 11.41).

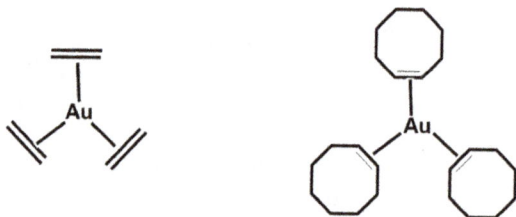

Fig. 11.41: Tris(olefin) gold complexes.

Substituted ethylenes, for example, stilbene (1,2-diphenylethylene) and the p-tolyl derivative also form similar compounds with nickel [78, 79] (Fig. 11.42). Complexes with four olefins are rare. A compound with four η^2-coordinated ethylene molecules is known as tetrakis(η^2-ethylene)-cobalt (2.2.2-cryptand)-potassium salt [80] (Fig. 11.42).

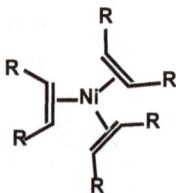

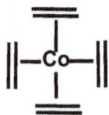

R = Ph, p-Tol   Fig. 11.42: Nickel– and cobalt–olefin complexes.

11.6.4 Bidentate diolefin complexes

The polyolefins containing isolated double bonds can act as bidentate ligands. Non-conjugated diolefins like Dewar benzene and cyclooctadiene act as bidentate ligands in metal complexes (Fig. 11.43).

Fig. 11.43: Bidentate diolefin complexes.

Group 16 metal hexacarbonyls form substitution products with cyclooctadiene by replacement of carbon monoxide in which the diene acts as a bidentate ligand.

Nonconjugated diolefins undergo isomerization in the presence of transition metals to become conjugated (as four-electron donors). However, cyclooctadiene and norbornadiene react with the dinuclear iron carbonyl, $Fe_2(CO)_9$, to form complexes of the bidentate diolefins (Fig. 11.44).

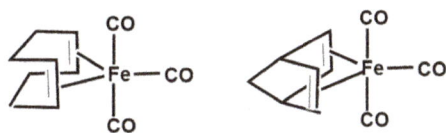

Fig. 11.44: Iron–carbonyl complexes of cyclooctadiene-1,5 and norbornadiene.

Reduction of nickel(II) acetylacetonate with Et_2AlOEt in the presence of cyclooctadiene produces the complex $[Ni(C_8H_{12})_2]$ (Fig. 11.45).

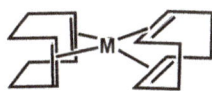

M = Ni, Pt   Fig. 11.45: Nickel complex of cyclooctadiene.

Palladium(II) chloride forms the diene complexes, $[Pd(diene)X_2]$, with cyclooctadiene and 1,5-hexadiene. The Kharasch reagent, $[Pd(PhCN)_2Cl_2]$, reacts with allyl chloride to yield a complex of hexadiene-1,5 formed by coupling of the olefin.

Platinum forms complexes with nonconjugated diolefins in both oxidation states, +2 and 0. Thus, $[Pt(diene)X_2]$ complexes have been obtained with cyclooctadiene, 1,5-hexadiene and cyclooctatetraene. Reduction of the complex $[Pt(cyclooctadiene-1,5)Cl_2]$ with isopropylmagnesium bromide in the presence of additional cyclooctadiene-1,5 gives $[Pt(C_8H_{12})_2]$.

Rhodium(III) chloride and the rhodium complex of butadiene react with cyclo-octadiene to form the binuclear dimer, [RhX(diene)]$_2$. Norbornadiene and cyclooctadiene-1,5 combine with the salt Na[IrCl$_6$] · 6H$_2$O to form [Ir(diene)Cl]$_2$ complexes. A norbornadiene complex [Cu(diolefin)Cl]$_2$ is formed by reduction of copper(II) chloride with sulfur dioxide in the presence of norbornadiene (Fig. 11.46).

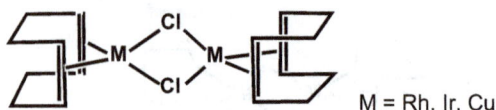

M = Rh, Ir, Cu

Fig. 11.46: Chlorine-bridged dinuclear complexes with cyclooctadiene-1,5 ligand.

Cyclooctadiene-1,5 and norbornadiene form [Ru(diolefin)X$_2$] compounds [81, 82] with ruthenium(II) halides and Ir(diolefin)[HC(CBu^{t}CO)$_2$] [83] in which the diolefin is attached as a bidentate ligand. Palladium also forms a complex with a bent cyclo-octa-1,5-diene molecule [84].

The larger ring, cyclodecadiene-1,6 derivative is obtained either directly from the diolefin and rhodium(III) chloride, or by isomerization of cyclodecadiene-1,5 (Fig. 11.47).

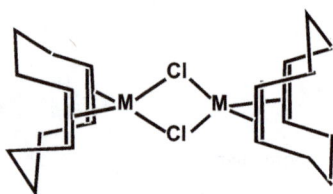

Fig. 11.47: Chlorine-bridged dinuclear complexes with cycloodecadiene-1,5 ligand.

11.6.5 Bridging diolefin complexes

In the previous examples, both double bonds of a diolefin were attached to the same metal atom. Even in some conjugated olefins the C=C groups can act independently, for example, in butadiene, cyclohexadiene-1,3 or fulvenes, to form bridges (Fig. 11.48). Bridging complexes are also known with norbornadiene and cyclooctatetraene. These are compounds could be described as inverse organometallic complexes since the metal is not the coordination center.

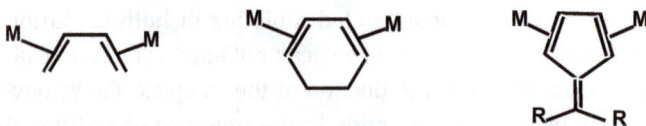

Fig. 11.48: Bridging coordination mode of conjugated diolefins.

Butadiene forms manganese and diiron derivatives with a bridged structure by replacement of carbon monoxide from the corresponding metal carbonyl complexes, and a manganese cyclohexadiene complex is also known (Fig. 11.49).

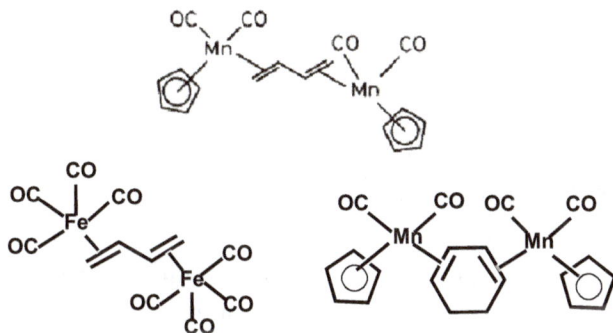

Fig. 11.49: Manganese– and iron–diene complexes.

The diolefins react with platinum chlorides to form bridged butadiene and cyclohexadiene-1,3 complexes (Fig. 11.50).

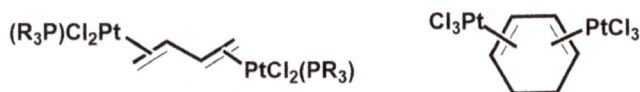

Fig. 11.50: Platinum–diene complexes.

The monovalent metals of the copper–silver–gold triad form bridged complexes, since they can coordinate only one double bond. Thus, copper(I) chloride reacts with butadiene to form $[C_4H_6(CuCl)_2]$, with cyclohexadiene-1,3 to form the unstable $[C_6H_{10}(CuCl)_2]$, and with norbornadiene to form $[C_7H_8(CuBr)_2]$ and $[C_7H_8(CuCl)_2]$ (Fig. 11.51).

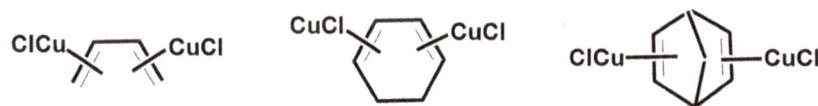

Fig. 11.51: Copper–diene complexes.

Silver ion forms bridged-butadiene complexes in 1:1 and 2:1 ratios by absorption of butadiene by aqueous silver nitrate (Fig. 11.52).

Gold forms the bridged complex $[C_8H_{12}(AuCl)_2]$ on ultraviolet irradiation of cyclooctadiene-1,5 with tetrachloroauric acid, $H[AuCl_4]$.

Molecules with several C = C bonds like cyclooctatetraene, can form a doubly bidentate bridge (Fig. 11.53).

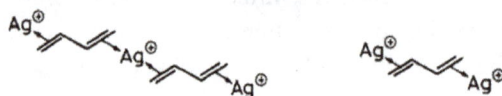

Fig. 11.52: Silver–diene complexes.

Fig. 11.53: Bridging coordination of cyclooctatetraene as tetraolefin.

Cyclooctatetraene reacts photochemically with the cyclopentadienyl-metal dicarbonyls, $(\eta^5\text{-}C_5H_5)M(CO)_2$ (M = Co, Rh), to form a cyclooctatetraene complex $[(\mu\text{-}C_8H_8)\{M(\eta^5\text{-}C_5H_5)\}_2]$ (Fig. 11.54).

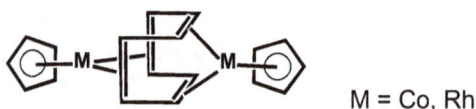

M = Co, Rh

Fig. 11.54: Cobalt and rhodium cyclooctatetraene complexes.

The nickel complex, $[Ni(C_{12}H_{18})]$, reacts with cyclooctatetraene to form polymeric $[Ni(C_8H_8)]_n$, with doubly bidentate-cyclooctatetraene bridges (Fig. 11.55).

Fig. 11.55: Nickel cyclooctatetraene complex.

11.6.6 Tridentate olefin complexes

Unconjugated triolefins can act as tridentate ligands: thus, cyclononatriene-1,4,7 reacts with molybdenum hexacarbonyl to form $[Mo(C_9H_{12})(CO)_3]$ (Fig. 11.56).

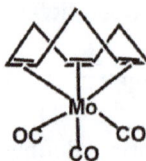

Fig. 11.56: Molybdenum complex of cyclononatriene-1,4,7.

The reduction of nickel(II) acetylacetonate with $Et_2Al–OEt$ in the presence of cyclododecatriene-l,5,9 results in the formation of $[Ni(C_{12}H_{18})]$ (Fig. 11.57). In this compound the nickel atom is coordinatively unsaturated and readily accepts a carbon monoxide molecule, to form $[Ni(C_{12}H_{18})(CO)]$.

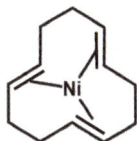

Fig. 11.57: Nickel complex of cyclododecatriene-l,5,9.

References

[1] Clark GR, Marsden K, Rickard CEF, Roper WR, Wright LJ Syntheses and structures of the chalcocarbonyl complexes $OsCl_2(CO)(CE)(PPh_3)_2$ (E = S, Se, Te). J Organomet Chem 1988, 338, 393–410.

[2] Albano VG, Monari M, Busetto L, Carlucci L, Zanotti V. Synthesis and molecular structure of the novel heterodinuclear thiocarbonyl complex $[FeMn(\mu-CO)(\mu-CS)(CO)_5Cp]$ (Cp = η^5-C_5H_5). Gazz Chim Ital 1992,122, 201–04.

[3] Takemoto S, Ohata J, Umetani K, Yamaguchi M, Matsuzaka H A Diruthenium μ-carbido complex that shows singlet-carbene-like reactivity. J Am Chem Soc 2014, 136, 15889–92.

[4] North TE, Thoden JB, Spencer B, Bjarnason A, Dahl LF Synthesis and stereophysical characterization of the Fischer-Palm-related 49-electron thiocarbonyl-capped triangular nickel clusters $Ni_3(\eta^5$-C_5H_5-$xMex)_3(\mu_3$-$X)(\mu_3$-$Y)$ (x = 0, 1), containing π-acceptor X and Y capping Ligands (X = Y = CS; X = CS. Organometallics 1992, 11, 4326–37.

[5] Saillard J-Y, Grandjean D. Structure du [1–6-η-(benzoate de méthyle)] dicarbonylsélénocarbonylchrome. Acta Crystallogr Sect B Struct Crystallogr Cryst Chem 1978, 34, 3772–75.

[6] Clark GR, James SM Coordinated CSe. The crystal and molecular structure of dichlorocarbonylselenocarbonylbis(triphenylphosphine)-ruthenium(II), $RuCl_2(CO)(CSe)$ $(PPh_3)_2$. J Organomet Chem 1977, 134, 229–36.

[7] Mutoh Y, Kozono N, Araki M, Tsuchida N, Takano K, Ishii Y. Ruthenium Seleno- and Tellurocarbonyl Complexes: Selenium and Tellurium Atom Transfer to a Terminal Carbido Ligand. Organometallics 2010, 29, 519–22.

[8] Frogley BJ, Hill AF, Watson LJ New binding modes for CSe: coinage metal coordination to a tungsten selenocarbonyl complex. Dalton Trans 2019, 48, 12598–606.

[9] Cade IA, Hill AF, McQueen CMA Isoselenocarbonyl complexes. Dalton Trans 2019, 48, 2000–12.

[10] Jones WD, Foster GP, Putinas JM Preparation and structural examination of a series of new, low-valent iron phosphine isocyanide complexes with bent carbon-nitrogen-carbon linkages. Inorg Chem 1987, 26, 2120–27.

[11] Warner S, Lippard SJ Synthesis, structure, and reactions of rhenium aminocarbyne complexes formed from $[ReCl_2(CNR)_3(PMePh_2)_2]^+$ (R = *tert*-Bu or Me) cations under reductive coupling conditions. Organometallics 1989, 8, 228–36.

[12] Haehnel M, Ruhmann M, Theilmann O, Roy S, Beweries T, Arndt P, Spannenberg A, Villinger A, Jemmis ED, Schulz A, Rosenthaler U Reactions of Titanocene Bis(trimethylsilyl)acetylene

Complexes with Carbodiimides. An Experimental and Theoretical Study of Complexation versus C–N Bond Activation. J Am Chem Soc 2012, 134, 15979–91.

[13] Semproni SP, McNeil WS, Baillie RA, Patrick BO, Campana CF, Legzdins P. Ground-State Electronic Asymmetry in Cp*W(NO)(η^1-isonitrile)$_2$ Complexes. Organometallics 2010, 29, 867–75.

[14] Scherer W, Tafipolsky M, Öfele K On the electron delocalization in cyclopropenylidenes - An experimental charge-density approach. Inorganica Chim Acta 2008, 361, 513–20.

[15] Caskey SR, Stewart MH, Johnson MJA, Kampf JW Carbon–Carbon Bond Formation at a Neutral Terminal Carbido Ligand: Generation of Cyclopropenylidene and Vinylidene Complexes. Angew Chem 2006, 118, 7582–84.

[16] Seidel G, Gabor B, Goddard R, Heggen B, Thiel W, Fürstner A Gold Carbenoids: Lessons Learnt from a Transmetallation Approach. Angew Chem Int Ed 2014, 53, 879–82.

[17] Werner H, Schwab P, Bleuel E, Mahr N, Windmüller B, Wolf J Carbenerhodium Complexes of the Half-Sandwich-Type: Synthesis, Substitution, and Addition Reactions. Chem Eur J 2000, 6, 4461–70.

[18] Kuhn N, Al-Sheikh A 2,3-Dihydroimidazol-2-ylidenes and their main group element chemistry. Coord Chem Rev 2005, 249, 829–57.

[19] Arnold PL, Liddle ST F-block N-heterocyclic carbene complexes. Chem Commun. 2006, 3959–71.

[20] Oldham WJ Jr., Oldham SM, Smith WH, Costa DA, Scott BL, Abney KD. Synthesis and structure of N-heterocyclic carbene complexes of uranyl dichloride. Chem Commun 2001, 1348–49

[21] Nakai H, Hu X, Zakharov LN, Rheingold AL, Meyer K. Synthesis and Characterization of N-Heterocyclic Carbene Complexes of Uranium(III). Inorg Chem 2004, 43, 855–57.

[22] Evans WJ, Kozimor SA, Ziller JW Bis(pentamethyl-cyclopentadienyl)U(III) oxide and U(IV) oxide carbene complexes. Polyhedron 2004, 23, 2689–94

[23] Schönherr H-J, Wanzlick H-W Chemie nucleophiler Carbene, XX HX-Abspaltung aus. 1.3-Diphenyl-imidazoliumsalzen. Quecksilbersalz-Carben-Komplexe. Chem Ber 1970, 103, 1037–46.

[24] Öfele K 1,3-Dimethyl-4-imidazolinyliden-(2)-pentacarbonylchrom ein neuer übergangsmetall-carben-komplex. J Organomet Chem 1968, 12, P42–3.

[25] Lappert MF Contributions to the chemistry of carbene metal chemistry. J Organomet Chem 2005, 690, 5467–73.

[26] Martín C, Muñoz-Molina JM, Locati A, Alvarez E, Maseras F, Belderrain TR, Pérez PJ Copper(I) –Olefin Complexes: The Effect of the Trispyrazolylborate Ancillary Ligand in Structure and Reactivity. Organometallics 2010, 29, 3481–89.

[27] Motloch P, Blahut J, Císařová I, Roithová J X-ray characterization of triphenylphosphine-gold (I) olefin π-complexes and the revision of their stability in solution. J Organomet Chem 2017, 848, 114–17.

[28] Dalla Riva Toma JM, Toma PH, Fanwick PE, Bergstrom DE, Byrn SR. Photochemical synthesis and crystal structure of two potentially useful metal carbonyl complexes: pentacarbonyl(η^2-cis-cyclooctene)-tungsten(0) and tetracarbonylbis(η^2-cis-cyclooctene)tungsten(0). J Crystallogr Spectrosc Res 1993, 23, 41–47.

[29] Otto S, Roodt A, Elding LI Bridge-splitting kinetics, equilibria and structures of trans-biscyclooctene complexes of platinum(II). Dalton Trans 2003, 2519–25

[30] Pampaloni G, Peloso R, Graiff C, Tiripicchio A. Synthesis, Characterization, and Olefin/CO Exchange Reactions of Copper(I) Derivatives Containing Bidentate Oxygen Ligands. Organometallics 2005, 24, 4475–82.

[31] Jayaratna NB, Pardue DB, Ray S, Yousufuddin M, Thakur KG, Cundari TR, Rasika Dias HV, Silver(I) complexes of tris(pyrazolyl)borate ligands bearing six trifluoromethyl and three additional electron-withdrawing substituents. Dalton Trans 2013, 42, 15399–410.

[32] Motloch P, Blahut J, Císařová I, Roithová J X-ray characterization of triphenylphosphine-gold (I) olefin π-complexes and the revision of their stability in solution. J Organomet Chem 2017, 848, 114–17.

[33] Cohen SA, Auburn PR, Bercaw JE Structure and reactivity of bis(pentamethylcyclopentadienyl) (ethylene)titanium(II), a simple olefin adduct of titanium. J Am Chem Soc 1983, 105, 1136–43.

[34] Hirsekorn KF, Hulley EB, Wolczanski PT, Cundari TR J Am Chem Soc. 2008, 130, 1183

[35] a) Takahashi T, Kasai K, Suzuki N, Nakajima K, Negishi E, Isolation and Characterization of the Ethylene-Bridged Zirconocene Complex $(Cp_2ZrM)_2(CH_2CH_2)$ Organometallics 1994, 13, 3413–14; (b) Fischer R, Gebhardt P, Görls H, Reactive Intermediates of the Catalytic Carbomagnesation Reaction: Isolation and Structures of $[Cp_2ZrEt]_2(\mu\text{-ethene})$, $[Cp_2Zr(ethene)$ (L)] (L = THF, Pyridine), and $[(indenyl)_2Zr(ethene)(THF)]$ and of Metallacycles with Norbornen, Organometallics 2000, 19, 13, 2532–40

[36] Cotton FA, Kibala PA A new type of metal-olefin complex. Synthesis and characterization of four compounds that contain an ethylene bridge perpendicularly bisecting a metal-metal axis. Inorg Chem 1990, 29, 3192–96.

[37] Choualeb A, Blacque O, Schmalle HW, Fox T, Hiltebrand T, Berke H Olefin Complexes of Low-Valent Rhenium. Eur J Inorg Chem 2007, 2007, 5246–61.

[38] Adamson TT, Kelley SP, Bernskoetter WH Iron-Mediated C–C Bond Formation via Reductive Coupling with Carbon Dioxide. Organometallics 2020, 39, 3562–71.

[39] Burcher B, Sanders KJ, Benda L, Pintacuda G, Jeanneau E, Danopoulos AA, Braunstein P, Olivier-Bourbigou H, Breuil PAR Straightforward Access to Stable, 16-Valence-Electron Phosphine-Stabilized Fe^0 Olefin Complexes and Their Reactivity. Organometallics 2017, 36, 605–13.

[40] Burrell AK, Clark GR, Rickard CEF, Roper WR, Ware DC Synthesis and structure of osmium and ruthenium complexes containing tetrafluoroethylene and maleic anhydride as ligands. J Organomet Chem 1990, 398, 133–58.

[41] Bender BR, Norton JR, Miller MM, Anderson OP, Rappé AK Structure of $Os(CO)_4(C_2H_4)$, an Osmacyclopropane. Organometallics 1992, 11, 3427–34

[42] Liu Y, Du J, Deng L. Synthesis, Structure, and Reactivity of Low-Spin Cobalt(II) Imido Complexes $[(Me_3P)_3Co(NAr)]$. Inorg Chem 2017, 56, 8278–86.

[43] Gil-Rubio J, Weberndörfer B, Werner H A series of new fluororhodium(I) complexes. J Chem Soc Dalton Trans 1999, 1437–44.

[44] Werner H, Papenfuhs B, Steinert P. Synthese, Struktur und Photochemie von Olefiniridium(I)-Komplexen mit Acetylacetonatoliganden. Z Anorg Allg Chem 2001, 627, 1807–14.

[45] Dreissig W, Dietrich H Die Kristallstruktur von Bis-triphenylphosphin-äthylennickel(0). Acta Crystallogr Sect B Struct Crystallogr Cryst Chem 1968, 24, 108–16.

[46] Beck R, Shoshani M, Krasinkiewicz J, Hatnean JA, Johnson SA Synthesis and chemistry of bis (triisopropylphosphine) nickel (I) and nickel(0) precursors. Dalton Trans 2013, 42, 1461–75.

[47] Scherer W, Eickerling G, Shorokhov D, Gullo E, McGrady GS, Sirsch P Valence shell charge concentrations and the Dewar-Chatt-Duncanson bonding model. New J Chem 2006, 30, 309–12.

[48] Mautz J, Heinze K, Wadepohl H, Huttner G Reductive Activation oftripod Metal Compounds: Identification of Intermediates and Preparative Application. Eur J Inorg Chem 2008, 2008, 1413–22.

[49] Fornies J, Martin A, Martin LF, Menjon B, Tsipis A All-organometallic analogues of Zeise's salt for the three group 10 metals. Organometallics 2005, 24, 3539–46.

[50] Ohashi M, Kambara T, Hatanaka T, Saijo HR, Ogoshi S, Palladium catalized coupling reactions of tetrafluoroethylene with arylzinc compounds. J Am Chem Soc 2011, 133, 3256–59.

[51] Shackleton Z, Kilner CA, Reid GD, Halcrow MA (η^2-Tetracyanoethene)-bis(triphenylphosphine) palladium–dichloromethane (1/0.7). Acta Crystallogr Sect C Cryst Struct Commun 2003, 59, m136–8.

[52] Otto S, Roodt A, Elding LI, Bridge splitting of *trans*-[PtCl$_2$(C$_2$H$_4$)]$_2$ by ethene using a simple combined NMR–UV/vis cell: Crystal and molecular structure of *cis*-[PtCl$_2$(C$_2$H$_4$)$_2$]. Inorg Chem Commun 2006, 9, 744–46

[53] Weller MT, Henry PF, Ting VP, Wilson CC. Crystallography of hydrogen-containing compounds: Realizing the potential of neutron powder diffraction. Chem Commun 2009, 2973–89.

[54] Dub PA, Rodriguez-Zubiri M, Daran J-C, Brunet -J-J, Poli R. Platinum-Catalyzed Ethylene Hydroamination with Aniline: Synthesis, Characterization, and Studies of Intermediates. Organometallics 2009, 28, 4764–77.

[55] Pryadun RS, Gerlits OO, Atwood JD Structural studies on platinum alkene complexes and precursors. J Coord Chem 2006, 59, 85–100.

[56] Wu J, Noonikara-Poyil A, Muñoz-Castro A, Dias HVR Gold(I) ethylene complexes supported by electron-rich scorpionates. Chem Commun 2021, 57, 978–81.

[57] Carmona E, Marin JM, Poveda ML, Atwood JL, Rogers RD Preparation and properties of dinitrogen trimethylphosphine complexes of molybdenum and tungsten. 4. Synthesis, chemical properties, and x-ray structure of cis-[Mo(N$_2$)$_2$(PMe$_3$)$_4$]. The crystal and molecular structures of trans-[Mo(C$_2$H$_4$)$_2$(PMe$_3$)$_4$]. J Am Chem Soc 1983, 105, 3014–22.

[58] Carmona E, Galindo A, Poveda ML, Rogers RD Synthesis and properties of cis-bis(dinitrogen) tetrakis(trimethylphosphine)- tungsten(0). Crystal and molecular structures of [W(N$_2$)(PMe$_3$)$_5$] and trans-[W(C$_2$H$_4$)$_2$(PMe$_3$)$_4$]. Inorg Chem 1985, 24, 4033–39.

[59] Szymańska-Buzar T, Kern K, Downs AJ, Greene TM, Morris LJ, Parsons S Crystal structure of *trans*-[W(CO)$_4$(η^2-C$_2$H$_4$)$_2$] and IR and 1H NMR studies of the reactions of this and related ethene carbonyl complexes of tungsten(0), [W(CO)$_n$(η^2-C$_2$H$_4$)$_{6-n}$] (n = 3–5). New J Chem 1999, 23, 407–16.

[60] Jonas K, Häselhoff -C-C, Goddard R, Krüger C. Manganese(II) cyclopentadienide and cyclopentadienylmanganese(biphenyl) as starting materials for the synthesis of carbonyl free organomanganese complexes. Inorg Chim Acta 1992, 198–200, 533–41.

[61] Choi J-C, Sakakura T. Structure and Reactivity of an Unusual Rhodium(I) Bis(ethylene) Complex, RhCl(CH$_2$CH$_2$)$_2$(PMe$_3$)$_2$. Organometallics 2004, 23, 3756–58.

[62] Price DW, Drew MGB, Hii KK, Brown JM The Chatt-Dewar-Duncanson Model Revisited: X-ray, DFT and NMR Studies of Rhodium-Alkene Binding—Deviations from Structural Ideality. Chem-Eur J 2000, 6, 4587–96.

[63] Bühl M, Håkansson M, Mahmoudkhani AH, Öhrström L X-ray Structures and DFT Calculations on Rhodium–Olefin Complexes: Comments on the [103]Rh NMR Shift–Stability Correlation. Organometallics 2000, 19, 5589–96.

[64] Bhirud VA, Uzun A, Kletnieks PW, Craciun R, Haw JF, Dixon DA, Olmstead MM, Gates BC J Organomet Chem. 2007, 692, 2107

[65] Moser E, Jeanneau E, Mézailles N, Olivier-Bourbigou H, Breuil PAR Simplified and versatile access to low valent Ni complexes by metal-free reduction of NiII precursor. Dalton Trans, 2019, 48, 4101–04.

[66] Otto S, Roodt A, Elding LI Bridge splitting of *trans*-[PtCl$_2$(C$_2$H$_4$)]$_2$ by ethene using a simple combined NMR-UV/vis cell: Crystal and molecular structure of *cis*-[PtCl$_2$(C$_2$H$_4$)$_2$]. Inorg Chem Commun 2006, 9, 764–66.

[67] Stricker M, Oelkers B, Rosenau CP, Sundermeyer J. Copper(I) and Silver(I) Bis (trifluoromethanesulfonyl)imide and Their Interaction with an Arene, Diverse Olefins, and an NTf 2–Based Ionic Liquid. Chem-Eur J 2013, 19, 1042–57.

[68] Dempsey JN, Baenziger NC The Crystal Structure of an Ethylene-Palladium Chloride Complex. J Am Chem Soc 1955, 77, 4984–87.

[69] Molinos E, Brayshaw SK, Kociok-Köhn G, Weller AS Sequential dehydrogenative borylation/ hydrogenation route to polyethyl-substituted, weakly coordinating carborane anions. Organometallics 2007, 26, 2370–82.

[70] Rifat A, Kociok-Köhn G, Steed JW, Weller AS Cationic iridium phosphines partnered with [$closo$-CB$_{11}$H$_6$Br$_6$]$^-$. Organometallics 2004, 23, 428–32.

[71] Howard JAK, Spencer JL, Mason SA Proc Roy Soc London SerA 1983 386,145.

[72] Santiso-Quiñones G, Reisinger A, Slattery J, Krossing I. Homoleptic Cu–phosphorus and Cu–ethene complexes. Chem Commun 2007, 5046.

[73] Fianchini M, Campana CF, Chilukuri B, Cundari TR, Petricek V, Dias HVR Use of [SbF$_6$]$^-$ to Isolate Cationic Copper and Silver Adducts with More than One Ethylene on the Metal Center. Organometallics 2013, 32, 3034–41.

[74] Reisinger A, Trapp N, Knapp C, Himmel D, Breher F, Raegger H, Krossing I Silver-Ethene Complexes [Ag(η^2-C$_2$H$_4$)$_n$][Al(R$_F$)$_4$] with n = 1, 2, 3 (R$_F$ = fluorine-substituted group). Chem-Eur J 2009, 15, 9505–20.

[75] Rasika-Dias HV, Fianchini M, Cundari TR, Campana CF Synthesis and Characterization of the Gold(I) Tris(ethylene) Complex [Au(C$_2$H$_4$)$_3$][SbF$_6$]. Angew Chem 2008, 120, 566–69.

[76] Schaefer J, Himmel D, Krossing I [Au(η^2-C$_2$H$_4$)$_3$]$^+$[Al(OR$_F$)$_4$]$^-$. A stable homoleptic (ethene)gold complex. Eur J Inorg Chem 2013, 2013, 2712–17.

[77] Hooper TN, Green M, Haddow M, McGrady J, Russell C. Synthesis, Structure and Reactivity of Stable Homoleptic Gold(I) Alkene Cations. Chem-Eur J 2009, 15, 12196–200

[78] Nattmann L, Saeb R, Nöthling N, Cornella J An air-stable binary Ni(0)–olefin catalyst. Nature Catal 2020, 3, 6–13.

[79] Nattmann L, Cornella J A Robust 16-Electron Ni(0) Olefin Complex for Catalysis. Organometallics 2020, 39, 3295–300.

[80] Brennessel WW, Ellis JE 2.2.2-Cryptand)potassium tetrakis(η^2-ethylene)cobaltate(–I). Acta Crystallogr, Sect E: Struct Rep Online 2012, 68, m1257–m1258.

[81] Chiririwa H, Meijboom R. Dichlorido(η^4-cycloocta-1,5-diene)bis(propanenitrile-κN)ruthenium (II). Acta Crystallogr, Sect E: Struct Rep Online 2011, 67, m1336–m1336.

[82] Hirano M, Asakawa R, Nagata C, Miyasaka T, Komine N, Komiya S Ligand displacement reaction of Ru(η^4-1,5-COD)(η^6-1,3,5-COT) with Lewis bases. Organometallics 2003, 22, 2378–86.

[83] Vikulova ES, Ilyin IY, Karakovskaya KI, Piryazev DA, Turgambaeva AE, Morozova NB Volatile iridium(I) complexes with β-diketones and cyclooctadiene: syntheses, structures and thermal properties. J Coord Chem 2016, 69, 2281–90.

[84] Gardiner MG, Ho CC, McGuinness DS, Liu YL Air and Moisture Tolerant Synthesis of a Chelated bis(NHC) Methylpalladium(ii) Complex Relevant to Alkyl Migration Processes in Catalysis. Aust J Chem 2020, 73, 1158–64.

12 Compounds with three-electron ligands

12.1 Allylic complexes and related compounds

The η^3-connectivity can be achieved in several ways. The prototype is the allyl group but cyclic olefins can have allylic fragments acting as three-electron ligands (Fig. 12.1).

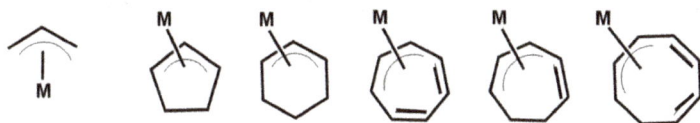

Fig. 12.1: Allylic coordination of various cyclic olefins.

The most important three-electron ligand is the π-allyl group (bonded trihapto), which has an open chain of three sp^2 hybridized carbon atoms, each having a π-electron available for metal–ligand bond formation (Fig. 12.2).

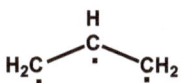

Fig. 12.2: The electronic structure of the π-allyl ligand.

The preparative methods for π-allylic complexes are as follows:
(a) Allyl halides or alcohols react with metal halides or metal carbonyls.
(b) Alkali metal salts of a metal carbonyl anion react with an allyl halide to give a σ-bonded derivative followed by UV irradiation to promote σ–π rearrangement (Fig. 12.3).

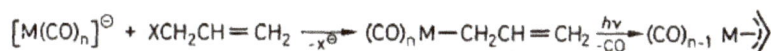

Fig. 12.3: Formation of allyl complexes by σ–π rearrangement.

(c) Allyl Grignard reagents react with metal halides:

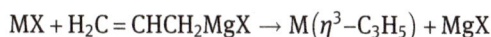

$$MX + H_2C = CHCH_2MgX \rightarrow M(\eta^3\text{–}C_3H_5) + MgX$$

(d) Metal hydrides add to dienes (Fig. 12.4).

https://doi.org/10.1515/9783110695274-013

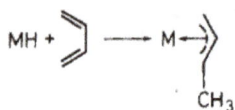

Fig. 12.4: Addition of dienes to metal hydrides.

12.1.1 Homoleptic allyl metal complexes

Homoleptic η^3-allyl metal complexes, that is, containing only allylic groups as ligands, have been described with two, three and four allyl ligands (Fig. 12.5).

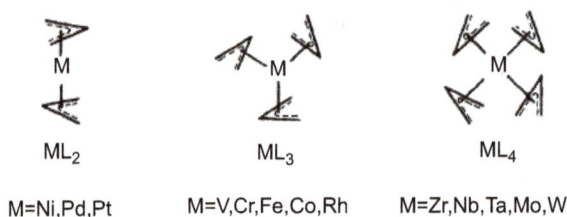

ML$_2$ ML$_3$ ML$_4$

M=Ni,Pd,Pt   M=V,Cr,Fe,Co,Rh   M=Zr,Nb,Ta,Mo,W

Fig. 12.5: Homoleptic η^3-allyl complexes.

Most of the binary metal–allyl complexes are mononuclear compounds (Tab. 12.1). No cluster compounds containing only π-allylic ligands are known.

Tab. 12.1: Binary π-allyl metal complexes.

Group IV	Group V	Group VI	Group VII	Group VIII
**	V(C$_3$H$_5$)$_3$	Cr(C$_3$H$_5$)$_3$ Cr$_2$(C$_3$H$_5$)$_8$	–	Fe(C$_3$H$_5$)$_3$ Co(C$_3$H$_5$)$_3$ Ni(C$_3$H$_5$)$_2$
Zr(C$_3$H$_5$)$_4$	Nb(C$_3$H$_5$)$_3$	Mo(C$_3$H$_5$)$_4$ Mo$_2$(C$_3$H$_5$)$_8$	–	Rh(C$_3$H$_5$)$_3$ Pd(C$_3$H$_5$)$_2$
Hf(C$_3$H$_5$)$_4$	Ta(C$_3$H$_5$)$_4$	W(C$_3$H$_5$)$_4$	Re$_2$(C$_3$H$_5$)$_4$	Ir(C$_3$H$_5$)$_3$ Pt(C$_3$H$_5$)$_2$

Recent examples include bis(allyl) metal complexes [M(Me$_3$Si-C$_3$H$_3$-SiMe$_3$)$_2$] with M = Co, Ni [1] and [M(Me$_3$Si-C$_3$H$_3$-SiMe$_3$)$_2$] M = Y, Tm [2] (Fig. 12.6).

Fig. 12.6: Bis(allyl) metal complexes.

The tris(allyl) iron complex $[Fe(\eta^3\text{-}C_3H_5)_3$ is formed in a Grignard reaction with iron(III) chloride at low temperature (−78 °C):

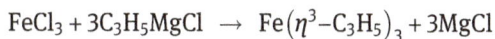

$$FeCl_3 + 3C_3H_5MgCl \;\rightarrow\; Fe(\eta^3\text{--}C_3H_5)_3 + 3MgCl$$

Pentadiene-1,4 forms a bis(η^3-allylic) compound with nickel chloride by reduction with triethylaluminum reacts. The dimer $[Rh(CO)_2Cl]_2$ with allylmagnesium chloride to form tris(η^3-allyl)rhodium, $Rh(\eta^3\text{-}C_3H_5)_3$:

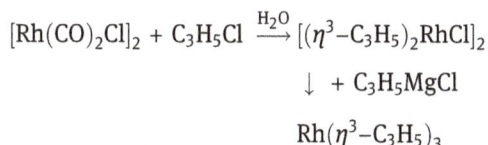

$$[Rh(CO)_2Cl]_2 + C_3H_5Cl \xrightarrow{H_2O} [(\eta^3\text{--}C_3H_5)_2RhCl]_2$$
$$\downarrow\; + C_3H_5MgCl$$
$$Rh(\eta^3\text{--}C_3H_5)_3$$

Tetrakis(allyl) metal complexes include $[Sm(\eta^3\text{-}C_3H_5)_4]$ [3], tetrakis(η^3-phenylpropargyl)-zirconium $[Zr(\eta^3\text{-}PhC_3H_4\text{-})_4]$ [4] and tetrakis(η^3-1-(trimethylsilyl)allyl)-thorium $[Th(Me_3\text{-}Si\text{-}C_3H_4\text{-}SiMe_3)_4]$ (Fig. 12.7) [5]. Thorium and uranium also form binary tetraallyl derivatives, $M(\eta^3\text{-}C_3H_5)_4$, from Grignard reactions.

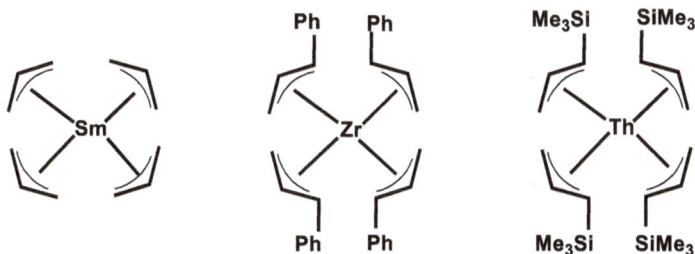

Fig. 12.7: Tetrakis(allyl) metal complexes.

12.1.2 Complexes with bridging allylic ligands

Bridging allylic ligands are present in some palladium complexes and in dichromium and dimolybdenum tetraallyls. Pentadiene-1,4 forms a bis(η^3-allylic) compound with nickel chloride by reduction with triethylaluminum and 1,6-diphenyl hexadiene forms a palladium complex [6] (Fig. 12.8).

12.1.3 Heteroleptic mixed allyl–ligand metal complexes

There is a large number and variety of heteroleptic, mixed ligand allyl metal complexes. A first family includes mixed allyl halide complexes, sometimes also containing additional carbon monoxide (Fig. 12.9).

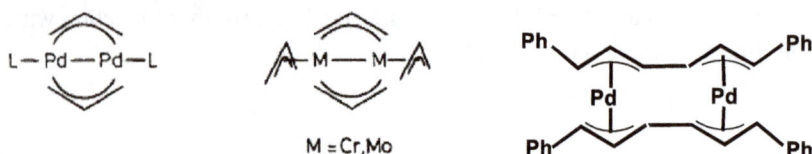

Fig. 12.8: Bridging allylic ligands.

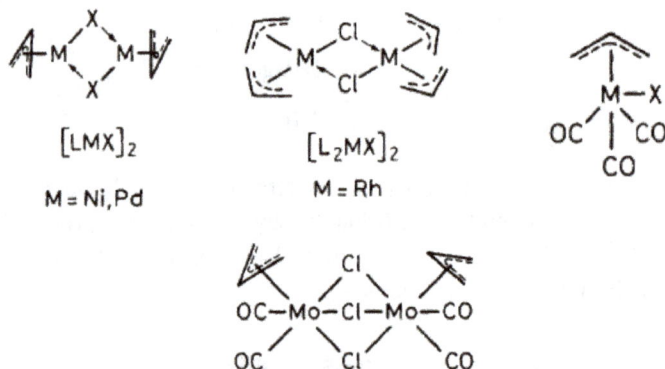

Fig. 12.9: Mixed allyl metal halide complexes.

The rhodium dimer, $[(\eta^3\text{-}C_3H)_2RhCl]_2$, has been obtained from the carbonyl chloride, $[Rh(CO)_2Cl]_2$, with allyl chloride.

Dimeric $[\eta^3\text{-}C_3H_5NiX]_2$ (X = halogen) forms on heating nickel tetracarbonyl with allyl halides which replace completely the carbon monoxide ligands.

Highly active forms of nickel and palladium, prepared by reduction of the anhydrous dihalides with potassium, react with allyl halides to produce $[(\eta^3\text{-}C_3H_5)MX]_2$ (M = Ni, Pd).

Palladium(II) chloride forms monomeric $(\eta^3\text{-}C_3H_5)PdCl(Cyh)$ [7]. Monoolefins react with palladium(II) chloride to form the η^3-allylic dimers, $[(\eta^3\text{-}C_3H_4R)PdCl]_2$. The dimeric compound $[(\eta^3\text{-}C_6H_9)PdCl]_2$ is also formed in the reaction of cyclohexene with $PdCl_2$ in acetic acid or from cyclohexadiene-1,3 with $[Pd(CO)Cl]_2$.

Numerous π-allyl derivatives are known for the metals of this triad. Iron pentacarbonyl forms $(\eta^3\text{-}C_3H_5)Fe(CO)_3X$ (X = halogen) with allyl halides.

Another family of mixed complexes are allyl–phosphine metal compounds with iridium [8], nickel [9] and platinum [10] (Fig. 12.10).

The association of allyl ligands with tetrahydrofuran produces a series of mixed complexes with scandium [11], yttrium [12] and lanthanides [12] (Fig. 12.11).

Allyl cyclopentadienyl metal complexes are numerous. The dichloride, $(\eta^5\text{-}C_5H_5)_2$-$TiCl_2$, reacts with allylmagnesium bromide to give $(\eta^5\text{-}C_5H_5)_2Ti(\eta^3\text{-}C_3H_5)$ [13]. The

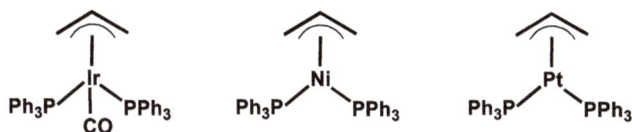

Fig. 12.10: Mixed allyl–phosphine metal complexes.

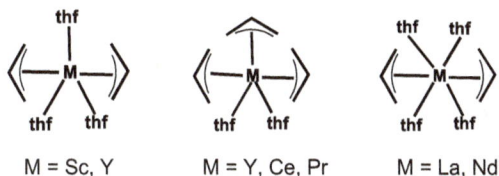

M = Sc, Y   M = Y, Ce, Pr   M = La, Nd

Fig. 12.11: Mixed allyl–tetrahydrofuran metal complexes.

analogous zirconium compound yields a diallyl derivative, $(\eta^5\text{-}C_5H_5)_2Zr(\eta^3\text{-}C_3H_5)_2$. A related binary cobalt complex, $(\eta^5\text{-}C_5Me_5)Co(\eta^3\text{-}C_3H_5)$, is also known [14] (Fig. 12.12).

The 18-electron compound, $Ni(\eta^5\text{-}C_5H_5)(\eta^3\text{-}C_3H_5)$ (Fig. 12.12), can be obtained by treatment of nickelocene with allylmagnesium chloride, treatment of $[(\eta^3\text{-}C_3H_5)NiBr]_2$ with sodium cyclopentadiene, or by treatment of nickel(II) chloride with allylmagnesium chloride and lithium cyclopentadienide:

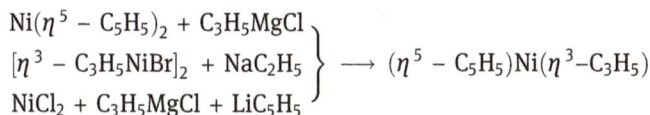

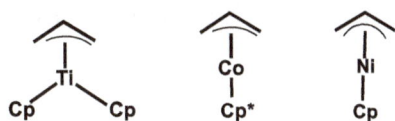

$$\left.\begin{array}{l} Ni(\eta^5-C_5H_5)_2 + C_3H_5MgCl \\[4pt] [\eta^3-C_3H_5NiBr]_2 + NaC_2H_5 \\[4pt] NiCl_2 + C_3H_5MgCl + LiC_5H_5 \end{array}\right\} \longrightarrow (\eta^5-C_5H_5)Ni(\eta^3-C_3H_5)$$

Fig. 12.12: Mixed allyl–cyclopentadienyl metal complexes.

A known scandium derivative is the mixed ligand complex, $[(\eta^5\text{-}C_5H_5)_2Sc(\eta^3\text{-}C_3H_5)]$, prepared from $(\eta^5\text{-}C_5H_5)_2ScCl$ and C_3H_5MgCl. Yttrium and lanthanide derivatives (Fig. 12.13) of the type $[(\eta^5\text{-}C_5H_5)_3Ln(\eta^3\text{-}C_3H_5)]$ [15–119] were also prepared from $((\eta^5\text{-}C_5H_5)_3)LnCl$ and C_3H_5MgCl.

Some cylopentadienyl complexes with two [20] and three [21] allyl ligands are also known with lanthanide central atoms (Fig. 12.14).

A number of substituted allyl ligands also form mixed cyclopentadienyl metal complexes. These include zirconium [22], tantalum [23], ruthenium [24] and palladium [25] complexes with various substituted allyl groups (Fig. 12.15).

M = Y, Nd, Sm, Gd, Tb, Yb, Lu,

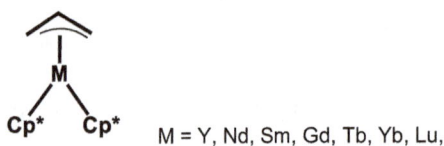

Fig. 12.13: Mixed allyl–cyclopentadienyl lanthanide complexes.

Fig. 12.14: Cyclopentadienyl mixed lanthanide complexes with two and three allyl ligands.

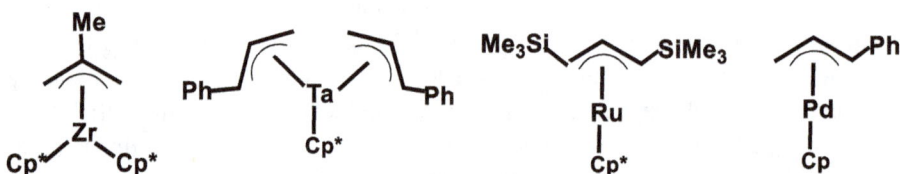

Fig. 12.15: Metal complexes with substituted allyl ligands.

Carbonyl ligands may accompany the coordinated allyl groups in a variety of compositions. Some examples are illustrated here. Thus, η^3-allylvanadium penta-carbonyl, η^3-$C_3H_5V(CO)$, is obtained from allyl chloride and sodium hexacarbonyl-vanadate. Sodium pentacarbonylmanganate reacts with allyl chloride under UV irradiation to form η^3-allymanganese tetracarbonyl:

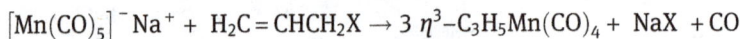

$$[Mn(CO)_5]^- Na^+ + H_2C = CHCH_2X \rightarrow 3\ \eta^3 - C_3H_5Mn(CO)_4 + NaX + CO$$

Manganese pentacarbonyl hydride adds to butadiene to form a η^3-allylic complex (Fig. 12.16).

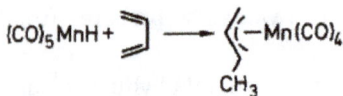

Fig. 12.16: Addition of butadiene to $(CO)_5MnH$.

The dimer $[(\eta^3$-$C_3H_5)Fe(CO)_3]_2$ is formed by elimination of iodine from $(\eta^3$-$C_3H_5)Fe(CO)_3I$ during chromatography over alumina, or in the reaction shown in Fig. 12.17.

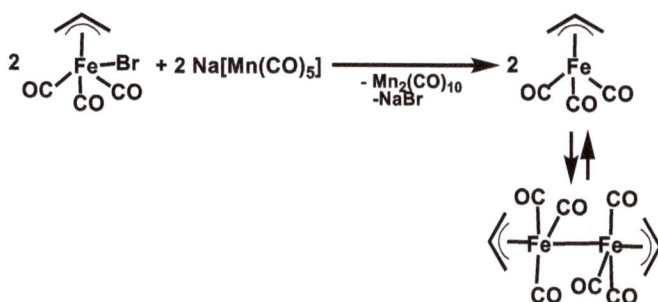

Fig. 12.17: Formation of the [(η³-C₃H₅)Fe(CO)₃]₂ dimer.

The reaction of Na[Co(CO)₄] with allyl halides produces an unstable η³-allyl derivative which readily isomerizes to the η³-allylic (η³-C₃H₅)Co(CO)₃, The η³-methylallyl, (η³-MeC₃H₄)Co(CO)₃, is formed by addition of the hydride HCo(CO)₄ to butadiene. The cation, [(η³-C₅H₅)Co(CO)(η³-C₃H₅)]⁺ is formed by reacting (η⁵-C₅H₅)Co(CO)₂ and an allyl halide. A binuclear compound 2,3-dimethylenebuta-1,4-diyl)-hexacarbonyl-dicobalt is also known [26] (Fig. 12.18).

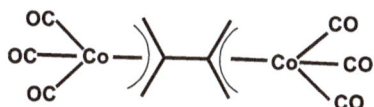

Fig. 12.18: A binuclear cobalt complex.

Many other mixed allyl metal carbonyl complexes of various compositions are known and some [27, 28] are illustrated in Fig. 12.19.

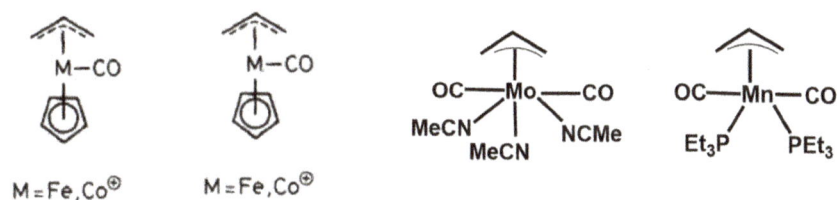

Fig. 12.19: More allyl metal complexes.

Finally, a rare uranium complex with cyclooctatetraene and 2-methylprop-1-en-3-yl [29] is mentioned (Fig. 12.20).

Fig. 12.20: A uranium complex.

12.1.4 Allylic fragments in cyclic polyolefins

Allyl groups can be fragments of various organic ligands, like η^3-cyclopentadienyl, η^3-cyclohexenyl, η^3-cycloheptatrienyl or, η^3-cycloheptadienyl groups, which can act as π-allylic ligands to form allylmetal complexes. Selected examples are presented here.

Chromocene is reduced by a mixture of hydrogen and carbon monoxide to a complex in which one of the five-membered rings becomes a η^3-ligand (Fig. 12.21).

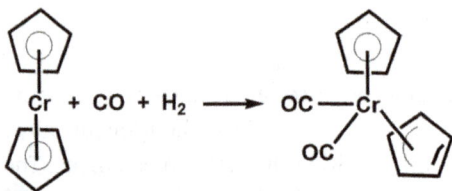

Fig. 12.21: Cyclopentene as allylic ligand in a chromium complex.

The tungsten complex, $(\eta^5\text{-}C_5H_5)_2W(CO)_2$, contains a bent trihapto-cyclopentadienyl ligand. The 18-electron rule requires one of the C_5H_5 groups to be bonded as an η^3-allylic fragment) (Fig. 12.22).

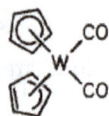

Fig. 12.22: Allylic η^3-coordination of a cyclopentadiene molecule.

The η^5-C_5H_5Ni group requires only three electrons to achieve a noble-gas configuration; consequently, it forms η^3-allylic complexes, as in the η^5-cyclopentadienyl-η^3-cyclopentenylnickel complex, Ni(η^3-C_5H_5)(η^3-C_5H_7) (Fig. 12.23).

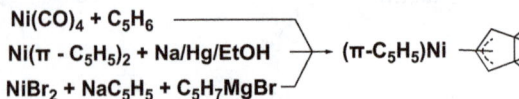

Ni(CO)$_4$ + C$_5$H$_6$

Ni(π - C$_5$H$_5$)$_2$ + Na/Hg/EtOH → (π-C$_5$H$_5$)Ni

NiBr$_2$ + NaC$_5$H$_5$ + C$_5$H$_7$MgBr

Fig. 12.23: Preparation of a nickel η^3-coordinated cyclopentadiene.

A compound of nickel with allylic coordination of the central unit is formed by cyclopentenyl [30] (Fig. 12.24).

Tris(η^3-cyclopentadienylmetal) compounds are rare but known for tin(II) and lead(II) as dinegative anions $[M(\eta^3\text{-}C_5H_5)_3]^{2-}$ [31] (Fig. 13.25).

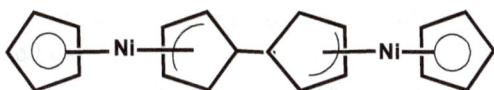

Fig. 12.24: Dinickel complex of cyclopentenyl.

Some six-membered rings can also act as η^3-ligands. Cyclohexa-1,3-diene reacts with molybdenum hexacarbonyl to form a complex in which one of the rings is a five-electron and

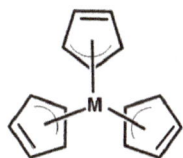

M = Sn, Pb   Fig. 12.25: Tris(η^3-cyclopentadienylmetal) compounds.

the other is a three-electron donor (Fig. 12.26). Cyclohexene forms monocyclic chromium [32], molybdenum [33] and ruthenium [34] complexes with η^3-connectivity (Fig. 12.26).

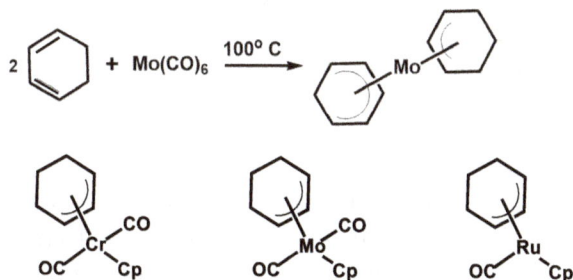

Fig. 12.26: Six-membered rings as η^3-ligands.

Seven-membered rings: As the (η^5-C$_5$H$_t$)Fe(CO) group requires only a three-electron ligand, a cycloheptatrienyl group attaches to it as a *trihapto* ligand. The dinuclear iron carbonyl, Fe$_2$(CO)$_9$, reacts with cycloheptatriene-1,3,5 to yield a product in which a dimetallic, Fe$_2$(CO)$_6$ unit is attached to the ring through two η^5-allylic fragments (Fig. 12.27).

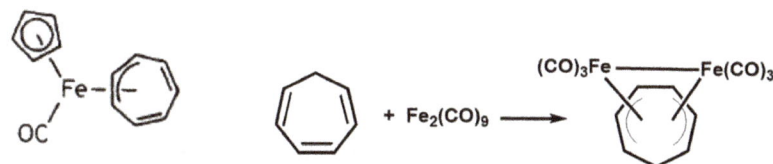

Fig. 12.27: Cycloheptatriene coordination as η^3-ligand in iron complexes.

Since the $Co(CO)_3$ group requires only three electrons to acquire a noble gas configuration in the cycloheptatriene complex, η^7-$C_7H_7Co(CO)_3$, prepared from $Co_2(CO)_8$ and cycloheptatriene under UV irradiation, the cyclic ligand must be bonded as a η^3-allylic fragment leaving a butadiene fragment in the ring not involved in coordination. Complexes with cycloheptatriene connected in η^3-allylic mode are also known for ruthenium, osmium [35], palladium [36], platinum [37] and palladium [38, 39] (Fig. 12.28).

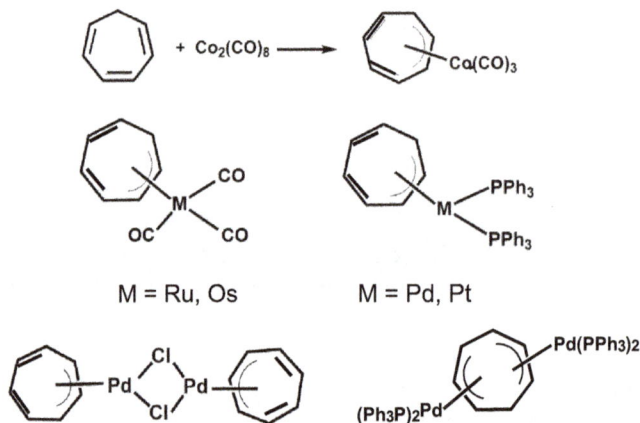

Fig. 12.28: Cycloheptatriene as η^3-ligand in various complexes.

Eight-membered rings. Niobium forms a mixed ligand complex, $(\eta^5$-$C_5H_5)_2Nb(\eta^3$-$C_8H_9)$, in which a cyclooctatrienyl group is bonded tri*hapto* to satisfy the 18-electron rule. Similar complexes are known for molybdenum [40] and palladium [41, 42] (Fig. 12.29).

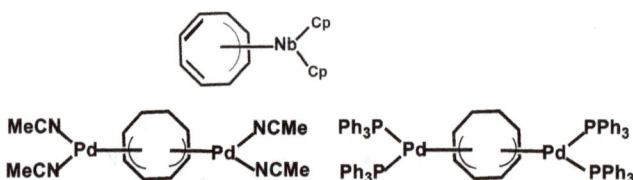

Fig. 12.29: Cyclooctatriene as η^3-ligand.

12.2 Cyclopropenyl complexes

The unsaturated three-membered cyclopropenyl ring can act as a three-electron ligand in transition metal complexes (Fig. 12.30). The starting materials are the trialkyl(aryl)-cyclopropenium salts, $[C_3R_3]^+X^-$.

Fig. 12.30: Cyclopropenyl as η^3-ligand.

Metal carbonyl or other anionic complexes react with cyclopropenium salts to give products other than η^3-cyclopropenyl complexes. Thus, triphenylcyclopropenium cation form a salt with the hexacarbonylvanadate anion $[(\eta^3\text{-}C_3Ph_3)V(CO)]^-$ which is converted by UV irradiation to a cyclopropenium complex (Fig. 12.31).

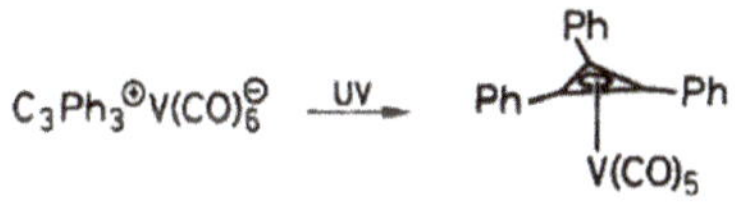

$$C_3Ph_3^{\oplus}V(CO)_6^{\ominus} \xrightarrow{\text{UV}} $$

Fig. 12.31: Vanadium complex of triphenylcyclopropenium.

A molybdenum-carbonyl derivative containing the triphenylcyclopropenyl ligand has been obtained from the chloride (Fig. 12.32).

$$Mo(CO)_3(MeCN)_3 + C_3Ph_3^+Cl^- \longrightarrow MoCl(MeCN)_2(CO)_2(\eta^3\text{-}C_3Ph_3)$$

$$\downarrow + LiC_5H_5$$

$$(\eta^5\text{-}C_5H_5)Mo(CO)_2(\eta^3\text{-}C_3Ph_3)$$

Fig. 12.32: Synthesis of a molybdenum complex.

Dicobalt octacarbonyl also reacts with a triphenylcyclopropenium salt (Fig. 12.33).

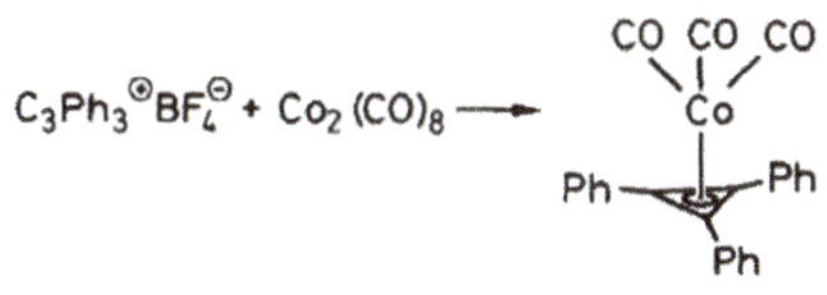

$$C_3Ph_3^{\oplus}BF_4^{\ominus} + Co_2(CO)_8 \longrightarrow$$

Fig. 12.33: Cobalt complex of triphenylcyclopropenium.

The dimeric nickel compound satisfies the 18-electron rule (Fig. 12.34).

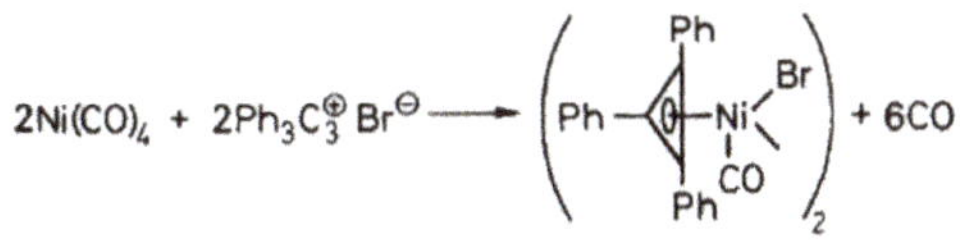

$$2Ni(CO)_4 + 2Ph_3C_3^{\oplus}Br^{\ominus} \longrightarrow$$

$$+ 6CO$$

Fig. 12.34: Dimeric nickel compound.

The η^3-bonding is found in $[Ni(\eta^3\text{-}C_3Ph_3)Cl(py)_2] \cdot py$ and $(\eta^5\text{-}C_5H_5)Ni(\eta^3\text{-}C_3Ph_3)$. The latter is prepared as shown in Fig. 12.35.

$$Ni(C_3Ph_3)Br(py)_2 \cdot py + TlC_5H_5 \xrightarrow{C_6H_6}$$

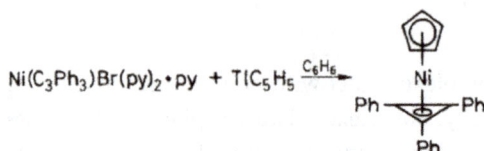

Fig. 12.35: Preparation of a mononuclear nickel triphenylcyclopropenyl complex.

Other nickel derivatives containing the trialkylcyclopropenyl ligand are similarly prepared by ligand replacement (Fig. 12.36).

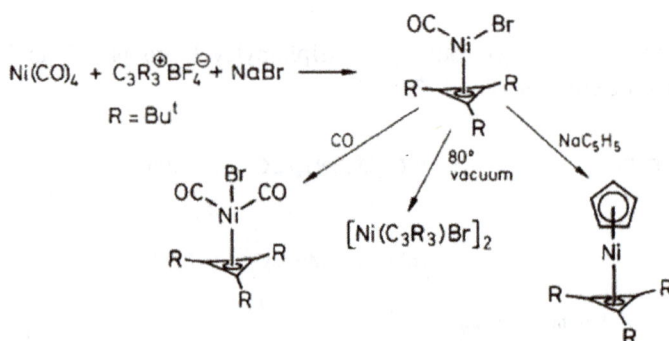

Fig. 12.36: Replacement reactions of nickel tri-*tert*-butylcyclopropenyl complexes.

Triarylcyclopropenium salts behave similarly with zero-valent platinum and palladium compounds.

References

[1] Schormann M, Garratt S, Bochmann M. Reactivity of Silyl-Substituted Allyl Compounds with Group 4, 5, 9, and 10 Metals: Routes to η^3-Allyls, Alkylidenes, and *sec*-Alkyl Carbocations. Organometallics 2005, 24, 1718–24.

[2] White RE, Hanusa TP, Kucera BE. Compositional variations in monomeric trimethylsilylated allyl lanthanide complexes. J Organomet Chem 2007, 692, 3479–85.

[3] Sanchez-Barba LF, Hughes DL, Humphrey SM, Bochmann M. Synthesis and Structures of New Mixed-Metal Lanthanide/Magnesium Allyl Complexes. Organometallics 2005, 24, 5329–34.

[4] Denomme DR, Dumbris SM, Hyatt IFD, Abboud KA, Ghiviriga I, McElwee-White L. Synthesis and Electronic Structure of Tetrakis(η^3-phenylpropargyl)zirconium. Organometallics 2010, 29, 5252–56.

[5]   Carlson CN, Hanusa TP, Brennessel WW. Metal Allyl Complexes with Bulky Ligands: Stabilization of Homoleptic Thorium Compounds, [(SiMe$_3$)$_n$C$_3$H$_{5-n}$]$_4$Th (n = 1, 2). J Am Chem Soc 2004, 126, 10550–51.

[6]   Murahashi T, Nakashima H, Nagai T, Mino Y, Okuno T, Jalil MA, Kurosawa H. Stereoretentive Elimination and Trans-olefination of the Dicationic Dipalladium Moiety [Pd$_2$Ln]$^{2+}$ Bound on 1,3,5-Trienes. J Am Chem Soc 2006, 128, 4377–88.

[7]   Faller JW, Sarantopoulos N. Retention of Configuration and Regiochemistry in Allylic Alkylations via the Memory Effect. Organometallics 2004, 23, 2179–85.

[8]   Godard C, Duckett SB, Henr C, Polas YCS, Toose R, Whitwood AC. Whitwood (η^3-Allyl)-carbonyl-bis(triphenylphosphine)-iridium. Chem Commun 2004, 1826–28.

[9]   Jiménez-Tenorio M, Carmen Puerta M, Salcedo I, Valerga P, De Los Ríos I, Mereiter K. Oligomerization of styrenes mediated by cationic allyl nickel complexes containing triphenylstibine or triphenylarsine. Dalton Trans 2009, 1842–52.

[10]   Crociani B, Benetollo F, Bertani R, Bombieri G, Meneghetti F, Zanotto L. Convenient synthesis of cationic allylplatinum(II) complexes with tertiary phosphines by oxidative allyl transfer from ammonium cations to platinum(0) substrates. Crystal and molecular structures of η^3-propenyl- and η^3-2-methylpropenyl-bis(triphenylphosphine. J Organomet Chem 2000, 605, 28–38.

[11]   Standfuss S, Abinet E, Spaniol TP, Okuda J. Allyl complexes of scandium: Synthesis and structure of neutral, cationic and anionic derivatives. Chem Commun 2011, 47, 11441–43.

[12]   Robert D, Abinet E, Spaniol TP, Okuda J. Cationic Allyl Complexes of the Rare-Earth Metals: Synthesis, Structural Characterization, and 1,3-Butadiene Polymerization Catalysis. Chem-Eur J 2009, 15, 11937–47.

[13]   Chen J, Kai Y, Kasai N, Yasuda H, Yamamoto H, Nakamura A. Steric effect of allyl substituent on the molecular structures of allyltitanium complexes. J Organomet Chem 1991, 407, 191–205.

[14]   Nehl H. Darstellung und Reaktionen von (η^3-Allyl)(η^5-pentamethyl-cyclopentadienyl)-cobalt(II) und seinen Derivaten. Chem Ber 1993, 126, 1519–27.

[15]   Evans WJ, Kozimor SA, Brady JC, Davis BL, Nyce GW, Seibel CA. et al., Metallocene Allyl Reactivity in the Presence of Alkenes Tethered to Cyclopentadienyl Ligands. Organometallics 2005, 24, 2269–78.

[16]   Woen DH, White JRK, Ziller JW, Evans WJ. Mechanochemical C–H bond activation: Synthesis of the tuckover hydrides, (C$_5$Me$_5$)$_2$Ln(μ-H)(μ-η^1:η^5-CH$_2$C$_5$Me$_4$)Ln(C$_5$Me$_5$) from solvent-free reactions of (C$_5$Me$_5$)$_2$Ln(μ-Ph)$_2$BPh$_2$ with K(C$_5$Me$_5$). J Organomet Chem 2019, 899, 120885.

[17]   Demir S, Zadrozny JM, Nippe M, Long JR. Exchange coupling and magnetic blocking in bipyrimidyl radical-bridged dilanthanide complexes. J Am Chem Soc 2012, 18546–49.

[18]   Evans WJ, Ulibarri TA, Ziller JW. Rectivity of (C$_5$Me$_5$)$_2$Sm and related species with alkenes: Synthesis and structural characterization of a series of organosamarium allyl complexes. J Am Chem Soc 1990, 112, 2314–24.

[19]   Göttker-Schnetmann I, Kenyon P, Mecking S. Coordinative chain transfer polymerization of butadiene with functionalized aluminum reagents. Angew Chem Int Ed 2019, 58, 17777–81.

[20]   Yu N, Nishiura M, Li X, Xi Z, Hou Z. Cationic Scandium Allyl Complexes Bearing Mono (cyclopentadienyl) Ligands: Synthesis, Novel Structural Variety, and Olefin-Polymerization Catalysis. Chem-Asian J 2008, 3, 1406–14.

[21]   Taube R, Maiwald S, Sieler J. XLVII. Darstellung und Charakterisierung einiger anionischer Allylneodym(III)-Komplexe als Katalysatoren für die stereospezifische Butadienpolymerisation: Li[Nd(η^3-C$_3$H$_5$)$_4$]·1,5Dioxan, Li[Nd(π-C$_5$H5)(η^3-C$_3$H$_5$)$_3$]·2Dioxan und Li [Nd(η^5-C$_5$Me$_5$)(η^3-C$_3$H$_5$)$_3$]·3D. J Organomet Chem 1996, 513, 37–47.

[22] Herberich GE, Kreuder C, Englert U. Bis(pentamethyl-cyclopentadienyl)(η^3-trimethylenemethane)zirconium: A New Mode of Bonding for the Trimethylenemethane Ligand. Angew Chem Int Ed English 1995, 33, 2465–66.

[23] Mashima K, Yamanaka Y, Gohro Y, Nakamura A. Preparation, characterization, and reactions of 16-electron Ta(η^5-C$_5$Me$_5$)(η^3-1-phenylallyl)$_2$. J Organomet Chem 1993, 459, 131–38.

[24] Peters M, Bannenberg T, Bockfeld D, Tamm M. Pentamethylcyclopentadienyl ruthenium "pogo stick" complexes with nitrogen donor ligands. Dalton Trans 2019, 48, 4228–38.

[25] Norton DM, Mitchell EA, Botros NR, Jessop PG, Baird MC. A superior precursor for palladium (0)-based cross-coupling and other catalytic reactions. J Org Chem 2009, 74, 6674–80.

[26] Aguirre-Etcheverry P, Ashley AE, Balázs G, Green JC, Cowley AR, Thompson AL, O'Hare D. Synthesis, structure, and ligand exchange reactions of tetramethyleneethane complexes of cobalt. Organometallics 2010, 29, 5847–58.

[27] Honzíček J, Kratochvíl P, Vinklárek J, Eisner A, Padělková Z. Effect of the substitution on the protonation of allyl cyclopentadienyl molybdenum(II) compounds. Organometallics 2012, 31, 2193–202.

[28] Kerscher T, Mihan S, Beck W. Addition of organometallic nucleophiles (carbonylmetallates) to the allyl ligand of [(η^3-C$_3$H$_5$)Pd(μ-Cl)]$_2$. Synthesis and structure of [(η^3-C$_3$H$_5$)Mn(CO)$_3$PEt$_3$. Z Naturforsch B 2011, 66, 861–64.

[29] Webster CL, Ziller JW, Evans WJ. Reactivity of U3+ Metallocene allyl complexes leads to a nanometer-sized uranium carbonate, [(C$_5$Me$_5$)$_2$U]$_6$(μ-$\kappa^{1:2}$-CO$_3$)$_6$. Organometallics 2013, 32, 4820–27.

[30] Pasynkiewicz S, Pietrzykowski A, Bukowska L, Słupecki K, Jerzykiewicz LB, Urbańczyk-Lipkowska Z. Reactions of nickelocene with diphenylmethyl- and triphenylmethyllithium. J Organomet Chem 2000, 604, 241–47.

[31] Armstrong DR, Duer MJ, Davidson MG, Moncrieff D, Russell CA, Stourton C, Steiner A, Stalke D, Wrigth DS. "Paddle-wheel" tris(cyclopentadienyl)tin (II) and -lead (II) complexes: Syntheses, structures, and model MO calculations. Organometallics 1997, 16, 3340–51.

[32] Norman DW, Ferguson MJ, Stryker JM. General synthesis of cyclopentadienylchromium(II) η^3-allyl dicarbonyl complexes. Organometallics 2004, 23, 2015–19.

[33] Faller JW, Murray HH, White DL, Chao KH. Stereoselective syntheses of some cyclohexene derivatives using complexes of molybdenum. Organometallics 1983, 2, 400–09.

[34] Crocker M, Green M, Morton CE, Nagle KR, Orpen AG. Reactions of co-ordinated ligands. Part 34. Synthesis, structure, and reactivity of cationic dieneruthenium complexes; crystal structures of [Ru(η^4-C$_6$H$_8$)(CO)(η-C$_5$H$_5$)][BF$_4$] and [Ru(η^3-C$_6$H$_9$)(CO)(η-C$_5$H$_5$)]. J Chem Soc, Dalton Trans 1985, 2145–53.

[35] Astley ST, Takats J, Huffman JC, Streib WE. Solid-state structure and fluxional solution behavior of the ambient organometallic nucleophiles (η^3-C$_7$H$_7$)MCO)$_3$ (M = Ru, Os). Organometallics 1990, 9, 184–89.

[36] Murahashi T, Usui K, Tachibana Y, Kimura S, Ogoshi S. Selective construction of Pd$_2$Pt and PdPt$_2$ triangles in a sandwich framework: carbocyclic ligands as scaffolds for a mixed-metal system. Chem-Eur J 2012, 18, 8886–90.

[37] Murahashi T, Usui K, Inoue R, Ogoshi S, Kurosawa H. Metallocenoids of platinum: Syntheses and structures of triangular triplatinum sandwich complexes of cycloheptatrienyl. Chemical Science 2011, 2, 117–22.

[38] Rosset J-M, Glenn MP, Cotton JD, Willis AC, Kennard CHL, Byriel KA, Riches BH, Kitching W. η^3-Allylpalladium Complexes from Medium-Ring Cycloalkenes. Organometallics 1998, 17, 1968–83.

[39] Yamamoto K, Teramoto M, Usui K, Murahashi T. Anti dinuclear adducts of cycloheptatriene and cycloheptatrienyl ligands: Anti-[Pd$_2$(μ-C$_7$H$_8$)(PPh$_3$)$_4$][BF$_4$]$_2$ and anti-[M$_2$(μ-C$_7$H$_7$)(PPh$_3$)$_4$] [BF$_4$] (M = Pd, Pt). J Organomet Chem 2015, 784, 97–102.

[40] Spencer DM, Beddoes RL, Dissanayake RK, Helliwell M, Whiteley MW. Synthesis and reactions of the cyclooctadienyl-molybdenum complexes [MoBr(CO)$_2$(NCMe)$_2$(η^3-C$_8$H$_{11}$)] (η^3-C$_8$H$_{11}$ = 1–3-$\eta^{4,5}$-C$_8$H$_{11}$ or 1–3-$\eta^{5,6}$-C$_8$H$_{11}$). J Chem Soc Dalton Trans 2002, 1009–19.

[41] Murahashi T, Kato N, Ogoshi S, Kurosawa H. Synthesis and structure of dipalladium complexes containing cyclooctatetraene and bicyclooctatrienyl ligands. J Organomet Chem 2008, 693, 894–98.

[42] Murahashi T, Kimura S, Takase K, Uemura T, Ogoshi S, Yamamoto K. Bis-cyclooctatetraene tripalladium sandwich complexes. Chem Commun 2014, 50, 820–22.

13 Compounds with four-electron ligands

Cyclic polyolefins having two or more double bonds can participate in tetra-hapto-bonding (η^4-connectivity). The four carbon atoms of an sp^2-hybridized butadiene fragment must be coplanar. Virtually all conjugated cyclic dienes are able to form η^4-complexes, contributing the four π-electrons of their butadiene fragment. Some seven- and eight-membered polyenes can behave in a similar manner when the metal atom requires four electrons (or a multiple of four) to achieve a noble gas configuration. Some typical four-electron ligands and their attachment to the metal atom are shown in Fig. 13.1.

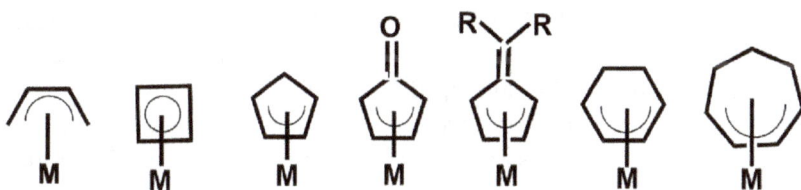

Fig. 13.1: Various ligands capable of tetrahapto-bonding.

13.1 Butadiene complexes and related compounds

Butadiene reacts as a chelating four-electron donor.

Vanadium forms a butadiene complex photochemically from cyclopentadienyl-vanadium tetracarbonyl, $(\eta^5\text{-}C_5H_5)V(CO)_4$, and butadiene by replacement of two carbonyl molecules to yield $(\eta^5\text{-}C_5H_5)V(CO)_2(\eta^4\text{-}C_4H_6)$.

Irradiation of cyclopentadienylvanadium tetracarbonyl with cyclohexadiene-1,3 yields $(\eta^5\text{-}C_5H_5)V(CO)_2(\eta^4\text{-}C_6H_8)$.

The unstable chromium complex, $(\eta^4\text{-}C_4H_6)Cr(CO)_4$, is obtained by reacting chromium vapor with butadiene and carbon monoxide (Fig. 13.2).

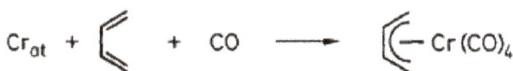

Fig. 13.2: Formation of a cobalt-butadiene complex.

Molybdenum hexacarbonyl reacts with butadiene by replacement of two carbonyl molecules to form $(\eta^4\text{-}C_4H_6)_2Mo(CO)_4$ and $(\eta^4\text{-}C_4H_6)Mo(CO)_2$. The metal-atom synthesis affords tris(η^4-butadiene) complexes of molybdenum and tungsten [1] (Fig. 13.3).

https://doi.org/10.1515/9783110695274-014

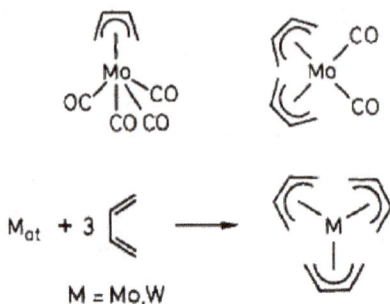

M_{at} + 3 [butadiene] ⟶ [complex]

M = Mo,W

Fig. 13.3: Molybdenum and tungsten complexes of butadiene.

Chromium, molybdenum and tungsten form the cyclobutadiene complexes $(\eta^4\text{-}C_4H_4)M(CO)_4$ by reaction of 1,2-dichlorocyclobutenes with the metal hexacarbonyls in the presence of sodium amalgam. Molybdenum hexacarbonyl reacts similarly to form $(\eta^4\text{-}C_4H_4)Mo(CO)_4$ and $(\eta^4\text{-}C_4H_4)_2Mo(CO)_2$.

Cyclopentadienylmanganese tricarbonyl reacts with butadiene photochemically to replace two carbon monoxide groups and form $[(\eta^5\text{-}C_5H_5)Mn(\eta^4\text{-}C_4H_6)(CO)]$ (Fig. 13.4).

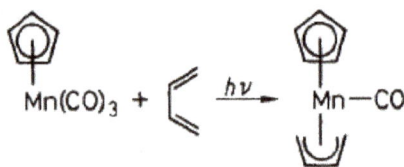

$Mn(CO)_3$ + [butadiene] $\xrightarrow{h\nu}$ Mn—CO

Fig. 13.4: Mixed manganese-butadiene cyclopentadienyl complex.

The electronic requirements of the $Fe(CO)_3$ group favor butadiene complexes, as in butadiene–iron tricarbonyl $\eta^4\text{-}C_4H_6Fe(CO)_3$ (Fig. 13.5). This was the first transition metal complex of butadiene, prepared in 1930 by the reaction of butadiene and iron pentacarbonyl. Cyclopentadienyliron dicarbonyl bromide, $(\eta^5\text{-}C_5H_5)Fe(CO)_2Br$, reacts with butadiene in the presence of aluminum bromide to form a butadiene-containing cation (Fig. 13.5).

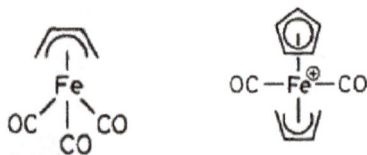

Fig. 13.5: Iron-butadiene complexes.

α,α'-Dibromoxylene reacts with $Fe_2(CO)_9$ to form a butadiene fragment following bromide elimination and π-electron redistribution (Fig. 13.6). Irradiation of the product in the presence of iron pentacarbonyl results in complex formation with the second butadiene fragment of the molecule to yield two isomers (*syn* and *anti*) (Fig. 13.6).

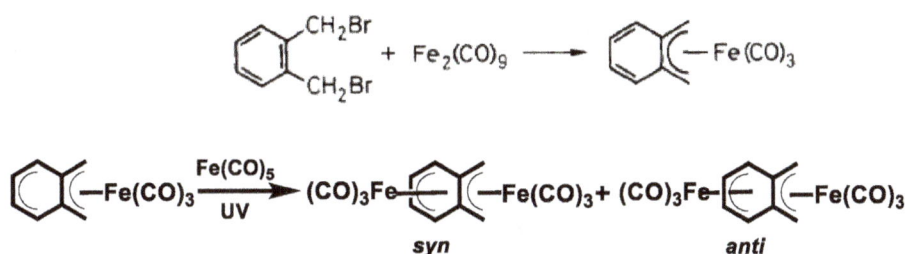

Fig. 13.6: Iron complexes containing butadiene moieties derived from a benzene ring compound.

Vinylbenzene and *para*-divinylbenzene form butadiene complexes with iron carbonyls by perturbing the aromatic conjugation in the six-membered ring on association of a ring double bond with the exocyclic vinyl group (Fig. 13.7).

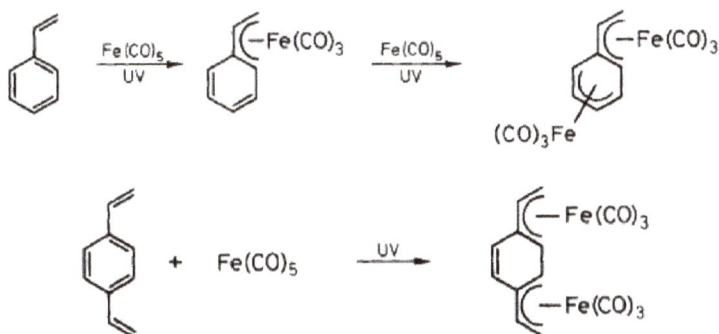

Fig. 13.7: Iron complexes derived from divinylbenzene.

Organic molecules which initially do not contain a butadiene fragment can react with iron carbonyls to form butadiene complexes as a result of rearrangement. Thus, tetramethylallene reacts with $Fe_2(CO)_9$ to form a trimethylbutadiene complex in addition to the expected tetramethylallene derivative (Fig. 13.8).

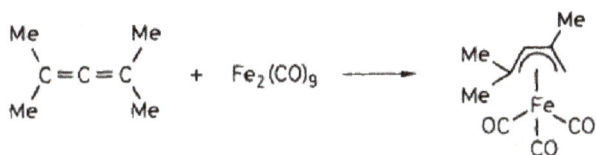

Fig. 13.8: Conversion of tetramethylallene into trbutylbutadiene by complexation.

Cobaltocene reacts with butadiene to form the complex $[(\eta^5-C_5H_5)Co(\eta^4-C_4H_6)]$, and dicobalt octacarbonyl forms the bimetallic complexes $[Co_2(CO)_6(\eta^4-C_4H_6)]$ and $[Co_2(CO)_4(\eta^4-C_4H_6)_2]$ (Fig. 13.9).

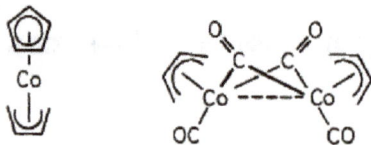

Fig. 13.9: Cobalt-butadiene complexes.

The reaction of rhodium(III) chloride with butadiene yields the compound $(\eta^4-C_4H_6)_2$-RhCl, and an analogous iridium complex is formed in the reaction of $Na_2[IrCl_4]$ with butadiene. The compound $(\eta^5-C_5H_5)Ir(\eta^4-C_4H_6)$ has been prepared by treating $(\eta^4-C_4H_6)_2IrCl$ with TlC_5H_5 (Fig. 13.10).

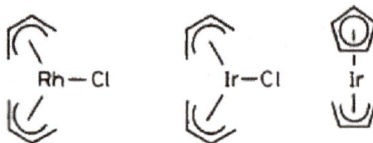

Fig. 13.10: Rhodium- and iridium-butadiene complexes.

Uranium forms a complex with 1,4-diphenylbutadiene [2] (Fig. 13.11).

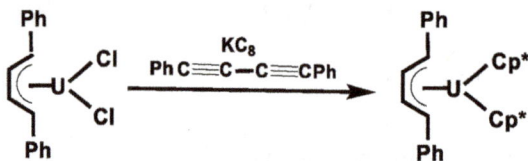

Fig. 13.11: A rare uranium-butadiene complex.

13.2 Cyclobutadiene complexes

Cyclobutadiene itself cannot be isolated but coordination to a transition metal stabilizes many derivatives. Because of the nonexistence of the free ligand, reactions in which the cyclobutadiene ligand is formed simultaneously with the complex must be used for the synthesis of cyclobutadiene complexes. The precursors are usually compounds containing a four-membered ring (e.g., 1,2-dichlorocyclobutene and photo-α-pyrone). The most convenient procedures start from acetylenes, since these readily available reagents can often be converted into cyclobutadiene complexes by reaction with transition metal carbonyls or halides.

Of the group 14 element triad, only the titanium cyclobutadiene complex is known. Titanium(III) chloride with cyclooctatetraene and diphenylacetylene in the presence of *iso*-PrMgCl gives a mixed ligand complex containing tetraphenylcyclobutadiene (Fig. 13.12).

$$TiCl_3 + C_8H_8 + C_2R_2 \xrightarrow{\text{iso-PrMgCl}}$$

R = Ph

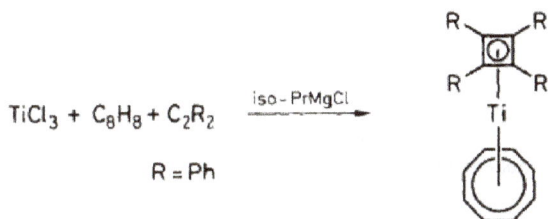

Fig. 13.12: Mixed titanium-cyclobutadiene cyclooctatetraene complex.

Vanadium and niobium form η^4-tetraphenylcyclobutadiene complexes starting from diphenylacetylene (R = Ph) (Fig. 13.13).

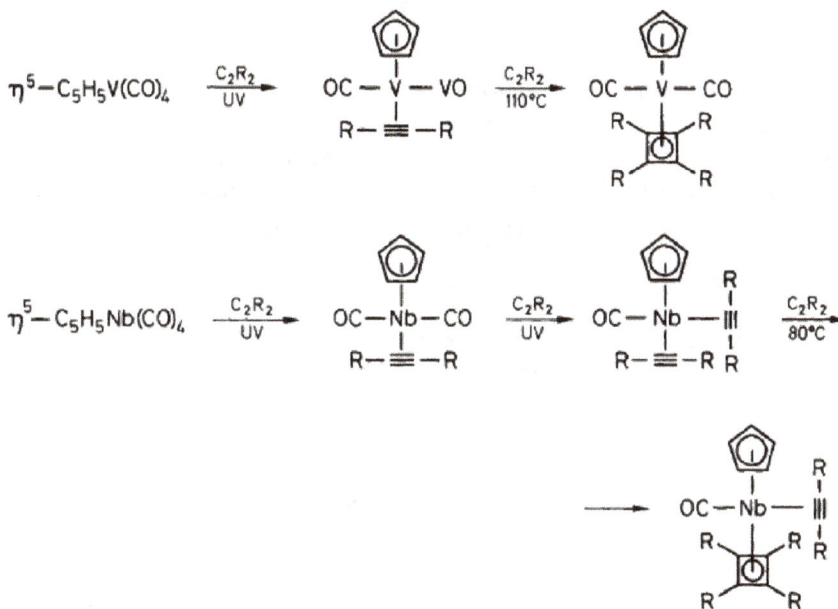

Fig. 13.13: Vanadium- and niobium-cyclobutadiene complexes.

Chromium, molybdenum and tungsten form the cyclobutadiene complexes, (η^4-C$_4$R$_4$) M(CO)$_4$, by reaction of 1,2-dichlorocyclobutenes with the metal hexacarbonylsand sodium amalgam.

A bis(η^4-tetraphenylcyclobutadiene) complex ofmolybdenum has been obtained from Mo(CO)$_6$ and diphenylacetylene (Fig. 13.14).

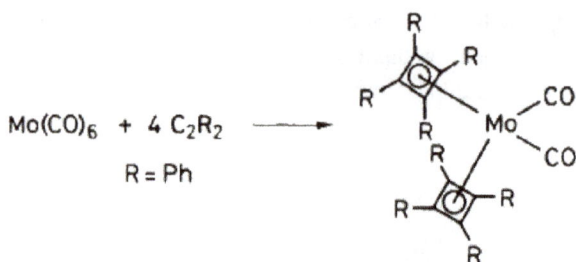

Fig. 13.14: Molybdenum complex with tetraphenylcyclobutadiene.

The first complex of unsubstituted cyclobutadiene was $[(\eta^4\text{-}C_4H_4)Fe(CO)_3$ pre-pared in 1965 by the action of 1,2-dichlorocyclobutene-3,4 on $Fe_2(CO)_9$ (Fig. 13.15).

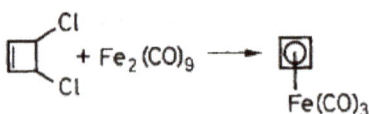

Fig. 13.15: Preparation of the cyclobutadiene iron complex $[(\eta^4\text{-}C_4H_4)Fe(CO)_3$.

In $[(\eta^4\text{-}C_4H_4)Fe(CO)_3]$, the planar four-membered ring is coordinated to iron. Many other cyclobutadiene complexes have now been prepared. This compound can also be prepared via photochemical transformation of α-pyrone, followed by coordina-tion to iron and elimination of carbon dioxide. The procedure has been extended to vanadium-, cobalt- and rhodium-cyclobutadiene complexes (Fig. 13.16).

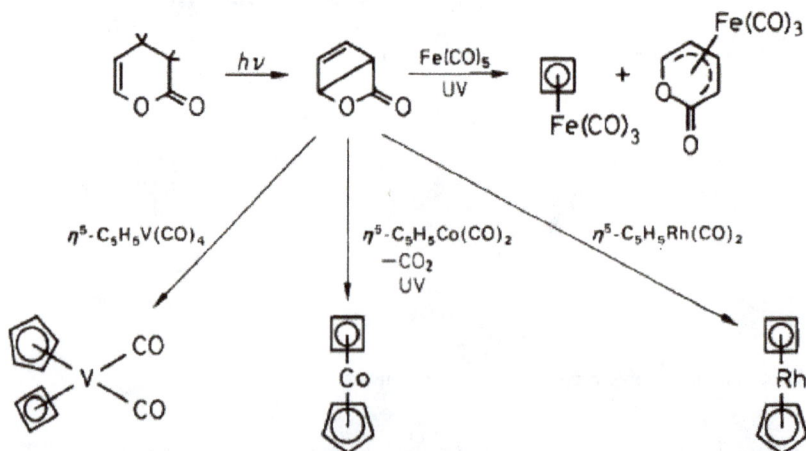

Fig. 13.16: Formation of cyclobutadiene complexes from α-pyrone.

Friedel–Crafts acetylation, aminomethylation, mercuration and other metallations, aromatic substitution reactions of the iron-coordinated cyclobutadiene ring
of η^4-$C_4H_4Fe(CO)_3$ reflect the aromatic character of this complex (Fig. 13.17).

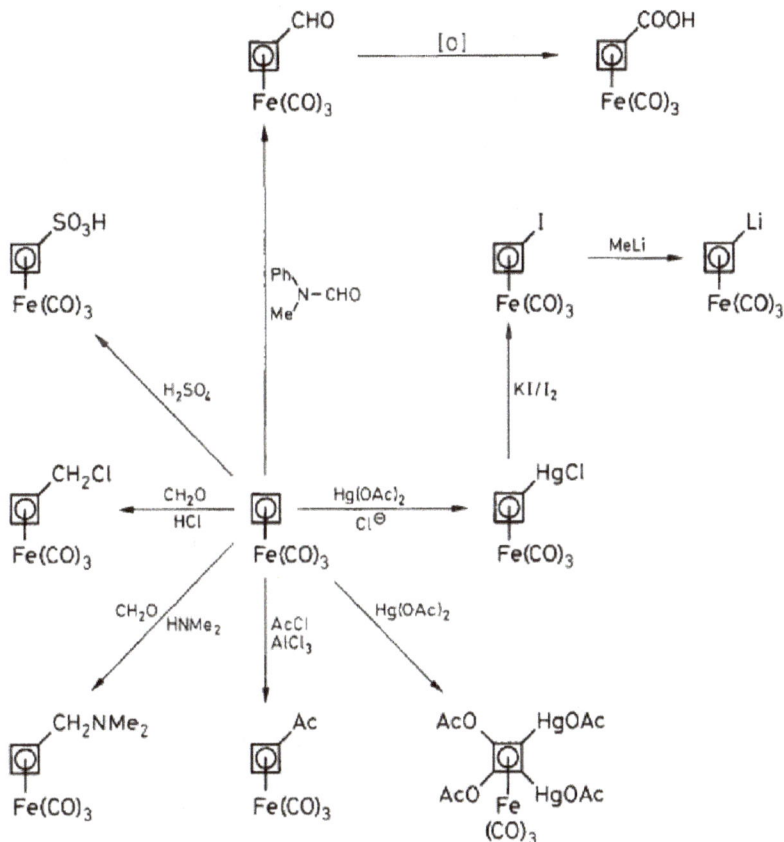

Fig. 13.17: Aromatic reactions of η^4-$C_4H_4Fe(CO)_3$.

Oxidation of $(\eta^4$-$C_4H_4)Fe(CO)_3$ with Fe^{3+}, Ce^{4+} or Ag^+ generates transient, free cyclobutadiene which can be trapped with various organic compounds. This is cleverly
exploited for the preparation of many unusual compounds which are not available
by other methods.

Diphenylacetylene acts on iron pentacarbonyl at 240 °C to yield a tetraphenylcyclobutadiene complex, $(\eta^4$-$C_4Ph_4)Fe(CO)_3$, in addition to a tetraphenylcyclopenta-
dienone complex (Fig. 13.18).

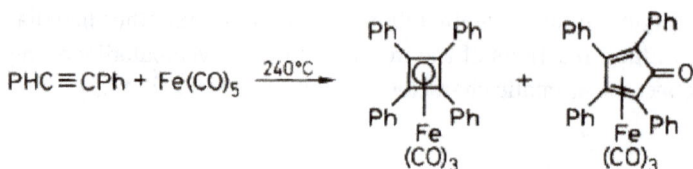

Fig. 13.18: Formation of an iron-cyclobutadiene complex from diphenylacetylene.

Macrocyclic diacetylenes also react with iron carbonyls to form cyclobutadiene complexes, among other products [3] (Fig. 13.19).

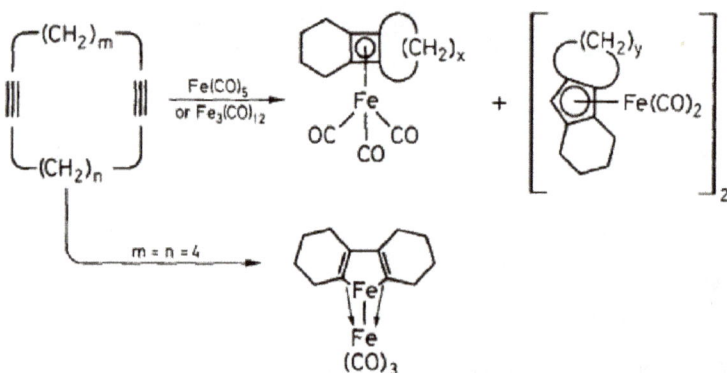

Fig. 13.19: Reactions of macrocyclic diacetylenes with iron pentacarbonyl.

Cobalt forms cyclobutadiene complexes, especially when the metal atom is part of a η^5-C_5H_5Co fragment which requires four electrons to fulfill a noble gas configuration. Thus, 1,2-dichlorocyclobutene forms a cyclobutadiene complex by reaction with $Na[Co(CO)_4]$; the primary product can be converted to a cyclopentadienylcobalt derivative (Fig. 13.20). A rhodium complex, $(\eta^5$-$C_5H_5)Rh(\eta^4$-$C_4Ph_4)$, has been prepared similarly.

With macrocyclic diacetylenes, transannular ring closure occurs to form cyclobutadiene derivatives (Fig. 13.21).

The binuclear nickel complex $[(\eta^4$-$C_4Me_4)NiCl_2]_2$ was the first-ever cyclobutadiene complex synthesized. It was obtained in 1959 by reacting 1,2-dichlorotetramethylcyclobutene with nickel tetracarbonyl. Bis(tetraphenylcyclobutadiene)nickel is obtained from $(\eta^4$-$C_4Ph_4)NiBr_2$ and dilithio-tetraphenylbutadiene (Fig. 13.22).

Disubstituted acetylenes form the palladium complexes $[(\eta^4$-$C_4Ph_4)PdBr_2]_2$ by reacting with the Kharasch reagent, which can transfer the cyclobutadiene ligand to other metals (Fig. 13.23).

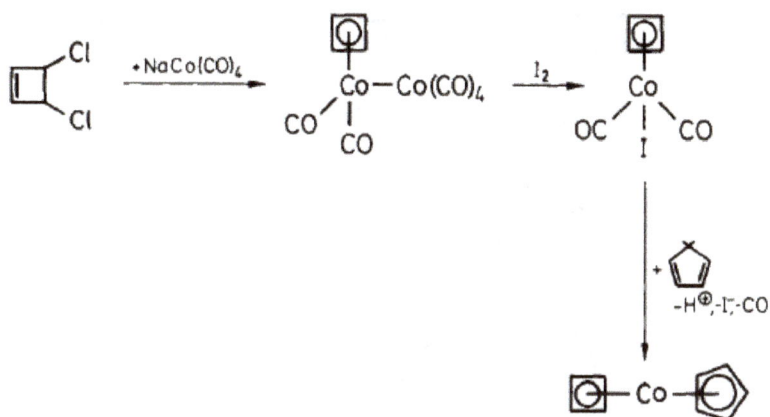

Fig. 13.20: Formation of cobalt-cyclobutadiene complexes.

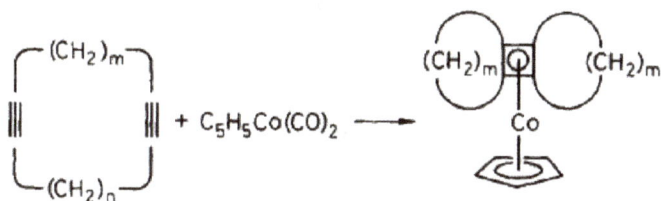

Fig. 13.21: Reaction of macrocyclic diacetylenes with a cobalt carbonyl compound.

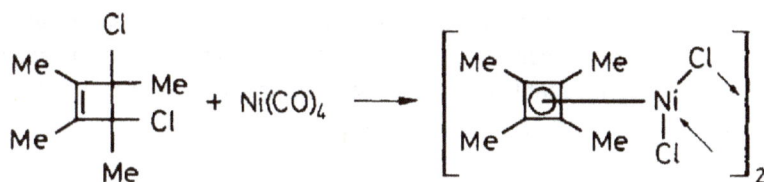

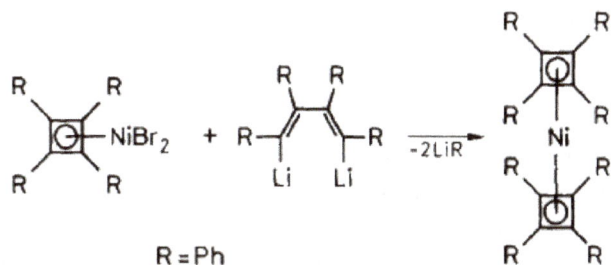

R = Ph

Fig. 13.22: Formation of nickel-cyclobutadiene complexes.

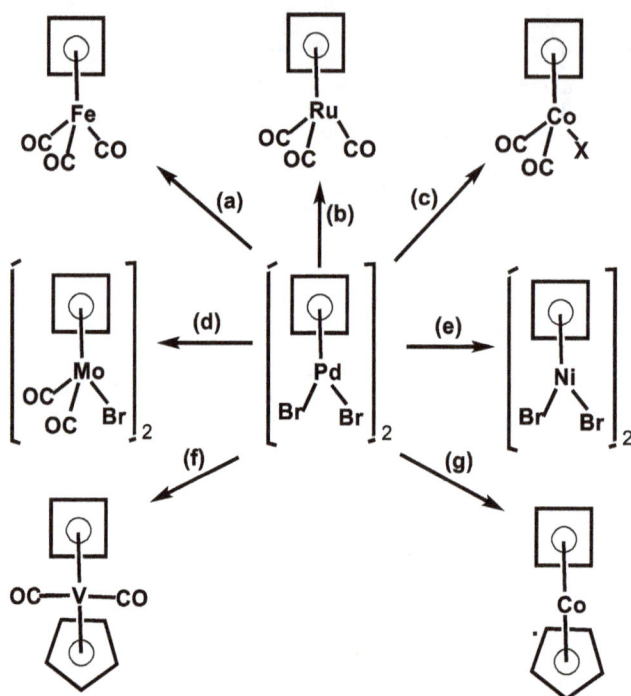

Fig. 13.23: Transfer of cyclobutadiene ligand from palladium to other metals. (a) $Fe(CO)_5$; (b) $Ru_3(CO)_{12}$; (c) $Co_2(CO)_8$; (d) $Mo(CO)_6$; (e) $Ni(CO)_4$; (f) $C_5H_5V(CO)_4$; (g) $Co(C_5H_5)_2$.

The chemistry of the cyclobutadiene complexes of transition metals is a beautiful illustration of the role which complexation by transition metals can play in stabilizing organic molecules incapable of independent existence, and in modifying the chemical reactivity of a coordinated organic group.

13.3 Complexes with cyclic dienes and polyenes

13.3.1 Cyclohexadiene

Since the $Fe(CO)_3$ group requires only four electrons, it readily forms η^4-complexes with cyclohexadienes. Thus, heating iron pentacarbonyl with cyclohexadiene-1,3 yields η^4-$C_6H_8Fe(CO)_3$. The same compound is formed even if a non-conjugated cyclohexadiene such as the 1,4-isomer is used. By heating with $Fe(CO)_5$, cyclohexadiene-1,4 undergoes isomerization, and the complex of the conjugated diene is formed. Thus, complex formation can be a driving force strong enough to produce isomerization of a cyclic diolefin. The bis(η^4-cyclohexadiene) complex (η^4-$C_6H_8)_2$Fe (CO) is also known (Fig. 13.24).

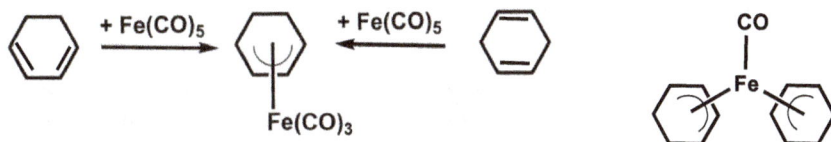

Fig. 13.24: Cyclohexadiene as η^4-ligand in iron complexes.

13.3.2 Cycloheptadiene-1,3

This diolefin can form the same type of complexes as cyclohexadiene-1,3 but few examples are known. With iron pentacarbonyl, cycloheptadiene-1,3 forms the expected (η^4-C_7H_{10})Fe(CO)$_3$ (Fig. 13.25).

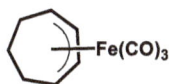

Fig. 13.25: Iron complex of cycloheptadiene-1,3.

Cycloheptatriene uses only two of its three double bonds to coordinate a Fe(CO)$_3$ fragment. The product can be hydrogenated in the presence of Raney nickel to form a tetrahapto cycloheptadiene-1,3 complex (Fig. 13.26).

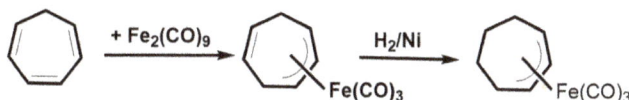

Fig. 13.26: Iron complexes from cycloheptatriene-1,3.

13.3.3 Cyclooctatriene

With cyclooctatriene-1,3,5-triene as a tetrahapto ligand, a ruthenium complex is known [4] (Fig. 13.27).

Fig. 13.27: Cyclooctatriene as η^4-connected ligand.

13.3.4 Cyclooctatetraene

The reaction between cyclooctatetraene and iron pentacarbonyl yields several compounds, depending upon the conditions, two of which are of the η^4-diene type. In η^4-$C_8H_8Fe(CO)_3$, two double bonds act as a conjugated-diene system, leaving the other two double bonds free. The complex $(\mu\text{-}C_8H_8)\{Fe(CO)_3\}_2$ obtained from cyclooctatetraene and $Fe_2(CO)_9$ contains two $Fe(CO)_3$ units attached to the two butadiene fragments. A η^4-ruthenium complex of cyclooctatetraene) has also been reported (Fig. 13.28).

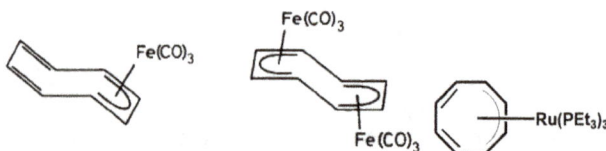

Fig. 13.28: Cyclooctatetraene as η^4-ligand in iron complexes.

13.3.5 Complexes with heteroatom molecules as four-electron ligands

Some divinylboranes, boron-containing heterocycles and carboranes can also act as four-electron ligands. Thus, alkoxydivinylboranes react with $Fe_2(CO)_9$ to form iron-tricarbonyl complexes on UV irradiation (Fig. 13.29).

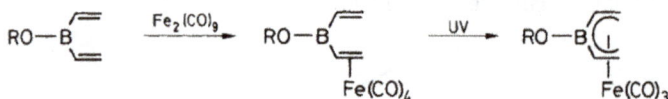

Fig. 13.29: Divinylborane moiety as a four-electron donor.

The five-membered heterocycle, thiadiborolene, also behaves as a four-electron ligand and forms an iron-tricarbonyl complex. Two electrons come from the olefinic double bond and another two from the sulfur atom, while the boron atoms having vacant p_z-orbitals seem to permit delocalization in the ring (Fig. 13.30).

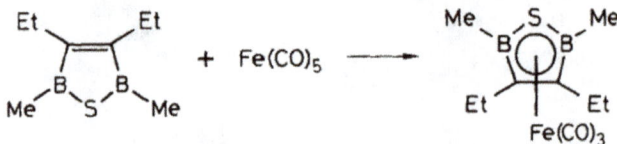

Fig. 13.30: Thiadiborolene heterocycle as four electron donor.

Several η^4-complexes derive from five-membered heterocyclic ligands containing a butadiene fragment. Typical is a group of silacyclopentadiene complexes of iron (Fig. 13.31).

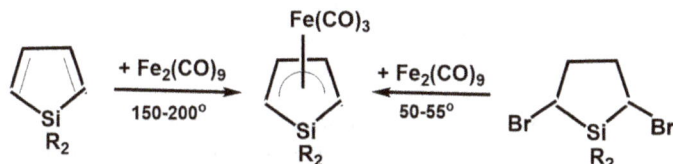

Fig. 13.31: Iron complex of silacyclopentadiene.

Thiophene dioxide forms a η^4-complex with iron on UV irradiation with iron carbonyl (Fig. 13.32).

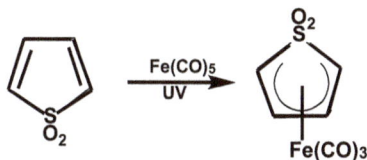

Fig. 13.32: Iron complex of thiophene dioxide.

13.4 Trimethylenemethyl radical as a four-electron ligand

In trimethylenemethyl free radical, each carbon atom is sp^2 hybridized and has a π-electron, available for donation. The four carbon atoms are nearly coplanar and all three C–C bonds are equivalent. Such a species cannot exist free, but by coordination to a transition metal it can act as a four-electron donor and can be stabilized.

Chromium- and molybdenum-carbonyl derivatives are formed from 1-chloro-2-chloromethylpropene-2 and an iron derivative has been prepared by reacting 1-chloro-2-chloromethylpropene-2 with $Fe_2(CO)_9$ (Fig. 13.33).

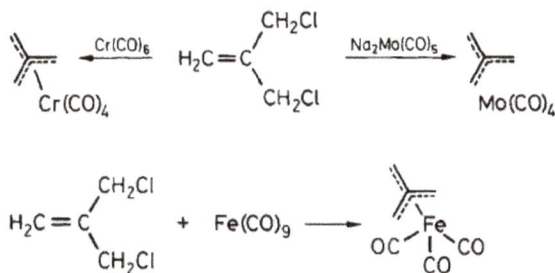

Fig. 13.33: Metal complexes with η^4-trimethylenemethyl.

References

[1]   Kaupp M, Kopf T, Murso A, Stalke D, Strohmann C, Hanks JR, Cloke FGN, Hitchcock PB.
      Trigonal prismatic structure of tris(butadiene)- molybdenum and related complexes revisited:
      diolefin or metallacyclopentene coordination? Organometallics 2002, 21, 5021–28.
[2]   Pagano JK, Erickson KA, Scott BL, Morris DE, Waterman R, Kiplinger JL. Synthesis and
      characterization of a new and electronically unusual uranium metallacyclocumulene,
      $(C_5Me_5)_2U(\eta^4$-1,2,3,4-PhC$_4$Ph). J Organomet Chem 2017, 829, 79–84.
[3]   King RB, Haiduc I, Eavanson CW, Reactions of transition metal compounds with macrocyclic
      alkadiynes. III. Intramolecular transannular cyclizations and related processes with iron. J Am
      Chem Soc 1973, 95, 2508–16.
[4]   Komiya S, Planas JG, Onuki K, Lu Z, Hirano M. Versatile coordination modes and
      transformations of the cyclooctatriene ligand in Ru(C$_8$H$_{10}$)L$_3$ (L = tertiary phosphine).
      Organometallics 2000, 19, 4051–9.

14 Compounds with five-electron ligands

There are several organic molecules able to furnish five π electrons. In addition to cyclopentadienyl, the acyclic pentadienyl, cyclohexadienyl, cycloheptadienyl and other groups can act as pentahapto five-electron ligands (Fig. 14.1).

Fig. 14.1: Various ligands capable of pentahapto-bonding.

The η^5-cyclopentadienyl group is the best one and the most popular. Alone or in association with other ligands, this ligand forms a large number of compounds with a variety of structures. Depending upon electronic requirements, a metal atom may bond one, two or sometimes more η^5-cyclopentadienyl groups (Fig. 14.2).

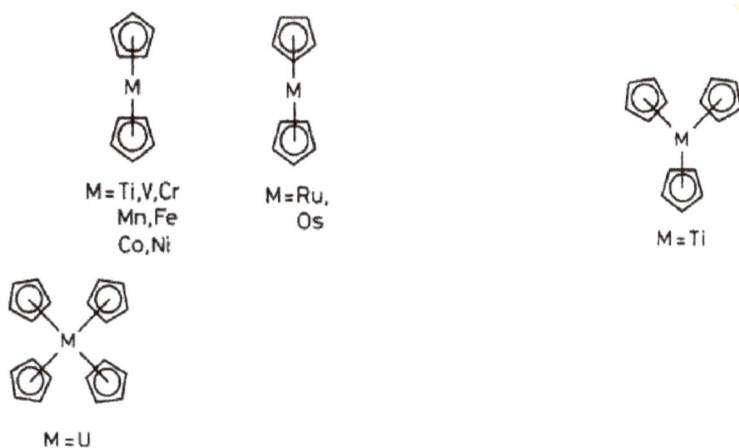

Fig. 14.2: Several types of cyclopentadienyl metal complexes.

Binary cyclopentadienyl–metal compounds, that is, derivatives containing only C_5H_5 groups and metal atoms in the molecule, are known for almost all transition metals. When several C_5H_5 groups are present in a molecule, they may be bonded in different ways (e.g., pentahapto and monohapto). Simple metallocenes of early transition metals are difficult to obtain, and for some (e.g., "titanocene"), the structure was found to be more complicated than believed initially.

The η^5-cyclopentadienyl derivatives of most transition metal are stable thermally, but their oxidative stability varies from metal to metal. Several derivatives,

https://doi.org/10.1515/9783110695274-015

$M(\eta^5\text{-}C_5H_5)_2$, are paramagnetic and do not obey the effective atomic number rule, having either an electron deficit or an electron excess. However, the compounds which have 18 electrons in their valence shell, for example, $Fe(C_5H_5)_2$, $Ru(C_5H_5)_2$ and $Os(C_5H_5)_2$, among binary compounds, are the most stable and exhibit aromatic character, obvious in numerous aromatic substitution reactions.

Many mixed-ligand complexes have also been prepared, and these include cyclopentadienylmetal carbonyls, cyclopentadienylmetal halide, cyclopentadienyl metal sulfides and several others.

14.1 Metallocene complexes

The most common metallocenes are bis(η^5-cyclopentadienyl) metal sandwich compounds, $M(\eta^5\text{-}C_5H_5)_2$, in which the two C_5H_5 rings are in parallel planes, in either eclipsed or staggered orientation. Thus, in solid $V(C_5H_5)_2$, $Cr(C_5H_5)_2$ and $Ru(C_5H_5)_2$, the parallel rings are eclipsed and $Fe(C_5H_5)_2$ and $Co(C_5H_5)_2$ are in staggered conformation.

14.1.1 Ferrocene

Bis(cyclopentadienyl)iron or ferrocene, $Fe(\eta^5\text{-}C_5H_5)_2$, is the subject of a huge volume of literature systematized in numerous reviews and monographs. It was the discovery of ferrocene that initiated an explosive growth in organic transition metal chemistry after the peculiar type of bonding in this compound was recognized.

The ferrocene molecule has an antiprismatic structure with parallel, staggered C_5H_5 rings with all carbon atoms equidistant from metal. In the ferrocenium salt [Fe$(\eta^5\text{-}C_5H_5)_2]^+[BiCl_4]^-$, the rings are eclipsed.

Almost all the ferrocene preparations have in common the conversion of cyclopentadiene to the $[C_5H_5]^-$ anion by a base (dimethylamine, sodium metal, Grignard reagents, etc.), followed by the reaction between the anion and an iron(II) salt, usually the chloride (Fig. 14.3).

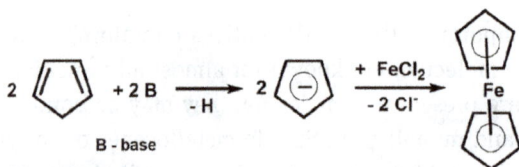

Fig. 14.3: Synthesis of ferrocene.

A simple preparation involves treatment of cyclopentadiene with KOH and $FeCl_2$ in tetrahydrofuran in the presence of [18]-crown-6 ether; the method can also be applied to substituted derivatives. Ferrocene is obtained industrially from cyclopentadiene with iron oxides at elevated temperatures.

Substituted ferrocenes are obtained either by substitution reactions on the ferrocene molecule, or by starting from substituted cyclopentadienes with iron carbonyls or iron(II) chloride.

Due to its aromatic character, observed shortly after its discovery, ferrocene is an extremely interesting compound. Ferrocene can be acylated in the presence of aluminum chloride, can be mercurated, sulfonated and, by indirect methods, nitrated or halogenated. These aromatic substitution reactions have been extensively investigated (Fig. 14.4).

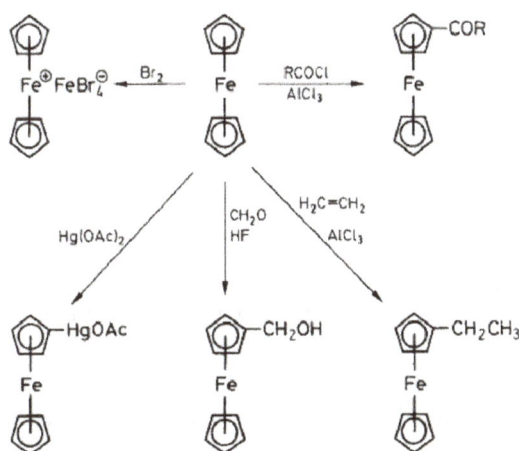

Fig. 14.4: Aromatic substitution reactions of ferrocene.

The electron density in ferrocene is higher than in benzene, and thus ferrocenylamine is a stronger base than aniline, and the ferrocenylcarboxylic acid is weaker than benzoic acid.

Ferrocenyllithium is a versatile starting material for the preparation of many substituted derivatives (Fig. 14.5).

Further reading

Okuda J. Ferrocene – 65 years after. Eur J Inorg Chem 2017, 217–19.

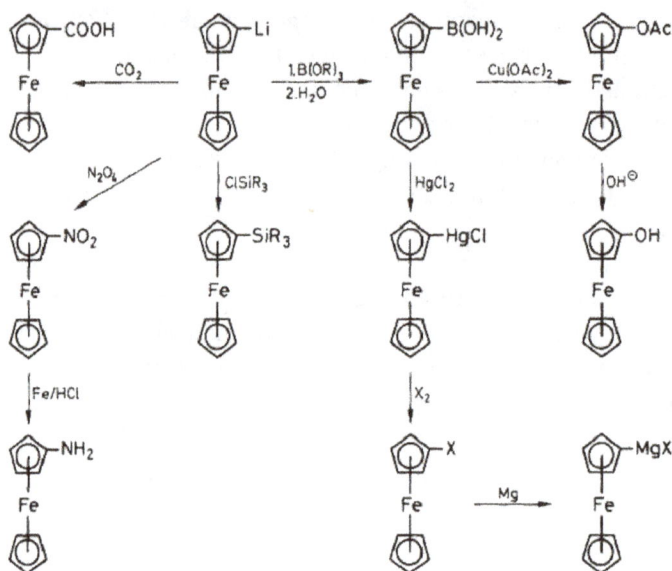

Fig. 14.5: Substitution reactions of lithioferrocene, a versatile reagent.

14.1.2 Other metallocenes

After the discovery of ferrocene attempts were made to prepare similar bis- and tris-(cyclopentadenyl) metal compounds with other metals. Many were successful, some of them being sandwich compounds, others having cyclopentadienyl ligand in bent structures, particularly when additional cooperating ligands (e.g., halogens, carbon monoxide) were present.

The compound first reported as "titanocene", $Ti(C_5H_5)_2$, prepared from $TiCl_4$ and sodium cyclopentadienide, by reducing the resulting $(\eta^5\text{-}C_5H_5)_2TiCl_2$, exists in several forms described as green, black and metastable titanocenes. Two of these are now known to be dimers. The green form is $\mu\text{-}(\eta^5\text{-fulvalene})$di-hydrido-bis$(\eta^5\text{-}$cyclopentadienyl) dititanium while the black form contains a $\eta^1{:}\eta^5$-bridge (Fig. 14.6). Zirconium and hafnium tetrachlorides reacting with $Na[C_5H_5]$ form bis(cyclopentadienyl) metal dichlorides, $(\eta^5\text{-}C_5H_5)_2MCl_2$, (M = Zr, Hf), which can be converted to tetrasubstituted derivatives, $M(C_5H_5)_4$, by reaction with excess $Na[C_5H_5]$. In Zr

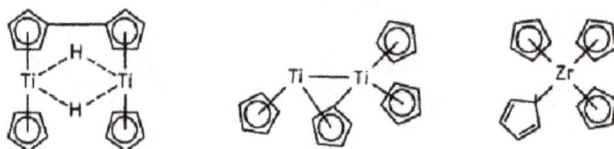

Fig. 14.6: Titanium an zirconium cyclopentadienyls.

$(C_5H_5)_4$ only three cyclopentadienyls are penta*hapto* while the fourth is monohapto (Fig. 14.6).

Vanadium tetrachloride and sodium cyclopentadienide form $(\eta^5\text{-}C_5H_5)_2VCl_2$, and with excess reagent, $V(\eta^5\text{-}C_5H_5)_2$ (vanadocene). The latter can be conveniently prepared from $VCl_2 \cdot 2THF$ with $Na[C_5H_5]$. Niobocene, $Nb(\eta^5\text{-}C_5H_5)_2$ can be obtained by reducing $(\eta^5\text{-}C_5H_5)_2NbCl_2$ but exists only in solution. A solid described earlier as "niobocene" was found to be a dimeric hydride with $(\eta^1{:}\eta^5\text{-}C_5H_4)$ bridges (Fig. 14.7).

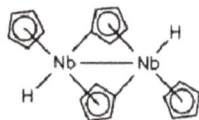

Fig. 14.7: Dimeric niobocene cyclopentadienyl compound.

Chromocene, $Cr(\eta^5\text{-}C_5H_5)_2$ is prepared by the reaction of anhydrous chromium(III) chloride with sodium cyclopentadienide, or better from $CrCl_2.THF$. With only 16 electrons in its valence shell, this compound is unstable and air-sensitive. Chromocene is also obtained from chromium hexacarbonyl and cyclopentadiene.

Molybdenum and tungsten analogues could not be prepared, forming instead the hydrides, $(\eta^5\text{-}C_5H_5)_2MH_2$.

Manganocene, $Mn(\eta^5\text{-}C_5H_5)_2$, from anhydrous manganese(II) chloride and sodium cyclopentadienide, is a supramolecular array formed of alternating Mn^{2+} cations and $[C_5H_5]^-$ anions in a chain. This is the only metallocene of a first row transition element having an ionic structure. The substituted 1,1-dimethylmanganocene contains two molecular forms: an ionic high-spin form (with a $Mn\text{-}C_5H_5$ distance of 2.433 A) and low-spin η^5-complex (with a $Mn\text{-}C_5H_5$ distance of 2.144 A) (Fig. 14.8).

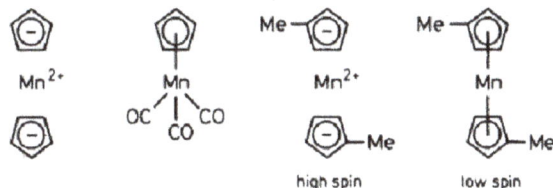

Fig. 14.8: A variety of manganese cyclopentadienyls.

Ruthenocene, $Ru(\eta^5\text{-}C_5H_5)_2$, which is prepared from ruthenium(III) chloride and sodium cyclopentadienide is even more stable thermally than ferrocene, but its aromatic-substitution reactions are more difficult to carry. Lithiation is possible with BuLi.TMEDA, and the product can be further converted to iodoruthenocene (Fig. 14.9).

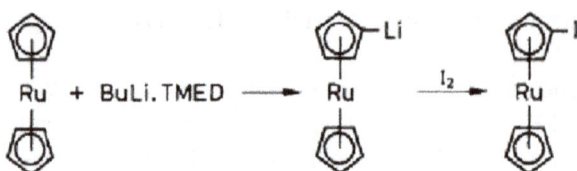

Fig. 14.9: Ruthenocene cyclopentadienyl complexes.

Mercuration products of ferrocene (e.g., mono- and dimercurated derivatives), ruthenocene and η^5-cyclopentadienylmanganese tricarbonyl can be obtained, indicating aromatic properties of these π-complexes (Fig. 14.10).

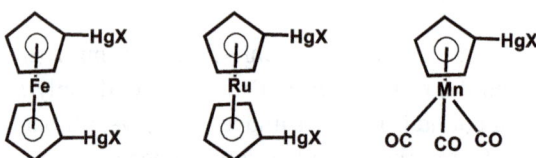

Fig. 14.10: Mercuration derivatives of some metallocenes.

Osmocene, $Os(\eta^5\text{-}C_5H_5)_2$, prepared form osmium tetrachloride and sodium cyclopentadienide, can be acetylated in a Friedel–Crafts reaction to form a monosubstituted derivative.

Cobaltocene, $Co(\eta^5\text{-}C_5H_5)_2$, is prepared form cobalt(II) chloride and sodium cyclopentadienide. With 19 electrons in the valence shell, one electron more than the noble-gas configuration, this compound is readily oxidized to the 18 electron cation, $[Co(\eta^5\text{-}C_5H_5)_2]^+$. This tendency to achieve the 18 electron configuration also manifests in the addition of carbon tetrachloride, reduction with lithium alanate and addition of alkyl halides, in which one of the cyclopentadienyl rings in the products becomes attached to the metal as a four-electron donor (Fig. 14.11).

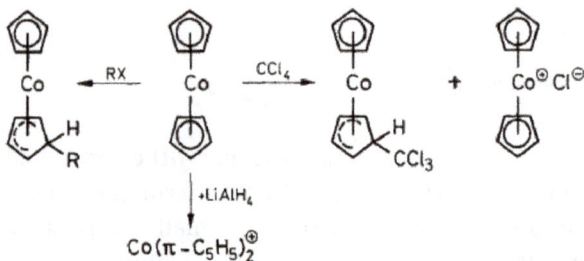

Fig. 14.11: Cobaltocene derivatives.

Rhodium and iridium behave otherwise. Thus, rhodium(III) and iridium(III) chlorides form compounds with $Na[C_5H_5]$ in which only one of the two rings is a five-electron donor (Fig. 14.12).

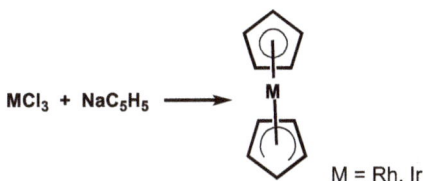

$MCl_3 + NaC_5H_5 \longrightarrow$

M = Rh, Ir

Fig. 14.12: Rhodium and iridium cyclopentadienyl compounds.

Nickelocene, $Ni(\eta^5\text{-}C_5H_5)_2$ can be made from anhydrous nickel bromide, cyclopenta-diene and diethylamine, or from the complex $[Ni(NH_3)_4]Cl_2$ and sodium cyclopenta-dienide. It has 20 electrons which makes it sensitive to oxidation to the cation, $[Ni(\eta^5\text{-}C_5H_5)_2]^+$, but removal of a second electron leads to decomposition, rather than to formation of an 18 electron dication.

Cyclopentadiene and nickel tetracarbonyl produce a dicyclopentadienyl derivative in which the second ring is bonded through a η^3-allylic fragment, since the η^5-C_5HNi group requires only three electrons to achieve a noble-gas configuration. Nickelocene itself can be reduced with sodium amalgam to form the same compound (Fig. 14.13).

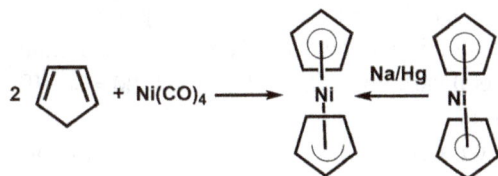

$2 \quad + Ni(CO)_4 \longrightarrow \quad Ni \quad \xleftarrow{Na/Hg} \quad Ni$

Fig. 14.13: Nickel cyclpentadienyl compounds.

The lanthanides form air-sensitive, but thermally stable tris(cyclopentadienyl) derivatives, $M(C_5H_5)_3$. Europium and ytterbium form bis(cyclopentadienyl) complexes, $M(C_5H_5)_2$, by the reaction of cyclopentadiene with the metals.

Cyclopentadienyl derivatives of actinides have also been prepared. Thorium tet-rachloride forms the tetrakis derivative, $Th(C_5H_5)_4$, with sodium cyclopentadienide. The uranium compounds $U(C_5H_5)_4$ and $U(C_5H_5)_3$ have similarly been prepared.

For the synthesis of transuranium cyclopentadienyls such as $(C_5H_5)_3NpCl$, $Pu(C_5H_5)_4$, $Am(C_5H_5)_3$, $Cm(C_5H_5)_3$ and $Bk(C_5H_5)_3$, the reaction between bis(cyclopentadienyl) beryllium and metal trichlorides has been used (sometimes on a microgram scale). For the preparation of $Cm(C_5H_5)_3$, the reaction of $CmCl_3$ with bis(cyclopentadienyl) magnesium can also be used.

Recent work describes compounds containing substituted cyclopentdienyl ligands, including derivatives of neptunium $[Np(\eta^5\text{-}C_5H_4SiMe_3)_3]^-$ [1], plutonium $[\{Pu^{II}\{\eta^5\text{-}C_5H_3(SiMe_3)_2\}_3]^-$ [2], americium $Am(\eta^5\text{-}C_5Me_4H)_3$ [3] and californium [Cf $(\eta^5\text{-}C_5Me_4H)_2Cl_2K(OEt_2)]_n$ [4].

14.1.3 Cyclopentadienylmetal carbonyls and carbonyl halides

There is a variety of compounds of this type (Fig. 14.14).

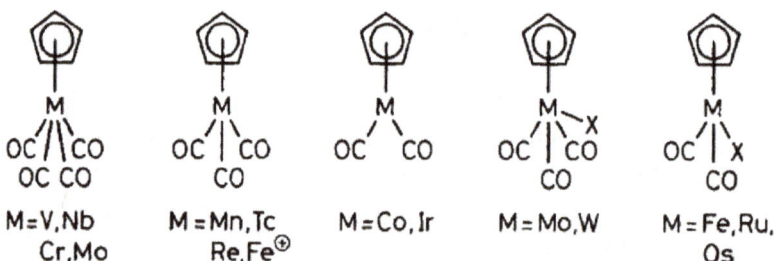

Fig. 14.14: Cyclopentadienylmetal carbonyl complexes.

The reaction of $TiCl_4$ with $Na[C_5H_5]$ yields $(\eta^5\text{-}C_5H_5)_2TiCl_2$. If the reaction between titanium tetrachloride and sodium cyclopentadienide is carried out under carbon monoxide, the product is $(\eta^5\text{-}C_5H_5)Ti(CO)_2$; the same sandwich compound is obtained by reducing $(\eta^5\text{-}C_5H_5)_2TiCl_2$ with aluminum powder in THF, in a carbon monoxide atmosphere. Cyclopentadienylmetal carbonyls, $(\eta^5\text{-}C_5H_5)_2M(CO)_2$ (M = Zr, Hf), are among the few carbonyl derivatives of these two elements.

Stable cyclopentadienyl vanadium tetracarbonyl, $(\eta^5\text{-}C_5H_5)V(CO)_4$, with a noble-gas electronic configuration, is formed in the reaction between sodium hexacarbonyl vanadate and cyclopentadienylmercury chloride.

$$Na[V(CO)_6] + \eta^1 - C_5H_5HgCl \longrightarrow \eta^5 - C_5H_5V(CO)_4 + Hg + 2CO + NaCl$$

Irradiation or heating of $(\eta^5\text{-}C_5H_5)V(CO)_4$ yields $(\eta^5\text{-}C_5H_5)_2V_2(CO)_5$, which is a prototypal example of a compound with semibridging carbonyl groups (Fig. 14.15). Refluxing in THF leads to a tetranuclear compound $[(\eta^5\text{-}C_5H_5)V(CO)]_4$.

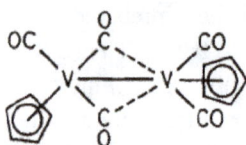

Fig. 14.15: The structure of $(\eta^5\text{-}C_5H_5)_2V_2(CO)_5$.

Cyclopentadienylniobium tetracarbonyl $(\eta^5\text{-}C_5H_5)Nb(CO)_4$ is obtained by the reaction of $Na[C_5H_5]$ with the $[Nb(CO)_6]^-$ anion, or by reduction of $(\eta^5\text{-}C_5H_5)_2NbCl_2$ with an Na/Cu/Al alloy under carbon monoxide.

Irradiation of $(\eta^5\text{-}C_5H_5)Nb(CO)_4$ by UV yields a trinuclear compound, $(\eta^5\text{-}C_5H_5)_3$ $Nb_3(CO)_7$, which contains a novel carbonyl bridge (Fig. 14.16).

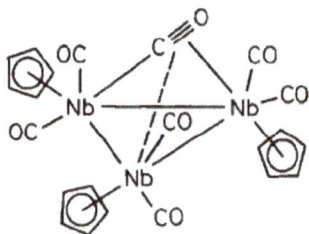

Fig. 14.16: Trinuclear niobium complex.

Chromium and molybdenum hexacarbonyl, heated with cyclopentadiene yield dimers, $[(\eta^5\text{-}C_5H_5)M(CO)_3]_2$, containing metal–metal bonds and no carbonyl bridges. Unlike cyclopentadiene, which forms the Mo–Mo bonded dimer, pentamethylcyclo-pentadiene forms with molybdenum hexacarbonyl, a dimer which contains a triple Mo≡Mo bond with two fewer carbonyl groups (Fig. 14.17).

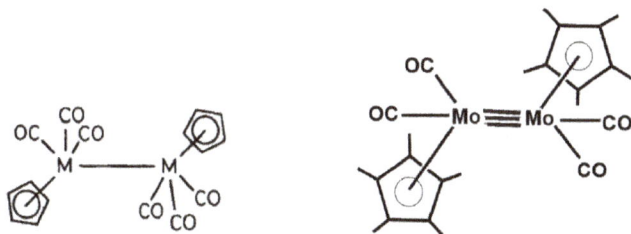

Fig. 14.17: Binuclear molybdenum cyclopentadienyls.

The metal–metal triple-bonded compounds, $[(\eta^5\text{-}C_5H_5)M(CO)_2]_2$ (M = Cr, Mo, W), are formed on refluxing the metal–metal single-bonded dimers $[(\eta^5\text{-}C_5H_5)M(CO)_3]_2$ in toluene. Both the chromium and molybdenum compounds contain metal–metal triple bonds, but the molecular geometry is different: the chromium derivative is a *trans*-isomer, while the molybdenum compound contains a linear $C_5H_5\text{-Mo}\equiv\text{Mo-}C_5H_5$ fragment (Fig. 14.18).

Carbon monoxide converts manganocene to cyclopentadienyl-manganese tri-carbonyl, $(\eta^5\text{-}C_5H_5)Mn(CO)_3$, which can also be obtained from sodium cyclopenta-dienide and $Mn_2(CO)_{10}$, or $[Mn(py)_2Cl_2]$ with cyclopentadiene, magnesium metal

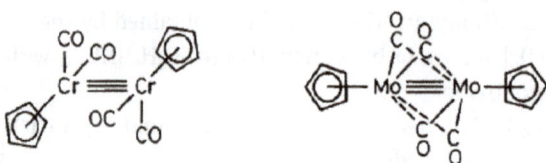

Fig. 14.18: Cyclopentadienyl compounds with metal-metal triple bonds.

and carbon monoxide. The compound, known as cymantrene, has a noble gas configuration for the metal, and undergoes aromatic substitution reactions (Fig. 14.19)

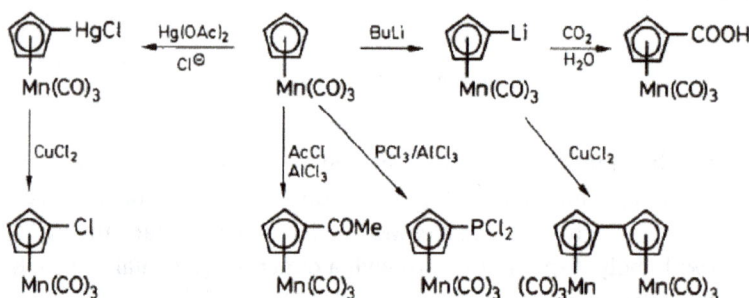

Fig. 14.19: Manganese cyclopentadenyl compounds.

Technetium and rhenium derivatives of the type $(\eta^5\text{-}C_5H_5)M(CO)_3$, prepared by coupling the pentacarbonylmetal chlorides with sodium cyclopentadienide:

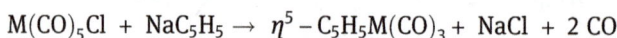

$$M(CO)_5Cl + NaC_5H_5 \rightarrow \eta^5 - C_5H_5M(CO)_3 + NaCl + 2\ CO$$

The dimer $[(\eta^5\text{-}C_5H_5)Fe(CO)_2]_2$ is reduced by sodium amalgam to give a strongly nucleophilic anion $[(\eta^5\text{-}C_5H_5)Fe(CO)_2]^-$ which is a useful starting material (Fig. 14.20).

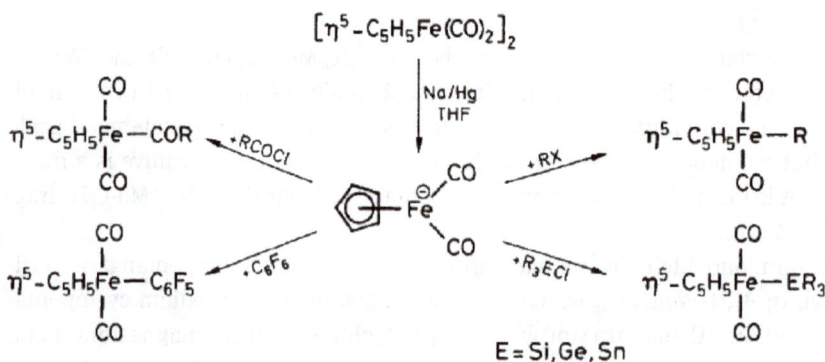

Fig. 14.20: Reactions of cyclopentadienyl iron dicarbonyl anion.

A dimer analogous to that of iron, $[(\eta^5\text{-}C_5H_5))Ru(CO)_2]_2$ has been prepared from the dimeric carbonyl chloride $[Ru(CO)_2Cl_2]_2$ and sodium cyclopentadienide.

The metal-metal bonded dimer, $[(\eta^5\text{-}C_5H_5)Os(CO)_2]_2$, prepared from $Os(CO)_3Cl$ and sodium cyclopentadienide, contains no carbonyl bridges. Thus there is a remarkable difference between the iron and osmium compounds.

Rhodium and iridium cyclopentadienylmetal complexes were reported as mononuclear $(\eta^5\text{-}C_5H_5)M(CO)_2$, binuclear $(\eta^5\text{-}C_5H_5)_2M(\mu_2\text{-}CO)(CO)_2$ (M = Rh, Ir) and trinuclear $[(\eta^5\text{-}C_5H_5)Ir(CO]_3$ compounds [5].

References

[1] Dutkiewicz MS, Apostolidis C, Walter O, Arnold PL, Reduction chemistry of neptunium cyclopentadienide complexes: from structure to understanding. Chem Sci 2017, 8, 2553–61.

[2] Windorff CJ, Chen GP, Cross JN, Evans WJ, Furche F, Gaunt, AJ, Janicke MT, Kozimor SA, Scott BL. Identification of the formal +2 oxidation state of plutonium: Synthesis and characterization of $[Pu^{II}[C_5H_3(SiMe_3)_2]_3]^-$. J Am Chem Soc 2017, 139, 3970–73.

[3] Conrad A. P. Goodwin CAP, Su J, Albrecht-Schmitt TE, Blake AV, Batista ER, Daly SR, Dehnen S, Evans WJ, Gaunt AJ, Kozimor SA, Lichtenberger N, Scott BL, Yang P. [Am(C_5Me_4H)_3]: An organometallic americium complex. Angew Chem Int Ed. 2019, 58.

[4] Goodwin CAP, Su J, Stevens LM, White FD, Anderson NH, Auxier II JD, Albrecht-Schönzart TE, Batista ER, Briscoe SF, Cross JN, Evans WJ, Gaiser AN, Gaunt AJ, James MR, Janicke MT, Jenkins TF, Jones ZR, Kozimor SA, Scott BL, Sperling JM, Wedal JC, Windorff CJ, Yang P, Ziller JW. Isolation and characterization of a californium metallocene, Nature 2021, 599, 421–24.

[5] Hashikawa Y, Murata Y. Synthesis and oligomerization of CpM(CO)_2. ACS Omega 2021, https://doi.org/10.1021/acsomega.1c05739

15 Six-electron ligands

Six π electrons, able to ensure η^6-hexahapto connectivity, are offered by benzene, cycloheptatriene, cyclooctene, cyclooctadiene (Fig. 15.1),

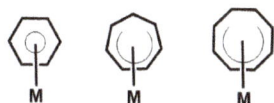

Fig. 15.1: Various ligands capable of hexahapto-bonding.

15.1 Homoleptic sandwich complexes

The most important six-electron ligand is benzene which forms homoleptic sandwich complexes (Fig. 15.2). Polyphenyls and condensed polyarenes can use one or more of their aromatic rings in metal bonding.

(a)

M = Ti, V, Cr, Mo, W, Fe, Ru, Os

Fig. 15.2: Homoleptic bis(benzene) sandwich complexes.

Bis(benzene)titanium is formed in the reaction of the metal vapor with benzene at 77 K, and the reaction can be extended to other arenes (toluene, mesitylene, etc.).

Paramagnetic bis(benzene)vanadium, $V(\eta^6\text{-}C_6H_6)_2$, obtained by the reaction of vanadium tetrachloride with benzene, in the presence of aluminum chloride and aluminum powder, followed by alkaline hydrolysis, has 17 electrons and can be readily reduced with alkali metals to the 18 electron anion, $[V(\eta^6\text{-}C_6H_6)_2]^-$. Vanadium hexacarbonyl reacts with benzene and its substituted derivatives to also form benzene complexes. Bis(benzene)vanadium can be metallated with n-BuLi, to give $[V(\eta^6\text{-}C_6H_6Li)_2]$. Niobium vapor and benzene, toluene or mesitylene give bis(arene) niobium derivatives.

The group 16 elements require 12 electrons, and can achieve a noble-gas configuration by coordinating two benzene molecules. Thus, bis(benzene)chromium Cr $(\eta^6\text{-}C_6H_6)_2$, the compound which historically opened this class, was obtained from a reductive Friedel–Crafts reaction with chromium chloride and benzene in the presence of aluminum chloride and aluminum powder, followed by reduction with sodium dithionite

https://doi.org/10.1515/9783110695274-016

$$3CrCl_3 + 2Al + AlCl_3 + 6C_6H_6 \rightarrow 3\left[\left(\eta^6 - C_6H_6Cr\right)\right]^+ AlCl_4^-$$

$$\xrightarrow[OH^-]{Na_2S_2O_4} Cr\left(\eta^6 - C_6H_6\right)_2$$

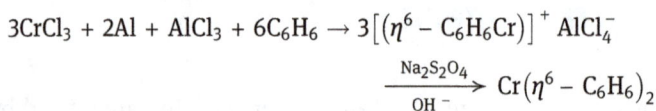

Substitution reactions are difficult, but metallation is possible, and further reactions of the metallated derivatives can lead to various products. The coordinated-benzene molecule retains its aromaticity as shown by substitution reactions. Bis(benzene) chromium can be metalated with the n-BuLi · TMEDA complex (Fig. 15.3).

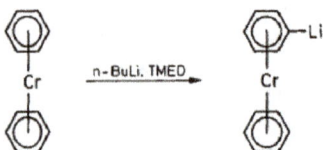

Fig. 15.3: Lithiation of bis(benzene) chromium.

The less stable bis(benzene) derivatives of molybdenum and tungsten, $M(\eta^6\text{-}C_6H_6)_2$ (M = Mo, W) are obtained by reductive Friedel–Crafts reactions or from metal atom vapor and benzene.

Iron, ruthenium and osmium chlorides heated with aromatic hydrocarbons in the presence of aluminum chlorides and aluminum powder, after hydrolysis, yield $[M(\eta^6\text{-}C_6H_6)_2]^{2+}$ cations, which can be precipitated as hexafluorophosphates. The hexamethylbenzene derivative can be reduced to a monopositive cation, $[Fe(\eta^6\text{-}arene)_2]^+$ and to the unstable, neutral $[Fe(\eta^6\text{-}arene)_2]$. Similarly, the complex salt, $[Fe(\eta^6\text{-}arene]^{2+}[PF_6]_2$ is reduced with $Na[BH_4]$ to a cyclohexadienyl complex, and with sodium dithionite to a monopositive cation (Fig. 15.4).

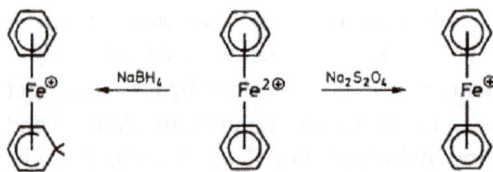

Fig. 15.4: Bis(benzene iron and its reactions.

Further reading

Pampaloni G. Aromatic hydrocarbons as ligands. Recent advances in the synthesis, the reactivity and the applications of bis(η^6-arene) complexes. Coord Chem Rev 2010, 254, 402–419.

Seyferth D. Bis(benzene)chromium. Its discovery by E.O. Fischer and W. Hafner and subsequent work by the research groups of E.O. Fischer, H. H. Zeiss, F. Hein, C. Elschenbroich, and others. Organometallics 2003, 21, 2800–2820.

Kündig EP. Synthesis of transition metal η^6-arene Complexes. Topics Organomet. Chem. 2004, 7, 3–20.

15.2 Heteroleptic, mixed ligand sandwich compounds

Numerous sandwich compounds are formed by associating different hydrocarbons as ligands. The number of possible combinations seems unlimited and only a selection is presented here, leaving the imagination of the reader to search for more.

Manganese(II) chloride reacts with sodium cyclopentadienide and phenyl magnesium bromide to give a mixed derivative, along with a bimetallic compound derived from biphenyl (Fig. 15.5).

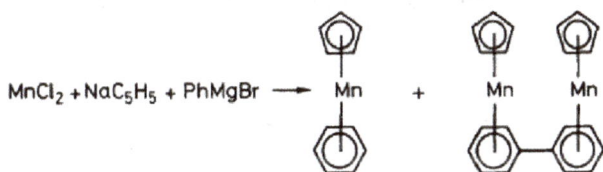

Fig. 15.5: Mixed manganese benzene cyclopentadienyl sandwich complex.

Similarly, $(\eta^5\text{-}C_5H_5)Re(\eta^6\text{-}C_6H_6)$ forms from rhenium(V) chloride, $C5H_nMgBr$ and cyclohexadiene, under UV irradiation.

The cation $[(\eta^5\text{-}C_5H_5)Fe(\eta^6\text{-}C_6H_6)]^+$ is obtained from $(\eta^5\text{-}C_5H_5Fe(CO)_2Cl$ and benzene with aluminum chloride, or from ferrocene and benzene in the presence of aluminum chloride and aluminum metal powder (Fig. 15.6).

Fig. 15.6: Mixed benzene cyclopenadienyl sandwich complex.

Like in ferrocene, one of the cyclopentadienyl rings of ruthenocene can be replaced by an aromatic ring (mesitylene, hexamethylbenzene, etc.) by heating with an aluminum chloride–aluminum metal powder mixture (Fig. 15.7).

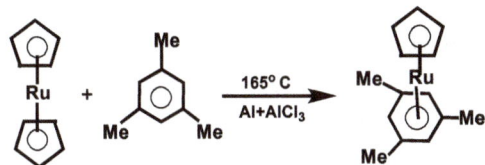

Fig. 15.7: Mixed ruthenium complex.

A cation of the type $[(\eta^5\text{-}C_5H_5)Co[(\eta^6\text{-}C_6H_6)]^{2+}$ can be obtained by hydride abstraction of the η^5-cyclopentadienyl-cobalt cyclohexadiene complex with trityl tetrafluoroborate (Fig. 15.8).

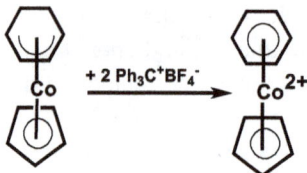

Fig. 15.8: Mixed cobalt benzene cyclopentadienyl sandwich.

15.3 Sandwich η^n-complexes of some heterocyclic ligands

Borabenzene ligands form several transition metal complexes. In these ligands, the boron atom contributes no electrons, but its vacant p_z orbital permits cyclic conjugation in the ring (Fig. 15.9).

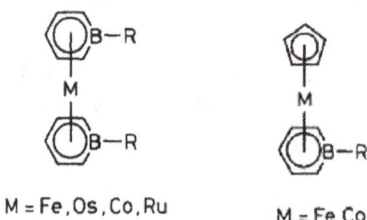

M = Fe, Os, Co, Ru

M = Fe, Co   Fig. 15.9: Borabenzene sandwich complexes.

A typical preparation is shown (Fig. 15.10).

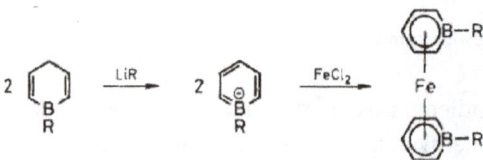

Fig. 15.10: Synthesis of an iron borabenzene sandwich complex.

The silacyclopentadiene heterocycles forms a mixed cobalt cyclopentadienyl complex (Fig. 15.11).

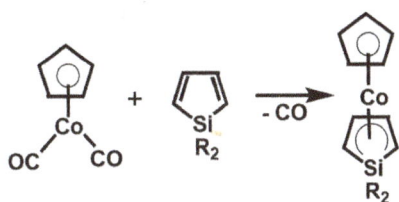

Fig. 15.11: Mixed silacyclopentadiene-cobalt cyclopentadienyl complex.

The pyrrole ring can form an azaferrocene and bis(pyridine) chromium complexes can be prepared by reactions with metal vapor (Fig. 15.12).

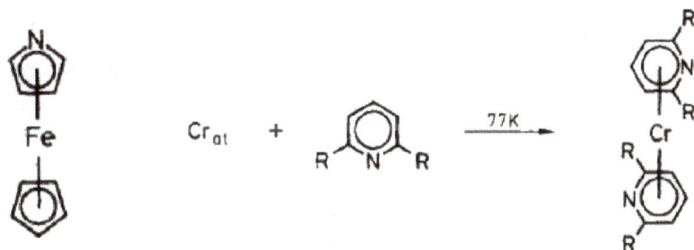

Fig. 15.12: Pyrrole and pyridine sandwich complexes.

Phosphaferrocenes iron complexes and arsole analoges are also known (Fig. 15.13).

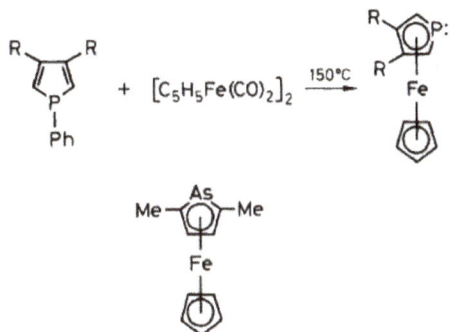

Fig. 15.13: Phospha- and arsaferrocenes.

Diphosphaferrocenes have been obtained with the aid of lithiophospholes. Iron complexes of arsoles are also known (Fig. 15.14).

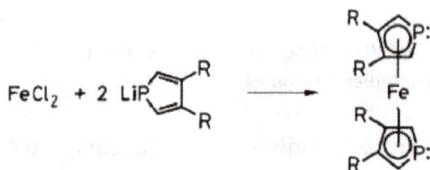

Fig. 15.14: Diphosphaferrocenes.

Thiophene behaves like benzene in its reaction with ferrocene and an aluminum metal–aluminum bromide mixture, to give a mixed cyclopentadienyliron thiophene complex (Fig. 15.15).

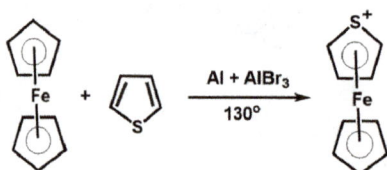

Fig. 15.15: Cyclopentadienyliron thiophene complex.

16 Complexes with seven-electron ligands

The η^7 connectivity is provided by cycloheptatriene. Homoleptic cycloheptatriene complexes can be obtained. Cycloheptatriene reacts with several vaporized metals (Ti, V, Fe, Co) to yield η^7-cycloheptatrienyl complexes with titanium, vanadium and chromium while iron gives $Fe(\eta^5\text{-}C_7H_7)(\eta^7\text{-}C_7H_9)$ imposed by the 18 electron rule (Fig. 16.1).

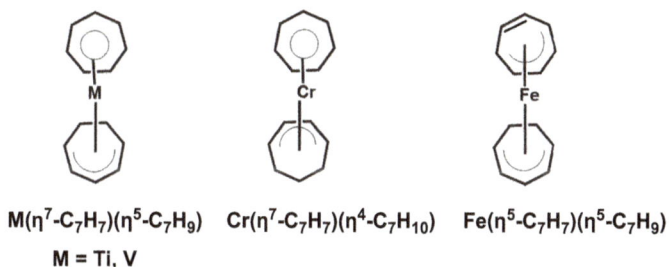

$M(\eta^7\text{-}C_7H_7)(\eta^5\text{-}C_7H_9)$ $Cr(\eta^7\text{-}C_7H_7)(\eta^4\text{-}C_7H_{10})$ $Fe(\eta^5\text{-}C_7H_7)(\eta^5\text{-}C_7H_9)$

M = Ti, V

Fig. 16.1: Homoleptic cycloheptatriene complexes.

Several cycloheptatriene complexes are heteroleptic compounds.

Cycloheptatriene reacts with cyclopentadienyl vanadium tetracarbonyl to yield the η^7-complex, $V(\eta^7\text{-}C_7H_7)(\eta^5\text{-}C_5H_5)$, but the reaction of vanadium metal vapor with cycloheptatriene produces η^6-complexes (Fig. 16.2).

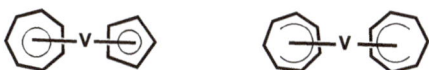

Fig. 16.2: Vanadium complexes withcycloheptatriene.

The mixed η^7-cycloheptatrienyl-η^5-cyclopentadienyl complex of chromium has been prepared by the reaction of anhydrous chromium(III) chloride withcyclopentadiene, cycloheptatriene and iso-PrMgBr or by treatment of $(\eta^5\text{-}C_5H_5)Cr(\eta^6\text{-}C_6H_6)$ with a cyclo-heptatrienyl salt, followed by reduction with an alkali metal dithionite. Related η^7-cycloheptatrienyl–molybdenum complexes with a η^6-ligand as second substituent, must be in cationic form in order to preserve the 18 electron configuration (Fig. 16.3).

https://doi.org/10.1515/9783110695274-017

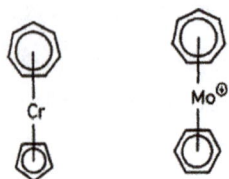

Fig. 16.3: Heteroleptic chromium and molybdenum complexes with cycloheptatriene.

17 Complexes with eight-electron ligands

Cyclooctatetraene is the eight-electron donor ligand capable of η^8-connectivity. Cyclooctatetraene forms η^8-complexes only with the early transition metals which require a large number of electrons to achieve a noble-gas electronic configuration. No η^8-cyclooctatetraene complex of a transition metal beyond group 16 is known.

In the titanium compound $Ti(\eta^8\text{-}C_8H_8)(\eta^5\text{-}C_5H_5)$, the cyclooctatetraene ring is bonded in octahapto-fashion and both rings are coplanar (Fig. 17.1).

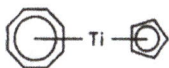

Fig. 17.1: Cyclooctatetraene as η^8 ligand.

The bis(cyclooctatetraene)vanadium complex $V(C_8H_8)_2$ was obtained by reacting K_2 C_8H_8 with $VCl_3.THF$. Only one of the ring is connected as $\eta^8\text{-}C_8H_8$ while the second one is in a $\eta^4\text{-}C_8H_8$ fashion [1] (Fig. 17.2)

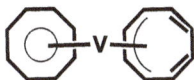

Fig. 17.2: Vanadium complex with two differently connected cyclooctatetraene ligands.

The mixed ligand complex of chromium with cyclooctatetraene and cyclopentadienyl exists in two equilibrium forms, and is readily converted to a cation, the eight-membered ring becoming a six-electron donor (Fig. 17.3).

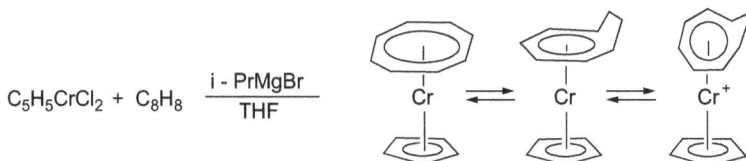

$$C_5H_5CrCl_2 + C_8H_8 \xrightarrow[\text{THF}]{i\text{ - PrMgBr}}$$

Fig. 17.3: Chromium complex with cyclooctatetraene.

Scandium derivatives of cyclooctatetraene can be prepared according to the sequence shown in Fig. 17.4.

$$ScCl_3 \cdot 3\,THF \xrightarrow{K_2C_8H_8} \eta^8\text{-}C_8H_8ScCl \cdot THF$$

$$\xrightarrow{K_2C_8H_8} K^+[Sc(\eta^8\text{-}C_8H_8)_2]^-$$

$$\xrightarrow{NaC_5H_5} (\eta^8\text{-}C_8H_8)Sc(\eta^5\text{-}C_5H_5)$$

Fig. 17.4: Formation of scandium cyclooctatetraene complexes.

https://doi.org/10.1515/9783110695274-018

The lanthanides and actinides (uranium thorium, protactinium, neptunium and plutonium) form η^8–cyclooctatetraene complexes, since the η^8-C_8H_8 ligand has molecular orbitals able to interact with the f-orbitals of the metals.

The pyrophoric bis(cyclooctatetraene) complex, $U(\eta^8$-$C_8H_8)_2$, ("uranocene") is obtained from uranium(IV) chloride and $K_2C_8H_8$. Both rings are coplanar and act as eight-electron donors; f-orbitals participate in the bonding. The uranium derivative can be anodically oxidized to air-stable cation $[U(\eta^8$-$C_8H_8)_2]^+$, but anions like $[M(\eta^8$-$C_8H_8)_2]^+$ with lanthanides M = La, Ce, Nd, Er and actinides M = Np, Pu, Am, have also been prepared.

The tetraphenylcyclooctatetraene complex, $U(\eta^8$-$C_8H_4Ph_4)_2$, prepared from UCl_4 and $K_2C_8H_4Ph_4$-1.3.5.7 is air-stable and sublimes at 400 °C (!) in vacuo. It has a nearly eclipsed configuration with the phenyl groups tilted away from the C_8-plane.

Reference

[1]   Gourier D, Samuel E, Bachmann B, Hahn F, Heck J. Bis(cyclooctatetraene)vanadium: X-ray structure and study of molecular motions by EPR and ENDOR spectroscopy in frozen solution. Inorg Chem 1992, 31, 86–95.

18 Inverse sandwich complexes

Traditional sandwich complexes refer to those organometallic complexes composed of one metal center and two planar conjugated ligands located in parallel on both sides of the metal center. Inverse sandwich complexes consist of one planar aromatic ligand and two metal centers binding on opposite sides of the ligand plane. This type of compound is also known under the term of "triple decker" sandwiches [1].

18.1 Cyclobutadiene center

Inverse sandwich structures with a cyclobutadiene center have been first reported with iron in [μ -C_4H_4))Fe{Cp(CO)$_2$}$_2$] [2] (Fig. 18.1)

Cp(CO)$_2$Fe $++$

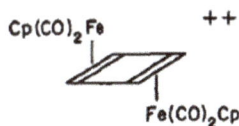

Fe(CO)$_2$Cp   Fig. 18.1: Inverse sandwich with cyclobutadiene center.

Dinickel inverse sandwich complexes with cyclobutadiene centers were prepared from $NiBr_2$ with appropriate dilithio reagents [3].

18.2 Cyclopentadiene center

In these *anti*-bimetallic compounds, the metals are found to be bridged by a single carbocyclic ring. Examples include [(THF)$_3$Ca]$_2$(1,3,5-triphenylbenzene) and [ArSn]$_2$(C_8H_8).

A typical inverse sandwich complex is the cyclopetadienyl triple decker cation [{μ-C_5C_5){Ni(C_5H_5)}$_2$]$^+$ [4, 5] (Fig. 18.2).

Ni

Ni

Fig. 18.2: Inverse sandwich with cyclopentadiene center.

https://doi.org/10.1515/9783110695274-019

Coordination of two metal atoms on opposite sides of the arene leads to the formation of quite unusual inverse sandwich complexes and was found to stabilize low oxidation states of magnesium and calcium in [(thf)$_2$Mg(Br)-C$_6$H$_2$-2,4,6-Ph$_3$] and [(thf)$_3$Ca{μ-C$_6$H$_3$-1,3,5-Ph$_3$}Ca(thf)$_3$] [6] and of gallium and indium [7] in compounds of the type [M(μ,η^5-C$_5$Me$_5$)M]$^+$ (M = Ga [8] and In [9]. Quantum chemical calculations also corroborated the particular stability of such "inverse" sandwich cations [10].

18.3 Benzene center

Titanium forms a benzene-centered inverse sandwich compound [{μ-C$_5$H(SiMe$_3$)$_4$} {Ti(η^5-C$_5$Me$_5$)$_5$}] [11], and a divanadium complex [(μ-C$_6$H$_6$){V(η^5-C$_5$H$_5$)}$_2$ [12] was prepared by reacting (η^5-C$_5$H$_5$)V(C$_3$H$_5$)$_2$ with 1,3-cyclohexadiene in refluxing heptane (Fig. 18.3).

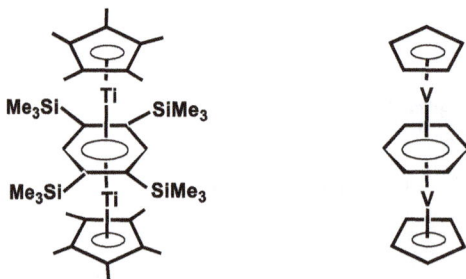

Fig. 18.3: Inverse sandwich with benzene ring as center.

It is worth mentioning that the titanium compound readily undergoes arene exchange reactions with other aromatics, for example, toluene and mesitylene, with retention of the triple-decker sandwich structure.

Another divanadium(I) inverse sandwich complex with a toluene bridge, [(μ-η^6: η^6-C$_7$H$_8$)[V(Nacnac)]$_2$, has a bulky ligand and was prepared by reduction of VCl$_2$ (Nacnac) (Nacnac=HC(C(Me)NC$_6$H$_3$-iPr$_2$)$_2$) with KC$_8$ in toluene [13]. Similar complexes with bulky ligands are known with chromium [14, 15] (Fig. 18.4).

An unprecedented monovalent transition metal complex, [(μ-η^6:η^6-C$_6$H$_6$)(MnAr)$_2$] (Ar = C$_6$H-2,6-(C$_6$H$_2$-2,4,6-Pr$^i_3$)$_2$−3,5-Pr$^i_2$), with η^6-arene coordination, was synthesized by reduction of the corresponding metal halide ArMn-X with potassium graphite in THF [16] (Fig. 18.5). Similar bulky ligands were needed to build dicobalt [17] and dirhodium [18] inverse sandwich compounds.

Nickelocene in benzene reacts with the Brønsted acid H$_2$O.B(C$_6$F$_5$)$_3$ to give salt of [(η^5-C$_5$H$_5$)Ni(η^6-C$_6$H$_6$)Ni(η^5-C$_5$H$_5$)]$^+$ which is the first example of a triple-decker nickel sandwich with a bridging benzene ligand [19] (Fig. 18.5).

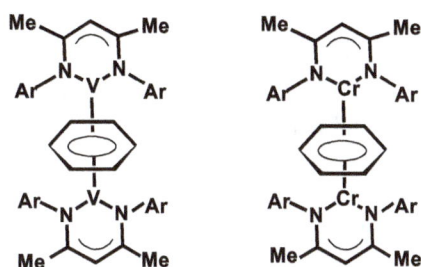

Fig. 18.4: Inverse sandwich complexes with bulky external ligands.

Several diuranium inverse sandwich compounds have been reported including [(μ-C$_6$H$_6$)(UCp*$_2$)$_2$] and [(μ-C$_6$H$_6$)(UCp*X)$_2$ where X = N(SiMe$_3$)$_2$, OC$_6$H$_4$Bu$^t_2$-2,6 and CH(SiMe$_3$)$_2$ [20, 21], [(μ-C$_6$H$_6$)(UX$_2$)$_2$] where X = N(SiMe$_3$)$_2$ [22], [(μ-C$_6$H$_6$)(UX$_3$)$_2$], with X = OSi(OBu^t)$_3$ [23] and others [24, 25] (Fig. 18.5).

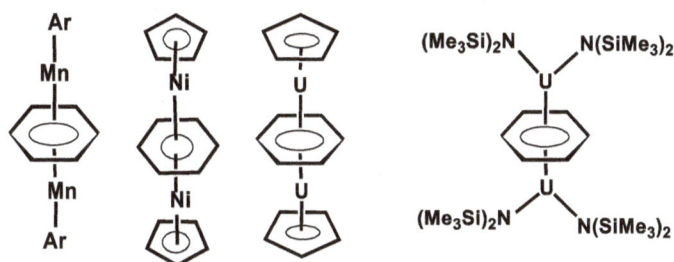

Fig. 18.5: More inverse sandwich complexes.

Unusual divalent lanthanides are stabilized in inverse sandwich complexes [(μ-C$_6$H$_6$)(LnCp^R)$_2$] (where Ln = La, Ce, R = C$_5$H$_4$SiMe$_3$) prepared by reduction of Ln(CH$_2$SiMe$_3$)$_3$ with potassium graphite in benzene. In these complexes, the bridging C$_6$H$_6$ center is non-planar (Fig. 18.6).

Fig. 18.6: Unusual lanthanide inverse sandwich.

18.4 Cyclooctatetraene center

Cycloctatetraene replaces the benzene bridge in the inverse sandwich complex [(μ-C_6H_6){U(C_5Me_5)$_2$}$_2$] to form [(μ-C_8H_8){U(C_5Me_5)$_2$}$_2$] and maintaining the triple decker structure (Fig. 18.7).

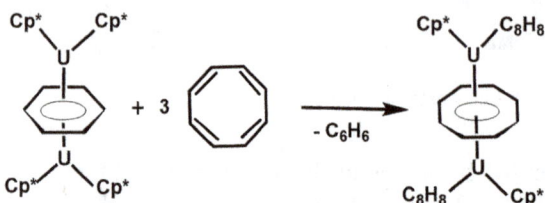

Fig. 18.7: Formation of a uranium inverse sandwich with cyclooctatetraene center.

A triple decker compound containing non-planar central cycloctatetraene molecule [(μ-C_8H_8){Ti(C_8H_8)}$_2$] has also been reported[26] (Fig. 18.8).

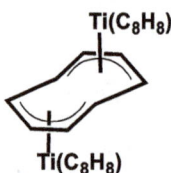

Fig. 18.8: Nonplanar cyclooctatetraene in a dititanium inverse sandwich.

18.5 Fused arene rings as centers

The vanadium inverse sandwich with naphalene bridge [(μ-$C_{10}H_8$){V(η^5-C_5H_5)}$_2$] was obtained from V(η^5-C_5H_5)$_2$ by ligand transfer from Yb($C_{10}H_8$)(THF)$_3$ [27]. A heterobinuclear inverse sandwich compound with naphthalene central bridge [(μ-$C_{10}H_8$){η^4-MnCO)$_3${Fe(η^5-C_5H_5)}] [28] was formed from [(η^6-$C_{10}H_8$)Mn(CO)$_3$]$^+$ with [(η^5-indenyl)Fe (CO)$_3$]$^+$ (Fig. 18.9).

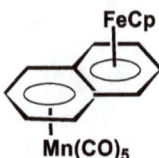

Fig. 18.9: A rare heterobinuclear inverse sandwich.

The reactions of [FeCl(η^5-C_5Me_5)(TMEDA)] with potassium arenes produced inverse sandwich compounds [(η-arene){Fe(η^5-C_5Me_5)}$_2$] (where arene = naphtalene and anthracene). With anthracene the sandwich with one $Fe(\eta^5\text{-}C_5Me_5)$ unit located over the central ring is formed first and this migrates to the outer ring on heating in the solid state [29] (Fig. 18.10).

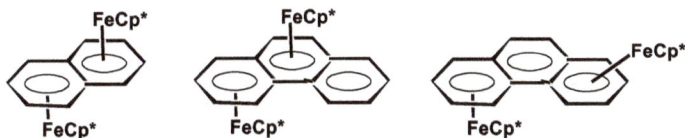

Fig. 18.10: Iron inverse sandwich compounds.

Inverse sandwiches with phenanthrene and pyrene decorated with $Cr(CO)_3$ units have also been obtained [30] (Fig. 18.11).

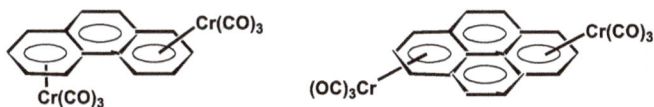

Fig. 18.11: Chromium inverse sandwich complexes.

The phenanthrene compound was formed in the reaction of $Cr(CO)_6$ with phenanthrene during a prolonged reaction time or with $[NH_4]_3[Cr(CO)_3]$. The pyrene complex was prepared in a similar way.

A series of inverse sandwich complexes have been reported with pentalene, *sym*-indacene and *asym*-indacene These include $[(\mu\text{-pentalene})\{M(\eta^5\text{-}C_5Me_5)\}_2$ with M = Fe,Co,Ni,Ru, RuandFe:Ru, Fe:Co pairs, $[(\mu\text{-}(sym\text{-indacene})\{M(\eta^5\text{-}C_5Me_5)\}_2$ and $[(\mu\text{-}(asym\text{-indacene})\{M(\eta^5\text{-}C_5Me_5)\}_2$ with M = Fe, Co, Ni (Fig. 18.12).

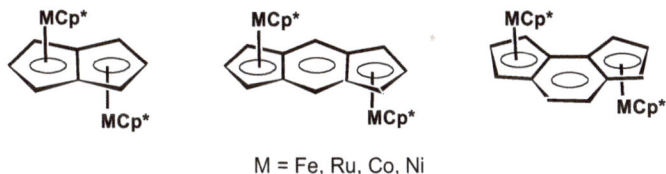

M = Fe, Ru, Co, Ni

Fig. 18.12: Inverse sandwich complexes with pentalene and isomeric indacenes.

The componds were prepared in reactions of $M(\eta^5\text{-}C_5Me_5(acac)$ (where M = Fe, Co, Ni or Ru) with LiC_5Me_5 and $Li_2(arene)$ [31, 32].

Alkaly metals form unexpected inverse sandwich complexes with tetraphenyl-pentalene [33] (Fig. 18.13).

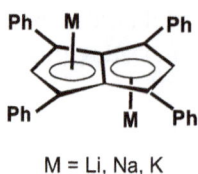

M = Li, Na, K

Fig. 18.13: Inverse sandwich complexes with alkali metals.

A last inverse sandwich to be mentioned here is a triindenyl trinuclear iron compound *syn,syn,anti*-[(μ-tri-indenyl){Fe(η^5-C$_5$H$_5$)}.] obtained by exchange reaction between K$_3$[tri-indenyl] and [Fe(η^5-C$_5$H$_5$)(fluorene)] [34] (Fig. 18.14).

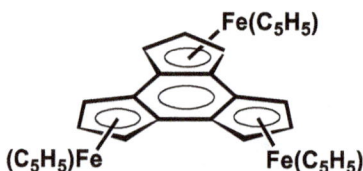

Fig. 18.14: A trimetallic inverse sandwich with tri-indenyl.

References

[1] Lauher JW, Elian M, Summerville RH, Hoffmann R. Triple-decker sandwiches. J Am Chem Soc 1976, 98, 3219–24.

[2] Sanders A, Giering WP. η^2-Cyclobutadienoid transition metal complexes. Preparation and characterization of a binuclear complex possessing a bridging cyclobutadiene ligand. J Am Chem Soc 1974, 96, 5247–48.

[3] Yu C, Wu B, Yang Z, Chen H, Zhang WX, Xi Z. Inverse-sandwich cyclobutadiene dinickel Complexes: Synthesis and structural characterization. Bull Chem Soc Japn 2020, 93, 1314–18.

[4] Salzer A, Werner H. A New Route to Triple-Decker Sandwich Compounds. Angew Chem Int Ed 1972, 11, 930–32.

[5] Dubler E, Textor M, Oswald HA, Salzer A. X-Ray Structure Analysis of the Triple-Decker Sandwich Complex Tris(η-cyclopentadienyl)dinickel Tetrafluoroborate. Angew Chem Int Ed 1974, 135–36.

[6] Krieck S, Görls H, Yu L, Reiher M, Westerhäusen M. (η^2,η^6,η^6-2,4,6-Triphenylbenzene)-hexakis(tetrahydrofuran)-dicalcium stable "inverse" sandwich complex with unprecedented organocalcium(I): Crystal structures of [(thf)$_2$Mg(Br)-C$_6$H$_2$-2,4,6-Ph$_3$] and [(thf)$_3$Ca{μ-C$_6$H$_3$-1,3,5-Ph$_3$}Ca(thf)$_3$]. J Am Chem Soc 2009, 131, 2977.

[7] Dagorne S, Atwood DA. Synthesis, Characterization, and Applications of Group 13 Cationic Compounds. Chem Rev 2008, 108, 4037–71.

[8] Buchin C, Gemel C, Cadenbach T, Schmid R, Fischer RA. The [Ga$_2$(C$_5$Me$_5$)]$^+$ Ion: Bipyramidal Double-Cone Structure and Weakly Coordinated, Monovalent Ga$^+$. Angew Chem, Int Ed 2006, 45, 1074–76.

[9] Jones JN, Macdonald CL, Gorden JD, Cowley AH. Use of a smaller counterion results in an 'inverse sandwich' diindium cation. J Organomet Chem 2003, 666, 3–5.

[10] Fernández I, Cerpa E, Merino G, Frenking G. Structure and Bonding of [E–Cp–E']+ Complexes (E and E' = B–Tl; Cp = Cyclopentadienyl). Organometallics 2008, 27, 1106–11.

[11] Gyepes R, Pinkas J, Cisarova I, Kubista J, Horacek M, Mach K. Synthesis, molecular and electronic structure of a stacked half-sandwich dititanium complex incorporating a cyclic p-faced bridging ligand. RSC Adv 2016, 6, 94149–59.

[12] Duff AW, Jonas K, Goddard R, Kraus HJ, Krueger C. The first triple-decker sandwich with a bridging benzene ring. J Am Chem Soc 1983, 105, 5479–80.

[13] Tsai Y-C, Wang PY, Lin K-M, S-a, Chen J-M. Synthesis and reactions of β-diketiminato divanadium(i) inverted-sandwich complexes. Chem.Commun. 2008, 205.

[14] Huang Y-S, Huang G-T, Liu Y-L, Yu J-SK, Tsai Y-C. Reversible Cleavage/Formation of the Chromium–Chromium Quintuple Bond in the Highly Regioselective Alkyne Cyclotrimerization. Angew Chem Int Ed 2017, 56, 15427–31.

[15] Monillas WH, Yap GPA, Theopold KH. A Tale of Two Isomers: A Stable Phenyl Hydride and a High-Spin (S=3) Benzene Complex of Chromium. Angew Chem Int Ed 2007, 46, 6692–94.

[16] Ni C, Ellis BD, Fettinger JC, Long GJ, Power PP. Univalent transition metal complexes of arenes stabilized by a bulky terphenylligand: Differences in the stability of Cr(i), Mn(i) or Fe(I) complexes. Chem Commun 2008, 1014–16.

[17] DeRosha DE, Mercado BQ, Lukat-Rodgers G, Rodgers KR, Holland PL. Enhancement of C-H oxidizing ability in CoO_2 complexes through an isolated heterobimetallic oxo intermediate. Angew Chem Int Ed 2017, 56, 3211.

[18] Zhang N, Sherbo RS, Bindra GS, Zhu D, Budzelaar PHM. Rh and Ir β-Diiminate Complexes of Boranes, Silanes, Germanes, and Stannanes. Organometallics 2017, 36, 4123–235.

[19] Priego JL, Doerrer H, Rees LH, Green MLH. Weakly-coordinating anions stabilise the unprecedented monovalent and divalent η-benzene nickel cations [(η-C_5H_5)Ni(η-C_6H_6)Ni(η-C_5H_5)]$^{2+}$ and [Ni(η-C_6H_6)$_2$]$^{2+}$. Chem.Commun. 2000, 779–80.

[20] Evans WJ, Kozimor SA, Ziller JW, Kaltsoyannis N. Structure, reactivity, and density functional theory analysis of the six-electron reductant, [(C_5Me_5)$_2$U]$_2$(µ-η^6:η^6-C_6H_6), synthesized via a new mode of (C_5Me_5)$_3$M reactivity. J Am Chem Soc 2004, 126, 14533–47.

[21] Evans WJ, Traina CA, Ziller JW. Synthesis of heteroleptic uranium (M-η^6:η^6-C_6H_6)$^{2-}$ sandwich complexes via facile displacement of (η^5-C_5Me_5)$^{1-}$by ligands of lower hapticity and their conversion to heteroleptic bis(imido) compounds. J Am Chem Soc 2009, 131, 17473–81.

[22] Arnold PL, Prescimone A, Farnaby JH, Mansell SM, Parsons, Kaltsoyannis N. Characterising pressure-induced uranium C-H agostic bonds. Angew Chem Int Ed 2015, 54, 6735–39.

[23] Camp C, Mougel V, Pecaut J, Maron L, Mazzanti M. Cation-mediated conversion of the state of charge in uranium arene inverted-sandwich complexes. Chem -Eur J 2013, 19, 17528.

[24] Mills P, Moro F, McMaster J, van Slageren W, Lewis W, Blake AJ, Liddle ST. A delocalized arene-bridged diuranium single-molecule magnet. Nat Chem 2011, 3, 454.

[25] Kotyk CM, Fieser ME, Palumbo CT, Ziller JW, Darago LE, Long JR, Furche F, Evans WJ. Isolation of +2 rare earth metal ions with three anionic carbocyclic rings: Bimetallic bis (cyclopentadienyl) reduced arene complexes of La^{2+} and Ce^{2+} are four electron reductants. Chem Sci 2015, 6, 7267–72.

[26] Breil H, Wiike G. Di(cyclooctatetraene)titanium and tri(cyclooctatetraene)-dititanium. Angew Chem Int Ed 1966, 5, 898–99.

[27] Bochkarev MN, Fedushkin IL, Schumann H. Loebel J Reactions of naphthaleneytterbium with bis(cyclopentadienyl) complexes of cobalt, nickel, chromium and vanadium. X-Ray crystal structure of the triple-decker CpV$C_{10}H_8$VCp. J.Organomet.Chem. 1991, 410, 321–26.

[28] Rheingold JA, Virkaitis KL, Carpenter GB, Sun S, Sweigart DA, Czech PT, Overly KR. Chemical and electrochemical reduction of polyarene manganese tricarbonyl cations: Hapticity changes and generation of *syn*- and *anti*-facial bimetallic η^4,η^6-naphthalene complexes. J Am Chem Soc 2005, 127, 11146–58.

[29] Hatanaka T, Ohki, Kamachi T, Nakayama T, Yoshizawa K, Katada M, Tatsumi K. Naphthalene and anthracene complexes sandwiched by two [(Cp*)FeI] fragments: Strong electronic coupling between the FeI centers. Chem Asian J 2012, 7, 1231–42.

[30] Peitz DJ, Palmer RT, Radonovich LJ, Woolsey NF. Preparation and structure of (μ-phenanthrene)- and (μ-pyrene)-bis(tricarbonylchromium). Organometallics 1993, 12, 4580–84.

[31] Manriquez JM, Ward MD, Reiff WM, Calabrese JC, Jones NL, Carroll PJ, Bunel EE, Miller JS. Structural and physical properties of delocalized mixed-valent [Cp*M(pentalene)M'Cp*]$^{n+}$ and [Cp*M(indacene)M'Cp*]$^{n+}$ (M, M' = Fe, Co, Ni; n = 0, 1, 2) complexes. J Am Chem Soc 1995, 117, 6182–93.

[32] Carey DM-L, Morales-Verdejo C, Muñoz-Castro A, Burgos F, Abril A, Adams C, Molins E, Cador O, Chavez I, Manriquez JM, Arratia-Perez R, Saillard JY. Polyhedron 2010, 29, 1137–43.

[33] Boyt SM, Jenek NA, Sanderson HJ, Kociok-Köhn G, Hintermair U. Synthesis of a Tetraphenyl-Substituted Dihydropentalene and Its Alkali Metal Hydropentalenide and Pentalenide Complexes. Organometallics 2021, 10.1021/acs.organomet.1c00495.

[34] Santi S, Orian L, Donoli A, Bisello A, Scapinello M, Benetollo F, Ganis P. Cecconi A Synthesis of the prototypical cyclic metallocene triad: Mixed-valence properties of [(FeCp)₃(trindenyl)] Isomers. Angew Chem Int Ed 2008, 47, 5331–34.

19 Organometallic compounds with σ-transition metal–carbon bonds

19.1 General

The organometallic compounds containing σ-transition metal–carbon bonds have been considered, for a long while, less stable than the main group metal organometallics. The presence of unoccupied d-orbitals in the valence shell is a source of kinetic lability. There was a much slower development of the chemistry of homoleptic σ-transition metal organometallics compared to other classes of compounds described in previous chapters. However, there are early examples of σ-transition metal organometallics: $(CH_3)_3PtI$ (1909) [1] (with the tetrameric structure $[(CH_3)_3PtI]_4$ reported later in 1998 [2]), and $Li[Cu(CH_3)_2]$, (1952) [3] (the crystal and molecular structure determined in 1998 along with other organocuprates), the anions $[CuMe_2]^-$, $[CuPh_2]^-$ and the intermediate, monosubstituted species $[Cu(Br)CH(SiMeI)]^-$ [4] obtained as salts with $Li(12$-crown-4$)^+$ counterion.

The kinetic lability of the σ-bonded transition metal organometallics can be caused by several mechanisms of decomposition. The most relevant are β-hydride elimination, β-alkyl elimination and α-hydrogen abstraction. The hydrogen atom in the β-position of an organic ligand is interacting with the empty d-orbitals of the transition metals (agostic bond), and the result is the formation of an olefin and transition metal hydride (Fig. 19.1).

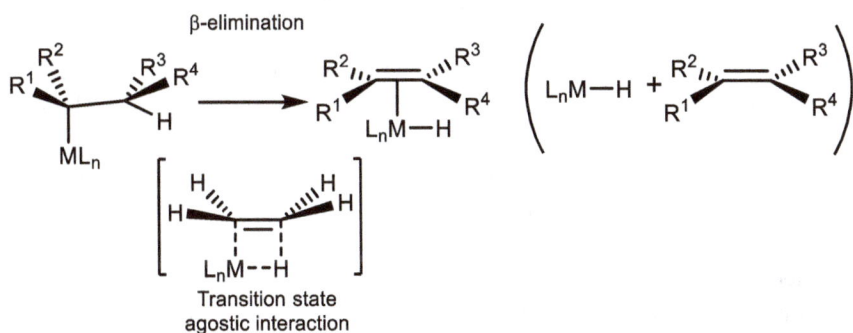

Fig. 19.1: β-Elimination.

Organic groups such as CH_2SiR_3, $CH(SiR_3)_2$, CH_2Ph, CH_2CMe_3 lacking a β-hydrogen, form rather stable σ-derivatives. Such groups are also bulky and, therefore, exhibit a favoring steric influence as well.

Another mechanism of decomposition, sometimes in competition with β-hydride elimination, is β-methyl elimination:

https://doi.org/10.1515/9783110695274-020

M = Ti, Zr, Nb, Hf

Fig. 19.2: β-Methyl elimination.

α-Abstraction might be viewed as the analogue of α-hydride elimination in instances where the alkyl group possesses no β-hydrogen atoms (Fig. 19.3).

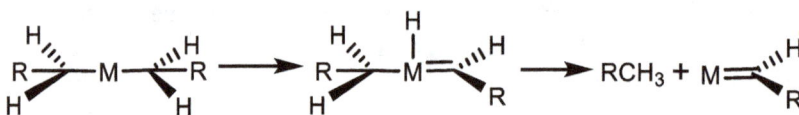

Fig. 19.3: α-Abstraction.

It is worth mentioning that all three mechanisms are relevant for synthetic applications in organic chemistry.

The stability of σ-bonded transition metal organic derivatives is increased when stabilizing factors are involved:

– by use of organic groups of appropriate structures to avoid β-elimination,
– steric protection with the aid of bulky substituents,
– chelate ring formation (participation of the transition metal as a heteroatom in a metallacycle),
– coordination of certain ligands to the transition metals to block the d-orbitals and prevent the decomposition via mechanisms involving empty d-orbitals.

The σ-bonded group R can be alkyl, aryl, σ-vinyl or σ-allyl, alkynyl, perhalogenated alkyls or aryls (CF_3, C_3F_7, C_6Cl_5, C_6F_5, etc.), acyl or σ-alkynyl groups (-C≡C-R). The most favored are those unable of β-elimination.

The electronegative character of the organic group also increases the stability of the σ-bonded compounds. Thus, aromatic derivatives and polyfluorinated or polychlorinated groups yield more stable compounds.

Ligands with π-donor properties facilitate the use of both occupied and vacant metal d-orbitals to achieve noble gas configurations. Thus, metal carbonyl or cyclopentadienyl metal moieties lacking only one electron form σ-bonded organic derivatives. Very often, the role of these ligands is only secondary in determining the structure and properties of the organometallic compound.

Based on the stabilizing factors the following types of compounds with σ-metal–carbon bonds have been described:

a) Homoleptic alkyl and aryl derivatives (neutral or ionic), that is, compounds containing exclusively σ-metal–carbon bonds;
b) Heteroleptic compounds, containing σ-transition metal–carbon bonds and additional ligands. These may include:
   - adducts of homoleptic derivatives with donor ligands, such as neutral CrPh$_3$.3THF, cationic [CrPh$_2$(bipy)$_2$]$^+$, or anionic [R$_3$AuX]$^-$;
   - compounds with monofunctional ligands, for example, halide R-MX, alkoxy R-M(OR)$_n$, amino R-M(NR'$_2$)$_n$ or mercapto derivatives, R-M(SR')$_n$;
   - metal carbonyl derivatives, R$_n$M(CO)$_x$;
   - cyclopentadienylmetal derivatives, (η^5-C$_5$H$_5$)$_m$MR$_n$ and cyclopentadienyl-metal carbonyl derivatives, (η^5-C$_5$H$_5$)MR$_n$(CO)$_x$.
c) Chelate rings and metallocycles with σ-carbon and M–X (X = O, N, S, P, As, etc.) bonds.

A comparison of the thermodynamic parameters like the metal–carbon bond energy (or bond dissociation enthalpy) suggests no significant differences between transition metal and main group organometallics. The bond strengths of second-row transition metal-carbon bonds was studied for different hybridizations on carbon using the set of molecules M-CH$_3$, M-CH = CH$_2$, and M-C $\equiv$ CH without additional ligands. The transition metal–carbon bond strength depends on the hybridization of the carbon atoms (decreases in the order sp > sp^2 > sp^3) and on the electronic structure of the metal (decreases from left to right in the periodic table). For alkyl chains of different lengths and with different numbers of substituents on the bonding carbon, M-CH$_3$, M-C$_2$H$_5$, M-n-C$_3$H$_7$, and M-iso-C$_3$H$_7$ (the same type of hybridization, sp^3) the strength of the transition metal–carbon bond decreases in the order M-methyl > M-ethyl > M- n-propyl > M-iso-propyl. The difference between the metal-alkyl bond strengths is larger to the left in the Periodic Table while the difference essentially disappears to the right [5]. There is a significant difference in the trend of the transition metal–carbon bond strength for transition metals compared to main group metals: the bond energy increases down the transition metal group, for example, Ti(CH$_2$CMe$_3$)$_4$ (185 kJ/mol), Zr (CH$_2$CMe$_3$)$_4$ (226 kJ/mol), Hf(CH$_2$CMe$_3$)$_4$ (243 kJ/mol) [6], while for the main group metals the bond energy is decreasing, for example, Si–Me (290 kJ/mol) and Pb–Me (130 kJ/mol). If we consider that the bond energy of Ti–Me bond is 200 kJ/mol and Ge–Me is 240 kJ/mol (same period), it is clear that the transition metal–carbon bonds are thermodynamically in the same range as the main group–carbon bonds.

The syntheses of σ-bonded derivatives are no different from those used in main group organometallic chemistry. Among the most important are:
- Reaction of a metal halide or halogeno complex with organolithium, organomagnesium or other organometallic reagents able to transfer an organic group:

$$M\text{-}X + M'R \longrightarrow M\text{-}R + M'X$$
$$M = Li, MgX, Al, Hg, Tl, etc.$$

– Reaction of a metal hydride with an olefin (addition). This reaction is the reverse of β-elimination:

$$M\text{-}H + H_2C\text{=}CHR \longrightarrow M\text{-}CH_2\text{-}CH_2R$$

– Reaction of a metal hydride with diazomethane to give σ-methyl derivatives:

$$M\text{-}H + CH_2N_2 \longrightarrow M\text{-}CH_3 + N_2$$

– Reaction of an anionic metal complex anion (metal carbonyl or cyclopentadienylmetal carbonyl) with a halogenated organic compound:

$$L_nM^- + RX \longrightarrow L_nM\text{-}R + X^-$$

– Oxidative addition of polar organic substrates to 16-electron metal complexes (Fig. 19.4).

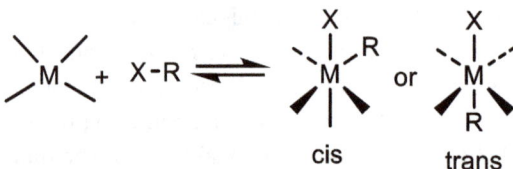

cis   trans

Fig. 19.4: Oxidative addition.

A particular oxidative addition specific for aryl derivatives is the ortho-metallation. A C–H bond of the phenyl group, part of a ligand coordinated to the metal, is cleaved by the metal to form a new M–C bond and a C–H bond:
– nucleophilic attack on coordinated ligands (Fig. 19.5);

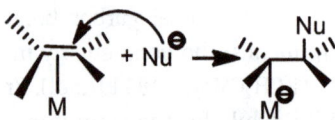

Fig. 19.5: Nucleophilic attack on coordinated olefins.

– reaction of metal vapor with organic halides:

$$M + RX \longrightarrow R\text{-}MX \text{ or } R\text{-}M + X$$

19.2 Homoleptic compounds

The use of organic ligands lacking hydrogens in the β-position led to first examples of homoleptic organometallic compounds with σ-transition metal–carbon bonds with R = CH(SiMe$_3$)$_2$ and R = CH$_2$SiMe$_3$, CH$_2$CMe$_3$ and CH$_2$SnMe$_3$ [7].

The perhalophenyl groups (C_6X_5, X = F, Cl) were proved to be the most suitable ligands to form homoleptic anions $[M(C_6X_5)_n]^{z-}$ with the first-row transition metals, as well as with a number of the heavier ones (M = Zr, Hf, Rh, Ir, Pd, Pt, Ag, Au). The stability, molecular geometry and other properties are determined by the nature of the ligand, the electronic configuration of the $M^{(n-z)+}$ ion and its size. Their stoichiometry is related to a compromise between electronic and steric factors. An important mechanism to gain stability is to reduce the electronic unsaturation of a metal ion (Lewis acid) by binding the highest number of C_6X_5 ligands (Lewis bases) allowed by interligand repulsions: maximum four pentachlorophenyl ligands, $[M(C_6Cl_5)_4]^{z-}$ but up to six for the less bulky pentafluorophenyl group in the case of larger sized and/or electron poorer metals: $[M(C_6F_5)_5]^{2-}$ (M = Ti, V, Cr, Rh) and $[M(C_6F_5)_6]^{2-}$ (M = Zr, Hf). Most of the $[M(C_6X_5)_n]^{z-}$ compounds have an open-shell electronic structure (<18 electrons, effective atomic number rule) [8]. A list of pentahalogenophenyl homoleptic species is provided below. Most are hypervalent anions, that is, compounds in which the number of M–C bonds is larger than the formal valence of the metal:

d^0 $[Ti^{IV}(C_6Cl_5)_4]$ (8 e$^-$), $[Zr^{IV}(C_6F_5)_6]^{2-}$ (12 e$^-$), $[Hf^{IV}(C_6F_5)_6]^{2-}$ (12 e$^-$);

d^1 $[Ti^{III}(C_6Cl_5)_4]^-$ (9 e$^-$), $[Ti^{III}(C_6F_5)_5]^{2-}$ (11 e$^-$), $[V^{IV}R_4]$ (9 e$^-$), R = C_6Cl_5, C_6F_5;

d^2 $[V^{III}(C_6Cl_5)_4]^-$ (10 e$^-$), $[Cr^{IV}(C_6Cl_5)_4]$ (10 e$^-$);

d^3 $[Cr^{III}(C_6Cl_5)_4]^-$ (15 e$^-$), $[Cr^{III}(C_6F_5)_5]^{2-}$ (13 e$^-$);

d^4 $[Cr^{II}(C_6Cl_5)_4]^{2-}$ (12 e$^-$), $[Cr^{II}(C_6F_5)_4]^{2-}$ (12 e$^-$), $[Mn^{III}(C_6F_5)_4]^-$ (12 e$^-$);

d^5 $[Fe^{III}(C_6Cl_5)_4]^-$ (13 e$^-$);

d^6 $[Fe^{II}(C_6F_5)_4]^{2-}$ (14 e$^-$), $[Co^{III}(C_6Cl_5)_4]^-$ (14 e$^-$) R = C_6Cl_5, C_6F_5, $[Rh^{III}(C_6Cl_5)_3]$ (18 e$^-$), $[Rh^{III}(C_6Cl_5)_4]^-$ (18e$^-$), $[Rh^{III}(C_6F_5)_4]$ (16 e$^-$), $[Pt^{IV}(C_6Cl_5)_4]$ (18 e$^-$);

d^7 $[Co^{II}R_4]^{2-}$ (15 e$^-$) R = C_6Cl_5, C_6F_5, $[Rh^{II}(C_6Cl_5)_4]^{2-}$ d^7 (15 e$^-$), $[Ir^{II}(C_6Cl_5)_4]^{2-}$ (15 e$^-$), $[Ni^{III}R_4]^-$ (15 e$^-$) R = C_6Cl_5, C_6F_5, $[Pt(C_6Cl_5)_4]^-$ (15 e$^-$);

d^8 $[Ni^{II}R_4]^{2-}$ (16 e$^-$) R = C_6Cl_5, C_6F_5, $[Pd^{II}R_4]^{2-}$ (16 e$^-$) R = C_6Cl_5, C_6F_5, $[Pt^{II}(C_6Cl_5)_4]^{2-}$, (16 e$^-$), $[Au^{III}(C_6Cl_5)_4]^-$, $[Au^{III}(C_6F_5)_4]^-$ (16 e$^-$);

d^9 $[Cu^IR_2]^-$ (14 e$^-$) R = C_6Cl_5, C_6F_5, $[Ag^IR_2]^-$ (14 e$^-$) R = C_6Cl_5, C_6F_5, $[Au^IR_2]^-$ (14 e$^-$) R = C_6Cl_5, C_6F_5.

A special case is the unsaturated empty-shell compound $[Ti^{IV}(C_6Cl_5)_4]$ with only eight valence electrons, prepared by oxidation of organotitanium(III) anion $[Ti^{III}(C_6Cl_5)_4]^-$ [9].

Unprecedented dimetallated benzene compounds have been obtained by stoichiometric 1,4-double deprotonation of the aromatic ring to form a peculiar type of inverse sandwich complexes in which the benzene rings are embedded in cyclic structures formed by tetramethylpyperidine ligands alternating with metal pairs of chromium–sodium [10], manganese–sodium [11] and iron–sodium [10] (e.g., Figure 19.6).

Similar complexes were obtained by manganation of naphthalenes, anthracene and phenathrene [12].

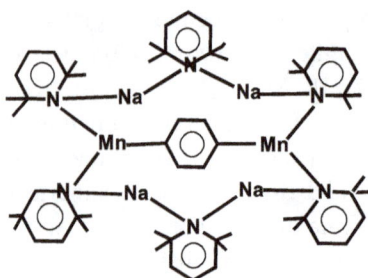

Fig. 19.6: Dimetallated benzene derivative.

19.2.1 Titanium, zirconium, hafnium

The homoleptic TiR_4 are prepared from titanium tetrachloride and alkyllithiums. The unstable tetramethyltitanium can be stabilized in the orthophenanthroline or bipyridyl complexes. The fully substituted $TiMe_4$ can be further converted to $Li[TiMe_5]$.

Tetrasubstituted derivatives, $M(CH_2SiMe_3)_4$ (M = Ti, Zr, Hf) and $Ti(CH_2Ph)_4$, are stable, and the trisubstituted complexes ($Ti(CH_2SiMe_3)_3$ and $Ti[CH(SiMe_3)_3]$) are also known. The compound, $Ti[CH(SiMe_3)_3$, forms from $TiCl_4$ and $LiCH(SiMe_3)_2$. Titanium tetrachloride and phenyllithium form tetraphenyltitanium, which polymerizes to give $(TiPh_2)_x$.

Tetrakis(pentafluorophenyl)zirconium, $Zr(C_6F_5)_4$, and tetrabenzylzirconium, $Zr(CH_2Ph)_4$, are known, but tetraphenylzirconium is unstable. Six Zr–C bonds are formed - in the hypervalent anion, $[ZrMe_6]^{2-}$.

Ylide derivatives are known for titanium and zirconium:

$$MCl_4 + nMe_3P{=}CH_2 \longrightarrow Cl_4M(CH_2PMe_3)_n$$

M = Ti, n = 2 and 3
M = Zr, n = 2, 3 and 4

$$(\eta^5\text{-}C_5H_5)_2MCl_2 + 2Me_3P{=}CH_2 \longrightarrow [(\eta^5\text{-}C_5H_5)_2M(\text{-}CH_2\text{-}PMe_3)_2]^{2+}\ 2Cl^-$$

M = Ti, Zr

19.2.2 Vanadium, niobium, tantalum

Vanadium trichloride forms the hypervalent anionic, hexasubstituted $Li_4[VPh_6]$, with phenyllithium and triphenylvanadium is obtained from $VCl_3.3THF$ and phenyllithium in THF. Both trimethylsilylmethyl derivatives, $V(CH_2SiMe_3)_n$ with $n = 3$ and 4 are known. 2,4,6-Trimethylphenyl (Mes) derivatives of the neutral MR_3 and anionic $[MR_4]^-$ are also known.

The reaction of $[VCl_3(thf)_3]^-$ with LiC_6Cl_5 in a 1:8 molar ratio followed by the appropriate treatment allows the isolation of $[NBu_4][V^{III}(C_6Cl_5)_4]$. This complex is air- and moisture-stable in the solid state. This behavior is in contrast with the observation of the ease with which the related compound [Li(thf)$_4$][V(Mes)$_4$] is air oxidized to give neutral [V(Mes)$_4$] [13].

The arylation of [VCl$_3$(thf)$_3$] with organolithium derivatives, LiR, of polychlorinated phenyl group [R = 2,4,6-trichlorophenyl or 2,6-dichlorophenyl gives four-coordinate, homoleptic organovanadium(III) derivatives of the [V^{III}R$_4$]$^-$ anions. The arylation of [VCl$_3$(thf)$_3$] with LiC$_6$F$_5$ also gives a homoleptic organovanadium(III) compound, but with a different stoichiometry: [V^{III}(C$_6$F$_5$)$_5$]$^{2-}$. In this five-coordinated species, the C$_6$F$_5$ groups define a trigonal bipyramidal environment for the vanadium atom [14].

High-spin and redox-active tetrahedral complexes of V(III), Fe(II) and Mn(II) were prepared with substituted phenylacetylide ligand 2,6-bis(trimethylsylyl)phenylacetylene, for example, the anion [(2,6-(Me$_3$Si)$_2$Ph-CC)$_4$V^{III}]$^-$.

The pentamethylniobium and -tantalum obtained from the corresponding methylmetal chlorides and methyllithium decompose by α-hydrogen abstraction. The hexamethyl derivative, TaMe$_6$, explodes even in vacuo. Highly substituted hypervalent phenyl anions, [NbPh$_6$]$^{4-}$, [TaPh$_6$]$^-$ and [NbPh$_7$]$^{3-}$ are also known.

19.2.3 Chromium, molybdenum, tungsten

Tetra-alkyl chromium compounds, CrR$_4$ (R = neopentyl, neophyl, tritylmethyl, and methyl), have been prepared by the interaction of the Grignard or lithium with the tetrahydrofuran adduct of chromic chloride, CrCl$_3$.3THF, or in the case of tetramethylchromium by an exchange reaction between methyllithium and chromium(IV) *tert*-butoxide. Chromium(III) chloride gives the trisubstituted CrPh$_3$ · 3THF from phenylmagnesium bromide in THF, which is readily converted to the η^6 -complexes of benzene and biphenyl. Excess of phenyllithium gives the hexasubstituted anion, Li$_3$ [CrPh$_6$] · nEt$_2$O. A tetraphenylchromium compound Li$_2$[CrPh$_4$] · 4THF is also known.

Disproportionation of Li$_3$[CrPh$_6$] · nEt$_2$O with CrCl$_3$ leads to Li[CrPh]$_3$ · nEt$_2$O or Li[CrPh$_4$],

The pentaphenylchromium derivative forms an adduct with Na(OEt$_2$)$_2$ of composition [CrPh$_5$]Na$_2$(Et$_2$O)$_3$.THF, which contains a trigonal bipyramidal CrPh$_5$ unit with interactions between sodium and the phenyl substituents (Fig. 19.7).

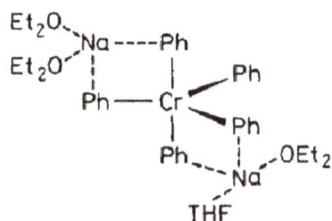

Fig. 19.7: Pentaphenylchromium complex.

Benzylmagnesium bromide forms with $CrCI_3$ a trisubstituted derivative, $Cr(CH_2Ph)_3$, which decomposes to form η^6-arene complexes.

A binuclear compound containing a chromium–chromium triple bond is obtained from $CrCl_2$ and methyllithium (Fig. 19.8).

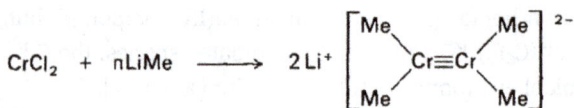

$$CrCl_2 \; + \; nLiMe \; \longrightarrow \; 2\,Li^+ \left[\begin{array}{c} Me \quad\quad Me \\ \diagdown \quad\quad \diagup \\ Cr\equiv Cr \\ \diagup \quad\quad \diagdown \\ Me \quad\quad Me \end{array} \right]^{2-}$$

Fig. 19.8: A tetramethyldichromium anion.

Multiple metal–metal bonding is also found in $Cri(CH_2SiMe_3)_5(PMe_3)_2$, which contains both bridging and terminal CH_2SiMe_3 groups (Fig. 19.9).

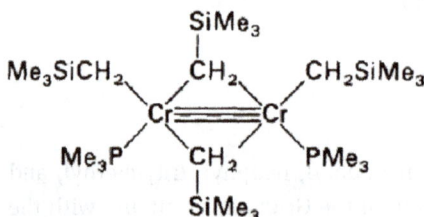

Fig. 19.9: A unique dichromium compound.

Six Mo–C σ-bonds are found in the $[MoPh_6]^{3-}$ anion, while tungsten forms $[WMe_8]^{2-}$ anions. The reaction of WCl_i with methyllithium or trimethylaluminum gives WMe_6 which is explosive. Excess methyllithium forms the $[WMe_8]^{2-}$ anion.

Molybdenum forms the binuclear, metal–metal triple-bonded compound $Mo_2(CH_2SiMe_3)_6$ (Fig. 19.10) from $MoCl_6$ and $Me_3Si–CH_2MgCl$.

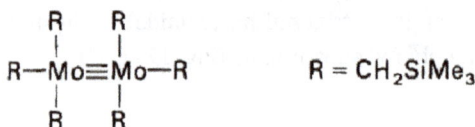

Fig. 19.10: A binuclear molybdenum compound.

A quadruple Mo–Mo bond is found in $Li_2[Mo_2Me_8] \cdot 4THF$. Related tungsten compounds are also known, including $Li_2[W_2Me_8] \cdot 4Et_2O$ and $W_2(CH_2SiMe_3)_6$, which contain multiple metal–metal bonds.

19.2.4 Manganese, technetium, rhenium

The thermally stable manganese(II) derivatives, MnR_2 ($R = CH_2SiMe_3$, CH_2Bu^i, CH_2-CMe_2Ph), are prepared using alkylmetal intermediates. The hypervalent anionic species, $[Mn^{II}R_3]^-$ and $[Mn^{II}R_4]^{2-}$, are obtained with $R = Me$, and $C \equiv C\text{-}R'$ ($R' = H$, Me, Ph).

The rhenium anion $[Re_2Me_8]^{2-}$ is prepared from $ReCl_5$ and LiMe and contains a quadruple Re–Re bond.

19.2.5 Iron, ruthenium, osmium

A rare homoleptic dimesityliron derivative is prepared from $FeCl_2$ by a Grignard route, and lithium tetrasubstituted ferrate anions and $[FeR_4]^-$ ($R = Me$, Ph) can be isolated. Six σ-Fe–carbon bonds are found in the hypervalent alkynyl–iron anions, $[Fe(C \equiv CR)_6]^{2-}$ ($R = H$, Me, Ph).

19.2.6 Cobalt, rhodium, iridium

With Co(II), only hypervalent anions are formed and tetrasubstituted cobalt anions, $[CoR_4]^{2-}$ with $R = CH_2SiMe_3$, Ph, C_6F_5, C_6Cl_5, Me are known. Hexasubstitution is achieved in the ethynyl derivatives, $[Co(C \equiv CR)_6]^{3-}$ and $[Co(C \equiv CR)_6]^{4-}$.

Apparently, rhodium and iridium compounds of this category are unknown.

19.2.7 Nickel, palladium, platinum

A six σ-alkyl platinum complex, the hexamethylplatinate, $[PtMe_6]^{2-}$, is obtained from $(Me_3PtI)_4$ or $(NR_4)_2[PtCl_6]$ with methyllithium.

19.2.8 Copper, silver, gold

With excess phenyllithium, copper(I) iodide forms the unstable tetrasubstituted anion as lithium salt $Li[CuPh_4] \cdot nEt_2O$.

Polymeric monovalent RCu compounds are obtained from copper(I) halides with organolithium, organozinc or Grignard reagents. The tetrameric pentafluorophenyl derivative, $(C_6F_5Cu)_4$ [15], is more stable than the alkyls. The trimethylsilylmethyl derivative is a tetramer and contains alkyl bridges and weak Cu–Cu bonding interactions.

For the applications of organocopper compounds in organic synthesis, the solubility is crucial [16] and it was found that mesitylcopper, $(MesCu)_n$ ($n = 4$, 5; Mes = 2,4,6-trimethylphenyl) is soluble in benzene, ether, THF and partially soluble in hexane. It can be readily prepared by a metathesis reaction between copper(I) chloride and mesitylmagnesium bromide in THF. Mesitylcopper represents an ideal approach to a readily soluble, stable, and versatile organocopper(I) synthon in organic synthesis. The tetrameric structure, similar with the structure of trimethylsilylmethyl derivative and the pentameric structure [17] are presented in Fig. 19.11. Bulkier complexes 2,4,6-triisopropylphenylcopper, 2,4,6-triethylphenylcopper, and 2,4,6-triethylphenylsilver have also tetrameric structures. Monomeric mesityl complexes were prepared, starting with the oligomeric framework of mesitylcopper in reaction with stronger σ-donor coligands or π-acceptors. Mesitylcopper with a N-heterocyclic carbene ligand yields a monomeric complex with a linearly coordinated Cu^I center (Fig. 19.11) [18].

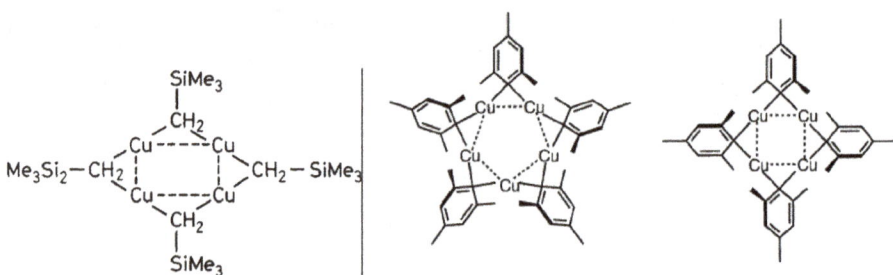

Fig. 19.11: Tetrameric and pentameric copper compounds.

Bis(alkynyl)titanocene π-accepting metalloligands (the so-called "organometallic π-tweezers") have the appropriate geometry to coordinate a [CuMes] unit within the Ti(C ≡ CSiMe$_3$)$_2$ binding pocket (Fig. 19.12) [19].

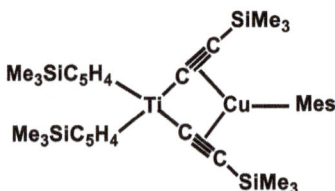

Fig. 19.12: Unusual mesitylcopper compound.

With crowded terphenyl ligand, Mes$_2$(C$_6$H$_3$), a organocopper dimer, was prepared and structurally characterized (Fig. 19.13) [20].

This complex shows two copper(I) centers with the formal coordination number 1 but each Cu^I ion lies close to the aromatic system of one mesityl substituent of the neighboring terphenyl ligand in a η^2-binding mode.

The disubstituted anionic complexes, [CuR$_2$]$^-$, serve as reagents in synthetic organic chemistry. Anionic alkynyl derivatives, [Cu^I(C = CR)$_2$]$^-$ (R = Me, Ph) and [Cu^I (C = CR)$_3$]$^{2-}$ (R = H, Me, Ph), are also known.

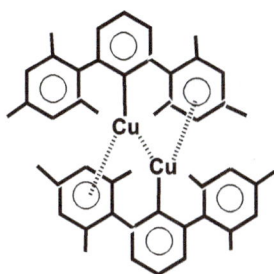

Fig. 19.13: Organocpper(I) dimer.

Organolead, tin and bismuth compounds form with silver nitrate the compound $[RAg]_2 \cdot AgNO_3$. The polymeric phenylsilver and other σ-aryl derivatives, $[AgR]_x$, are prepared from silver(I) salts and organozinc reagents. The dimeric, disubstituted silver derivatives, which contain aryl groups bridging silver and lithium atoms, are obtained from lithium aryls.

Di– and tetramethylaurate anions, $[AuMe_2]^-$ and $[AuMe_4]^-$, are prepared from organolithium reagents:

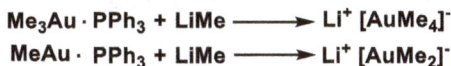

$$Me_3Au \cdot PPh_3 + LiMe \longrightarrow Li^+ [AuMe_4]^-$$
$$MeAu \cdot PPh_3 + LiMe \longrightarrow Li^+ [AuMe_2]^-$$

The former is linear, while the latter is square planar. The anions are more thermally stable than neutral species but less stable to oxygen.

Anionic species $[Au(C_6F_5)_n]^-$ ($n = 2$ or 4) are also known.

19.3 Heteroleptic compounds

The σ-bond organic derivatives of transition metals may be stabilized by a variety of additional ligands (co-ligands like halogens, amines, phosphines, halogens, etc.) which form heteroleptic compounds.

19.3.1 Organometallic halides

A class of heteroleptic compounds includes organometallic halides, known for several transition metals.

Titanium tetrachloride forms organotitanium halides, $MeTiCl_3$ and Me_2TiCl_2, in the reaction with organoaluminum and organolead reagents.

Zirconium and hafnium tetrachloride with $LiCH(SiMe_3)_2$ yield the triorganometal chlorides, $[(Me_3Si)_2CH_2]_3MCl$ (M = Zr, Hf).

Niobium and tantalum pentachlorides react with dimethylzinc to form the unstable trimethyl dichlorides, Me_3MCl_2.

The tetrasubstituted anions, $[NiR_4]_2$ (with R = Me, Ph), and halogen-bridged di-nuclear anions, $[(C_6F_5)_2NiX_2Ni(C_6F_5)_2]^{2-}$ (X = Cl, Br, also CN), are also known.

The square planar palladium compounds $(PR_3)_2PdRX$, and $(PR_3)_2PdR_2$ are prepared from the dihalides with Grignard reagents as both *cis*- and *trans*-isomers.

Platinum forms cubic, tetrameric $(Me_3PtX)_4$ (X = Cl, I) derivatives (Fig. 19.14) from the reaction of platinum(II) chloride with Grignard reagents.

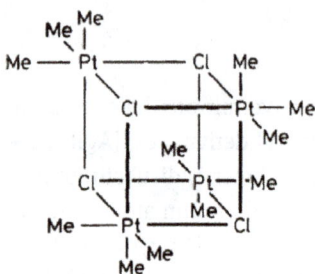

Fig. 19.14: Tetrameric organoplatinum chloride.

The dimeric compounds, $(R_2AuX)_2$, are prepared from $[Au(py)Cl_3]$ and methylmagnesium iodide.

19.3.2 Nitrogen donors

Efficient co-ligands are the nitrogen donors. The reaction of $VCl_3(THF)_3$ with one equivalent of R_2NLi (R = *iso*-Pr, Cy) formed the tetravalent $(R_2N)_2VCl_2$ which can be alkylated with RLi to yield the corresponding $(R_2N)_2VR_2$ derivatives in good yield [21].

Stable CoR_2(bipy) and $Ni(CH_2SiMe_3)_2$ (Fig. 19.15) are prepared from the metal(II) acetylacetonates, bipyridyl and aluminum trialkyls.

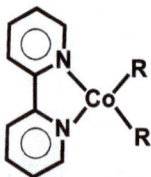

Fig. 19.15: Bipyridyl adducts.

Dimethylglyoximato chelates of cobalt, CoRL(DMG) (L = pyridine, H_2, etc.), have been investigated as B12 vitamin models, since the coenzyme of this vitamin also contains a Co-C bond in a similar coordinative environment (Fig. 19.16).

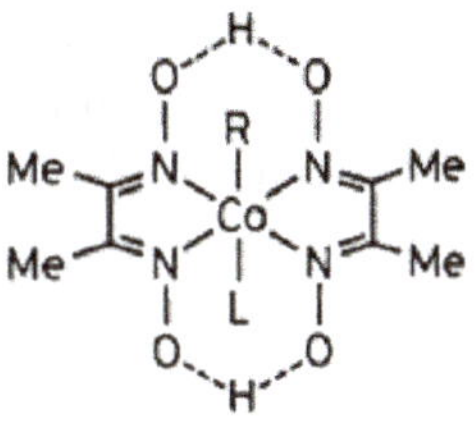

Fig. 19.16: Dimethylglioximato organocobalt complex.

19.3.3 Phosphines

Other versatile co-ligands are the phosphines. The adduct $WMe_6 \cdot PMe_3$ obtained from the components decomposes thermally or on photolysis to give $trans$-$[WMe_2(PMe_3)_4]$ (Fig. 19.17).

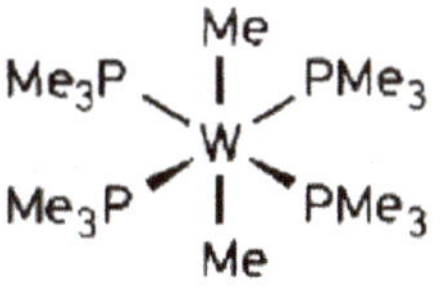

Fig. 19.17: Orgnotungsten tetraphosphine adduct.

Stable di- and tetraphenyl derivatives of cobalt, $CoPh_2(PEt_3)_2$, are known. Phosphino derivatives of cobalt $CoR_2(PR_3)_2$ (Fig. 19.18) are formed when the phenyl group attached to cobalt is $ortho$-substituted.

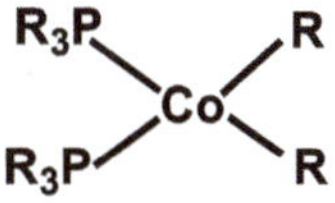

Fig. 19.18: Organocobalt diphosphine adducts.

– The treatment of a phosphine rhodium or iridium halide with Grignard reagents:

$$(PR'_3)_3MBr_3 \ + \ 2\ RMgBr \longrightarrow (PR'_3)_3MR_2Br$$

M = Rh, Ir

Stable bis(phosphine)metal dialkyls, $MR_2(PR_3)_2$, can be isolated from the corresponding dihalides with organolithium reagents.

The pentafluorophenyl derivative, $Ni(C_6F_5)(PPh_3)_2$, is a very stable compound.

Gold(I) derivatives, stabilized by complexation with tertiary phosphines, $AuR(PR_3)$, are obtained from the halides, $R_3P \cdot AuX$, with organolithium or Grignard reagents. Trimethylgold, prepared from gold(III) bromide and methyllithium, is stabilized by complexation with amines or tertiary phosphines, as $Me_3Au \cdot L$. The pentafluorophenyl derivatives of gold include the neutral species $Au(C_6F_5)_n(PR_3)$ ($n = 1$ or 3). Other examples are $(Me_3Si)_2CH$-AuL (L = PPh_3, $AsPh_3$).

- Titanium complexes of bis(dimethylphosphino)ethane are known as *trans*-TiMeCl(dmpe)$_2$ and *trans*-TiMe$_2$(dmpe)$_2$. A *trans*-geometry was evidenced by spectroscopic methods and X-ray diffraction, *trans*-MMe$_2$(dmpe)$_2$ for M = V, Cr or M, while the iron species obtained is *cis*-FeMe$_2$(dmpe)$_2$ [19].

19.3.4 Metal carbonyl derivatives, (CO)$_m$MR$_n$

Anionic species, [R-M(CO)$_5$]$^-$ (R = Me, Et, PhCH$_2$), are formed in the reaction of Na$_2$[Cr(CO)$_5$] with the corresponding organic halides.

For group VII metals the pentacarbonyl alkyls are typical. The sodium salt of the anion [Mn(CO)$_5$]$^-$ with methyl iodide forms the carbonyl compound (CO)$_5$Mn-CH$_3$. This absorbs carbon monoxide reversibly, to form an acetyl derivative, also available by an alternative route (Fig. 19.19).

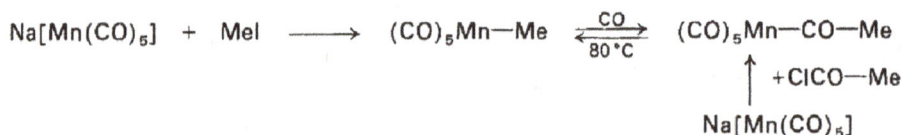

$$Na[Mn(CO)_5] + MeI \longrightarrow (CO)_5Mn-Me \underset{80\,°C}{\overset{CO}{\rightleftharpoons}} (CO)_5Mn-CO-Me$$

$$\uparrow +ClCO-Me$$

$$Na[Mn(CO)_5]$$

Fig. 19.19: Formation of a manganese acetyl derivative.

Other alkylmetal carbonyls, (CO)$_5$M-R (M = Mn, Re), are formed by the decarbonylation of acyl derivatives, (CO)$_5$M-CO-R (R = Ph, perfluoroalkyl).

Hydrometallation, as the addition of perfluoroethylene to a carbonyl hydride, can also be used:

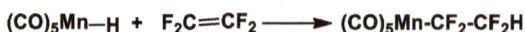

$$(CO)_5Mn-H + F_2C=CF_2 \longrightarrow (CO)_5Mn-CF_2-CF_2H$$

Tetracarbonyliron diiodide, Fe(CO)$_4$I$_2$, reacts with pentafluorophenyllithium to give (η^5-C$_6$H$_5$)Fe(CO)$_4$I, and analogous compounds are formed from iron pentacarbonyl with perfluoroalkyl iodides.

19.3.5 Cyclopentadienylmetal derivatives, (η^5-C$_5$H$_5$)$_m$MR$_n$, and cyclopentadienylmetal carbonyl derivatives, (η^5-C$_5$H$_5$)MR$_n$(CO)$_m$

The stable bis(cyclopentadienyl) titanium derivatives, (η^5-C$_6$H$_5$)$_2$TiR$_2$, are prepared from the corresponding dichloride with organolithium reagents. [TiPh$_2$]$_x$ reacts with cyclopentadiene to give (η^5-C$_6$H$_5$)$_2$TiPh$_2$.

Bis(cyclopentadienyl)vanadium chloride with phenyllithium yields (η^5-C$_6$H$_5$)$_2$V-Ph.

Li$_3$[CrPh$_6$] · nEt$_2$O reacts with cyclopentadiene to give the complex Li[(η^5-C$_5$H$_5$)CrPh$_3$].

The cyclopentadienylmetal tricarbonyl alkyls of molybdenum and tungsten are prepared from the corresponding anions and alkyl halides:

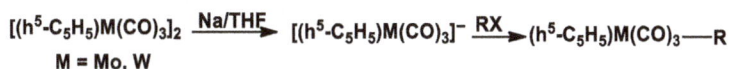

$$[(h^5\text{-}C_5H_5)M(CO)_3]_2 \xrightarrow{\text{Na/THF}} [(h^5\text{-}C_5H_5)M(CO)_3]^- \xrightarrow{\text{RX}} (h^5\text{-}C_5H_5)M(CO)_3\text{—R}$$
$$M = Mo, W$$

Cyclopentadienyliron dicarbonyl derivatives are obtained by the reaction of the nucleophilic anion, $[(\eta^5\text{-}C_5H_5)Fe(CO)_2]^-$, with alkyl halides, hexafluorobenzene or substituted perfluorobenzenes (Fig. 19.20). This anion also reacts with acyl halides to form the acyliron derivatives, $[(\eta^5\text{-}C_5H_5)Fe(CO)_2]CO\text{-}R$, which can be decarbonylated to $(\eta^5\text{-}C_5H_5)Fe(CO)_2R$ (R = perfluoroalkyl, Ph, etc.).

$$\text{Fe(CO)}_2^{\ominus} + RX \longrightarrow \overset{\displaystyle CO}{\underset{\displaystyle CO}{\mid}}\!\!\text{Fe—R} + X^{\ominus}$$

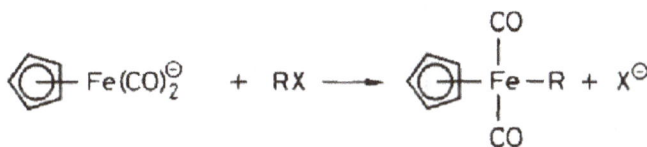

Fig. 19.20: Preparation of an iron compound.

The fluxional mixed $(\eta^5\text{-}C_5H_5)(\eta^1\text{-}C_5H_5)Fe(CO)_2$ derivative is prepared from the corresponding halide and sodium cyclopentadienide (Fig. 19.21).

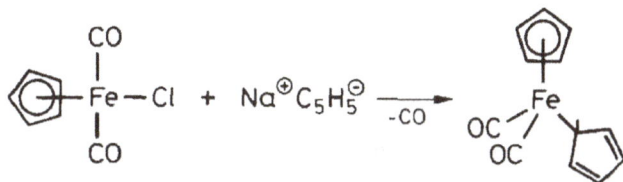

$$\overset{\displaystyle CO}{\underset{\displaystyle CO}{\mid}}\!\!\text{Fe—Cl} + Na^{\oplus}C_5H_5^{\ominus} \xrightarrow{-CO} \begin{array}{c} Fe \\ OC^{\diagup}\diagdown \\ OC \end{array}$$

Fig. 19.21: Preparation of fluxional $(\eta^5\text{-}C_5H_5)(\eta^1\text{-}C_5H_5)Fe(CO)_2$.

Metal-hydride addition to olefins leads to σ-derivatives of iron (Fig. 19.22).

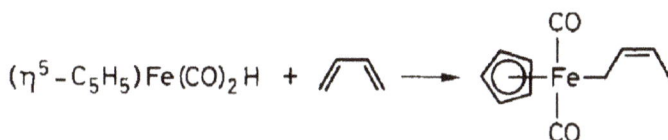

$$(\eta^5\text{-}C_5H_5)Fe(CO)_2H + \text{butadiene} \longrightarrow \overset{\displaystyle CO}{\underset{\displaystyle CO}{\mid}}\!\!\text{Fe—}$$

Fig. 19.22: Metal hydride addition to butadiene.

Cyclopentadienylnickel derivatives are obtained from the dimer with perfluoroalkyl iodides:

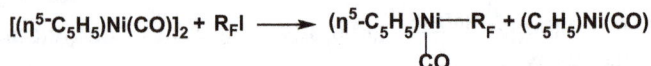

$$[(\eta^5\text{-}C_5H_5)Ni(CO)]_2 + R_Fl \longrightarrow (\eta^5\text{-}C_5H_5)Ni\text{---}R_F + (C_5H_5)Ni(CO)$$
$$\underset{CO}{|}$$

Ferrocene and other η^5-cyclopentadienylmetal derivatives form compounds in which a ring carbon atom is bonded to two gold atoms. These structures involve polycenter Au . . . Au . . . C-bonds. Their relation to σ-bonded compounds is illustrated by the interconversions shown in Fig. 19.23.

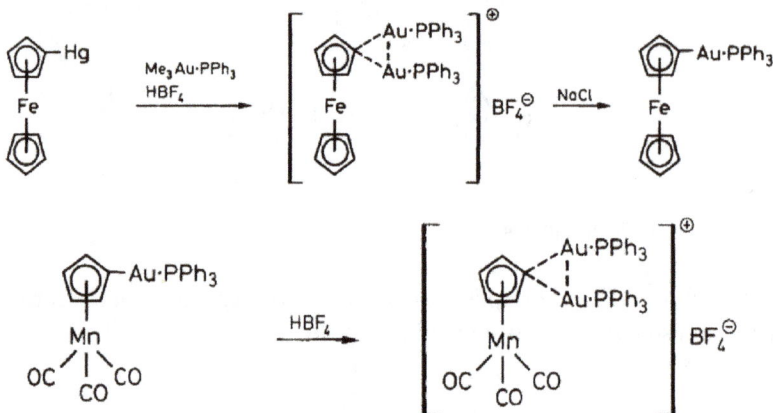

Fig. 19.23: Aurated ferrocene compounds.

The cyclopentadienylmetal derivatives of lanthanides (η^5-C$_5$H$_5$)$_2$M-R (M = Y, Dy, Ho, Er, Yb, Gd, Tm; R = Me, Ph, C≡CPh) are prepared from the corresponding halides and LiR, while the dimeric halides, (η^5-C$_5$H$_5$)$_2$MCl]$_2$, react with Li[AlR$_4$] or Mg[AlR$_4$]$_2$ to give the alkyl-bridged compounds, (η^5-C$_5$H$_5$)$_2$M(μ-R)$_2$AlR$_2$ with M = Sc, Y, Dy, Ho, Er, Tm, Yb; R = Me and M = Sc, Y, Ho, R = Et (Fig. 19.24).

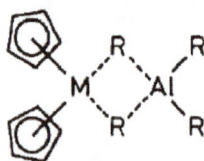

Fig. 19.24: Alkyl bridged heterobimetallic compounds.

Uranium forms the σ-derivatives (η^5-C$_5$H$_5$)$_3$U-R (R = -CC, etc.).

In this context, some arene metal complexes can be also mentioned. Thus, cobalt and nickel vapor react with bromopentafluorobenzene to afford Co(C$_6$F$_5$)$_2$ and unstable Ni(C$_6$F$_5$)Br, respectively. These interact further with toluene to give π-arene complexes (Fig. 19.25).

$$Co_{at} + C_6F_5Br \longrightarrow Co(C_6F_5)_2 \xrightarrow{PhMe}$$

$$Ni_{at} + C_6F_5Br \xrightarrow{-196\,°C} C_6F_5NiBr \xrightarrow{PhMe}$$

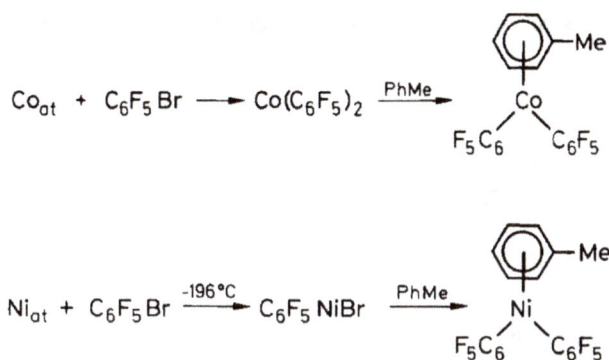

Fig. 19.25: Formation of cobalt and nickel π-arene complexes.

19.3.6 Metallacycles and chelate rings

Titanium forms a chelate ring with a bis(dimethylphosphino)amine ligand (Fig. 19.26).

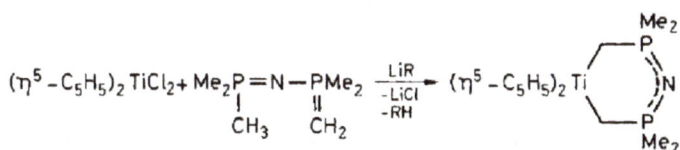

$$(\eta^5\text{-}C_5H_5)_2\,TiCl_2 + Me_2P\!=\!N\!-\!PMe_2 \xrightarrow[\substack{-LiCl\\-RH}]{LiR} (\eta^5\text{-}C_5H_5)_2\,Ti$$

Fig. 19.26: Titanium chelate ligand.

Titana metallocycles are prepared from diphenylacetylene and organodilithium compounds (Fig. 19.27).

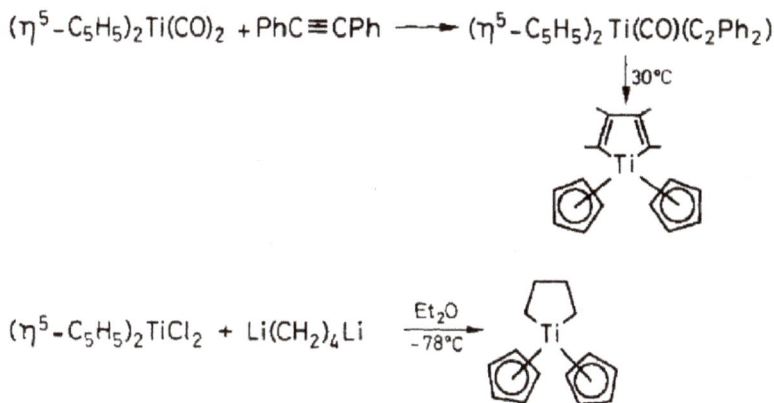

$$(\eta^5\text{-}C_5H_5)_2Ti(CO)_2 + PhC\!\equiv\!CPh \longrightarrow (\eta^5\text{-}C_5H_5)_2\,Ti(CO)(C_2Ph_2)$$

$$(\eta^5\text{-}C_5H_5)_2TiCl_2 + Li(CH_2)_4Li \xrightarrow[-78\,°C]{Et_2O}$$

Fig. 19.27: Titanium heteroatoms in five-membered rings.

Chromium–carbon σ-bonds can be part of a metallocycle in the following two structures shown in Fig. 19.28.

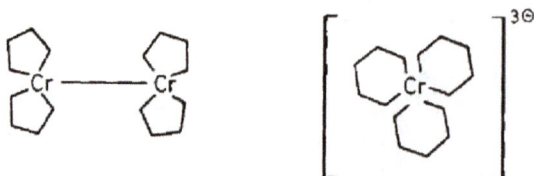

Fig. 19.28: Rings with chromium heteroatoms.

Chromium compounds containing four-membered chelate rings derived from phosphorus ylides are also obtained (Fig. 19.29).

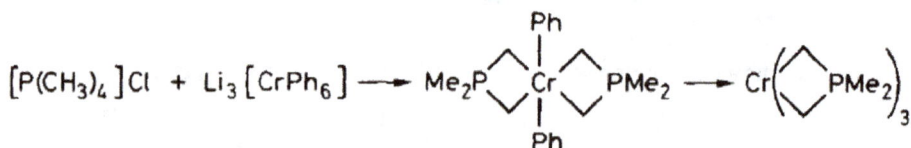

$$[P(CH_3)_4]Cl + Li_3[CrPh_6] \longrightarrow Me_2P \underset{Ph}{\overset{Ph}{\underset{|}{\overset{|}{Cr}}}} PMe_2 \longrightarrow Cr\left(\begin{array}{c} \\ \end{array}PMe_2\right)_3$$

Fig. 19.29: Four-membered rings with chromium heteroatoms.

Cyclic σ-carbon ruthenium metallacycles are derived from phosphorus ylides and polymethylene reagents (Fig. 19.30).

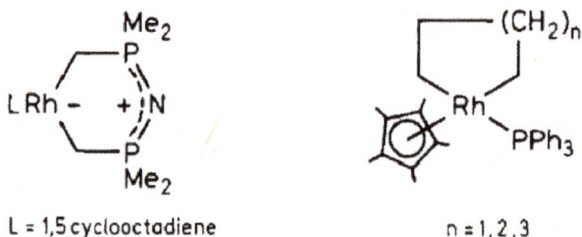

L = 1,5 cyclooctadiene   n = 1,2,3

Fig. 19.30: Rhodium heteroatom in metallacyles.

With the phosphorus ylide [Me P(CH$_2$)$_z$]$^-$ two Ni$_2$[Me$_2$P(CH$_2$)$_z$]$_4$ isomers have been prepared (Fig. 19.31).

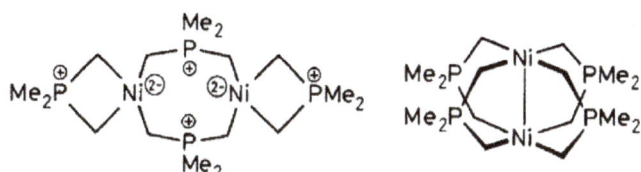

Fig. 19.31: Two nickel-containing isomers.

Double ylides of phosphorus react with metal halides to form nickel, palladium and platinum spirocyclic compounds (Fig. 19.32).

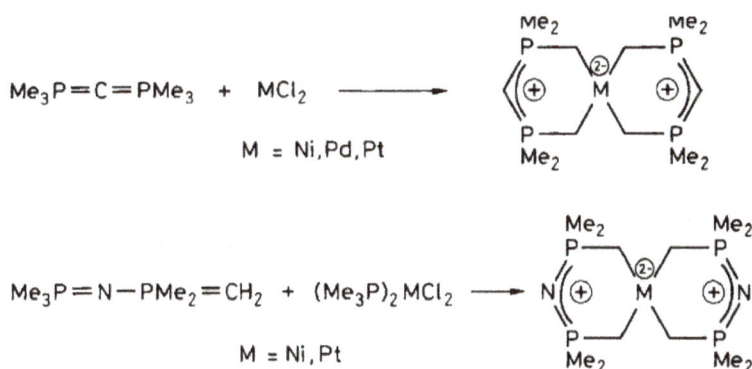

$Me_3P=C=PMe_3$ + MCl_2 ⟶

M = Ni, Pd, Pt

$Me_3P=N-PMe_2=CH_2$ + $(Me_3P)_2 MCl_2$ ⟶

M = Ni, Pt

Fig. 19.32: Spirocyclic nickel, palladium and platinum compounds.

Several metallacyclic compounds containing gold have been described (Fig. 19.33).

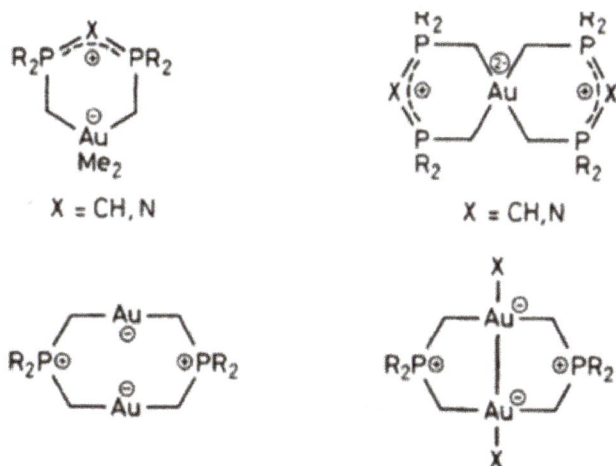

X = CH, N

X = CH, N

Fig. 19.33: Gold-containing ring compounds.

Heterocyclic gold compounds can be obtained by replacement of tin from a tetraphenylstannole with $AuCl_3$ (Fig. 19.34).

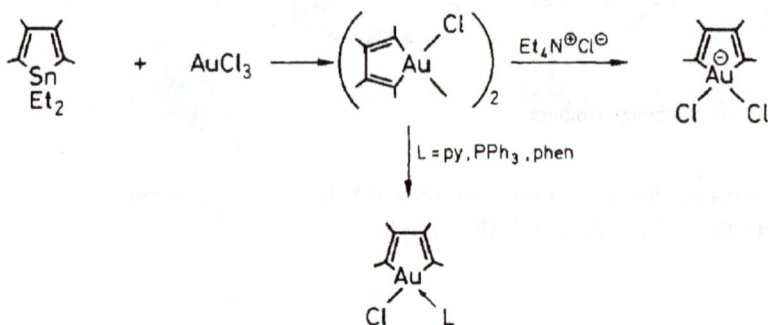

Fig. 19.34: Formation of auracyclopentadiene compounds.

Rare earth cationic complexes, $[M(CH_2)_2PMe_2)_3]Cl$ (M = La, Pr, Nd, Sm, Gd, Ho, Er, Lu), have been reported, in which the positive charge is localized at phosphorus rather than at the metal; these deprotonate to give neutral compounds containing chelate rings (Fig. 19.35).

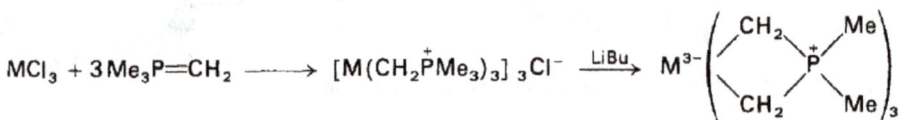

Fig. 19.35: Lanthanide ring compounds.

References

[1]   Pope WJ, Peachey SJ. The alkyl compounds of platinum. J Chem Soc Trans 1909, 95, 571–76.
[2]   Ebert K, Massa W, Donath H, Lorberth J, Seo B-S, Herdtweck E. Organoplatinum compounds: VI. Trimethylplatinum thiomethylate and trimethylplatinum iodide. The crystal structures of [(CH₃)₃PtS(CH₃)]₄ and [(CH₃)₃PtI]₄·0.5CH₃I. J Organomet Chem 1998, 559, 203–07.
[3]   Gilman H, Jones RG, Woods LA. The preparation of methylcopper and some observations on the decomposition of organocopper Compounds. J Org Chem 1952, 17, 1630–34.
[4]   Hope H, Olmstead MM, Power PP, Sandell J, Xu X. Isolation and X-ray crystal structures of the mononuclear cuprates [CuMe₂]⁻, [CuPh₂]⁻, and [Cu(Br)CH(SiMe₃)₂]⁻. J Am Chem Soc 1985, 107, 4337–38.
[5]   Siegbahn PEM. Trends of metal-carbon bond strengths in transition metal complexes. J Phys Chem 1995, 99, 12723–29.
[6]   Lappert MF, Patil DS, Pedley JB. Standard heats of formation and M–C bond energy terms for some homoleptic transition metal alkyls MRₙ. J Chem Soc Chem Commun 1975, 830–31.

[7]   Davidson PJ, Lappert MF, Pearce R. Stable homoleptic metal alkyls. Acc Chem Res 1974, 7, 209–17.
[8]   García-Monforte MA, Alonso PJ, Forniés J, Menjón B. New advances in homoleptic organotransition-metal compounds: The case of perhalophenyl ligands. Dalton Trans 2007, 3347–59.
[9]   Ara I, Forniés J, García-Monforte MA, Martín AB, Menjón B. Synthesis and characterization of pentachlorophenyl–metal derivatives with d^0 and d^{10} electron configurations. Chem–Eur J 2004, 10, 4186–97.
[10] Albors P, Carrella LM, Clegg W, Garcia-lvarez P, Kennedy AR, Klett J, Mulvey RE, Rentschler E, Russo L. Direct CH metallation with chromium(II) and iron(II): Transition metal host-benzenediide guest magnetic inverse-crown complexes. Angew Chem 2009, 121, 3367–71.
[11] Carrella LM, Clegg W, Graham DV, Hogg LM, Kennedy AR, Klett J, Mulvey RE, Rentschler E, Russo L. Inverse-crown complexes. Sodium-mediated manganation: Direct mono- and dimanganation of benzene and synthesis of a transition-metal inverse-crown complex. Angew Chem Int Ed 2007, 46, 4662–66.
[12] Blair VL, Clegg W, Mulvey RE, Russo L. Alkali-metal-mediated manganation(II) of naphthalenes: Constructing metalla-anthracene and metalla-phenanthrene structures Inorg. Chem 2009, 48, 8863–70.
[13] Alonso PJ, Fornié J, García-Monforte MA, Martín A, Menjón B. The first structurally characterised homoleptic organovanadium(III) compound. Chem Commun 2001, 2138–39.
[14] Alonso PJ, Fornies J, Garcia-Monforte MA, Martin A, Menjon B. New homoleptic organometallic derivatives of vanadium(III) and vanadium(IV): Synthesis, characterization, and study of their electrochemical behavior. Chem-Eur J 2005, 4713–24.
[15] Jäkle F. Pentafluorophenyl copper: Aggregation and complexation phenomena, photoluminescence properties, and applications as reagent in organometallic synthesis. Dalton Trans 2007, 2851–58.
[16] Stollenz M, Mesitylcopper MF. A powerful tool in synthetic chemistry. Organometallics 2012, 31, 7708–27.
[17] Eriksson H, Hakansson M. Mesitylcopper: Tetrameric and pentameric. Organometallics 1997, 16, 4243–44.
[18] Niemeyer M. Reaktion von Kupferarylen mit Imidazol-2-ylidenen oder Triphenylphosphan – Bildung von 1:1-Addukten mit zweifach koordinierten Kupferatomen. Z Anorg Allg Chem 2003, 629, 1535–40.
[19] Janssen MD, Köhler K, Herres M, Dedieu A, Smeets WJJ, Spek AL, Grove DM, Lang H, van Koten GJ. Monomeric bis(η^2-alkyne) complexes of copper (I) and silver(I) with η^1-bonded alkyl, vinyl, and aryl Ligands L. J Am Chem Soc 1996, 118, 4817–29.
[20] Niemeyer M. σ-Carbon versus π-arene interactions in the solid-state structures of trimeric and dimeric copper aryls $(CuAr)_n$ (n = 3, Ar = $2,6-Ph_2C_6H_3$; n = 2, Ar = $2,6-Mes_2C_6H_3$). Organometallics 1998, 17, 4649–56.
[21] Desmangles N, Gambarotta S, Bensimon C, Davis S, Zahalka H. Preparation and characterization of $(R_2N)_2VCl_2$ [R = Cy, i-Pr] and its activity as olefin polymerization catalyst. J Organomet Chem 1998, 562, 53–60.

Part IV: **Application of organometallics in organic synthesis**

20 Polar organometallics in organic syntheses

The contribution of organometallic chemistry to organic synthesis was open by polar organometallics and, in time, they became key reagents for the preparation of practically all classes of organic compounds. As it was described in the chapter dedicated to the formation of metal–carbon bonds, this class of compounds includes the organometallic derivatives of group 1 and 2 elements, together with the elements of group 12, mainly organozinc. The organomagnesium (Grignard) and organolithium reagents were the stars for a long while [1–3] but, in time, other polar organometallics joined the two in providing, in most of the cases, new and surprising synthetic paths [4–8]. The following selection of examples is intended to illustrate the significance of the polar organometallics in organic synthesis.

20.1 Reactivity of polar organometallics

20.1.1 General

The reactivity of polar organometallics is strongly related to the degree of polarity. Both the metal and the organic moiety containing the carbon involved in the formation of the organometallic species influence the degree of polarity. The nucleophilicity and/or basicity, on the other hand, need to be evaluated for each polar organometallics in relation not only with the two components mentioned above but the organic substrate to be reacted with. A wise choice, based on previous results, will help to fine-tune the organometallic contributions to organic synthesis.

In a polar organometallic, the organic backbone is not behaving as a free carbanion. The bond can be regarded as a combination of a covalent bond and an ionic interaction. This can explain the attenuation of the carbanion basicity by the metal [9]. Therefore, to select the most suitable reagent for a given transformation, the relative chemical potential of any individual organometallic reagent must be evaluated. The degree of polarity/ionicity can be derived based on Pauling's formula and electronegativities [10]: C–H 4%, C–Hg 10%, C–Zn 18%, C–Mg 30%, C–Li 43%, C–Na 47%, C–K 51%. We can assume, at this stage, that the carbanionic reactivity is somewhere around these values.

As already mentioned, the reactivity potential of an organometallic reagent depends also on the organic backbone and is causally related to the stability of the metal–carbon bond. The stability of a polar organometallic compound is strongly dependent on the type of the carbon involved in the bonding to the metal. The most stable compounds are formed by alkynes (sp C) followed by alkenes (sp^2 C) and alkyls (sp^3 C), in accordance with the acidity of the respective C–H bond. Less thermodynamic stability means a high reactivity potential. The intrinsic stability of the

https://doi.org/10.1515/9783110695274-021

organic moiety due to extended conjugation (the negative charge dispersed over several carbon atoms) (i.e., benzyl, cyclopentadienyl) is another factor to be considered. A good example is the reaction of phenylsodium (colorless) with toluene to give bright red benzylsodium and benzene [11]:

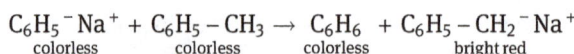

$$C_6H_5{}^-Na^+ + C_6H_5-CH_3 \rightarrow C_6H_6 + C_6H_5-CH_2{}^-Na^+$$

colorless   colorless   colorless   bright red

A more complex example of the influence of the degree of polarity on the resonance stabilization is the rearrangement of diphenylcyclopropylcarbinyl-/Y,Y-diphenylallylcarbinyl-lithium, sodium, potassium and magnesium organometallic compounds (Fig. 20.1) [12, 13].

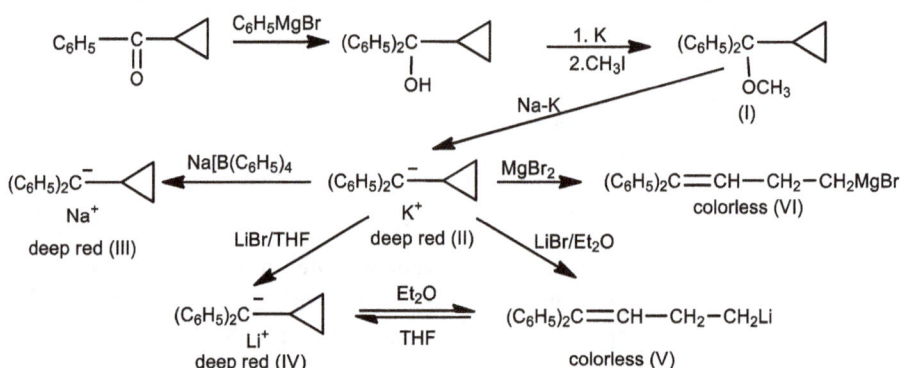

Fig. 20.1: Rearrangement of diphenylcyclopropylcarbinyl-/Y,Y-diphenylallylcarbinyl-metals (Li, Na, K, MgBr).

The stabilization of diphenylcyclopropylcarbinyl anion increases with the ionicity of the carbon–metal bond as the gain in stability brought by resonance overcompensates the strain of the cyclopropyl. The higher the separation of the carbanion and the metal cation, the more stable is the cyclic isomer. The diphenylcyclopropylcarbinyl methyl ether (I) (Fig. 20.1) was readily cleaved by sodium–potassium alloy in diethyl ether with the formation of a deep red precipitate almost quantitatively. No rearranged products could be detected. The reactions of diphenylcyclopropylcarbinylpotassium (II) with Na[BPh₄] and the reaction of diphenylcyclopropylcarbinylmethyl ether (I) (Fig. 20.1) with metallic sodium in ether gave the same product, diphenylcyclopropylcarbinylsodium (III) (Fig. 20.1). When diphenylcyclopropylcarbinylpotassium (II) was treated with lithium bromide in diethylether, the deep red color disappeared as the open-chain isomer Y,Y-diphenylallylcarbinyllithium (V) (Fig. 20.1) was formed, while the same reaction in THF led to the deep red diphenylcyclopropylcarbinyllithium (IV) (Fig. 20.1). When tetrahydrofuran was added to the colorless open-chain lithium compound prepared in diethyl ether, the deep red color of the cyclic anion immediately reappeared. A retro-cyclopropylcarbinyl-allylcarbinyl rearrangement was achieved simply through a solvent change. In a 2:1 mixture ether:tetrahydrofuran, the equilibrium between the closed and open forms lies more than 90% on the side of the cyclic anion. The reaction of diphenylcyclo-propylcarbinylpotassium (II) with magnesium bromide in tetrahydrofuran gave exclusively an open-chain product (VI) [12, 13].

It is often observed that reactions with polar organometallic reagents may take totally different courses with different metal counterions, even with the alkali metals [9]. The reaction of phenyl-M (M = Li, Na, K, MgBr) derivatives with acetophenone can yield two different products, the carbinolates as a result of the nucleophilic addition to the carbon–oxygen double bond and enolates by α-deprotonation (Fig. 20.2) [11].

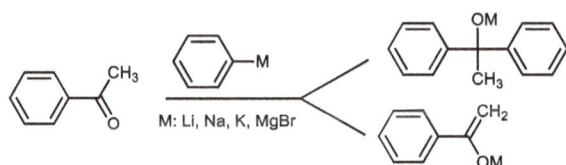

Fig. 20.2: Metallation of acetophenone with phenylmetal compounds.

The enolate:carbinolate ratio strongly depends on the metal. The enolate:carbinolate ratio found experimentally was 10:1 for phenylpotassium (mainly results in enolate formation), a 2:1 mixture was formed in reaction with phenylsodium, a 1:23 mixture was formed with phenyllithium (mainly the carbinolate formation), and for Grignard reagent, the carbinolate was obtained almost quantitatively. The regiospecificity can be related with the higher polarity of the C–M bond of the heavier alkali metals and thus stronger basicity than the Li and Mg derivatives [11].

The driving force of the reaction for the chemical transformations described in this chapter is the conversion of a polar organometallic compound into an essentially covalent hydrocarbon and a salt-like metal derivative, a process accompanied by a substantial gain in free reaction enthalpy.

20.1.2 Ortho-metallation

A reaction with significant applications in organic synthesis is *directedortho-metallation* (DoM), the deprotonation of a site *ortho* to a heteroatom-containing functional group in the presence of a strong base. The first reports on *ortho*-metallation go back to the works of Gilman [14] and Wittig [15] and refer to the lithiated intermediate obtained on the treatment of anisole with *n*-BuLi in ether. The use of *ortho*-metallation became very important and, although organolithium bases are still the most used, other organometallic systems have been applied [16].

Organolithium derivatives, especially alkyllithiums, are known for their reactivity, and, as a consequence, for many applications in organic synthesis, including *ortho*-lithiation, as well as in the synthesis of many other organometallic compounds. Most of the organolithium compounds are aggregated in solution, and the degree of aggregation is strongly dependent on carbanion structure, solvent polarity and the presence of donor ligands like *N,N,N′,N′*-tetramethylethylenediamine (TMEDA), *N,N, N′,N″,N″*-pentamethyldiethylenetriamine (PMDTA) or hexamethylphosphoramide (HMPA) [17]. Sometimes the observed aggregates are the actual reactive species; at other times, lower aggregates seem to be active. These observations raised the interesting question as to the role the various aggregates and mixed aggregates play in reactivity and selectivity. The substituents able to orient the metallation in the *ortho*-position are known as direct metallation groups (DMG). A DMG is usually a Lewis basic group that interacts with the Lewis acidic lithium cation through a heteroatom with coordinating ability to form the adduct II (Fig. 20.3). This step is helpful in planning organic syntheses, and it was treated as such and complex-induced proximity effect (CIPE) in deprotonation of aryl and heteroaryl organic substrates [18]. The lithium-proton exchange is facilitated by the proximity of the basic alkyllithium to the proton in *ortho*-position. An agostic metal–hydrogen interaction facilitates the proton removal [19]. The metallated intermediaries are usually reacted with an electrophile to afford the final products (**IV**).

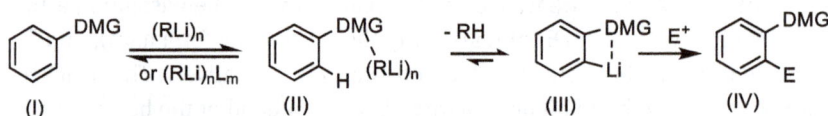

Fig. 20.3: Directed *ortho-metallation*, DoM.

Strong alkyllithium bases are needed for these metallations, the most common being MeLi, *n*-BuLi, *sec*-BuLi and *tert*-BuLi. Taking into account the result of the *ortho*-lithiation followed by the reaction with an electrophile, the product is the same as that of a traditional electrophilic substitution. The particularity of using this sequence of reactions is the regioselectivity: only the *ortho*-substitution is achieved compared to the mixture of *ortho*- and *para*-substitution formed in the aromatic electrophilic substitution.

The DMGs can be classified, according to their strength in directing metallation, as strong, moderate and weak. Examples of carbon- or heteroatom-based strong DMGs are CON-R, CSN-R, $CONR_2$, CH = NR, N-COR, $N-CO_2R$, $OCONR_2$ and OCH_2OMe; moderate DMGs are CF_3, NC, OMe, NR_2, F, Cl, $O-(CH_2)_2-OMe$ and $O-(CH_2)_2-NMe_2$; and weak DMGs are $CH(OR)_2$, CH_2O^-, O^- and S^- [20].

Inductive and steric effects as well as other functional groups on the arenes may also influence, in some cases, the reactivity of the proton in the *ortho*-position or even the site of metallation. CIPE can be used to control the regioselectivity of

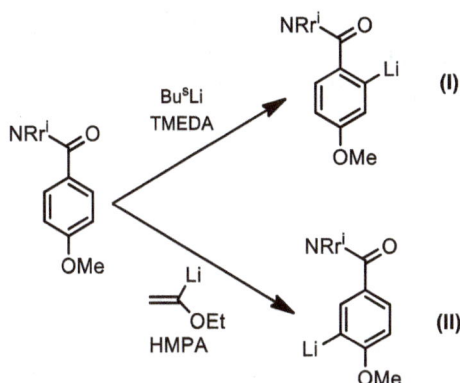

TMEDA=N,N,N',N'-tetramethylethylenediamine.
HMPA=Hexamethylphosphoramide

Fig. 20.4: Regioselective DoM with Bu^{s}Li in TMEDA (*N,N,N',N'*-tetramethylethylenediamine) (I) and α-ethoxyvinyllithium in HMPA (hexamethylphosphoramide) (II).

DoMs by altering the balance of inductive and association effects. The directed lithiation of *p*-methoxy carboxamide (Fig. 20.4) with two different lithiation reagents is Bu^{s}Li in TMEDA (Fig. 20.4 (**I**)) or α-ethoxyvinyllithium in HMPA (Fig. 20.4 (**II**)) [21].

The reaction with Bu^{s}Li/TMEDA provides the product of lithiation adjacent to the strongly complexing carboxamide (**I**), while α-ethoxyvinyl lithium/HMPA affords the kinetic product of lithiation adjacent to the methoxy group (**II**). Formation of (**II**) is related to a favorable inductive effect of the methoxy group.

Following the same concept, a benzylic position may be metallated more rapidly even in the presence of a DMG. An example is the lithiation reactions of tertiary benzylic esters (Fig. 20.5) and carbamates, where lithium precomplexation – the rate-determining step – precedes the proton transfer and the lithiation occurs at the benzylic methylene [22].

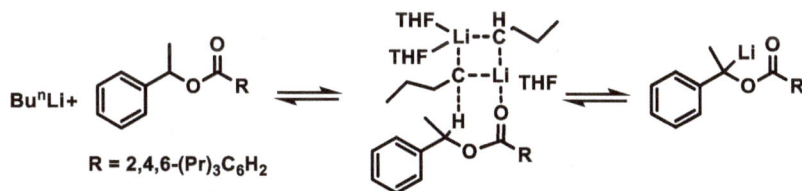

R = 2,4,6-(Pr)$_3$C$_6$H$_2$

Fig. 20.5: Lithium precomplexation providing selective DoM of tertiary benzylic esters with Bu^{n}Li.

When two DMGs are in a 1,3-relative position on the arene, the lithiation will be directed to the position between them through a cooperative coordination of the alkyllithium. In case of a 1,4-disposition of two DMGs, the metallation will be directed to the *ortho*-position closer to the stronger DMG. The lithiation of *N,N*-dimethyl-*p*-anisidine can, theoretically, take place in the adjacent position to either of the substituents but the decisive factor is the stronger dipolar interaction of oxygen with lithium in (**A**), more favorable than the dipolar interaction between nitrogen and lithium in (**B**) (Fig. 20.6) [19].

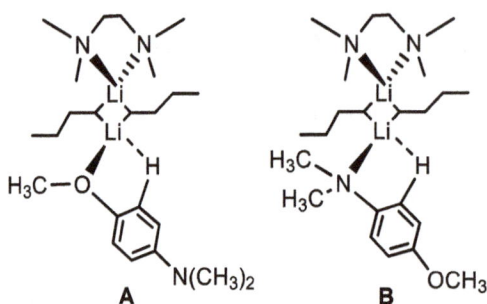

Fig. 20.6: Lithium precomplexation as a determining step in DoM of *N,N*-dimethyl-*p*-anisidine.

If the two DMGs have comparable (close or similar) strengths, a mixture is to be expected.

The heteroatoms in the heterocycles act as a directed metallation group. For the synthesis of 2-substituted saturated nitrogen heterocycles, the deprotonation of a sp^3 C–H bond next to nitrogen by *ortho*-lithiation is one of the methods (Fig. 20.7). To get very good results, additional functionalities were added, which proved effective in directed metallation adjacent to nitrogen in heterocycles [23].

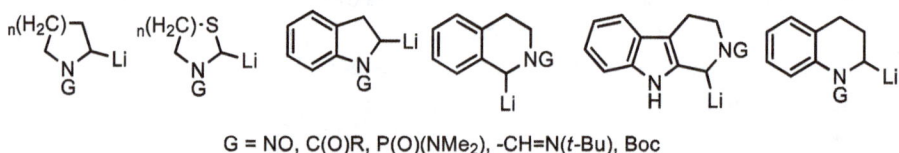

G = NO, C(O)R, P(O)(NMe₂), -CH=N(*t*-Bu), Boc

Fig. 20.7: Directed metallation of 2-substituted saturated nitrogen heterocycles.

The subsequent reaction of the *ortho*-lithiated compounds with electrophiles is usually straightforward. In case the electrophile contains an acidic proton, complications can occur due to the possible deprotonation as a competitive reaction to the nucleophilic attack.

Another process where the *ortho*-lithiated intermediates are used is the transmetallation reaction with transition metals, resulting in compounds having wide applications in catalysis.

Nature of base and solvent. As mentioned before, the alkyllithiums exist as aggregates in solution. For most of the reactions, the breakup of the aggregates by a strong donor, mainly an amine, is necessary to accelerate reactivity by an increased basicity. TMEDA is an excellent ligand and is therefore more commonly employed. A mechanistic approach based on experiments and computations is presented in Fig. 20.8 [19]. The first step is the breaking of the hexamer $(Bu^nLi)_6$ and the coordination of a Bu^nLi tetramer by anisole (1). Next, the TMEDA breaks the tetramer with the formation of the dimer (2) and free anisole, followed by the loss of one molecule of TMEDA in two steps, through dimer (3), leaving two free coordination sites open at one lithium (4). These could be coordinated by the anisole oxygen and by agostic Li–H interaction as a chelating ligand. The lithium-activated *ortho*-proton is removed subsequently by the adjacent strongly basic α-carbon atom of *n*-BuLi [19].

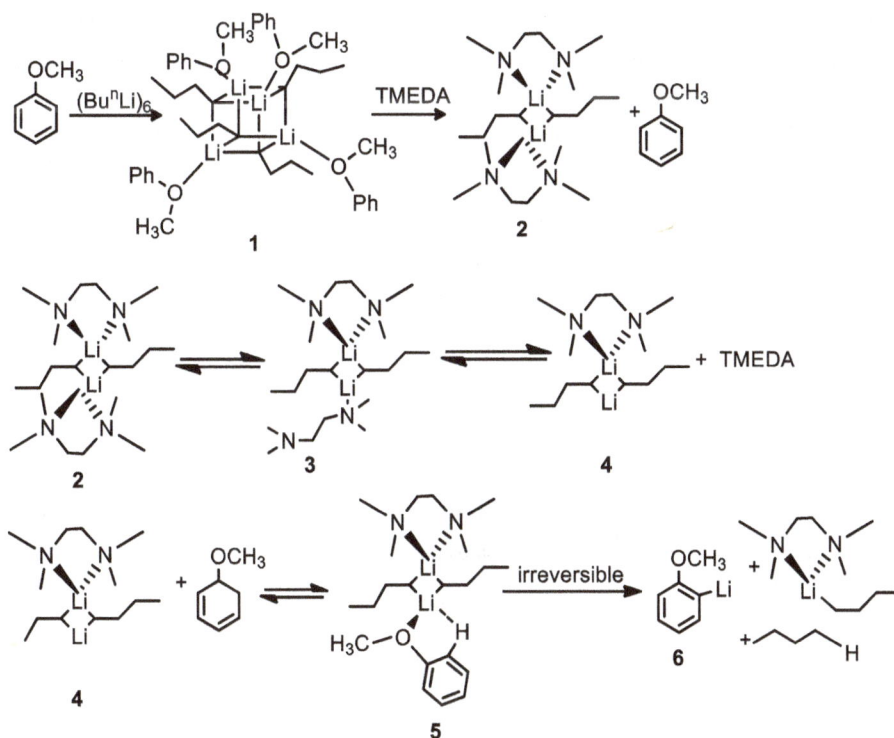

Fig. 20.8: Breakup of the organolithium aggregates by a strong donor to perform *ortho*-lithiation.

The choice of the appropriate solvent (ethers or amines) for a given reaction will take into account not only the lithiation but also the substrate for the subsequent reactions. In search for better yields and friendlier reaction conditions, experimental protocols were permanently improved [24].

To get optimal results, it is important to control the degree of aggregation and the structure of the solvates formed in a particular solvent and in the company of a given organic substrate [17, 25]. Although most of the reactions were performed in ethers during the years, attempts have been made to replace them, even if not entirely, with hydrocarbon solvents to avoid the problems linked to the sensibility of organolithiums to the water traces in ethers, their reaction with some of the ethers or the incompatibility of some organic substrates with the ethers. As the alkyllithium reagents are highly associated in hydrocarbons, a compromise has been found to improve the yields and to reduce the inconvenience of using ethers. The addition of measured amounts of ethers (usually THF) or bis-chelating amines (like TMEDA) able to activate alkyl-lithium reagents by promoting the disassociation of the aggregates to more reactive species in hydrocarbon solutions afforded good reagents for DoM reactions [24].

Hydrocarbon-based media metallation procedures involving "deficiency catalysis" can be applied for the *ortho*-lithiation of properly substituted aryls. To maximize the extent of metallation, a controlled deoligomerization of the n-BuLi hexamer found in hydrocarbons was achieved by the use of substoichiometric ratio of equivalent TMEDA to n-BuLi (0.1–0.2:1.0). In some cases, ether was necessary to obtain the expected results (Fig. 20.9). The proposed unsaturated TMEDA dimer has structure **4** in Fig. 20.8. If ether is added in the reaction mixture, the TMEDA can be replaced and a new unsaturated dimer can be formed. The two molecules of ether can coordinate to the same lithium atom in the dimer or one to each lithium atoms. As the generation of the coordinately unsaturated intermediates with either one molecule of TMEDA or two molecules of ether at the same lithium, it is crucial to maximize their concentration in the doped hydrocarbon media to achieve the greatest metallation efficiency (Fig. 20.9) [24].

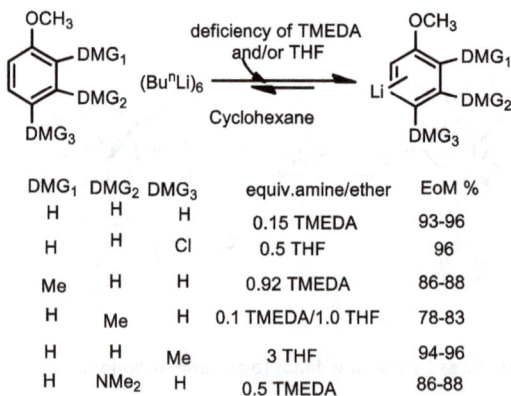

DMG$_1$	DMG$_2$	DMG$_3$	equiv.amine/ether	EoM %
H	H	H	0.15 TMEDA	93-96
H	H	Cl	0.5 THF	96
Me	H	H	0.92 TMEDA	86-88
H	Me	H	0.1 TMEDA/1.0 THF	78-83
H	H	Me	3 THF	94-96
H	NMe$_2$	H	0.5 TMEDA	86-88

Fig. 20.9: *Ortho*-lithiation in hydrocarbon-based media.

20.1.3 Organomagnesium reagents

Organomagnesium halides, RMgX, known as Grignard reagents were named after Victor Grignard who received Nobel Prize in 1912 for the contribution to their synthesis and the use as synthetic reagents in organic synthesis. Grignard reagents are solvated by ethers or, sometimes, by amines not only in solution but also in solid state (examples in [26]). There is a rapidly established equilibrium between the organomagnesium halide, RMgX, and the corresponding dialkyl-magnesium, R_2Mg (Schlenk's equilibrium):

$$2RMgCl \rightleftharpoons MgR_2 + MgCl_2$$

The Schleck equilibrium plays an important role in the reactivity of Grignard reagents not only in the classic nucleophilic addition to double carbon–oxygen bond [3, 27]. Both the substrate and the nucleophile are in the coordination sphere of Mg centers during the reaction. Different forms of the Grignard reagents may be involved in the process as it was found that the mononuclear and dinuclear species react with very similar activation energies. Also, the solvent has an important role in the reaction: the more solvated Mg species are more reactive, probably due to their flexibility which allows the structural reorganization from the reactant to the transition state [3, 27].

The well-known applications of Grignard reagents are permanently enriched with new synthetic protocols or new types of substrates to react with.

Nucleophilic addition of Grignard reagents to ketones in combination with various additives with catalytic properties is one possibility. Tertiary alcohols can be prepared in good-to-excellent yields in THF with bis(2-methoxyethyl) ether (diglyme) as an additive and tetrabutylammonium chloride ([NBu$_4$]Cl) as a catalyst (Fig. 20.10). The additive is expected to increase the nucleophilic reactivity of Grignard reagents by coordination to magnesium, while the catalyst is shifting the Schlenk equilibrium toward the dimeric Grignard reagents able to favor a six-membered transition state and to form tertiary alcohols [28].

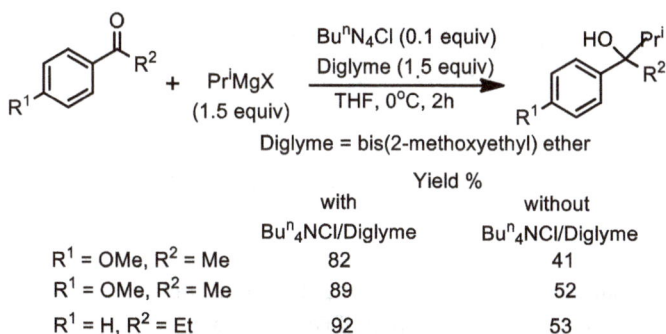

	with Bu$^n_4$NCl/Diglyme	without Bu$^n_4$NCl/Diglyme
	Yield %	
R^1 = OMe, R^2 = Me	82	41
R^1 = OMe, R^2 = Me	89	52
R^1 = H, R^2 = Et	92	53

Fig. 20.10: Nucleophilic addition of Grignard reagents to carbonyls in the presence of the additive (diglyme) and catalyst (Bu$^n_4$NCl).

Generation of aryl ketones without transition metal catalysts by reacting acid chlorides with Grignard reagents in the presence of bis[2-(N,N-dimethylamino)ethyl] ether was achieved in high yields (Fig. 20.11). The role of the tridentate additives (i.e., bis[2-(N,N-dimethylamino)ethyl] ether or PMDTA) is to moderate the nucleophilicity of the Grignard reagents, preventing its addition of newly formed ketones by coordinating the magnesium [29].

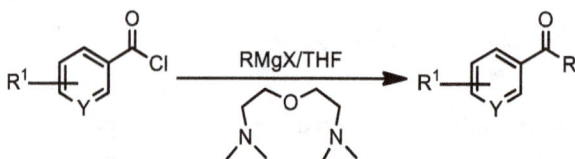

Y = C, N
R = alkyl, benzyl (1.1-1.3 equiv), -60°C), aryl, 1 equiv, -5°C - 0°C, 1-1.5 h
R¹ = halides, ether, ester, nitro, cyano,...
Yield 71-92 %

Fig. 20.11: Preparation of functionalized ketones using Grignard reagents.

The reaction of sodium methyl carbonate (SMC) (obtained from MeONa and CO_2) with primary and secondary aliphatic or alkynyl Grignard reagent is a source of carboxylic acids in excellent yields (Fig. 21.12). The reaction conditions afford pure carboxylic acids that require no further purification. These results demonstrate SMC as an effective CO_2 surrogate electrophile (the carboxylations with CO_2 often require low-temperature (−78 or −45 °C) conditions [30].

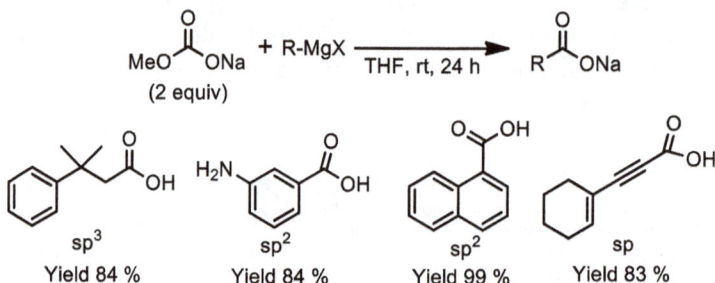

Fig. 20.12: Preparation of carboxylic acids from the reaction of sodium methyl carbonate with Grignard reagents.

An alternative to conventional transformations of Grignard reagents into esters was developed: the one-carbon homologative esterification of Grignard reagents with O-alkyl S-pyridin-2-yl thiocarbonates. The first step in the synthesis of esters using Grignard reagents is the formation of chelation-stabilized intermediates (Fig. 20.13) [31].

The reactivity of organolithium and Grignard reagents, important for organic synthesis, was extended by building new metallation systems described in the next paragraph.

R = alkyl, aryl, alkenyl, alkynyl
R^1 = benzyl, t-butyl,, methyl

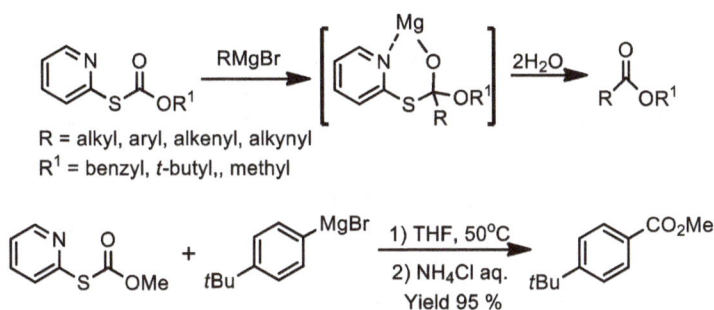

Fig. 20.13: Synthesis of esters using Grignard reagents.

20.1.4 Alkali-metal-mediated reactions

The monometallic organometallic compounds described in the previous paragraphs fail to be active in some reactions like the metal–hydrogen exchange by deprotonation of the C–H bonds of aliphatic compounds or even in aromatic molecules. Even the metal–halogen exchange reactions are not always successful for functionalized or less reactive substrates. A tool to widen the use of polar organometallics is the combination of an alkali or alkaline earth metal compound with another alkali metal compound (Li, Na, K) or with a compound containing a group 2 (Mg, Ca), group 4 (Ti), group 5 (V), group 6 (Cr), group 7 (Mn), group 8 (Fe), group 9 (Co), group 10 (Ni), group 11 (Cu), group 12 (Zn) or group 13 (Al) element. Important progress was made in the use of polar organometallics in metal–hydrogen exchange by using the alkali-metal-mediated metallation [32]. This category of compounds can cover many of the requirements of organic synthesis: high reactivity (in most of the cases avoiding low-temperature reactions), high selectivity and high functional group tolerance in the metallation step and in the following transformations. As monometallic compounds already described, most of these mixed compounds are found in solution as aggregates. The metallating agent can contain only one metal but different ligands, such as those employed for enantioselective ligand transfers or as unimetal superbases, like "Caubere reagent," the complex alkyllithium–lithium aminoalkoxide, nBuLi–Me$_2$N(CH$_2$)$_2$OLi [33], or at least two different metals. The best known examples are mixed alkali metal superbases and ate compounds (ate complexes are salts formed from the stoichiometric reaction of a Lewis base and Lewis acid, wherein the acidic moiety formally increases its valence and becomes anionic, i.e., Na[ZnR$_3$], lower ate complexes, or Na$_2$[ZnR$_4$], higher ate complexes) [4, 5, 32, 34–36]. The bimetallic complexes thus formed often exhibit unique chemistry that can be interpreted in terms of synergistic effects [32, 34]. Although the alkali metal is essential in most cases, the second metal performs the synthetic transformation. The reactivity of a given metallate complex depends on the involvement of the two metals in the transition states of the reaction intermediates as contacted ion pair structures or in separated charge structures. The charge separation is achieved by the transfer of the valence electron of the monovalent alkali metal to the more electronegative softer metal. To assess the synergistic effects in the reactivity of such complexes, comparisons between the behavior of the bimetallic compound and the parent mono-metallic compounds from which the bimetallic compound is constructed are useful. Different reactivity was noticed depending on the number of ligands around the non-alkali metals such as Mg, Al, Fe, Co or Zn present in a lower or higher order ate com-pounds. The higher order (or highly coordinated) ate compounds are generally more reactive than the lower order ones. An example is the deprotometallation of toluene by Na magnesates, a reaction possible using diazabicycloctane-activated $Na_2[Bu_4Mg]$ but not using diazabicycloctane-activated Bu_3MgNa (ion-pair structure) [37].

The reactivity of the mixed polar organometallics covers deprotometallation, halogen/metal exchange, nucleophilic transfer of a ligand from the organometal-lic complex to a carbon site or oxidation/reduction processes (with one- or two-electron transfer).

The metallation reactions (through metal–hydrogen or metal–halogen exchange) leading to the formation of carbon–magnesium or carbon–zinc bonds are an impor-tant step for the preparation of functionalized organic compounds. The transmetalla-tion reactions of organomagnesium and organozinc derivatives with catalytically active transition metal species (i.e., Pd, Ni, Ir, Cu) afford transition metal intermedi-ates relevant in cross-coupling with the formation of new carbon–carbon bonds (Kumada–Tamao–Corriu cross-coupling of organomagnesium compounds (see Sec-tion 21.2.2), and Negishi cross-coupling of organozinc compounds (see Section 21.2.3)):

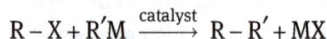

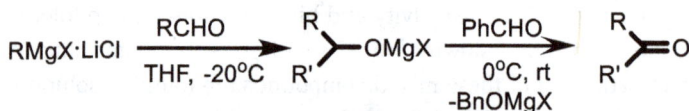

$$R - X + R'M \xrightarrow{\text{catalyst}} R - R' + MX$$

R = Ph, R'= Ph, Yield: 98 %
R = 4-BrC$_6$H$_4$, R'2-furyl, Yield: 95 %
R = 3-benxofuryl, R'= ferrocenyl, Yield 79 %
R = C$_6$H$_5$, R'= (CO)$_3$MnCp, Yield: 80 %

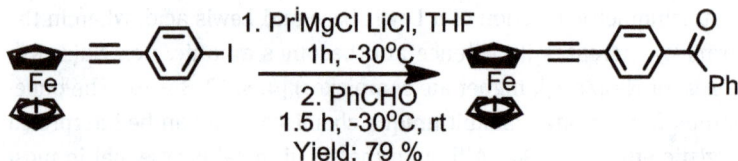

Fig. 20.14: Mixed metal RMgCl•LiCl-mediated preparation of magnesium alkoxides followed by the reaction with benzaldehyde.

where R, R′ are organic fragments; M is Mg, Zn, or Mg- or Zn-containing groups; X is halogen or other leaving groups.

The nucleophilic addition promoted by mixed metal RMgCl•LiCl systems led to magnesium alkoxides, further oxidized in the presence of benzaldehyde, as hydride acceptor, to form aryl and metallocenyl ketones (Fig. 20.14). The good results were correlated with the enhanced solubility of Mg alkoxides [38, 39].

The treatment of benzaldehyde with chiral BINOL-derived Li/Mg reagents allowed the enantioselective alkylation (Fig. 20.15). The dilithium (S)-binolate and Et$_2$Mg in 1:1 THF DME (1,3-dimethyl-2-imidazolidinone) gave the expected (S)-alcohol in an enantiomeric excess (ee) of 92% [40].

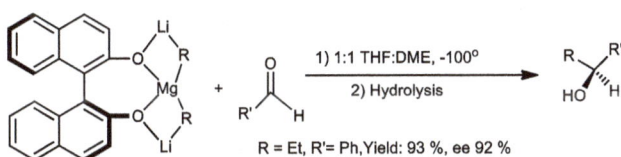

Fig. 20.15: Enantioselective alkylation with chiral BINOL-derived Li/Mg reagent.

A comparison of the reactivity of Bu^{n}Li, Bu^{n}MgCl and Bu$^n_2$Mg with Bu$^n_3$MgLi and Bu$^n_3$MgLi.LiCl toward acetophenone and benzophenone is a good example of the increased nucleophilicity of the mixed reagents (Fig. 20.16). The method was extended to various ketones and proved suitable for mixed lithium triorganomagnesates [41].

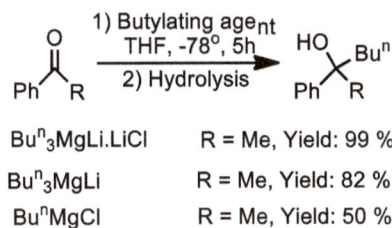

Bu$^n_3$MgLi.LiCl	R = Me, Yield: 99 %;
Bu$^n_3$MgLi	R = Me, Yield: 82 %;
Bu^{n}MgCl	R = Me, Yield: 50 %;

Fig. 20.16: Relative reactivity of Bu^{n}MgCl, Bu$^n_3$MgLi and Bu$^n_3$MgLi.LiCl toward acetophenone.

Application in the chemistry of heterocycles is also described in the literature. An example is the preparation of the symmetrically 3,3-dialkylated derivatives of 3,6-dihydro-1H-pyridin-2-one in a one-pot and a single-step procedure using magnesium "ate" complexes. One equivalent of [Bu$_3$Mg]Li used as the base allowed double proton abstraction from 3,6-dihydro-1H-pyridin-2-one. Deprotonation in the conditions described in Fig. 20.17 yielded stable magnesiates which on treatment with more than 2 equiv of alkyl halides provided 3,3-dialkylated products in good yield. In some cases, minor 3,5-dialkylated lactams were formed due to allylic conjugation [42].

Group 1/group 3 metallate complexes, lithium alkynyl trimethylaluminates in the presence of BF$_3$.OEt$_2$, have been used to stereospecifically and regioselectively

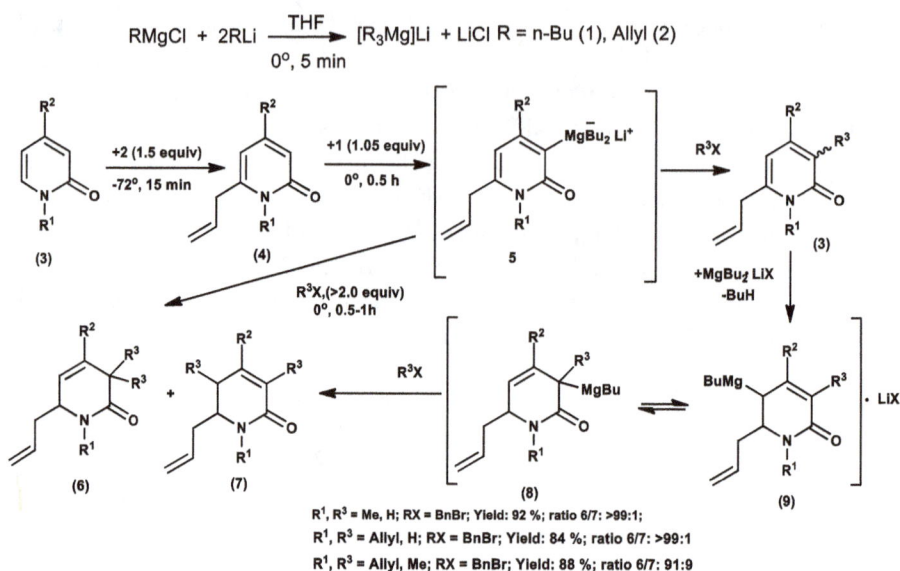

RMgCl + 2RLi →(THF, 0°, 5 min) [R₃Mg]Li + LiCl R = n-Bu (1), Allyl (2)

R¹, R³ = Me, H; RX = BnBr; Yield: 92 %; ratio 6/7: >99:1;
R¹, R³ = Allyl, H; RX = BnBr; Yield: 84 %; ratio 6/7: >99:1
R¹, R³ = Allyl, Me; RX = BnBr; Yield: 88 %; ratio 6/7: 91:9

Fig. 20.17: One-pot single-step procedure for the preparation of the symmetrically 3,3-dialkylated derivatives of 3,6-dihydro-1H-pyridin-2-one (6).

alkynylate trisubstituted epoxides at the more hindered carbon, introducing an alkyne and an alkyl substituent in the same reaction (Fig. 20.18) [43].

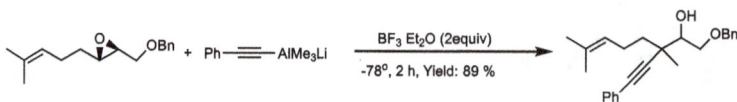

Fig. 20.18: Alkynylation of epoxide using lithium alkynyl trimethylaluminates.

20.1.4.1 Metal–hydrogen exchange reactions: superbases

In the early search for better deprotometallating agents, the reactivity of mixed Li, Na and Li, K bases was the answer. An alternative to the activation of organolithiums by using polar solvents is the addition of stoichiometric amounts of K or Na alkoxide. A classic example of a mixed alkali metal synergistic reagent obtained starting from polar organometallic is the Lochmann–Schlosser superbase, a binary mixture of n-butyllithium and potassium t-butoxide [44, 45]. The binary mixture [{(BuⁿLi)(tBuOK)}ₙ] (commonly written as LIC-KOR) designated "superbase" exhibits a reactivity intermediate between that of n-butyllithium and n-butylpotassium: enhanced reactivity compared to n-butyllithium but not so aggressive as n-butylpotassium. Changing n-Bu with tert-Bu, a solvent-free heterometallic tetralithium–tetrapotassium alkoxide, [(Butᵗ O)₈Li₄K₄], was obtained and structurally characterized, although its chemistry is not as impressive as that of LIC-KOR [46].

An example of site selectivity control using LIC-KOR is the metallation of *N*-pivolyl-2-(3-methoxyphenyl)ethylamine (Fig. 20.20). Different reaction products are obtained depending on the metallation reagent used: with *t*-butyllithium, the ben-zylic position is deprotonated; with Bu^{n}Li, the classic *ortho-metallation* is observed while with the "superbase" mixed metal reagent LIC-KOR, the hydrogen–metal exchange occurs at the aromatic *para*-position which is adjacent to the methoxy group but distant from the alkyl side chain (Fig. 20.19) [47].

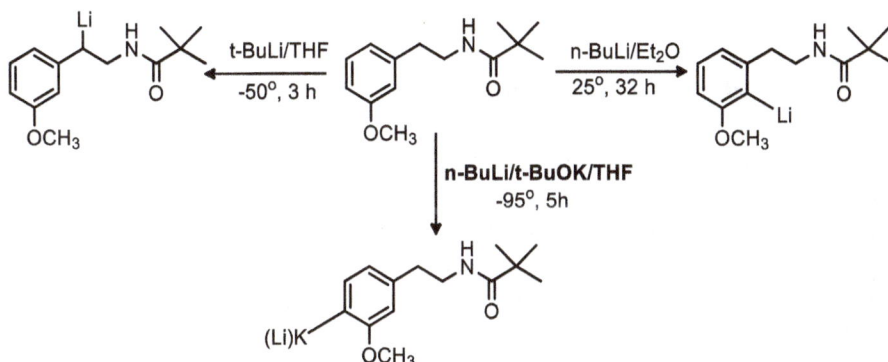

Fig. 20.19: Reagent-controlled site selectivity in metallation of *N*-pivolyl-2-(3-methoxyphenyl) ethylamine.

It was shown that it is not always necessary to prepare the LIC-KOR separately: mixing of equimolar amounts of B^{t}uOK, Bu^{n}Li and TMEDA (the order is not essential) in hexane or pentane at temperatures below –40 °C gives an extremely efficient metallating reagent. This mixture was successfully used for the metallation of ethene [48].

This new combination readily deprotonates weakly or nonactivated benzene derivatives, while exhibiting exceptional regioselectivity [49–51].

20.1.5 Turbo-Grignard reagents and related salt-supported complexes

The best introduction to the next paragraph is the quotation: "One class of reagent that stands head and shoulders above all others is the so-called 'turbo-reagents'. Bona fide synergistic reagents, their utility is so vast, greater than all others combined" – R.E. Mulvey [5].

Combinations of magnesium and zinc bases, (R$_2$NMX), or bis-amides [(R$_2$N)$_2$M] (M = Mg, Zn) having a limited solubility with molecules of lithium chloride result in soluble species with very good properties as selective deprotonating agents for a huge variety of aromatic and heterocyclic substrates or for metal–halogen exchange [7, 36, 52–55]. The turbo-Grignard reagents, the prototype being Pr^{i}MgCl · LiCl, were found to give excellent results in metal–halogen exchange reactions [6]. The use of halomagnesium amides as metallating agents goes back to the works of Hauser [56, 57]. The further development of this class of compounds is related to the complexes of sterically demanding amido ligands, most notably 2,2,6,6-tetramethylpiperidide (TMP). The TMP-active Hauser bases like TMPMgX (X = Cl, Br) with extension to "turbo-Hauser bases" opened the way for a reach and beautiful chemistry [58]. The LiCl-solubilized 2,2,6,6-tetramethylpiperidyl, TMPMCl.LiCl, and TMP$_2$M.2LiCl (M = Mg, Zn) metal amides do not affect sensitive functional groups (ester, nitrile, aldehyde, aryl or methyl ketone, azide, nitro) and react with heterocycles; hence, the preparation of polyfunctional aryl and heteroaryl organometallic species is possible.

To understand the reactivity of the new reagents and their use in organic synthesis, molecular structures offer valuable information like connectivity or aggregation [4, 5, 35]. For example, the molecular structure of the Hauser base, TMPMgCl, and the corresponding turbo-Hauser base TMPMgCl·LiCl [58]. The Hauser base, TMPMgCl, was prepared from Bu^{n}MgCl and TMP(H) in THF [59] and the turbo-Hauser base, TMPMgCl · LiCl, was obtained in the reaction of iPrMgCl · LiCl with TMP(H) in THF (Fig. 20.20) [60]. The position of the ligands was determined after the appropriate workup procedures when crystals suitable for the determination of the molecular structure were obtained: Hauser base, [TMP(THF)Mg(μ-Cl)$_2$Mg(THF)TMP], and turbo-Hauser base, [(THF)$_2$Li(μ-Cl)$_2$Mg(THF)TMP] [58].

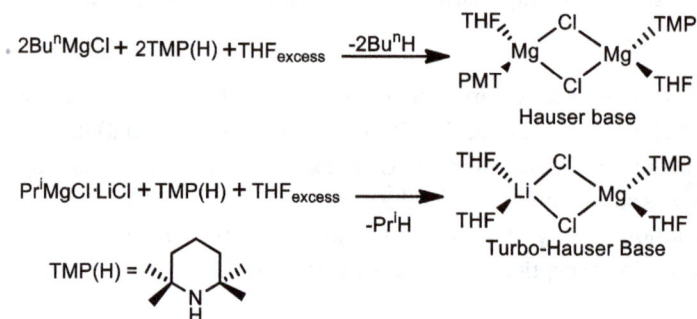

Fig. 20.20: Synthesis of Hauser base and turbo-Hauser base.

20.1.5.1 Metal–hydrogen exchange

The selectivities (chemo-, regio-, stereo-) and the reaction conditions of the metal–hydrogen exchange using turbo-reagents allowed the metallation of a huge variety of substrates.

The solubility of various turbo-reagents is important for their use in organic synthesis not in terms of yield but the reaction conditions. The reaction of the soluble TMPMgCl · LiCl compared to the less soluble Pr$^i_2$NMgCl · LiCl with isoquinoline (Fig. 20.21) led to the organomagnesiate with almost the same yield (~90%) but much faster in case of TMPMgCl·LiCl [60].

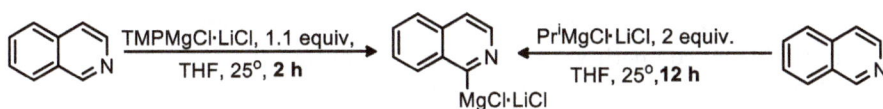

Fig. 20.21: Comparative magnesiation of isoquinoline reactions with TMPMgCl · LiCl and Pr$^i_2$NMgCl · LiCl.

The metallation of 3,5-dibromopyridine with lithium diisopropylamide proceeds selectively at the 4-position [61], while with TMPMgCl · LiCl, regioselectively orients the metallation in the 2-position with a high yield (Fig. 20.22) [60].

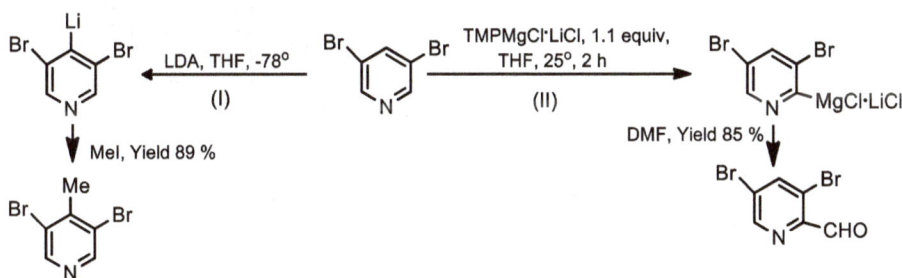

Fig. 20.22: Regioselective metallation of 3,5-bibromopyridine with lithium diisopropylamide (I) and TMPMgCl · LiCl (II) and subsequent reactions with electrophiles.

In some cases, the synthetic protocol is relevant for the process. For the metallation of pyrimidines, even the halogen-substituted ones, the inverse addition of the pyrimidine to the THF solution of TMPMgCl · LiCl afforded the magnesiated intermediates with complete regioselectivity (Fig. 20.23) [60].

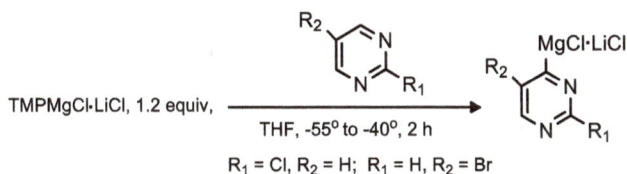

Fig. 20.23: The reaction of TMPMgCl · LiCl with substituted pyrimidines: the pyrimidines are added to the magnesium reagent.

Heterocycles, even those bearing more acidic protons such as thiazole, thiophene, furan, benzothiophene and benzothiazole, are magnesiated in mild conditions [60].

The combination of mild reaction conditions and appropriate basicity of TMPMgCl · LiCl allowed the metallation of all the available positions of a benzene ring by consecutive metallations. A hexasubstituted benzene was obtained with ~30% overall yield (Fig. 20.24) [62]. For the metallation of **2** with TMPMgCl ·

LiCl, a less polar solvent was necessary; therefore, a 1:2 mixture of THF:Et$_2$O was used to avoid the competitive deprotonation of proton H$_2$. The mixture of solvents changed the ratio of about 90:10 to 98.5:1.5 (Fig. 20.24) [62].

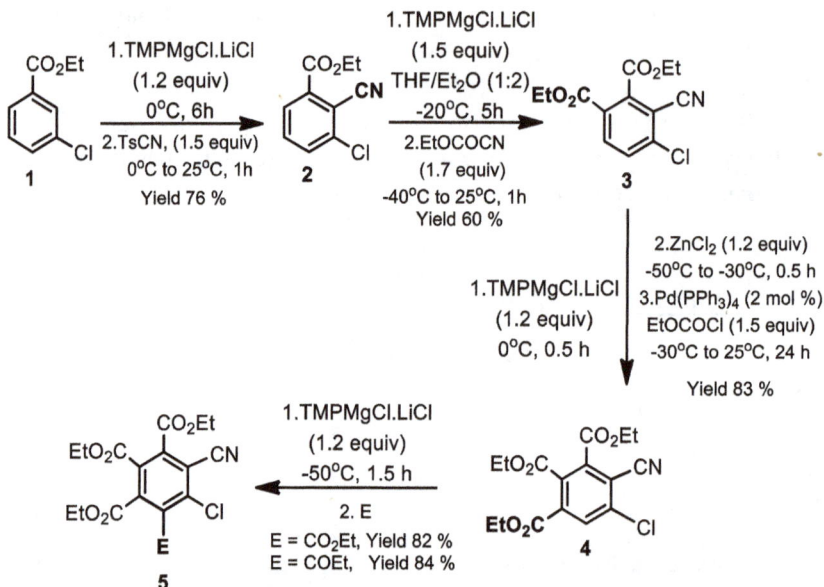

Fig. 20.24: Preparation of a hexasubstituted benzene derivative by a quadruple consecutive magnesiation with TMPMgCl · LiCl followed by reactions with electrophiles.

For reactions with less reactive substrates or aromatic compounds substituted with electron-donating groups or weakly electron-withdrawing groups, a stronger base, TMP$_2$Mg · 2LiCl, is used. The treatment of TMPMgCl · LiCl with TMPLi in THF affords TMP$_2$Mg · 2LiCl in very good yield [63]. The magnesiation of dimethyl-1,3-benzodioxan-4-one, an electron-rich aromatic ring, can be successfully performed and subsequently transformed in 6-hexylsalicilic acid, a natural product (Fig. 20.25) [64].

Sensitive functional groups like ketone, carbonate (OBoc. Boc = *tert*-butoxycarbonyl) or bis(dimethylamino) phosphonate group (OP(O)(NMe$_2$)$_2$) are not affected in the reaction with TMP$_2$MgCl.2LiCl (Fig. 20.26). Using Boc group as a directing group and to enhance the metallation, unsymmetrical benzophenone (1) was magnesiated (2) and further transformed in the 1,2-diketone (3) in 72% yield [64].

Another sensitive group, OP(O)(NMe)$_2$, orients the metallation in the 4-position, and it is not affected during the reaction with TMP$_2$MgCl.2LiCl (Fig. 20.27) [64].

The bulky bis(dimethylamino)phosphonate selectively directs the metallation to position 4, leading to the magnesiated reagent and after the reaction with iodine to aryl iodide in 91% yield [64].

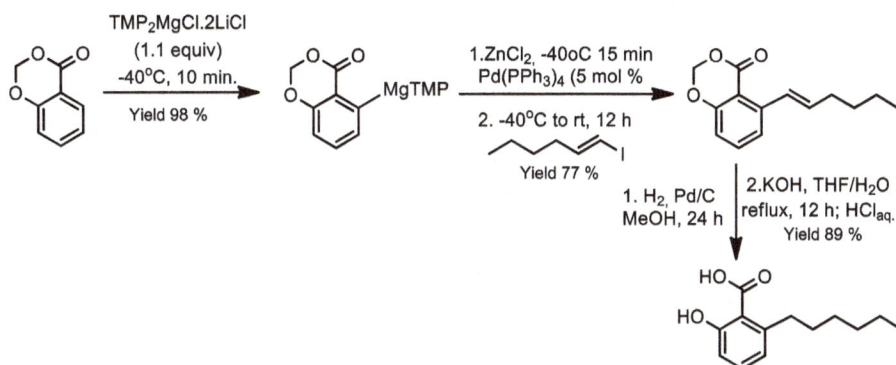

Fig. 20.25: Selective *ortho*-metallation of dimethyl-1,3-benzodioxan-4-one followed by transmetallation with ZnCl₂, Pd-catalyzed cross-coupling with (*E*)-1-hexenyl iodide, hydrogenation of double bond and dioxanone cleavage.

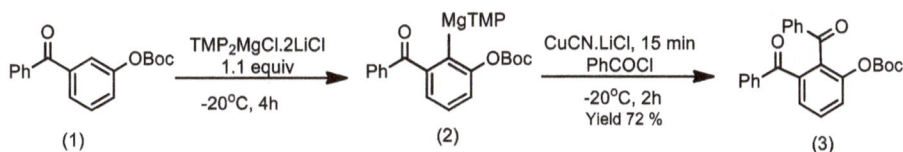

Fig. 20.26: Metallation of Boc-protected (3-hydroxy) benzophenone followed by catalytic coupling with benzoyl chloride.

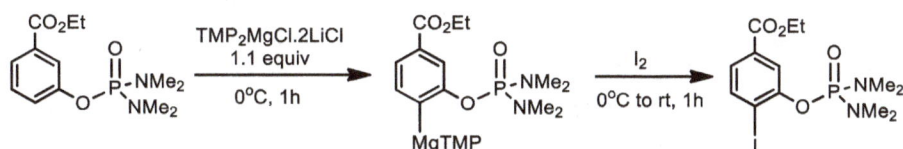

Fig. 20.27: Selective metallation followed by quenching of the magnesiate with iodine.

The selective deprotonative generation of the strained cyclohexynes from a cyclohexenyl triflate using (TMP)₂Mg · 2LiCl is illustrated in Fig. 20.28 [65]. The success of this transformation is the law nucleophilicity of the turbo-base.

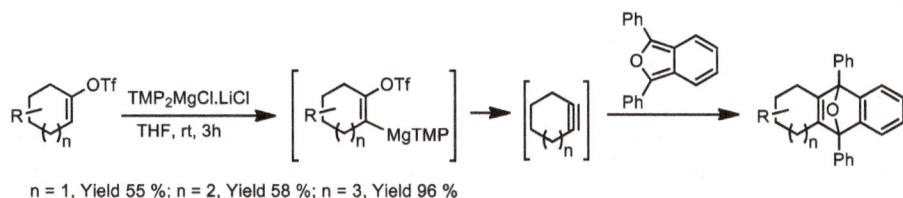

n = 1, Yield 55 %; n = 2, Yield 58 %; n = 3, Yield 96 %

Fig. 20.28: Deprotonative generation of cyclohexine.

The best result was, as expected, in the generation of cyclooctyne, the less strained cycles included in the experiment. The yields refer to the cycloaddition of the transient cycloalkyne with 1,3-diphenyl benzofuran [65].

The base-induced halogen migration, referred to as halogen dance, was found to occur by sequential deprotonation and halogen–metal exchange [66]. For aromatic/heteroaromatic chemistry, the halogen dance/Negishi coupling reactions allow the formation of two chemical bonds in one pot as an alternative to electrophilic aromatic substitution. The magnesium amide-mediated halogen dance (not effective with lithium amides) of bromothiophenes under mild reaction conditions is presented in Fig. 20.29.

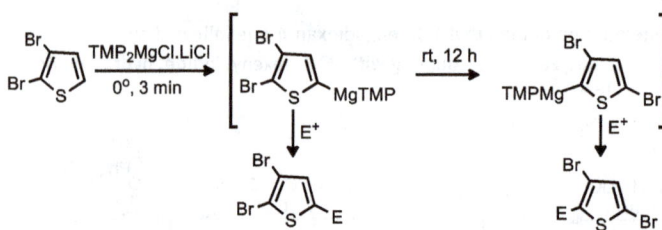

Fig. 20.29: The magnesium amide-mediated halogen dance of dibromothiophene.

A suggested mechanism implies not only the magnesium–hydrogen exchange but also magnesium–bromine exchange [66].

In a route to metal–hydrogen exchange on substrates containing a Lewis basic group or atom, the assistance of another Lewis acid can change the regioselectivity of a reaction. Following this idea, the C–H activation of various polyfunctional pyridines and related heterocycles by a stepwise activation with $BF_3 \cdot OEt_2$ followed by metallation with the appropriate TMP base was experimented successfully. The reactions in Fig. 20.30 are examples for the change in regioselectivity of 3-fluoropyridine (1) and the electron-deficient 3-bromo-4-cyanopyridine (2). To assess the regioselectivity, the metallated intermediaries are transmetallated with $ZnCl_2$ and crosscoupled with an electrophile (Negishi cross-coupling – see Section 21.2.3). The reaction of 3-fluoropyridine and 3-bromo-4-cyanopyridine with TMPMgCl.LiCl affords the magnesiated compounds at position 2 (A) and (F). Precomplexation with $BF_3 \cdot OEt_2$ (C) and (H) and metallation with TMPMgCl $\cdot$ LiCl provide different metallated pyridines (D) and (I). The coordination of $BF_3 \cdot OEt_2$ sterically blocks the 2-position, directing the metallation to positions 4 and 5, respectively [67].

The zinc-containing turbo-reagents allowed the metallation of more sensitive substrates than already selectively described for the magnesium base. High tolerance to nitro, aldehyde, methyl ketone or electron-poor N-heterocycles was achieved with TMPZnCl $\cdot$ LiCl and $(TMP)_2ZnCl \cdot 2LiCl$. Treatment of TMPLi with $ZnCl_2$ in THF produces the LiCl-solubilized base TMPZnCl $\cdot$ LiCl in quantitative yield [68]. One of

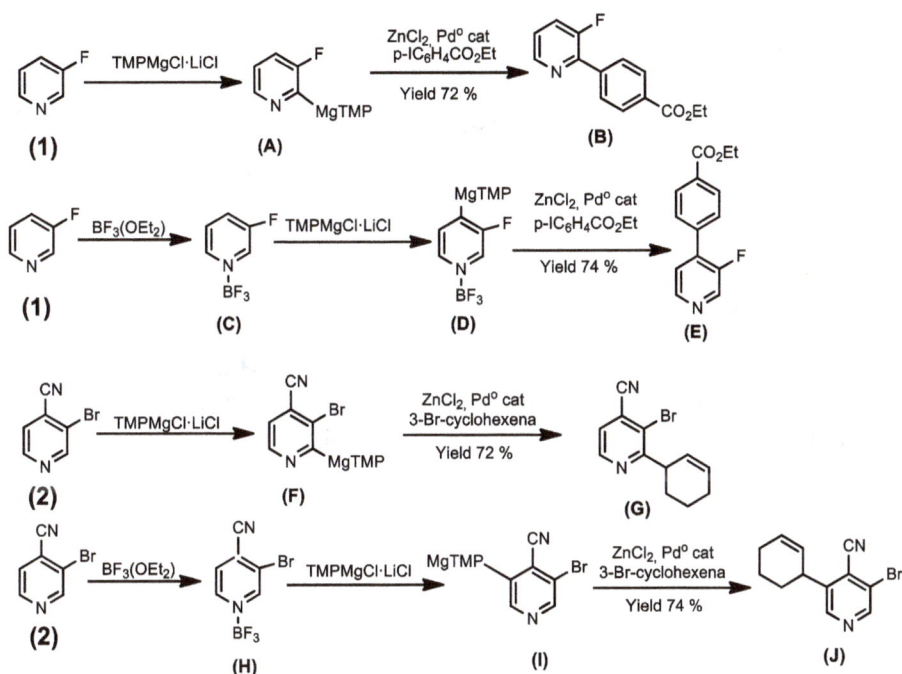

Fig. 20.30: Regioselective metallation of substituted pyridines with and without protecting/directing group (BF₃) on nitrogen.

the advantages brought by the zinc reagents is the possibility to perform metallations at elevated temperatures [69]. To support the last statement, the selective zincation of the dichloropyrimidine in position 5 followed by a copper(I)-catalyzed allylation with cyclohexenyl bromide leading to the fully substituted pyrimidine is presented in Fig. 20.31 [69].

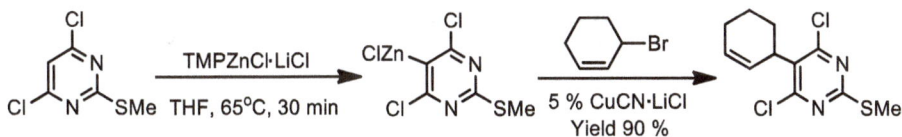

Fig. 20.31: Metallation of dichloropyrimidine with TMPZnCl · LiCl.

Direct zincation of 1-morpholino-6-chlorophthalazine using TMPZnCl · LiCl requires 48 h at 25 °C and produced the zincated species in low yield. The microwave-assisted procedure (a green chemistry approach) led to a complete zincation within 45 min (Fig. 20.32). The metallated chlorophthalazine derivative was further treated with 2-iodothiophene in the Negishi cross-coupling conditions [70].

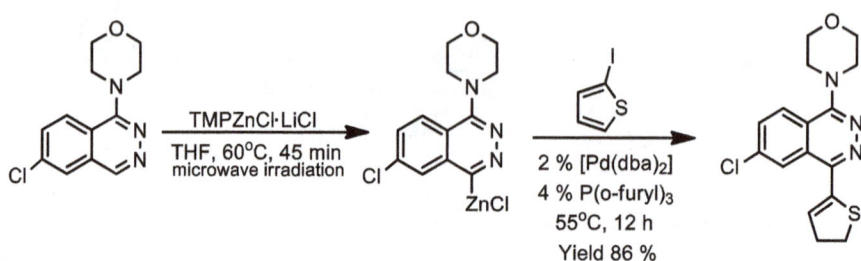

Fig. 20.32: Microwave-assisted zincation of 1-morpholino-6-chlorophthalazine.

The presence of a nitro group, very sensitive, is tolerated when TMPZnCl · LiCl was used for metallation, an example being the preparation of 2-zincated benzothiazole starting from 6-nitrobenzothiazole (Fig. 20.33) [68].

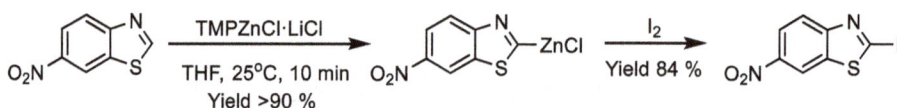

Fig. 20.33: Zincation of 6-nitrobenzothiazole followed by trapping with iodine.

A more powerful base able to zincate relatively unreactive unsaturated substrates was prepared starting from TMPMgCl · LiCl with ZnCl$_2$ in THF resulting in TMP$_2$-Zn · 2MgCl$_2$ · 2LiCl. As for the other turbo-reagents, LiCl ensures a good solubility of the base. The additional presence of MgCl$_2$ (2 equiv) considerably enhances its kinetic basicity. The zincation of 1,3,4-oxadiazole and the 1,2,4-triazole, sensitive heterocycles are prone to undergo fragmentation during the metallation process, is described in Fig. 20.34 [71].

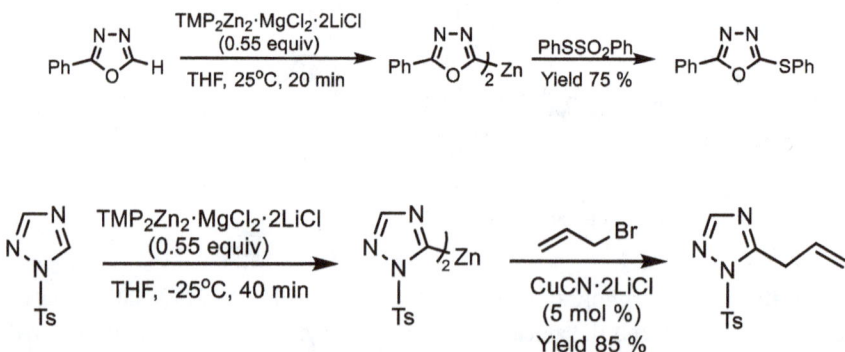

Fig. 20.34: Metallation of 1,3,4-oxadiazole and 1,2,4-triazole with TMP$_2$Zn · 2MgCl$_2$ · 2LiCl.

The reaction of TMP$_2$Zn · 2MgCl$_2$ · 2LiCl with 2-pyridone and 2,7-naphthyridone (heterocycles with pharmaceutical relevance) succeeded to prevent their decomposition or the complex mixture of products observed during the lithiation or magnesiation (Fig. 20.35). Zincations of the methoxyethoxymethyl-protected compounds followed by trapping with electrophiles provided functionalized 2-pyridones and 2,7-naphthyridones [72].

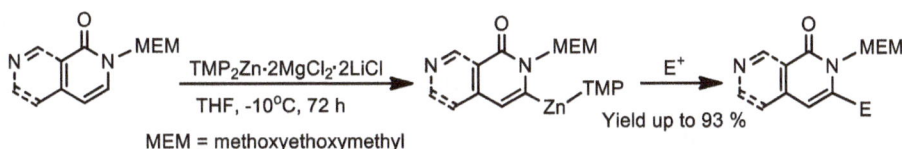

MEM = methoxyethoxymethyl

Fig. 20.35: Metallation of protected 2-pyridone and 2,7-naphthyridone with TMP$_2$Zn · 2MgCl2 · 2LiCl.

Air-stable solid reagents, RZnOPiv · Mg(OPiv)X · nLiCl (where OPiv = OCOtBu; R = aryl, heteroaryl or benzyl; X = Cl, Br or I), can be prepared by the transmetallation of a range of organomagnesium species with zinc pivalate which, after solvent removal, displays significantly improved air and moisture stability [36, 73–76]. The solid aryl-, heteroaryl- and benzylic zinc pivalates show very similar reactivity in cross-couplings, allylations or acylations compared with standard organozinc halides. Example for the preparation of air-stable solid-functionalized aryl, heteroaryl and benzyl organozinc reagents is presented in (FG = functional group) Fig. 20.36 [73].

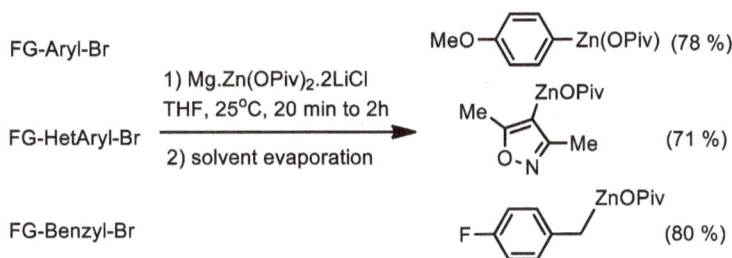

Fig. 20.36: Metallation of aryl, heteroaryl and benzyl with air-stable solid reagents, RZnOPiv · Mg (OPiv)X · nLiCl (OPiv = OCOtBu).

Solid allylic zinc reagents obtained from allylic chlorides or bromides with zinc dust in the presence of lithium chloride and magnesium pivalate (Mg(OCOBu^t)$_2$) in THF, after evaporation of the solvent, also display excellent thermal stability ($t_{1/2}$ is the half-lives when the reagents were stored at −24 °C, Fig. 20.37) [74].

The zinc allylic reagents, (R*)ZnOPiv · Mg(OPiv)X · nLiCl (where R* = various allylic groups; X = Cl, Br or I), can add readily to aldehydes and methyl ketones with high diastereoselectivity (Fig. 20.38).

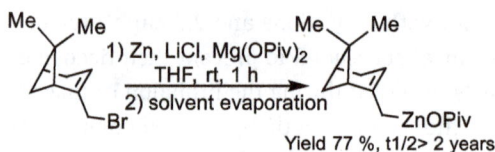

Fig. 20.37: Preparation of solid allylic zinc reagents.

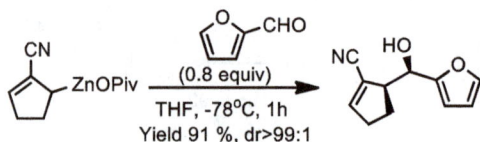

Fig. 20.38: Addition of allylic zinc pivalate to formyl group.

In reaction with acid chlorides, β,γ-unsaturated ketones are formed with high regio-selectivity [75].

Another base, TMPZnOPiv · Mg(OPiv)Cl · LiCl, compatible with nitro group, aldehyde or sensitive heteroaromatic rings, was prepared in solid state after removal of the solvent with significant tolerance toward hydrolysis or oxidation after air exposure. TMPZnOPiv · LiCl (Mg(OPiv)Cl is omitted for clarity) is prepared by the addition of solid Zn(OPiv)$_2$ to a solution of TMPMgCl · LiCl in THF [76]. In most cases, the metallation proceeded with excellent regio- and chemoselectivity. N-Methyl-3-formylindole was successfully zincated at position 2 (25 °C, 30 min) providing indolylzinc pivalate (Fig. 20.39) [76].

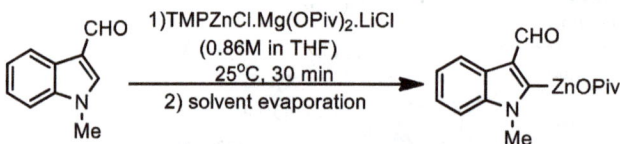

Fig. 20.39: Metallation of N-methyl-3-formylindole with TMPZnOPiv · LiCl (Mg(OPiv)Cl.

The synergistic reactivity of these salt-supported zinc reagents is due to the presence of magnesium pivalate and lithium chloride in the reaction mixtures. The solubility is explained by the molecular structure of [(THF)$_2$Li$_2$(μ-Cl)$_2$(μ-OPiv)$_2$Zn] obtained when the Zn(OPiv)$_2$ barely soluble in THF was dissolved on addition of 1 equiv of LiCl and crystals were deposited (Fig. 20.40). Solubility can, therefore, be attributed to form this molecular complex through the amphoteric Lewis acidic–Lewis basic resource of the salt, which completes the coordination of both the Lewis basic OPiv group and Lewis acidic Zn atom in the presence of lithium chloride [75].

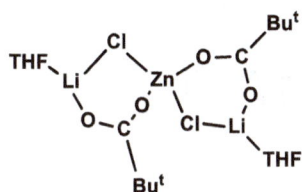

As mentioned before, the contribution of turbo reagents in metal–hydrogen ex-change reactions is more than impressive. To illustrate their performances, a limited number of examples have been chosen for the scope of this book but more examples can be found in the cited literature.

20.1.5.2 Magnesium–halogen exchange

Another path to organometallic compounds useful in organic synthesis is the metal–halogen exchange. For the preparation of substrates bearing highly sensitive func-tional groups, the less polar Mg–C bonds and the covalent Zn–C bonds were ex-ploited with good results as a first step in preparative chemistry in combination with the reactions mediated by transition metals. The traditional Grignard reagents are still important and are effective in Mg–I exchange, faster than Mg–Br exchange. The challenge is to extend the applicability of M–X exchange to Mg–Cl exchange in mild conditions and to find the conditions to perform this type of reaction in nonpolar non-ethereal solvents (as described earlier for the lithiation reactions [24, 77]).

For metal–halogen exchange turbo-Grignard reagents, the representative is Pr^{i}MgCl · LiCl [78]. Other combinations of ligands and ratios of the metals led to very effective reagents for magnesium–halogen exchange such as sBu$^s_2$Mg · 2LiCl, Bu^{s}Mg(OR) · LiOR and Bu$^s_2$Mg · 2LiOR [6].

The metallation capacity of Pr^{i}MgCl · LiCl to discriminate the most electron-poor bromine substituent in polybromides is an example of its regioselectivity (Fig. 20.41). The next step is a classic addition of organomagnesium compound to carbonyl deriva-tives (pivaldehyde in this case) to afford alcohols [78, 79].

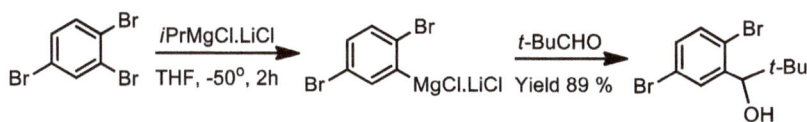

Fig. 20.41: Regioselective magnesiation of 1,2,4-tribromobenzene with Pr^{i}MgCl · LiCl followed by nucleophilic addition to pivaldehyde.

The treatment of 1-bromo-3,5-difluorobenzene with Pr^{i}MgCl · LiCl leads to a complete Mg–Br exchange within 1 h (Fig. 20.42). Transmetallation with ZnCl₂, followed by the

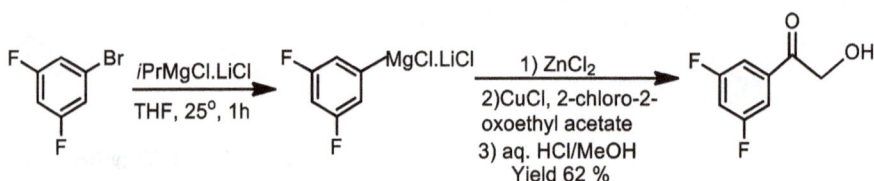

Fig. 20.42: Selective magnesiation of 1-bromo-3,5-difluorobenzene with Pr^{i}MgCl · LiCl, intermediate in the synthesis of α-hydroxyacetophenone.

addition of CuCl and acetoxyacetyl chloride and acidic deprotection, leads to α-hydroxyacetophenone in 62% yield on a 100 g scale [80].

Cycloalkenyl bromides such as 1,2-dibromocyclopentene react with Pr^{i}MgCl · LiCl to provide β-bromocyclopentenylmagnesium by a single Mg–Br exchange (Fig. 20.43). In the presence of a secondary alkylmagnesium halide and Li$_2$CuCl$_4$, these 2-bromoalkenylmagnesium compounds undergo bromine substitution and can then further react with electrophiles to give 1,2-difunctionalized cyclopentenes [81].

Fig. 20.43: Preparation of β-bromocyclopentenylmagnesium by selective reaction with Pr^{i}MgCl · LiCl.

The triazine group is compatible with the magnesium–halogen exchange conditions. An example is the selective Mg–I exchange, leading to the polyfunctionalized Grignard reagent, which by heating cyclizes, leading to carbazole in 75% yield (Fig. 20.44) [82].

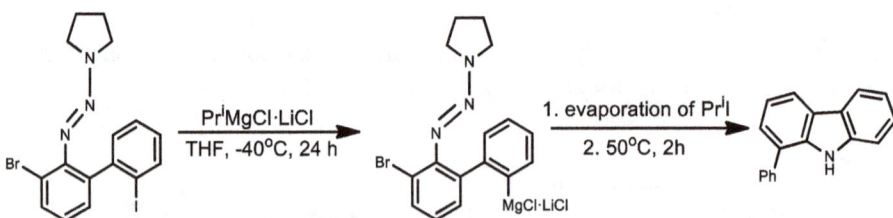

Fig. 20.44: Selective Mg–I exchange followed by cyclization.

Better exchange rates are obtained with dialkylmagnesium complexed with 2 equiv of LiCl, the first of the series being Pr$^i_2$Mg · LiCl. The dialkyl magnesiates proved to have a general synthetic value [79]. The displacement of the Schlenk equilibrium toward the formation of Pr$^i_2$Mg · LiCl was achieved by the treatment of 2 equiv of Pr^{i}MgCl · LiCl chelating additives (1,4-dioxane, PEG250, dimethoxyethane (DME) [15], crown-5 ether, N,N'-dimethyl-N,N'-propyleneurea (DMPU) or TMEDA). A 100% conversion was obtained with the addition of 1,4-dioxane to a solution of iPrMgCl · LiCl in THF [79].

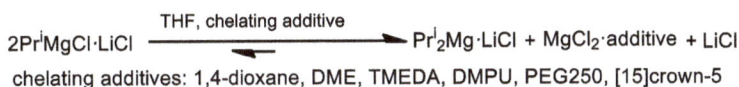

$$2Pr^iMgCl \cdot LiCl \xrightleftharpoons[\text{THF, chelating additive}]{} Pr^i_2Mg \cdot LiCl + MgCl_2 \cdot additive + LiCl$$

chelating additives: 1,4-dioxane, DME, TMEDA, DMPU, PEG250, [15]crown-5

The presence of the lithium chloride was found to be essential for achieving high exchange rates. The influence of the chelating additives was tested in the reaction of electron-rich 4-bromoanisole (Fig. 20.45). The Mg–Br exchange was 100% with 1,4-dioxane or [15] crown-5 ether, 77% with [18] crown-6 ether or TMEDA, 70% with DME, 60% with DME and DMPU, 55% with PEG250 while with the corresponding Grignard reagent, only 31% conversion was obtained [80].

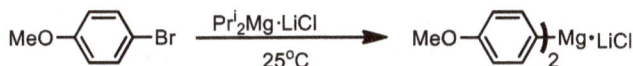

Fig. 20.45: Chelating additive-assisted magnesium bromide exchange in the reaction of 4-bromo-anisole with $Pr^i_2Mg \cdot LiCl$.

The reaction of 1-bromo-4-trimethylsilylbenzene with $Pr^i_2Mg \cdot LiCl$ (generated by adding 10% 1,4-dioxane to $Pr^iMgCl \cdot LiCl$) is completed within 24 h (for the same reaction with $Pr^iMgCl \cdot LiCl$ in THF, the conversion was only 36%; Fig. 20.46 (**I**)). The classical Grignard addition to furfural led to the corresponding alcohol in 92% yield [79].

An even more impressive increase in Mg–Br conversion, from 16% to more than 96%, was obtained for aryls substituted with strong electron-donating groups like N,N-dimethylamine (Fig. 20.46 (**II**)) [79].

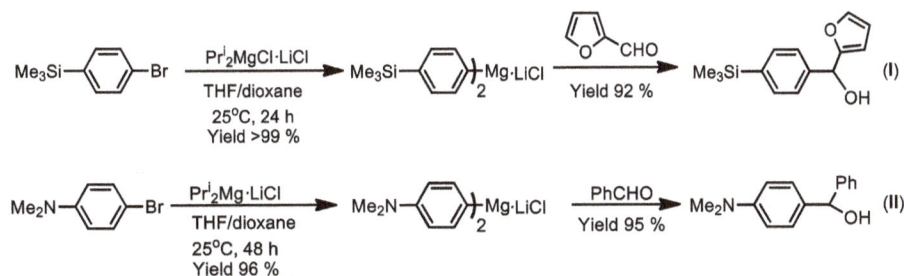

Fig. 20.46: Magnesium bromide exchange in reaction of 1-bromo-4-trimethylsilylbenzene (I) and 1-bromo-4-(dimethylamino)benzene (II) with $Pr^i_2Mg \cdot LiCl$.

The change of organic group in dialkyl magnesiates was also experimented. The reagent $Bu^s_2Mg \cdot LiCl$ was prepared from Bu^sLi (1 equiv) and Bu^sMgCl (1 equiv) in THF and used as 1 M solution in THF. A fast Mg–Br exchange was reported for highly electron-rich 1-bromo-3,4,5-trimethoxybenzene (Fig. 20.47) [79].

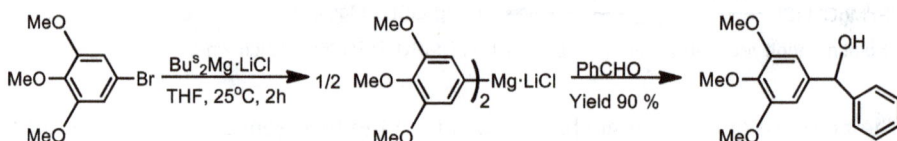

Fig. 20.47: Magnesium bromide exchange in the reaction of 1-bromo-3,4,5-trimethoxybenzene with $Bu^s_2Mg \cdot LiCl$ followed by nucleophilic addition to benzaldehyde.

For substrates sensitive to ethers or to avoid the reaction of the metallation reagents with the ethers, compounds soluble in nonpolar solvents were prepared as shown for *ortho*-lithiation reactions [24]. The preparation of aryl- and heteroaryl-magnesium reagents soluble in toluene can be realized by adding a long aliphatic chain in magnesium reagents. The replacement of chloride in readily available organometallics such as (Bu^s_2Mg, Bu^sLi) with alcoholates led to $Bu^sMg(OR) \cdot Li(OR)$ and $Bu^s_2Mg(OR) \cdot 2Li(OR)$ (Fig. 20.48), which are considerably more active than $Pr^iMgCl \cdot LiCl$ or Bu^s_2Mg LiCl [77].

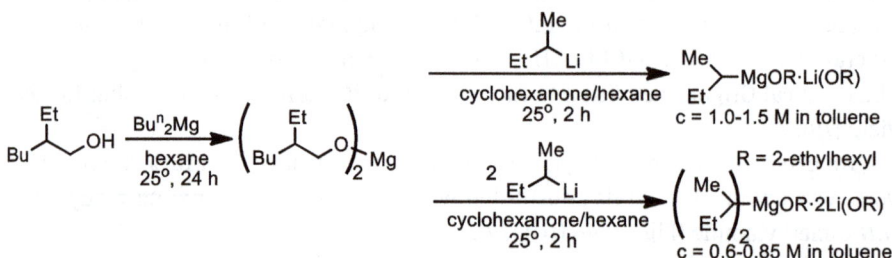

Fig. 20.48: Preparation of alkoxy-substituted organomagnesium reagents.

The mild conditions for the reaction of $Bu^sMg(OR).Li(OR)$ with functionalized aryl bromides allow magnesiation of a variety of substrates. The TMEDA added in 1:1 ratio with the metallating reagent is, most probably, coordinating lithium in the metallation species (I) in Fig. 20.49.

Various heterocyclic bromides are readily converted to toluene-soluble Grignard reagents for subsequent carbon–carbon cross-coupling transformations (Fig. 20.50) [77].

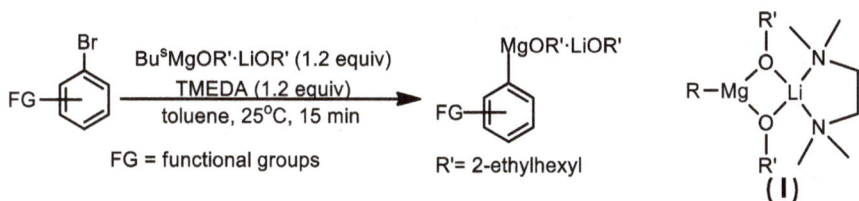

Fig. 20.49: Metallation of functionalized aryl halides with BusMg(OR).Li(OR).

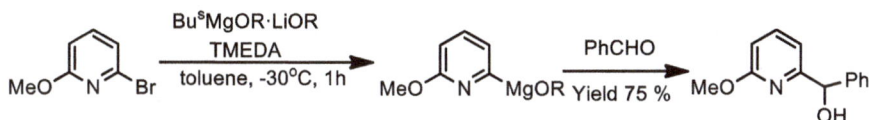

Fig. 20.50: Mg–Br exchange on heterocyclic bromides using Bu^{s}Mg(OR) · Li(OR) in toluene.

The Mg–Cl exchange is performed with the more powerful Bu$^s_2$Mg(OR) 2Li(OR) reagent in reaction with aryl chlorides bearing an *ortho*-chlorine substituent, providing the soluble Grignard reagents (Fig. 20.51) [77].

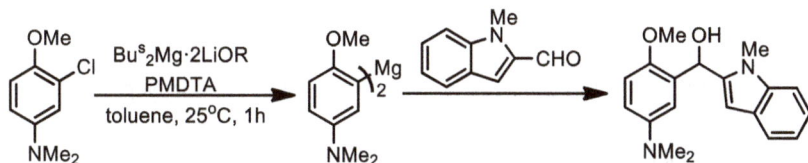

Fig. 20.51: Magnesium–chlorine exchange using Bu$^s_2$Mg(OR).2Li(OR) and subsequent nucleophilic addition to the carbonyl group.

20.1.5.3 Zinc–halogen exchange

The zinc–halogen exchange is another important tool for the preparation of alkyl, alkenyl, aryl and heteroaryl zinc organometallics bearing highly sensitive functional groups. Several types of zinc reagents are used with good results for a given substrate: R$_2$Zn, R$_2$Zn.2Li(OR), R$_3$ZnLi, R$_4$ZnLi$_2$ [7]. The use of dialkylzinc, although useful for metallation of substrates as secondary alkyls, is highly pyrophoric and somehow limited. The iodine–zinc exchange of secondary alkyl iodides proceeds using Pr$^i_2$Zn but when the reagent is prepared from 2 Pr^{i}MgBr and ZnBr$_2$, leading to Pr$^i_2$Zn · 2MgBr$_2$, the exchange reaction proceeds up to 200 times faster due to the presence of this magnesium salt [83]. This may be explained by the formation of the dibromozincate [Pr$^i_2$ZnBr$_2$]$^{2-}$[Mg$_2$Br$_2$]$^{2+}$. Another reactive combination is Pr$^i_2$Zn and Li(acac) (acac = acetylacetone) (10 mol%) in Et$_2$O:NMP (NMP = 1-methyl-2-pyrrolidinone), which is efficient for Zn–I exchange on aryl and heteroaryl and heterocyclic substrates bearing sensitive functional groups such as isothiocyanates or aldehydes (Fig. 20.52) [84].

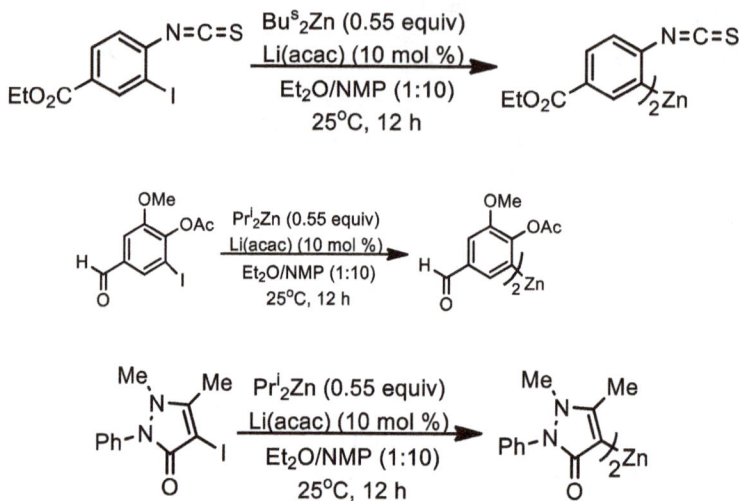

Fig. 20.52: Preparation of functionalized aryl and heterocyclic diorgano zinc compounds using R_2Zn and Li(acac)[84].

It is important to reiterate that zinc reagents serve as reagents for Negishi cross-coupling reactions (Section 21.2.3) [85].

Bimetallic combination of $Bu^s_2Zn \cdot 2LiOR$ is a useful reagent for Zn–X (X = Br, I) exchange in toluene. The applicability was experimented for the preparation of a wide range of polyfunctional aryl- and heteroaryl compounds. The preparation of $Bu^s_2Zn \cdot 2LiOR$ is resumed in Fig. 20.53. The solution of (ROZnEt · ROH) in toluene reacted with Bu^sLi (2.0 equiv, in cyclohexane) to get, after removal of the solvents and subsequent redissolution in toluene a very stable solution of $Bu^s_2Zn \cdot 2LiOR$ [86].

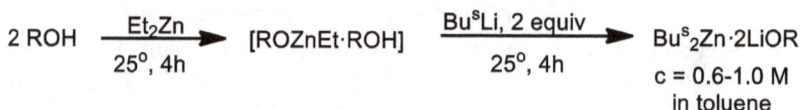

Fig. 20.53: Preparation of $Bu^s_2Zn \cdot 2LiOR$.

The alcohol accompanying the two metals plays a very important role in the reactivity of the reagent. The efficiency of the Zn–I exchange is improved by alcohols bearing N-coordination site as found for the metallation of 3-iodoanisole with Bu^s_2 Zn · 2LiOR: with $HO-CH(CH_3)-(CH_2)_5-CH_3$, 23% yield in 30 min; with $HO-CH_2CH_2 N(Et)_2$, 95% yield in 30 min; and with $HO-CH_2CH_2N-(CH_3)CH_2CH_2N(CH_3)_2$, 99% yield in 1 min. The presence of nitrogen atoms is important, and the coordination to lithium atoms prevents the formation of higher oligomeric lithium zincate. In the molecular structure of [Me$_2$Zn · LiOR] (R = CH$_2$CH$_2$N-(CH$_3$)CH$_2$CH$_2$N(CH$_3$)$_2$) (Fig. 20.54), lithium atoms are coordinated by two oxygen and two nitrogen atoms of the alkoxide while zinc atoms are tricoordinated by the two methyl groups and the oxygen of the alkoxy moiety [86].

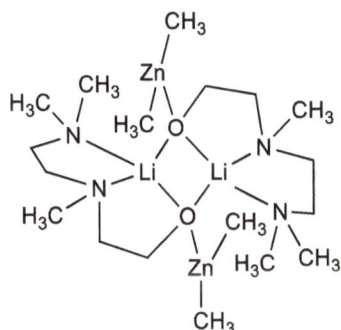

Fig. 20.54: The molecular structure of [Me$_2$Zn · LiOR] (R = CH$_2$CH$_2$N-(CH$_3$)CH$_2$CH$_2$N(CH$_3$)$_2$).

Functionalized aryl iodides are easily zincated using Bu$^s_2$Zn · 2LiOR, in mild conditions, with very good yields and the metallated species reacted with electrophiles under transition metal catalytic conditions (Fig. 20.55) [86].

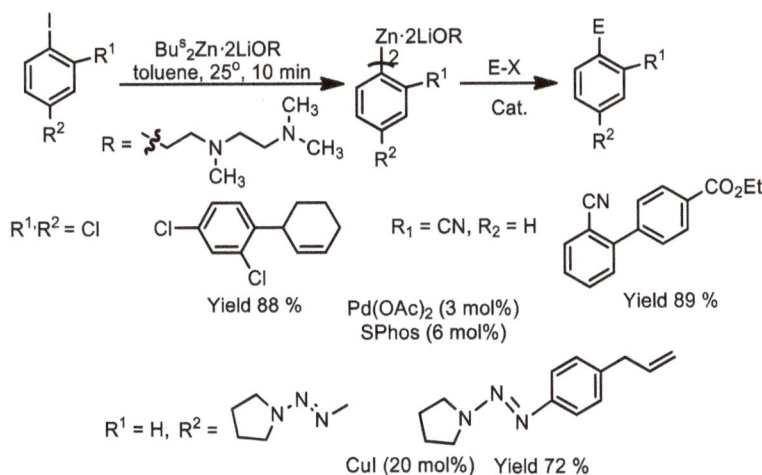

Fig. 20.55: Zinc iodide exchange using Bu$^s_2$Zn · 2LiOR.

Both zincate-type compounds, lower order triorganozincate (R$_3$ZnLi) and higher order tetraorganozincates (R$_4$ZnLi$_2$) are efficient reagents for zinc–halogen exchange. Saturated [87] and unsaturated [88] geminal dibromoderivatives are selectively metallated in good yield. Thus, when 1,1-dibromoalkenes are treated with triorganozincate, a bromine–zinc exchange takes place, leading to alkenylzinc reagents [88]. After hydrolysis, monobromoalkenes are obtained in 82–97% yield (Fig. 20.56).

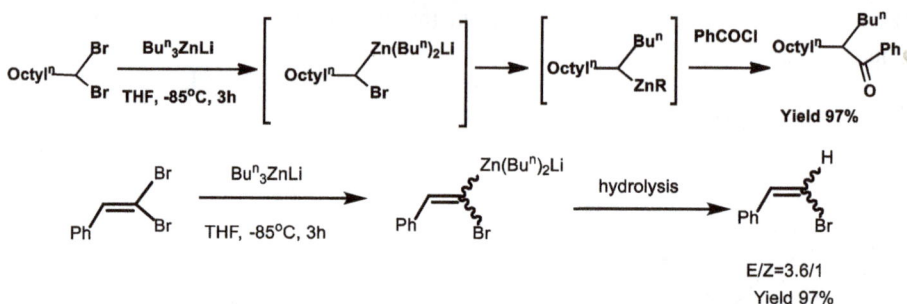

Fig. 20.56: Zn–Br exchange on dibromoalkanes and dibromoalkenes.

Highly reactive zincates, R_4ZnLi_2, are functional group tolerant and allow smooth zinc–halogen exchange reactions in the presence of a chiral acetal or an unprotected hydroxyl group if an excess of reagent is used in the last case (Fig. 20.57) [89].

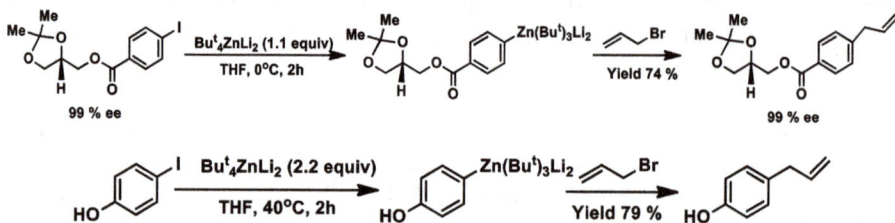

Fig. 20.57: Zn–I exchange using R_4ZnLi_2.

20.2 Organotitanium reagents in organic synthesis

The organotitanium reagents are an alternative to polar organometallics in terms of chemoselectivity and tolerance of functional groups in C–C bond formation in addition (Grignard, aldol and Michael, including enantioselective C–C bond-forming reactions induced by stoichiometric amounts of chiral titanium compounds) or substitution reactions. The capacity of Ti(IV) to accommodate up to six substituents/ligands was exploited in organic synthesis. Examples of Wittig-type or Knoevenagel olefination reactions will be further presented.

The are several types of titanium and organotitanium reagents with relevance in organic syntheses [1]:
- chlorotitanium reagents, $RTiCl_3$ and R_2TiCl_2;
- alkoxytitanium reagents, $RTi(OR')_3$ and titanium ate complexes, $RTi(OR)_4Li$;

– aminotitanium reagents, RTi(NR'₂)₃;
– cyclopentadienyltitanium reagents containing Cp group(s) in combination with chlorine, alkyl, alkoxy or amino group(s), organoaluminum fragments.

There is a general scheme for the preparation of organotitanium reagents involving the conversion of classical carbanions such as RMgX, RLi and R₂Zn into titanium analogues:

$$RLi + TiCl_4 \rightarrow RTiCl_3$$

$$RMgX$$

$$RZnX$$

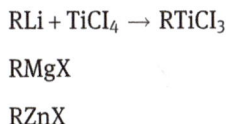

The most common organotitanium reagents are prepared via the intermediates obtained by the redistribution reaction (alkoxy) or the reaction of TiCl₄ with lithium amides [90]:

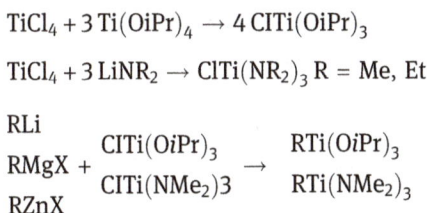

$$TiCl_4 + 3\,Ti(OiPr)_4 \rightarrow 4\,ClTi(OiPr)_3$$

$$TiCl_4 + 3\,LiNR_2 \rightarrow ClTi(NR_2)_3\ R = Me,\ Et$$

$$\begin{matrix} RLi \\ RMgX\ + \\ RZnX \end{matrix} \quad \begin{matrix} ClTi(OiPr)_3 \\ ClTi(NMe_2)3 \end{matrix} \quad \rightarrow \quad \begin{matrix} RTi(OiPr)_3 \\ RTi(NMe_2)_3 \end{matrix}$$

It is important to mention that, for some reactions, the isolation of RTiCl₃ is not always necessary: a stoichiometric mixture of TiCl₄ and RLi, for example, added to the reaction with the organic substrate is acting as the organotitanium reagent. The standard alkoxy titanating agent is chlorotriisopropoxytitanium, ClTi(OPrⁱ)₃, its reaction with polar organometallics affording a huge variety of alkoxy organotitanium reagents of the type RTi(OiPr)₃: R = alkyl (Me, Et, n-Bu), cyclopropyl, alkynyl, aryl, benzyl, vinyl, allyl, 3-furyl, etc.

20.2.1 Reactivity of organotitanium reagents

The steric and electronic properties (mainly the Lewis acidity) of the reagents can be adjusted in a predictable way by an interplay of the ligand X and the R group at titanium. The type of ligand is important for the stereoselectivity control. The alkoxy or amino groups are decreasing the Lewis acidity due to their electron donation effect while steric properties can be tuned by the appropriate choice of their size.

One of the first and significant contributions of a organotitanium reagents was the chemoselectivity in nucleophilic additions of triisopropoxymethyltitanium to carbonyl compounds. The chemoselective addition of triisopropoxymethyltitanium to aldehydes in the presence of ketones compared to reactivity of methyl lithium is illustrated by their reactions with benzaldehyde and acetophenone in a 1:1:1 ratio (Fig. 20.58). In the reaction with triisopropoxymethyltitanium, only the aldehyde adduct was detected while the methyllithium afforded a 1:1 mixture of the adducts to both carbonyl compounds [90].

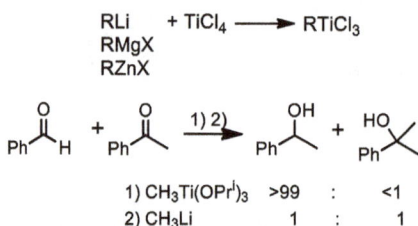

Fig. 20.58: Chemoselective addition of $CH_3Ti(OPr^i)_3$ and CH_3Li to a 1:1 mixture of adehyde:ketone.

The same chemoselectivity was observed when both the aldehyde and ketone are in the same substrate (Fig. 20.59) [90].

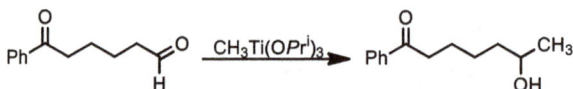

Fig. 20.59: Chemoselective addition of $CH_3Ti(OPr^i)_3$ within the same molecule.

Functional groups such as alkyl and aryl halides, esters, amides as well as nitro and cyano moieties are tolerated.

20.2.2 Titanium-based reagents for carbonyl methylenation and alkylidenation

The organotitanium reagents described in the previous paragraph are better alternatives for the polar organometallics. There is another class of reactions – alkylidenations – where the organotitanium reagents make a difference. The best known methylenation methods of aldehydes and ketones – the Wittig reaction – fail to react with the carbonyl group sterically hindered or base-sensitive substrates. Tebbe's reagent and other Cp_2Ti-containing reagents succeed not only in these cases but also have the ability to alkylidenate carboxylic and carbonic acid derivatives [91–93].

The titanium reagents are chemoselective and have a good functional group tolerance. Chemoselectivity is related to what functional groups are tolerated in the substrates containing the carbonyl group and to what functional groups are tolerated in the titanium reagents.

Tebbe's reagent, $Cp_2TiCH_2AlClMe_2$, a titanium–aluminum metallacycle [94], the Petasis reagent, Cp_2TiMe_2 [95], and Grubbs' reagents, dicyclopentadienyltitana-cyclobutane (including substituted at C2) [96], are olefination reagents for carbonyl groups in organic synthesis. The first two can be prepared in toluene according to Fig. 20.60.

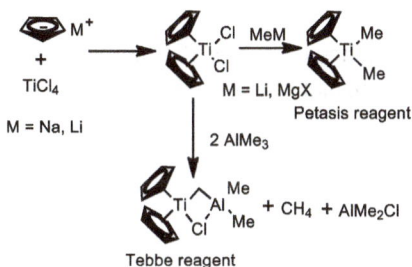

Fig. 20.60: Preparation of the Tebbe and Petasis reagents.

The molecular structure of the "illustrious Tebbe's reagent" was determined by X-ray diffraction as a cocrystal of $[Cp_2Ti(\mu_2\text{-}Cl)(\mu_2\text{-}CH_2)AlMe_2]$ and $[Cp_2Ti(\mu_2\text{-}Cl)_2AlMe_2]$ [97].

To achieve a good reactivity with the less reactive substrates, Tebbe's reagent is treated with a Lewis base like pyridine or THF to generate reactive titanocene methylidene, $Cp_2Ti = CH_2$. Dicyclopentadienyltitanacyclobutane provides the same intermediate, $Cp_2Ti = CH_2$, on heating (Fig. 20.61) [92]. Titanocene methylidene is a typical Schrock, electron-deficient (16e), carbene characteristic for the early transition metals in a high formal oxidation state.

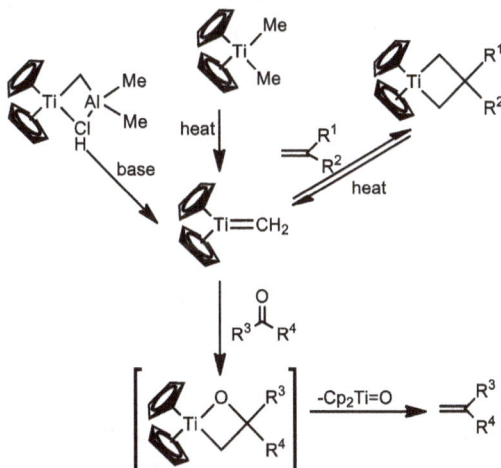

Fig. 20.61: Generation of the active $Cp_2Ti = CH_2$ intermediate and the olefination of the C = O bond.

The methylenation takes place via oxatitanacyclobutane to give alkenes in several minutes at room temperature or even below. It is accepted that the driving force is the formation of the strong titanium oxygen double bond. The Schrock carbenes are nucleophilic at the carbene carbon atom and electrophilic at titanium, and their reactivity toward carbonyl groups is dominated by their high-energy HOMOs. Thus, titanium alkylidenes would be expected to react readily with the most electrophilic carbonyl groups.

Tebbe's reagent methylenated esters and lactones to give enol ethers. In some cases, the ester and ketone in the same molecule are methylenated (Fig. 20.62) [98].

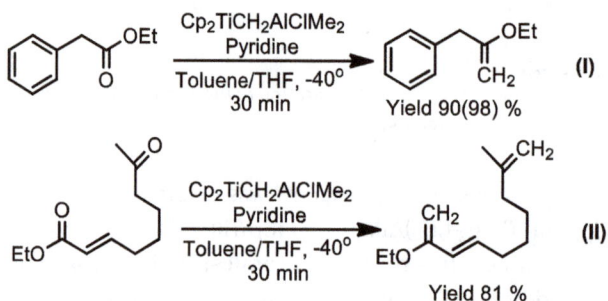

Fig. 20.62: Methylenation of esters (I) and a ketone-containing ester (II) with Tebbe's reagent.

Using the appropriate protocol, Tebbe's selective methylenation of aldehydes and ketones in the presence of esters or amides is possible, and it is used for the preparation of complex molecules. Examples include methylenation of the ketone in Fig. 20.63 in high yield, and stirred in toluene in the dark at 75 °C for 5 days[99].

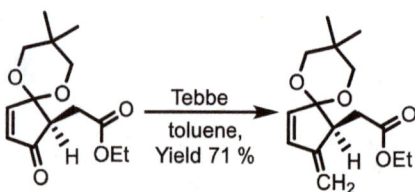

Fig. 20.63: Selective olefination of the ketone in the presence of an ester.

Methylenation of the less hindered methyl ester group in the diester containing a tertiary hydroxy group and a carboxylic acid protected with *tert*-butyldimethylsilyl (TBDMS) triflate using only 1 equiv of Tebbe's reagent at −78 °C followed by warming up to room temperature afforded the targeted enol ether (Fig. 20.64). This was the first example of a regioselective diester olefination [100].

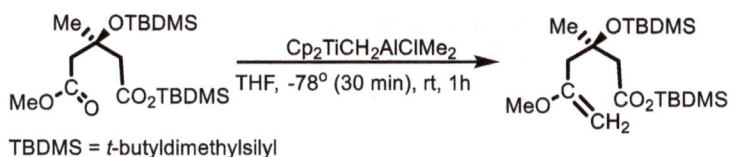

TBDMS = t-butyldimethylsilyl

Fig. 20.64: Methylenation of less hindered ester group in a diester compound.

As would be expected from a nucleophilic reagent, aldehydes and ketones can be selectively methylenated in the presence of less electrophilic carbonyl groups such as esters and amides.

The Petasis reagent, Cp_2TiMe_2, more stable and easier to handle, is used in the selective olefination reactions of highly functionalized substrates. An example is the selective methylenation of the formate ester, leaving the sterically hindered ethyl ester unchanged (Fig. 20.65) [101].

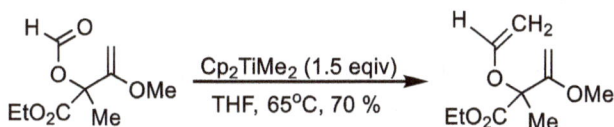

Fig. 20.65: Olefination of formate ester in the presence of ethylester using the Petasis reagent.

Petasis methylenation of highly strained β-lactones proceeds in 20–86% yield with excellent chemoselectivity (Fig. 20.66), while Tebbe's methylenation is unsuccessful [102].

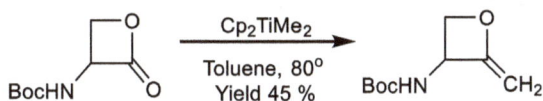

Fig. 20.66: Methylenation of strained β-lactones using the Petasis reagent.

The change from methyl to methylene-substituted groups in the Petasis reagent allowed the preparation of substituted olefins, enol ethers and enamines (Fig. 20.67) [103].

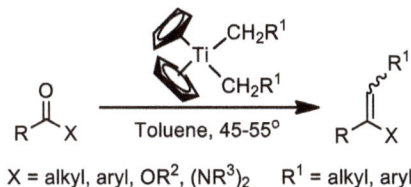

X = alkyl, aryl, OR^2, $(NR^3)_2$ R^1 = alkyl, aryl

Fig. 20.67: Olefination using the modified Petasis reagent.

An example is dibenzyltitanocene, a stable compound readily prepared from titano-cene dichloride and benzylmagnesium chloride, which reacts with carbonyl com-pounds to give phenyl-substituted olefins (Fig. 20.68), in some cases with quantitative conversion as a single geometrical isomer [103].

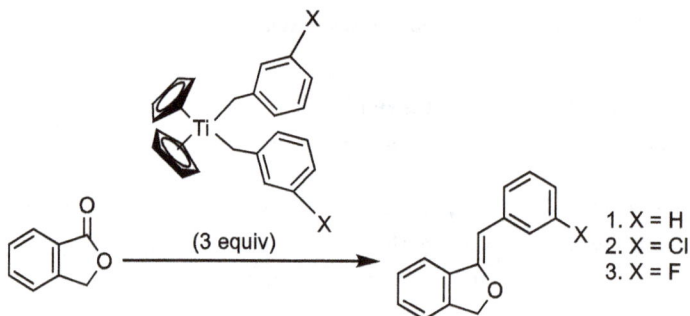

Fig. 20.68: Olefination using dibenzyltitanocene.

Another stereoselective method for the alkylidenation of esters to prepare Z-enol ethers is the reagent prepared from 1,1-dibromoalkane, zinc, titanium(IV) chloride and TMEDA in THF [104]. All reactions are Z-selective, and stereoselectivities are generally over 89%. An example is the alkylidenation of an α,β-unsaturated ester, as shown in Fig. 20.69.

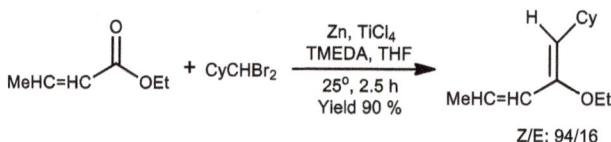

Fig. 20.69: Preparation of Z-enol ethers by olefination of the ester with 1,1-dibromoalkane, zinc, TiCl$_4$ and TMEDA in THF.

Bulky groups in the ester are reducing the stereoselectivity, while a branched one in the α-position to the carbonyl group ensures total Z-selectivity. The reaction is not very sensitive to the bulk of the group on the RCHBr$_2$ reactant. For example, *tert*-butyl ester gives modest selectivity for the corresponding Z-enol ether but *iso*-butyrate gives solely Z-enol ether (Fig. 20.70) [104].

The alkylidenation catalyzed by PbCl$_2$ is another way to enol ether prepara-tion (Fig. 20.71) [105].

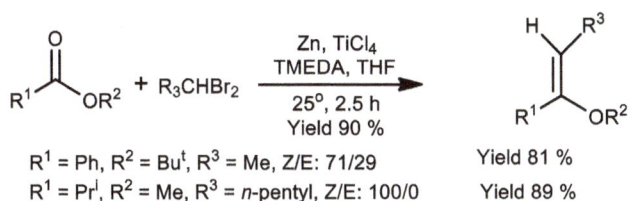

Fig. 20.70: Z-Selective alkylidenation using 1,1-dibromoalkane, zinc, TiCl₄ and TMEDA in THF.

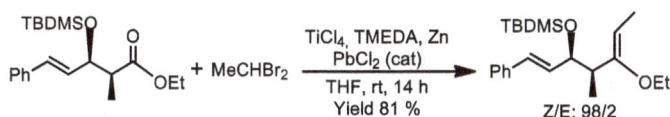

Fig. 20.71: PbCl₂-catalyzed alkylidenation.

References

[1]   Schlosser M. Organometallics in Synthesis: A Manual [Internet]. Schlosser M. editor Organometallics in Synthesis: A Manual, Hoboken, NJ, USA: John Wiley & Sons, Inc, 2001, 1–352.

[2]   Schlosser M. Organoalkali Reagents. Organometallics in Synthesis. Hoboken, NJ, USA: John Wiley & Sons, Inc., 2013, 1–166.

[3]   Peltzer RM, Gauss J, Eisenstein O, Cascella M. The Grignard reaction-unraveling a chemical puzzle. J Am Chem Soc 2020, 142, 2984–94.

[4]   Mongin F, Harrison-Marchand A. Mixed ggregAte (MAA): A single concept for all dipolar organometallic aggregates. 2. Syntheses and reactivities of homo/heteromaas, Chem Rev, 2013, 113, 7563–727.

[5]   Robertson SD, Uzelac M, Mulvey RE. Alkali-metal-mediated synergistic effects in polar main group organometallic chemistry. Chem Rev 2019, 119, 8332–405.

[6]   Ziegler DS, Wei B, Knochel P. Improving the halogen–magnesium exchange by using new turbo-Grignard reagents. Chem Eur J 2019, 25, 2695–703.

[7]   Balkenhohl M, Knochel P. Recent advances of the halogen–zinc exchange reaction. Chem Eur J Eur J 2020, 26, 3688–97.

[8]   Gentner TX, Mulvey RE. Alkali-metal mediation: diversity of applications in main-group organometallic chemistry. Angew Chem Int Ed 2021, 60, 9247–62.

[9]   Lambert C, Von Ragué Schleyer P. Are polar organometallic compounds "carbanions"? The Gegenion effect on structure and energies of alkali-metal compounds. Angew Chem Int Ed English 1994, 33, 1129–40.

[10] Pauling L. The nature of the chemical bond. IV. The energy of single bonds and the relative electronegativity of atoms, J Am Chem Soc, 1932, 54, 3570–82.

[11] O'Sullivan WI, Swamer FW, Humphlett WJ, Hauser CR. Influence of the metallic cation of certain organometallic compounds on the courses of some organic reactions. J Org Chem 1961, 26, 2306–10.

[12] Howden MEH, Maercker A, Burdon J, Roberts J. Small-ring compounds. XLV. Influence of vinyl and phenyl substituents on the interconversion of allylcarbinyl-type Grignard reagents 1. J Am Chem Soc 1966, 88, 1732–42.

[13] Maercker A, Roberts JD. Small-ring compounds. XLVI. Stabilized cyclopropylcarbinyl anions. A retro cyclopropylcarbinyl-allylcarbinyl rearrangement. J Am Chem Soc 1966, 88, 1742–59.

[14] Gilman H, Bebb RL. Relative reactivities of organometallic compounds XX. Metallation. J Am Chem Soc 1939, 61, 109–12.

[15] Wittig G, Pieper G, Fuhrmann G. Über die Bildung von Diphenyl aus Fluorbenzol und Phenyl-lithium (IV Mitteil. über Austauschreaktionen mit Phenyl-lithium). Berr Dtsch Chem Ges 1940, 73, 1193–97.

[16] Knochel P, Cole KP. Directed ortho metallation in 2021: A Tribute to Victor Snieckus (August 1, 1937–December 18, 2020). Org Process Res Dev 2021, 25, 2188–91.

[17] Reich HJ. Role of organolithium aggregates and mixed aggregates in organolithium mechanisms. Chem Rev 2013, 113, 7130–78.

[18] Whisler MC, MacNeil S, Snieckus V, Beak P. Beyond thermodynamic acidity: a perspective on the complex-induced proximity effect (CIPE) in deprotonation reactions. Angew Chem Int Ed 2004, 43, 2206–25.

[19] Bauer W, Ragué Schleyer PV. Mechanistic evidence for ortho-directed lithiations from one- and two-dimensional NMR spectroscopy and MNDO calculations. J Am Chem Soc 1989, 111, 7191–98.

[20] Snieckus V. Directed ortho metallation. Tertiary amide and o-carbamate directors in synthetic strategies for polysubstituted aromatics, Chem Rev, 1990, 90, 879–933.

[21] Shimano M, Meyers AI. Ethoxyvinyllithium-HMPA. Surprising regiochemistry in aromatic metallation. J Am Chem Soc 1994, 116, 10815–16.

[22] Fernández-Nieto F, Paleo MR, Colunga R, Raposo ML, Garcia-Rio L, Sardina FJ. The two alternative rate-determining steps in benzylic lithiation reactions of esters and carbamates. Org Lett 2016, 18, 5520–23.

[23] Campos KR. Direct sp3 C–H bond activation adjacent to nitrogen in heterocycles. Chem Soc Rev 2007, 36, 1069–84.

[24] Slocum DW, Reinscheld TK, White CB, Timmons MD, Shelton PA, Slocum MG, Sandlin RD, Holland EG, Kusmic D, Jennings J, Tekin KC, Nguyen Q, Bush SJ, Keller JM, Whitley PE. Ortho-lithiations reassessed: The advantages of deficiency catalysis in hydrocarbon media. Organometallics 2013, 32, 1674–86.

[25] Reich HJ. What's going on with these lithium reagents?. J Org Chem 2012, 77, 5471–91.

[26] Sakamoto S, Imamoto T, Yamaguchi K. Constitution of Grignard reagent RMgCl in tetrahydrofuran. Org Lett 2001, 3, 1793–95.

[27] Peltzer RM, Eisenstein O, Nova A, Cascella M. How solvent dynamics controls the Schlenk equilibrium of Grignard reagents: a computational study of CH3MgCl in tetrahydrofuran. J Phys Chem B 2017, 121, 4226–37.

[28] Zong H, Huang H, Liu J, Bian G, Song L. Added-metal-free catalytic nucleophilic addition of Grignard reagents to ketones. J Org Chem 2012, 77, 4645–52.

[29] Wang XJ, Zhang L, Sun X, Xu Y, Krishnamurthy D, Senanayake CH. Addition of Grignard reagents to aryl acid chlorides: An efficient synthesis of aryl ketones. Org Lett 2005, 7, 5593–95.

[30] Hurst TE, Deichert JA, Kapeniak L, Lee R, Harris J, Jessop PG. et al. Sodium methyl carbonate as an effective C1 Synthon. Synthesis of carboxylic acids, benzophenones, and unsymmetrical ketones. Org Lett 2019, 21, 3882–85.

[31] Usami S, Suzuki T, Mano K, Tanaka K, Hashimoto Y, Morita N. et al., Chelation-based homologation by reaction of organometallic reagents with O-alkyl S-pyridin-2-yl thiocarbonates: Synthesis of esters from Grignard reagents. Synlett 2019, 30, 1561–64.

[32] Mulvey RE. Avant-garde metalating agents: Structural basis of alkali-metal-mediated metallation. Acc Chem Res 2009, 42, 743–55.

[33] Caubère P. Unimetal super bases. Chem Rev 1993, 93, 2317–34.

[34] Mulvey RE. Modern ate chemistry: Applications of synergic mixed alkali-metal – magnesium or – zinc reagents in synthesis and structure building. Organometallics 2006, 25, 1060–75.

[35] Harrison-Marchand A, Mongin F. Mixed AggregAte (MAA): A single concept for all dipolar organometallic aggregates. 1. Structural data. Chem Rev 2013, 113, 7470–562.

[36] Benischke AD, Ellwart M, Becker MR, Knochel P. Polyfunctional zinc and magnesium organometallics for organic synthesis: some perspectives. Synth 2016, 48, 1101–07.

[37] Andrikopoulos PC, Armstrong DR, Hevia E, Kennedy AR, Mulvey RE, O'Hara CT. Stoichiometrically-controlled reactivity and supramolecular storage of butylmagnesiate anions. Chem Commun 2005, 1131–33.

[38] Becker MR, Knochel P. High-temperature continuous-flow zincations of functionalized arenes and heteroarenes using $(Cy_2N)_2Zn\cdot2LiCl$. Org Lett 2016, 18, 1462–65.

[39] Kloetzing RJ, Krasovskiy A, Knochel P. The Mg-oppenauer oxidation as a mild method for the synthesis of aryl and metallocenyl ketones. Chem Eur J 2007, 13, 215–27.

[40] Noyori R, Suga S, Kawai K, Okada S, Kitamura M. Enantioselective alkylation of carbonyl compounds. From stoichiometric to catalytic asymmetric induction. Pure Appl Chem 1988, 60, 1597–606.

[41] Hatano M, Ishihara K. Recent progress in the catalytic synthesis of tertiary alcohols from ketones with organometallic reagents. Synthesis (Stuttg) 2008, 1647–75.

[42] Sośnicki JG, Struk Ł. Single-step symmetrical double alkylation of β,γ-unsaturated δ-lactams via magnesium "ate" complexes. Synlett 2009, 1812–16.

[43] Zhao H, Engers DW, Morales CL, Pagenkopf BL. Reactions of alanes and aluminates with tri-substituted epoxides. Development of a stereospecific alkynylation at the more hindered carbon. Tetrahedron 2007, 63, 8774–80.

[44] Schlosser M. Zur aktivierung lithiumorganischer reagenzien. J Organomet Chem 1967, 8, 9–16.

[45] Lochmann L, Pospíšil J, Lím D. On the interaction of organolithium compounds with sodium and potassium alkoxides. A new method for the synthesis of organosodium and organopotassium compounds. Tetrahedron Lett 1966, 7, 257–62.

[46] Clegg W, Drummond AM, Liddle ST, Mulvey RE, Robertson A. A novel heterometallic alkoxide: Lithium-potassium tert-butoxide [(BU(t)O)8Li₄K₄]. Chem Commun 1999, 1569–70.

[47] Schlosser M, Simig G. 8-Methoxyisoquinoline derivatives through ortho-selective metallation of 2-(3-methoxyphenyl)ethylamine. Tetrahedron Lett 1991, 32, 1965–66.

[48] Brandsma L, Verkruijsse HD, Schade C, Schleyer PVR. The first successful direct metallation of ethene. J Chem Soc Chem Commun 1986, 260–61.

[49] Schlosser M, Jung HC, Takagishi S. Selective mono- or dimetallation of arenes by means of superbasic reagents. Tetrahedron 1990, 46, 5633–48.

[50] Lochmann L, Janata M. 50 Years of superbases made from organolithium compounds and heavier alkali metal alkoxides. Cent Eur J Chem 2014, 12, 537–48.

[51] Lochmann L. Reaction of organolithium compounds with alkali metal alkoxides – A route to superbases. Eur J Inorg Chem 2000, 1115–26.

[52] Haag B, Mosrin M, Ila H, Malakhov V, Knochel P. Regio- and chemoselective metallation of arenes and heteroarenes using hindered metal amide bases. Angew Chem Int Ed 2011, 50, 9794–824.

[53]   Rohbogner CJ, Clososki GC, Knochel P, General A. Method for meta and para functionalization of arenes using TMP2Mg-2 LiCl. Angew Chem Int Ed 2008, 47, 1503–07.

[54]   Li-Yuan Bao R, Zhao R, Shi L. Progress and developments in the turbo Grignard reagent i-PrMgCl·LiCl: A ten-year journey. Chem Commun 2015, 51, 6884–900.

[55]   Tolman CA. Steric effects of phosphorus ligands in organometallic chemistry and homogeneous catalysis. Chem Rev 1977, 77, 313–48.

[56]   Hauser CR, Walker HG. Condensation of certain esters by means of diethylaminomagnesium bromide. J Am Chem Soc 1947, 69, 295–97.

[57]   Frostick FC, Hauser CR. Condensations of esters by diisopropylaminomagnesium bromide and certain related reagents. J Am Chem Soc 1949, 71, 1350–52.

[58]   García-Álvarez P, Graham DV, Hevia E, Kennedy AR, Klett J, Mulvey RE. et al., Unmasking representative structures of TMP-active Hauser and turbo-Hauser bases. Angew Chem Int Ed 2008, 47, 8079–81.

[59]   Schlecker W, Huth A, Ottow E, Mulzer J. Regioselective metallation of pyridinylcarbamates and pyridinecarboxamides with (2, 2, 6, 6-tetramethylpiperidino)magnesium chloride. J Org Chem 1995, 60, 8414–16.

[60]   Krasovskiy A, Krasovskaya V, Knochel P. Mixed Mg/Li amides of the type $R_2NMgCl\cdot LiCl$ as highly efficient bases for the regioselective generation of functionalized aryl and heteroaryl magnesium compounds. Angew Chem Int Ed 2006, 45, 2958–61.

[61]   Gu YG, Bayburt EK. Synthesis of 4-alkyl-3,5-dibromo-, 3-bromo-4,5-dialkyl-and 3,4,5-trialkylpyridines via sequential metallation and metal-halogen exchange of 3,5-dibromopyridine. Tetrahedron Lett 1996, 37, 2565–68.

[62]   Lin W, Baron O, Knochel P. Highly functionalized benzene syntheses by directed mono or multiple magnesiations with TMPMgCl·LiCl. Org Lett 2006, 8, 5673–76.

[63]   Ihor E, Kopka, IZAF, Rathke MW. Preparation of hindered lithium amide bases and rates of their reaction with ether solvents. J Org Chem 1987, 52, 448–50, 1987, 52, 448–50.

[64]   Clososki GC, Rohbogner CJ, Knochel P. Direct magnesiation of polyfunctionalized arenes and heteroarenes using (TMP)2Mg·2 LiCl. Angew Chem Int Ed 2007, 46, 7681–84.

[65]   Hioki Y, Mori A, Okano K. Steric effects on deprotonative generation of cyclohexynes and 1,2-cyclohexadienes from cyclohexenyl triflates by magnesium amides. Tetrahedron 2020, 76, 131103.

[66]   Yamane Y, Sunahara K, Okano K, Mori A. Magnesium bisamide-mediated halogen dance of bromothiophenes. Org Lett 2018, 20, 1688–91.

[67]   Jaric M, Haag BA, Unsinn A, Karaghiosoff K, Knochel P. Highly selective metallations of pyridines and related heterocycles using new frustrated Lewis pairs or TMP-zinc and TMP-magnesium bases with $BF_3\cdot OEt_2$. Angew Chem Int Ed 2010, 49, 5451–55.

[68]   Mosrin M, Knochel P. TMPZnCl.LiCl: A new active selective base for the directed zincation of sensitive aromatics and heteroaromatics. Org Lett 2009, 11, 1837–40.

[69]   Bresser T, Mosrin M, Monzon G, Knochel P. Regio-and chemoselective zincation of sensitive and moderately activated aromatics and heteroaromatics using TMPZnCl·LiCl. J Org Chem 2010, 75, 4686–95.

[70]   Crestey F, Knochel P. Regioselective functionalization of chlorophthalazine derivatives. Synthesis (Stuttg) 2010, 1097–106.

[71]   Wunderlich SH, Knochel P. $(tmp)_2Zn\cdot 2MgCl_2\cdot 2LiCl$: A chemoselective base for the directed zincation of sensitive arenes and heteroarenes. Angew Chem Int Ed 2007, 46, 7685–88.

[72]   Ziegler DS, Greiner R, Lumpe H, Kqiku L, Karaghiosoff K, Knochel P. Directed zincation or magnesiation of the 2-pyridone and 2,7-naphthyridone scaffold using TMP bases. Org Lett 2017, 19, 5760–63.

[73] Bernhardt S, Manolikakes G, Kunz T, Knochel P. Preparation of solid salt-stabilized functionalized organozinc compounds and their application to cross-coupling and carbonyl addition reactions. Angew Chem Int Ed 2011, 50, 9205–09.

[74] Ellwart M, Knochel P. Preparation of solid, substituted allylic zinc reagents and their reactions with electrophiles. Angew Chem Int Ed 2015, 54, 10662–65.

[75] Hernán-Gómez A, Herd E, Hevia E, Kennedy AR, Knochel P, Koszinowski K. et al., Organozinc pivalate reagents: Segregation, solubility, stabilization, and structural insights. Angew Chem Int Ed 2014, 53, 2706–10.

[76] Stathakis CI, Manolikakes SM, Knochel P. TMPZnOPiv·LiCl: A new base for the preparation of air-stable solid zinc pivalates of sensitive aromatics and heteroaromatics. Org Lett 2013, 15, 1302–05.

[77] Ziegler DS, Karaghiosoff K, Knochel P. Generation of aryl and heteroaryl magnesium reagents in toluene by Br/Mg or Cl/Mg exchange. Angew Chem Int Ed 2018, 57, 6701–04.

[78] Krasovskiy A, Knochel P. A LiCl-mediated Br/Mg exchange reaction for the preparation of functionalized aryl- and heteroarylmagnesium compounds from organic bromides. Angew Chem Int Ed 2004, 43, 3333–36.

[79] Krasovskiy A, Straub BF, Knochel P. Highly efficient reagents for Br/Mg exchange. Angew Chem Int Ed 2005, 45, 159–62.

[80] McLaughlin M, Belyk KM, Qian G, Reamer RA, Chen CY. Synthesis of α-ydroxyacetophenones. J Org Chem 2012, 77, 5144–48.

[81] Despotopoulou C, Bauer RC, Krasovskiy A, Mayer P, Stryker JM, Knochel P. Selective mono- and 1,2-difunctionalisation of cyclopentene derivatives via Mg and Cu intermediates. Chem Eur J 2008, 14, 2499–506.

[82] Liu CY, Knochel P. Preparation of polyfunctional arylmagnesium reagents bearing a triazene moiety. A new carbazole synthesis. Org Lett 2005, 7, 2543–46.

[83] Micouin L, Knochel P. Iodine-zinc exchange reactions mediated by i-Pr₂Zn. A new preparation of secondary zinc reagents. Synlett 1997, 1997, 327–28.

[84] Kneisel FF, Dochnahl M, Knochel P. Nucleophilic catalysis of the iodine-zinc exchange reaction: Preparation of highly functionalized diaryl zinc compounds. Angew Chem Int Ed 2004, 43, 1017–21.

[85] Negishi E. Magical power of transition metals: past, present, and future (Nobel Lecture). Angew Chem Int Ed 2011, 50, 6738–64.

[86] Balkenhohl M, Ziegler DS, Desaintjean A, Bole LJ, Kennedy AR, Hevia E. et al. Preparation of polyfunctional arylzinc organometallics in toluene by halogen/zinc exchange reactions. Angew Chem Int Ed 2019, 58, 12898–902.

[87] Harada T, Kotani Y, Katsuhira T, Oku A. Novel method for generation of secondary organozinc reagent: Application to tandem carbon-carbon bond formation reaction of 1,1-dibromoalkane. Tetrahedron Lett 1991, 32, 1573–76.

[88] Harada T, Katsuhira T, Hattori K, Oku A. Stereochemistry in carbenoid formation by bromine/ lithium and bromine/zinc exchange reactions of gem-dibromo compounds. Tetrahedron 1994, 50, 7987–8002.

[89] Uchiyama M, Furuyama T, Kobayashi M, Matsumoto Y, Tanaka K. Toward a protecting-group-free halogen–metal exchange reaction: Practical, chemoselective metallation of functionalized aromatic halides using dianion-type zincate, tBu₄ZnLi^{2-}. J Am Chem Soc 2006, 128, 8404–05.

[90] Reetz MT, Westermann J, Steinbach R, Wenderoth B, Peter R, Ostarek R, Maus S. Chemoselective addition of organotitanium reagents to carbonyl compounds. Chem Ber 1985, 118, 1421–40.

[91] Pine SH. Carbonyl Methylenation and Alkylidenation Using Titanium-Based Reagents. Organic Reactions. Hoboken, NJ, USA: John Wiley & Sons, Inc, 1993, 1–91.

[92] Petasis NA, Hu Y-H. cyclopentadienyl titanium derivatives in organic synthesis. Curr Org Chem 1997, 1, 249–86.

[93] Hartley RC, McKiernan GJ. Titanium reagents for the alkylidenation of carboxylic acid and carbonic acid derivatives. J Chem Soc Perkin Trans 2002, 1, 2, 2763–93.

[94] Tebbe FN, Parshall GW, Reddy GS. Olefin homologation with titanium methylene compounds. J Am Chem Soc 1978, 100, 3611–13.

[95] Petasis NA, Bzowej EI. Titanium-mediated carbonyl olefinations. 1. Methylenations of carbonyl compounds with dimethyltitanocene. J Am Chem Soc 1990, 112, 6392–94.

[96] Brown-Wensley KA, Buchwald SL, Cannizzo L, Clawson L, Ho S, Meinhardt D, Stille JR, Straus D, Grubbs RH. Cp$_2$TiCH$_2$ complexes in synthetic applications. Pure Appl Chem 1983, 55, 1733–44.

[97] Thompson R, Nakamaru-Ogiso E, Chen C-H, Pink M, Mindiola DJ. Structural elucidation of the Illustrious Tebbe reagent. Organometallics 2014, 33, 429–32.

[98] Pine SH, Zahler R, Evans DA, Grubbs RH. Titanium-mediated methylene-transfer reactions direct conversion of esters into vinyl ethers. J Am Chem Soc 1980, 102, 3270–72.

[99] Göres M, Winterfeldt E. Enantiopure didemnenones via kinetic resolution. J Chem Soc Perkin Trans 1, 1994, 3525–31.

[100] Müller M, Lamottke K, Löw E, Magor-Veenstra E, Sieglich W. Stereoselective total syntheses of atrochrysone, torosachrysone and related 3,4-dihydroanthracen-1(2H)-ones. J Chem Soc Perkin Trans 2000, 1, 1, 2483–89.

[101] Vedejs E, Duncan SM. A synthesis of C(16),C(18)-bis-epi-cytochalasin D via Reformatsky cyclization. J Org Chem 2000, 65, 6073–81.

[102] Ndakala AJ, Hashemzadeh M, So RC, Howell AR. Synthesis of D – erythro - dihydrosphingosine and D – xylo -phytosphingosine from a template. Org Lett 2002, 4, 1719–22.

[103] Petasis N A, Bzowej E I. Titanium-mediated carbonyl olefinations. 2. Benzylidenations of carbonyl compounds with dibenzyl titanocene. J Org Chem 1992, 57, 1327–30.

[104] Okazoe T, Takai K, Oshima K, Utimoto K. Alkylidenation of ester carbonyl groups by means of a reagent derived from RCHBr$_2$, Zn, TiCl$_4$, and TMEDA. Stereoselective preparation of (Z)-alkenyl ethers. J Org Chem 1987, 52, 4410–12.

[105] Rutherford AP, Hartley RC. Unexpected stereoselectivity in the anionic oxy-Cope rearrangement of acyclic enol ethers. Tetrahedron Lett 2000, 41, 737–41.

21 Transition metal organometallics in organic syntheses

The transition metal organometallics are involved in organic synthesis in two ways: as reactants (i.e., nucleophilic attack to π-unsaturated ligands coordinated to transition metals and olefination of carbonyl derivatives) and in catalytic cycles (i.e., cross-coupling reactions, Heck reactions, homogeneous hydrogenation, carbonylation, olefin metathesis and polymerization). Most of the transition metal catalysts which changed for the better organic synthesis are not organometallic but coordination complexes. However, the key of the catalytic processes we will further discuss is the binding of the organic substrate to the transition metal with the formation of a transition metal–carbon bond, hence, they are within the scope of this book. The selection of reactions is meant to give an image on the huge contribution the organometallic chemistry has on organic chemistry and it is not comprehensive.

The transition metal–carbon bonds are formed during the catalytic cycles by specific reaction types not found in organic chemistry like oxidative addition, migratory insertion or nucleophilic attack to coordinated ligands.

21.1 Specific reaction types involving transition metal organometallics

21.1.1 Oxidative addition–reductive elimination

Oxidative addition is, usually, the first step of the catalytic cycle while the reverse reaction – reductive elimination – is the last step resulting in the release of the reaction product along with the catalyst in its active form. Oxidative addition of an organic substrate to a transition metal is one of the most important ways to build reactive intermediates for further transformations. At least one carbon–metal σ-bond is formed in the process (Fig. 21.1) [1]. These reactions are possible due to the ability of the transition metals to exist in different oxidation states and to change the coordination number.

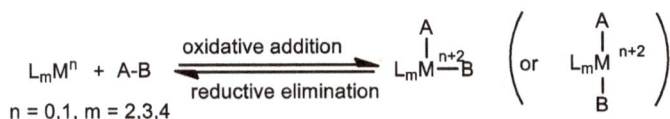

$$L_mM^n + A\text{-}B \underset{\text{reductive elimination}}{\overset{\text{oxidative addition}}{\rightleftharpoons}} L_mM\overset{A}{\underset{}{\overset{|}{-}}}B^{n+2} \quad \left(\text{or} \quad L_mM\overset{A}{\underset{B}{\overset{|}{\underset{|}{}}}}^{n+2} \right)$$

$n = 0,1, m = 2,3,4$

Fig. 21.1: Oxidative addition/reductive elimination.

https://doi.org/10.1515/9783110695274-022

The transition metal in a low oxidation state (0, + 1) and coordinatively unsaturated reacts with a A–B substrate to form two new bonds, M–A and M–B, with the increase of the oxidation state and the coordination number [2]. Electron-rich metals in the low oxidation state, electron-poor organic compound with low A–B dissociation energy and stable compound in the new oxidation state are important for the success of the oxidative addition. To fulfill the requirements for oxidative addition – coordinative unsaturation and low oxidation state – it is important to correlate the coordination numbers with the electronic structure of the metal: six-, five- and four-coordinate complexes are usually saturated, while five-, four-, three- and two-coordinate complexes are unsaturated for d^6, d^8 or d^{10} electronic configurations [2]. Eighteen electron complexes do not undergo oxidative addition. The systems going from d^{10} to d^8 (Ni(0), Pd(0) to Ni(II), Pd(II)), or d^8 to d^6 (Rh(I), Ir(I) to Rh(III), Ir(III)) are used more commonly for oxidative addition.

As the transition metals in low oxidation state act as nucleophiles, the ligands have an important role: the σ-donor ligands enhance the nucleophilicity, increasing the reaction rate, while the π-acceptor ligands decrease the electron density on the metal. The ease of dissociation of the ligand in order to free the coordination sites for the new M–A and M–B bonds is also to be taken into consideration. Some ligands should be avoided, irrespective of their donor properties, due to their propensity to form bridged compounds blocking the potential reactive site (i.e., hydroxide or sulfur ligands).

The best examples for coordinative unsaturation are square planar complexes (Fig. 21.2) [2]:

Fig. 21.2: Oxidative addition to a square planar complex.

Complexes of the type $M(PPh_3)_4$ (M = Ni(0), Pd(0), Pt(0)) are good catalysts in processes where oxidative addition is the first step (e.g., cross-coupling reactions). The active catalytic species, usually $M(PPh_3)_2$, are formed via dissociative mechanisms. Phosphine ligands are good σ-donor, increasing the electron density on the metal and at the same time are prone to dissociation. In most of the cases, the steric effects are much more important than electronic effects in determining the dissociation of phosphine ligands from transition metal complexes. The greater the size of the ligand cone, the greater is the tendency for dissociation [3].

Oxidative addition can proceed through a variety of mechanisms. The systems participating in the oxidative addition reactions to transition metal complexes are organic, organometallic or inorganic molecules. The polarity of the A–B bond is responsible for the relative position, cis or trans, of A and B in the oxidative addition product. The nonpolar bonds like hydrogen–hydrogen (H_2), halogen–halogen (X_2) and carbon–hydrogen bonds (including aldehydes) form mainly cis-products, although sometimes other electronic or steric factors can orient the two fragments in the trans-relative position. The polar A–B bonds (polar electrophiles) like hydrogen–halogen (H–X) and carbon–halogen (organic halides and acyl halides) usually form the trans-isomers. The substrates containing multiple bonds, like O_2, S_2, carbon–carbon double or triple bonds and ortho-diketones, form cyclic compounds.

For the nonpolar systems, the mechanism of the oxidative addition suggests the prior attachment of A–B bond to the metal through an agostic interaction followed by insertion (Fig. 21.3). The oxidative additions of H–H, C–H or Si–H bonds are the most common examples.

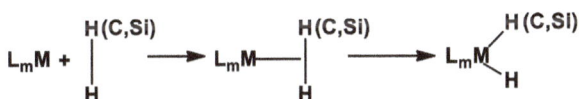

Fig. 21.3: Oxidative addition of nonpolar bonds via agostic interaction.

Oxidative addition is one of the best methods for C–H activation.

Oxidative additions of aryl and alkenyl halides or triflates proceed through concerted mechanisms analogous to oxidative additions of nonpolar systems (Fig. 21.4). The equivalent of the agostic interaction is the η^2-coordination intermediate.

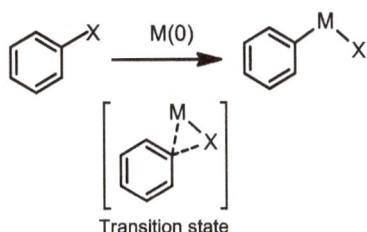

Transition state

Fig. 21.4: Oxidative addition of polar bonds via η^2-coordination intermediate.

The oxidative addition of polar systems is following S_N2 and ionic mechanisms in two steps as shown in Fig. 21.5: (1) the nucleophilic metal center reacts with the electrophilic atom, displacing the halide (A), and (2) the halide bonds to the metal (B):

As in classical S_N2 reactions, the primary halides are the most reactive, followed by secondary and tertiary halides and the order of reactivity of halides is I > Br > Cl. The phosphine ligands are good ligands in oxidative addition of polar electrophiles, that is, $L = P(Et_3)_3 > P(Et_2)Ph > PEt(Ph)_2 > PPh_3$ Negatively charged metal complexes will react very fast.

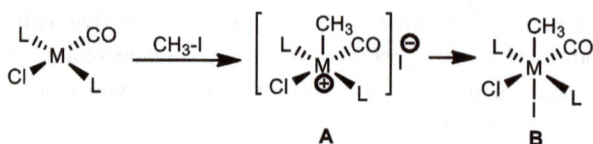

Fig. 21.5: Two-step oxidative addition of CH_3I to a square planar complex: A – the electrophile (CH_3^+) attacks the metal center to form an ionic complex; B – the iodine coordinates the metal to form the final neutral complex.

Radical mechanisms are also possible for the oxidative addition (Fig. 21.6). Nonchain radical mechanism involves single-electron transfer from the metal complex to the organohalide (the homolytic dissociation of C–X bond) followed by combination of the resulting radicals. For the radical reactions, solvents that do not react with the intermediate radicals should be used.

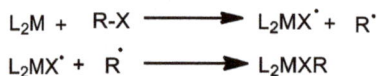

$$L_2M + R\text{-}X \longrightarrow L_2MX^{\cdot} + R^{\cdot}$$

$$L_2MX^{\cdot} + R^{\cdot} \longrightarrow L_2MXR$$

Fig. 21.6: Radical oxidative addition.

The rate of reactivity depends on the stability of the intermediate radical species: tertiary > secondary > primary carbon. Electron-rich metal centers react more rapidly since they can more easily donate the electron to the organic substrate. Chain radical mechanisms involve reactions between radical intermediates and even-electron starting materials resulting in the continuous regeneration of radicals as products.

Reductive elimination is an important step in many catalytic cycles, leading to the formation of the final product, in some cases, even the turnover-limiting step. Reductive elimination is the reverse of oxidative addition (Fig. 21.7). The oxidation state of the metal decreases by two units as the new bond in the product is formed, and two new open coordination sites become available.

$$L_mM\overset{A}{\underset{B}{\overset{n+2}{—}}} \longrightarrow L_mM^n + A\text{-}B$$

Fig. 21.7: Reductive elimination.

A *cis*-disposition of the eliminating ligands is an absolute requirement for reductive elimination. As expected, the reductive elimination is favored by factors opposite to those mentioned before for oxidative addition: electron-rich ligands bearing electron-donating groups, electron-poor metal centers bearing π-acidic ligands and/or ligands with electron-withdrawing groups and bulky ancillary ligands. The rates of reductive eliminations of alkanes are parallel to the steric demands of the eliminating ligands: C–C > C–H > H–H. The mechanisms of the reductive elimination are the same as for oxidative addition: nonpolar and moderately polar ligands react by concerted or radical mechanisms; highly polarized ligands and/or very electrophilic metal complexes react by ionic (S_N2) mechanisms.

21.1.2 Migratory insertion and β-hydride elimination

Another mechanism important for organic synthesis is the migratory insertion. This reaction involves the formal insertion of a neutral ligand (usually unsaturated) into another metal–ligand bond on the same complex (Fig. 21.8). The two groups involved in the migratory insertion must be *cisoidal* to one another. The empty coordination site left behind from where the fragment A originally was located is fastly occupied by another ligand.

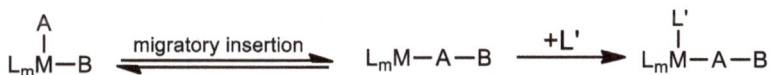

Fig. 21.8: Migratory insertion followed by complexation.

Common examples of ligands that can do migratory insertion reactions with one another are:

Neutral (A) = CO, CO, alkenes, alkynes, carbenes, NO, CR_2, CNR, RCN, O_2, CO_2
Anionic (B): H^-, R^- (alkyl), Ar^- (aryl), acyl-, RO-, R_2N-, O^{2-} (oxo) . . .

The M–B bond is generally polarized toward the ligand, making it nucleophilic and prone to interact with the electrophilic unsaturated ligand in the organometallic complex. There is no change in formal oxidation state of the metal (unless the ligand is an alkylidene/alkylidyne) but the total electron count of the complex decreases by 2 during the insertion. The coordination of an added ligand to the empty coordination site generated in the process is important to avoid the back-elimination reaction. Thermodynamically, the newly formed A–B and covalent M–A bonds must be more stable than the broken M–B and dative M–A bonds for insertion to be favored or the reverse reaction will prevail. Starting from a given organometallic compound, the reverse reaction is also important for organic synthesis: an example we can mention is the elimination of an olefin – the process known as β-hydride elimination.

There are two types of migratory insertions which differ in the number of atoms in the unsaturated ligand involved. Insertions of η^1-unsaturated ligands (i.e., CO, or carbenes) are referred to as 1,1-insertions because the anionic ligand moves from its current location one bond further from the metal. For η^2-ligands (i.e., alkenes, alkynes), one considers a 1,2-insertion as the anionic ligands are bonded two atoms further the metal to the distal atom of the unsaturated ligand.

Migratory insertion of CO in a metal–carbon bond (a 1,1-insertion) is a useful way to prepare various classes of carbonyl derivatives, depending on the neighboring ligands in the molecule: aldehydes, ketones or carboxylic acid derivatives (esters and amides) (Fig. 21.9). Also, with the appropriate combination of metals and ligands, the acyl can be decarbonylated.

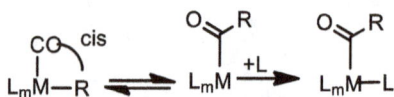

Fig. 21.9: Migratory insertion of CO into a metal–alkyl bond.

The classical example of the migratory insertion of CO into a metal–alkyl bond is presented in Fig. 21.10 [4].

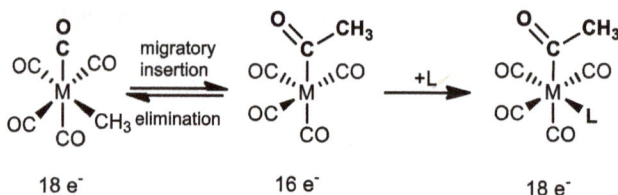

Fig. 21.10: Migratory insertion of CO into M–CH_3 bond followed by coordination of a ligand to the vacanted site.

Electron-deficient metals are recommended to increase the electrophilicity of the CO and increase the susceptibility to nucleophilic attack. The donor properties of the newly formed acyl ligand are not as good as the ones of the alkyl group; therefore, to shift the equilibrium toward the migratory insertion product, a better donating ligand than the CO should block the emptied coordination site. Phosphines are the ligands of choice to fulfill this task.

Migratory insertions of a π-system into M–B bonds (B = hydride, alkyl) results in the formation of two new σ-bonds in one step, in a stereocontrolled manner (Fig. 21.11). An alkene and a hydride usually react via migration of the hydride to the coordinated alkene ligand in a *syn* fashion. The transition state for this process is essentially an agostic interaction of the hydride with the emerging alkyl:

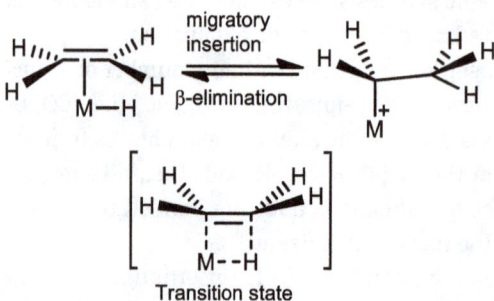

Transition state

Fig. 21.11: Migratory insertion of the olefin into a metal–hydride bond.

The 1,2-migratory insertion thermodynamic, in this case, depends strongly on the relative strength of the alkene–metal bond compared to the alkyl–metal bond. The driving force is the stronger alkyl–metal bond formed after migratory insertion. The presence of electron-withdrawing substituents, such as carbonyls or fluorine, known to stabilize the M–C bond, will have a positive influence on the migratory insertion. The migratory insertions of alkenes into M–H bond is faster than the migratory insertion in an M–R (R = alkyl) bond (both thermodynamically favored). The same order of reactivity was observed for the reversed reaction, β-elimination: β-hydride elimination is much faster than the alkyl elimination. For β-elimination, the eliminating moiety and the metal must have the ability to align in a *syn*-fashion.

Although insertion into M–R is relatively slow, this elementary step is critical for olefin polymerizations (Ziegler–Natta polymerization).

21.1.3 Nucleophilic attack on coordinated substrates

The formation of M–C bonds brings fundamental changes in the reactivity of the carbon containing moieties opening synthetic possibilities unknown in classic organic chemistry. Direct nucleophilic attack is possible to unsaturated ligands – carbon monoxide, alkenes, alkynes, arenes – coordinated by transition metals. Factors favoring nucleophilic attack at coordinated ligands are the coordinately saturated metal center, π-accepting ancillary ligands (e.g., CO), electron-poor or cationic metal centers and soft nucleophiles (hard nucleophiles usually attack the metal first).

The coordination of carbon monoxide or isonitriles to transition metals will allow the nucleophilic attack on the carbon atom involved in the M–C bond. The carbon monoxide reacts with strong nucleophiles like organolithium compounds to afford the two structures, **A** and **B** (Fig. 21.12).

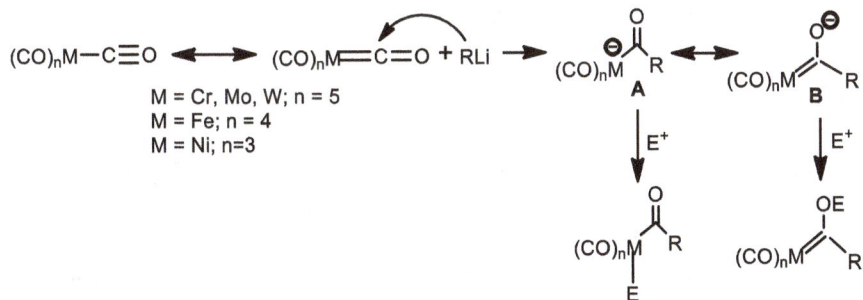

Fig. 21.12: The nucleophilic attack of organolithium compounds on coordinated carbon monoxide.

The subsequent reactions of **A** with electrophiles lead to precursors for carbonyl-containing organic compounds (aldehydes, ketones or derivatives of carboxylic acids). Reaction of **B** with electrophiles is a way to Fischer carbenes (Fig. 21.12).

The attack of softer nucleophiles, like hydroxide and alkoxides, to coordinated carbon monoxide (Fig. 21.13) affords transition metal hydrides and precursors for the synthesis of ester derivatives via reductive elimination, respectively.

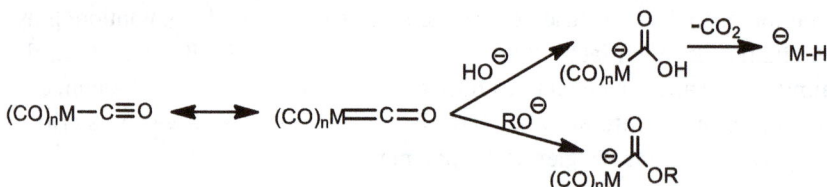

Fig. 21.13: The nucleophilic attack of hydroxide and alkoxide anions on coordinated carbon monoxide.

The elimination of carbon dioxide from the product of the nucleophilic attack of hydroxy anion to form hydrides is an important step in the water-gas shift reaction (the conversion of a CO/H_2O mixture to CO_2/H_2 mixture).

Nucleophilic attack on coordinated π-ligands is a perfect example for the contribution of organometallic chemistry to organic synthesis. Systems normally susceptible to electrophilic attack, π-unsaturation containing substrates, change their behavior when coordinated to a transition metal. The order of reactivity and the stereoselectivity were generally correlated with the number of carbon atoms in the π-system (even and odd). Also there are differences in reactivity closed-loop or open-ended π-systems. It was observed that the even unsaturated ligands are more reactive than odd, while open structures are more reactive than the closed ones:

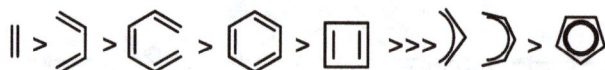

Regarding the stereochemistry, the open even polyenes are attacked at the terminal position while open odd polyenes are not normally attacked at a terminal carbon atom unless the metal fragment is strongly electron withdrawing. Cyclopentadienyl is not very reactive; therefore, it is used as a spectator ligand in organometallic compounds involved in organic synthesis.

The nucleophilic attack on the complexed monoolefins from the face opposite to the metal results in the formation of a new carbon–carbon bond and a carbon–metal bond (Fig. 21.14).

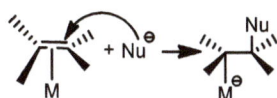

Fig. 21.14: Nucleophilic attack on coordinated monoolefins.

In the reaction of η^3-allyl complexes with nucleophiles (Fig. 21.15), the terminal carbon forms a new carbon–carbon bond with the nucleophile accompanied by the formation of a π-olefin complex.

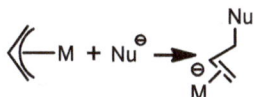

Fig. 21.15: Nucleophilic attack on coordinated η^3-allyl.

The two chosen examples illustrate the principles followed by all π-coordinated ligands when attacked by nucleophiles.

Examples of nucleophilic attack on π-coordinated alkenes, alkynes and arenes can be found in [5].

21.2 Carbon–carbon bond formation reactions

The formation of new C–C bonds is of outmost importance in organic and organometallic synthesis for the preparation of both new (like pharmaceuticals) as well as known compounds (like natural products). In the last four decades, significant progress has been made in this field mainly due to the wise use of the transition metals. The contributions of researchers to the development of synthetic methods afford now reactions difficult to imagine in what we can call the classical organic chemistry. To cite one of the Nobel Prize winner, Ei-ichi Negishi, now the ultimate goal of organic synthesis is "to be able to synthesize any desired and fundamentally synthesizable organic compounds (a) in high yields, (b) efficiently (in as few steps as possible, for example), (c) selectively, preferably all in >98–99% selectivity, (d) economically, and (e) safely, abbreviated as the $y(es)^2$ manner"[6].

There are several ways to build new carbon–carbon bonds we address in this paragraph: cross-coupling reactions and Heck reactions. Due to the crucial contribution to the synthetic organic chemistry, the Nobel Prize in Chemistry was awarded to Akira Suzuki, Ei-ichi Negishi and Richard Heck in 2010. There is an important number of reviews covering the cross-couplings in the preparation of complex organic molecules, including complicated natural products (see a selection [6–14]).

21.2.1 Cross-coupling reactions – general

The most relevant cross-coupling reactions are nickel or palladium catalyzed and involves a nucleophile, R′M, and an electrophile, R–X:

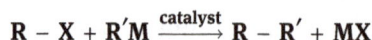

$$R - X + R'M \xrightarrow{\text{catalyst}} R - R' + MX$$

where R, R′ are organic fragments; M is the metal or metal-containing group; X is the halogen or other leaving groups; catalyst is mainly Ni(0) or Pd(0) complexes.

In the coupling processes, there are three components: an organometallic compound in stoichiometric amount (the nucleophile), an organic compound containing a leaving group (Cl, Br, I, Otf, OSO$_2$R, etc.) as a coupling partner (the electrophile) and the transition metal catalyst (mainly Ni or Pd complexes either prepared separately or in situ).The role of the transition metal catalyst is to activate less reactive organic substrates and to prevent homocoupling of the organic halides.

Based on the general reaction, various procedures were developed using specific elements in the organometallic reactant: Mg (Kumada–Tamao–Corriu (KTC) coupling), Al, Zn, Zr (Negishi cross-coupling), B (Suzuki and Suzuki–Myaura cross-couplings), Sn (Stille cross-coupling), Si (Hiyama–Denmark cross-coupling) and Li (Murahashi–Feringa cross-coupling). All these cross-coupling procedures offer a very good synthetic alternative for the coupling of less reactive electrophiles: aryl, benzyl, alkenyl, alkynyl, allyl and propargyl.

The catalytic cycle for all these cross-coupling reactions is presented in Fig. 21.16.

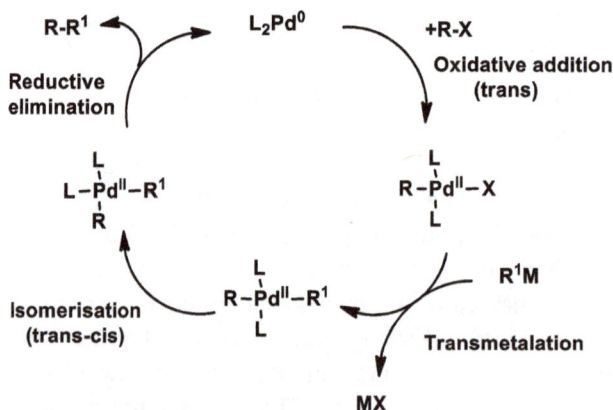

Fig. 21.16: General catalytic cycle for C–C cross-coupling.

The first step is the oxidative addition of R–X to the catalyst, a compound containing the transition metal in a low oxidation state (the metals of choice for cross-coupling are, in most of the cases, nickel or palladium). There are two possibilities to introduce the catalyst in the reaction: preformed, that is, M(PPh$_3$)$_4$, M = Ni(0) or Pd(0) or a precursor containing Ni(II) or Pd(II) with the appropriate ligands and re-ducing agents. The oxidative addition will result in the formation of a metal–carbon bond (Ni–C or Pd–C) and the metal-leaving group (mainly in *trans*-position relative to each other). The next step is the transmetallation, the exchange of the leaving group with the nucleophile. The driving force of this step is usually the formation of the inorganic salt or a more thermodynamically stable compound of the less elec-tronegative metal. The reductive elimination requires the *cis*-orientation of the two organic fragments; therefore, the next step is the *trans*/*cis*-isomerization of the com-plex. In the final step, the reductive elimination, the cross-coupling compound is released along with the catalyst which will start a new cycle.

21.2.2 The Kumada–Tamao–Corriu cross-coupling reactions

The first organometallic nucleophiles used to build new carbon–carbon bonds by cross-coupling reactions were organomagnesium reagents. The Grignard reagents are easily available either from direct exchange reaction of the halide with metallic magnesium or from commercial sources. The development of new synthetic meth-ods for the preparation of highly functionalized Grignard reagents from aryl, heter-oaryl, alkenyl and alkyl halides has expanded the scope of reaction methodologies with Grignard nucleophiles [15–19]. Another advantage of using Grignard reagents is the mild reaction condition, as will be shown further. The transition metal cata-lysts prevent the formation of unwanted homocoupling products.

The contributions describing nickel-catalyzed cross-coupling of alkenyl or aryl hal-ides with aryl or alkylmagnesium halides in 1972 by Kumada and Tamao (Fig. 21.17 (**1**)) [20, 21], as well as Corriu and Masse, (Fig. 21.17 (**2**),(**3**)) [22] are considered the ground-breaking discovery of a novel carbon–carbon bond formation reaction.

Progress was made by using palladium complexes instead of nickel compounds as catalysts with very good results in terms of yield and the substrate scope of the KTC reaction. In 1975, Murahashi [23] reported for the first time the coupling of Grignard reagents under palladium catalysis (Fig. 21.18). The palladium-catalyzed KTC coupling showed increased stereocontrol and broader substrate scope of the organometallic coupling partner.

The palladium-catalyzed cross-coupling reactions usually proceed best in polar nonprotic solvents such as dimethylformamide or *N*-methyl-2-pyrrolidinone; therefore, reaction conditions have to accommodate this observation with the known fact that Grignard reagents are prepared usually in strong coordination solvents to magnasium (II): diethylether, tetrahydrofuran (THF), dioxane or diethyleneglycol dimethylether (diglyme). A mixture of solvents is the answer to the abovementioned problem [23].

The efficiency of the catalyst is related to the first step in the catalytic cycle, oxidative addition of the electrophile (often the rate-determining step). The higher

(1)

dppe =1,2-bis(diphenylphosphino)ethane

(2)

(3)

R = 4-MeOC$_6$H$_4$, 4-MeC$_6$H$_4$, 3-MeBrC$_6$H$_4$, 4-BrC$_6$H$_4$, 2,4-Me$_2$C$_6$H$_3$

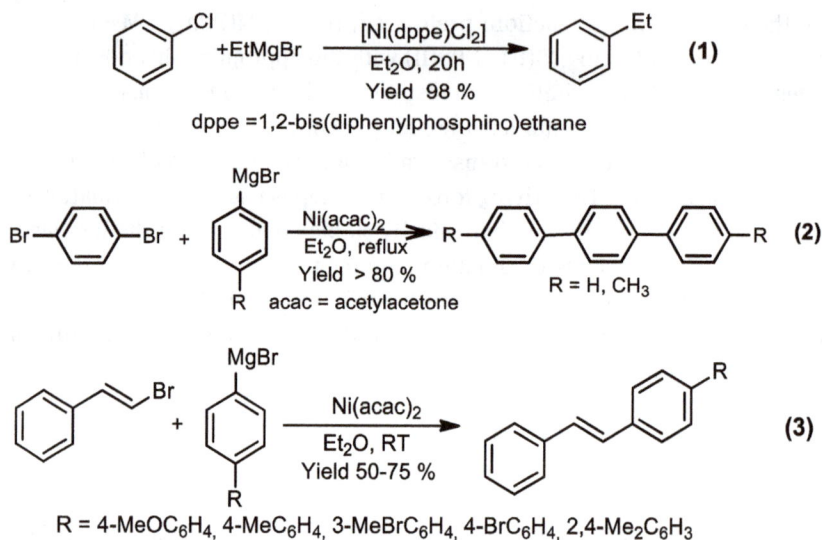

Fig. 21.17: Nickel-catalyzed cross-coupling reactions using organomagnesium reagents: (1) [20, 21], (2), (3) [22].

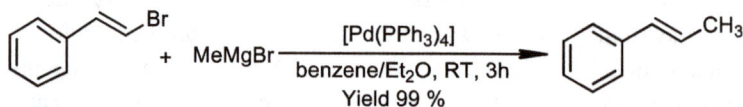

Fig. 21.18: Palladium-catalyzed cross-coupling reactions using organomagnesium reagents.

selectivity obtained with the palladium catalysts compared to nickel was the incentive for the further development of cross-coupling reactions. In cases where nickel and palladium catalysts have been shown to perform with similar activity, nickel has been preferably used due to its lower cost. Several examples of catalytic systems useful in building complex organic molecules were selected.

For functionalized substrates with more than one possible coupling reaction centers (like C–Br, C–Cl or C–OTf), a catalyst based on Pd(I) gave good results in chemoselective Csp^2–Csp^2 KTC couplings (Fig. 21.19). Exclusive bromoselectivity was observed in the presence of C–Cl and/or C–OTf bonds, regardless of the electronic or steric properties of the substrate. The C–C bond formations are extremely rapid (<5 min at RT) and are catalyzed by an air- and moisture-stable PdI dimer under open flask conditions [24].

The method proved to be compatible with heterocycles and functional groups, tolerating not only C–Cl, C–OTf, C–F but also C–CN, aldehydes, esters and sterically demanding groups (*ortho*-adamantyl). The larger scale applicability – on 1 g scale with 1 mol% catalyst loading – was successfully experimented [24].

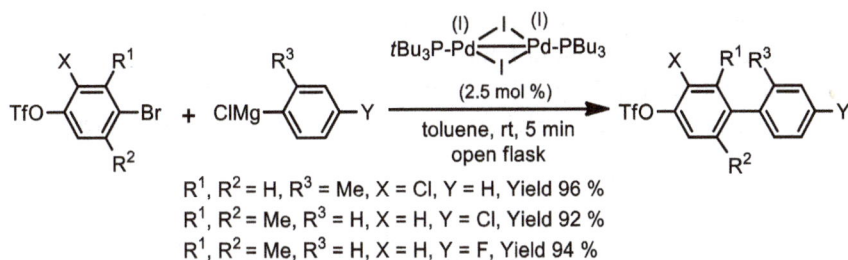

R^1, R^2 = H, R^3 = Me, X = Cl, Y = H, Yield 96 %
R^1, R^2 = Me, R^3 = H, X = H, Y = Cl, Yield 92 %
R^1, R^2 = Me, R^3 = H, X = H, Y = F, Yield 94 %

Fig. 21.19: Selective cross-coupling reactions of arylmagnesiumchlorides with functionalized arylbromides.

In the quest for catalytic systems to conduct C–C KTC and Negishi cross-coupling for a wider range of substrates bearing sensitive substituents with functionalized Grignard reagents (KTC) or without special handling of the organozinc reagents (Negishi), a versatile class of air-stable, highly active, well-defined precatalysts was prepared from PdCl$_2$, 2,6-disubstituted phenylimidazolium chloride stabilized by σ-donating 3-chloropyridine ligand, known as PEPPSI – **p**yridine-**e**nhanced **p**recatalyst **p**reparation **s**tabilization and **i**nitiation (Fig. 21.20) [25].

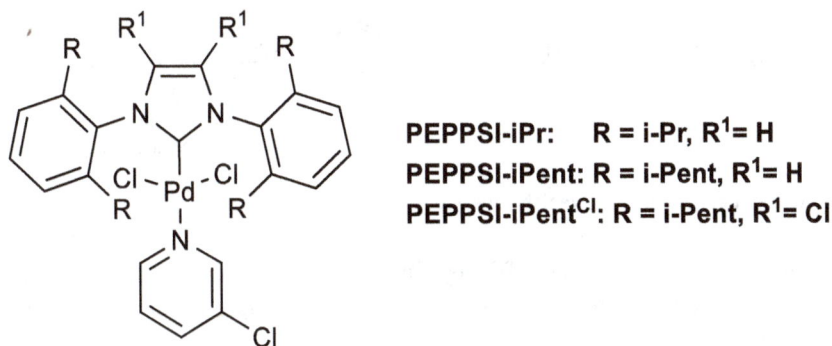

PEPPSI-iPr:　R = i-Pr, R^1= H
PEPPSI-iPent: R = i-Pent, R^1= H
PEPPSI-iPent^{Cl}: R = i-Pent, R^1= Cl

Fig. 21.20: Palladium catalysts based on NHC backbone.

The chemoselectivity, especially when nucleophiles or electrophiles (or both) contain Grignard-sensitive functional groups (-CN, -COOR, etc.), is often critical in the synthesis of complex organic molecules (pharmaceutical, natural products, etc.). The dinuclear palladium(I) complex [{(PtBu$_3$)PdI}$_2$] already mentioned is very effective in coupling bromides at room temperature, but not chlorides or triflates [24]. A step further is the rapid chemoselective KTC cross-coupling of aryl bromides in the presence of chlorides or triflates using Pd-PEPPSI-IPent^{Cl} in one-pot sequential KTC/KTC cross-couplings (Fig. 21.21). The same procedure can be applied for one-pot sequential KTC/Negishi cross-couplings (Fig. 21.22).The functionalized Grignard reagent is added to the solution of bromo-chloro arene/catalyst at 0 °C. After stirring for 15 min, the second Grignard reagent was added, and the resulting mixture stirred for 30 min yielding functionalized triaryl compounds in good yields [26].

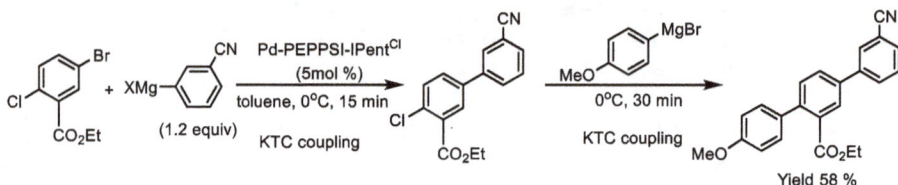

Fig. 21.21: One-pot sequential KTC/KTC cross-coupling for the preparation of functionalized triaryls.

As Pd-PEPPSI-IPent[Cl] is a highly reactive catalyst, the one-pot sequential KTC/Negishi cross-couplings of bromo-chloro/triflate-arenes was experimented, first with functionalized Grignard reagents followed by reaction with alkyl or aryl zinc reagents (Fig. 21.22). This procedure provided substituted triaryls, including heterocycles.

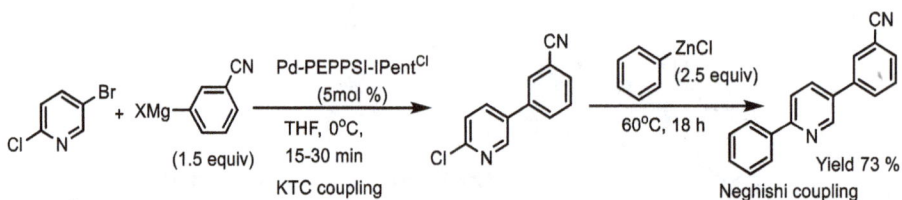

Fig. 21.22: One-pot sequential KTC/Negishi cross-coupling.

For both sequential KTC/KTC and KTC/Negishi couplings, no additional catalyst or special handling is required as products were obtained by simply adding Grignard or alkyl/aryl zinc reagents, respectively[26].

The prevention of the competing β-hydride elimination in the cross-coupling of alkenyl halides with alkyl Grignard reagents bearing β-hydrogens and the retention of stereochemistry is a challenge for the catalytic systems. Using the sterically hindered bidentate diphosphine ligands 1,1′-bis(di-tert-butylphosphino)ferrocene (dtbpf) and [oxydi(2,1-phenylene)]bis(diphenylphosphane) (DPEPhos) in the presence of tetramethylethylenediamine, the KTC cross-couplings of alkenyl halides with alkyl, alkenyl, aryl or heteroaryl substrates take place under ambient conditions and minimize the side product formation (Fig. 21.23). Notably, mild reaction conditions allow for the synthesis of alkenes bearing sensitive functionality, such as cyano and ester groups [27].

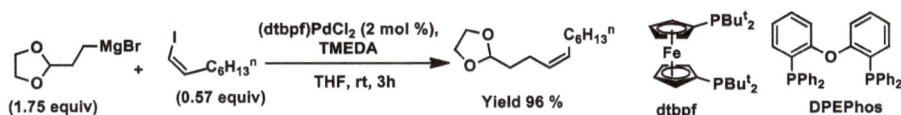

Fig. 21.23: KTC cross-coupling of alkenyl halides with sp^3-Grignard.

21.2.3 The Negishi cross-coupling reactions

The Negishi cross-coupling reactions are using organozinc reagents as nucleophiles. The cross-coupling reactions using more electronegative metals in the nucleophile reagent, Al, Zn, Zr and nickel or palladium catalyst afforded excellent results even with less reactive electrophiles [28] and were reported for the first time by Negishi in 1976 using a palladium(0) complex as catalyst and organoaluminum compound as nucleophile (Fig. 21.24) [29]:

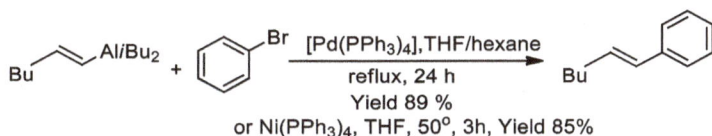

Fig. 21.24: Negishi cross-coupling of arylbromide with organoaluminum reagent.

Another premiere, the cross-coupling of aryls to get unsymmetrical biaryls using organozinc compounds (Fig. 21.25) was reported by Negishi in 1977 [30].

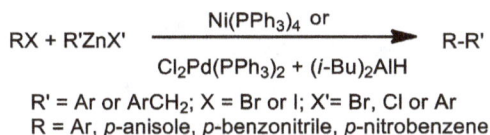

R' = Ar or ArCH₂; X = Br or I; X'= Br, Cl or Ar
R = Ar, p-anisole, p-benzonitrile, p-nitrobenzene

Fig. 21.25: Preparation of unsymmetrical biaryls by cross-coupling.

The nickel catalyst was preformed or prepared in situ by the reaction of Ni(acac)$_2$, PPh$_3$ and (Bu^i)$_2$AlH (1:4:1). Good results – yields over 85% – were reported in mild reaction conditions (cat. 5 mol%, 25 °C, 1–2 h), irrespective of the electronic properties of the substituents on the electrophile, electron-donating (–CH$_3$ and –OCH$_3$) or electron-withdrawing (CN or NO$_2$). Approximate relative order of reactivity of organic halides in oxidative addition to palladium is: allyl, propargyl > benzyl, acyl > alkenyl, alkynyl > aryl ≫ simple alkyl.

Highly general stereo-, regio-, and chemo-selective synthesis of terminal and internal conjugated enynes by the Pd-catalyzed reactions of alkynylzinc reagents with alkenyl halides are also very useful (Fig. 21.26) [31]. The combination of alkynylzinc reagents with arylhalides provided terminal and internal arylalkynes at room temperature in THF using $Pd(PPh_3)_4$ or $Cl_2Pd(PPh_3)_2 + Bu^i_2AlH$ as catalysts in very high yield (Fig. 21.26) [32].

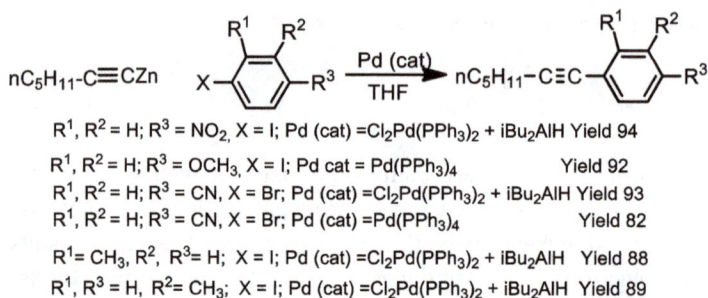

$$nC_5H_{11}-C\equiv CZn \quad X- -R^3 \xrightarrow[THF]{Pd\ (cat)} nC_5H_{11}-C\equiv C- -R^3$$

R^1, R^2 = H; R^3 = NO$_2$, X = I; Pd (cat) =Cl$_2$Pd(PPh$_3$)$_2$ + iBu$_2$AlH Yield 94
R^1, R^2 = H; R^3 = OCH$_3$, X = I; Pd cat = Pd(PPh$_3$)$_4$ Yield 92
R^1, R^2 = H; R^3 = CN, X = Br; Pd (cat) =Cl$_2$Pd(PPh$_3$)$_2$ + iBu$_2$AlH Yield 93
R^1, R^2 = H; R^3 = CN, X = Br; Pd (cat) =Pd(PPh$_3$)$_4$ Yield 82

R^1= CH$_3$, R^2, R^3= H; X = I; Pd (cat) =Cl$_2$Pd(PPh$_3$)$_2$ + iBu$_2$AlH Yield 88
R^1, R^3 = H, R^2= CH$_3$; X = I; Pd (cat) =Cl$_2$Pd(PPh$_3$)$_2$ + iBu$_2$AlH Yield 89

Fig. 21.26: Negishi cross-coupling of arylhalides with alkynylzinc reagents.

The preservation of both E- and Z-olefin geometry in the products of two-step Negishi zinc-mediated reactions in ethereal media can be achieved using zinc dust, TMEDA as additive, $PdCl_2(Amphos)_2$ (dichlorobis(p-dimethylaminophenyl-π-di-$tert$-butylphosphine)palladium(II) and the amphiphile PTS (the diester made from PEG-600, R-to-copherol and sebacic acid) which presumably supplies the hydrophobic pocket in which the in situ-generated water-sensitive organozinc halide reacts in water at room temperature (Fig. 21.27). Complete retention of E-stereochemistry was observed for the cross-couplings of stereoisomerically pure E-alkenyl halides (Fig. 21.27 (1)) while for Z-stereoisomers the results are in the range of 77:23 < Z/E < 99:1 (Fig. 21.27 (2)) [33].

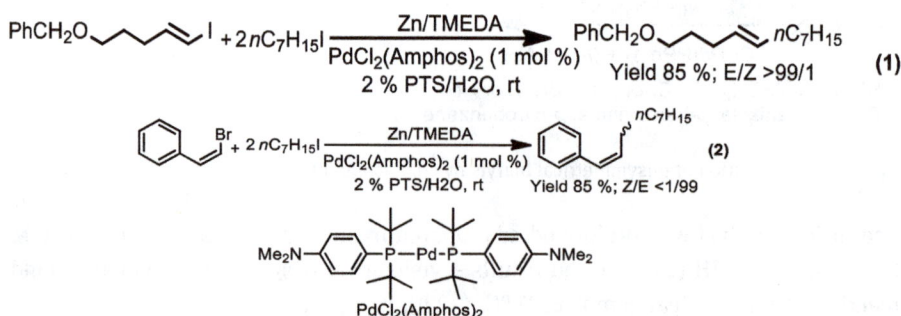

$$PhCH_2O\diagup\diagdown\diagup I + 2nC_7H_{15}I \xrightarrow[\substack{PdCl_2(Amphos)_2\ (1\ mol\ \%)\\2\ \%\ PTS/H2O,\ rt}]{Zn/TMEDA} PhCH_2O\diagup\diagdown\diagup nC_7H_{15}$$

Yield 85 %; E/Z >99/1   (1)

$$ Br + 2nC_7H_{15}I \xrightarrow[\substack{PdCl_2(Amphos)_2\ (1\ mol\ \%)\\2\ \%\ PTS/H2O,\ rt}]{Zn/TMEDA} nC_7H_{15}$$

Yield 85 %; Z/E <1/99   (2)

$$Me_2N- -P-Pd-P- -NMe_2$$

PdCl$_2$(Amphos)$_2$

Fig. 21.27: Cross-coupling of E-alkenyl (1) and Z-alkenyl (2) halides with alkyl halides.

This new micellar technology is promising as the water is the only medium.

Using TMEDA or N-methylimidazole (N-MeIm) and $PdCl_2(PPh_3)_2$, under standard Negishi conditions, virtually complete stereoretention and high yields were realized in the cross-coupling of (Z)-1-bromooct-1-ene and both primary and secondary alkyl iodides. Enhanced overall efficiency as well as the coupling of functionalized Z-alkenes was possible using bidentate flexible ligand-containing catalysts, PdCl$_2$(DPEPhos), PdCl$_2$ (Amphos)$_2$ or PdCl$_2$(dppf)$_2$ and N-MeIm additive, and the tolerance and mild reaction conditions of organozinc reagents (Fig. 21.28). The known drawbacks of couplings between alkenyl halides and alkylzinc reagents, like the formation of undesired by-products as well as the potential erosion of stereochemistry in the case of a Z-alkenyl halide, are avoided by applying the abovementioned procedure (Negishi-Plus couplings)[34].

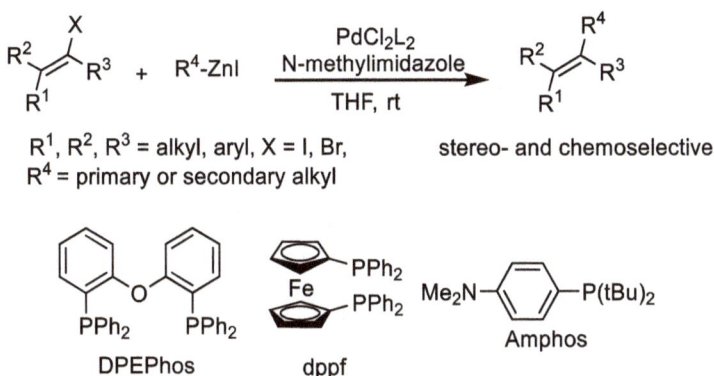

R^1, R^2, R^3 = alkyl, aryl, X = I, Br,   stereo- and chemoselective
R^4 = primary or secondary alkyl

DPEPhos   dppf   Amphos

Fig. 21.28: Cross-coupling of (Z)-1-bromooct-1-ene and primary and secondary alkyl iodides.

The ligands, the ones already described as well as others, proved to have a significant influence on Negishi cross-coupling of functionalized reaction partner, alkenyl halides and organozinc reagents [35].

The regio- and stereoselective cross-coupling reaction between 2-phenyl-N-tosylaziridine and organozinc reagents using air-stable Ni(II) source and dimethyl fumarate as ligand, in mild conditions, is a way to β-substituted amines (Fig. 21.29). The stereoselectivity of the reaction is related to a stereoconvergent mechanism, wherein the sulfonamide directs the C–C bond formation [36]:

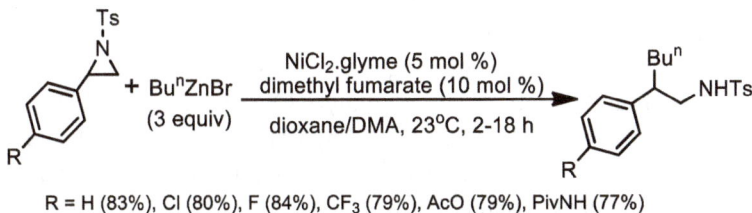

R = H (83%), Cl (80%), F (84%), CF$_3$ (79%), AcO (79%), PivNH (77%)

Fig. 21.29: Cross-coupling reaction between 2-phenyl-N-tosylaziridine and n-butylzinc bromide.

Both electron-deficient and electron-rich *para*- and *meta*-substituted styrenyl aziridines reacted with high efficiency, and a wide variety of functional groups were tolerated.

The palladium-catalyzed cross-coupling of silyl electrophiles with secondary zinc organometallics (silyl–Negishi cross-coupling) provides direct access to alkyl silanes. The ligand(s) in the structure of the palladium catalyst, tris[3,5-bis(1,1-di-methylethyl)phenyl]phosphine (DrewPhos) and bis[3,5-bis(1,1-dimethylethyl)phenyl] (1,1-dimethylethyl)phosphine (JessePhos) (Fig. 21.30), display the appropriate steric and electronic parameters and the ability to suppress isomerization and promote efficient and selective cross-coupling (the yields in brackets are obtained without catalysts):

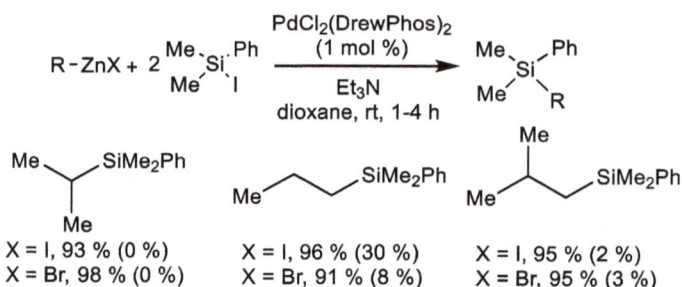

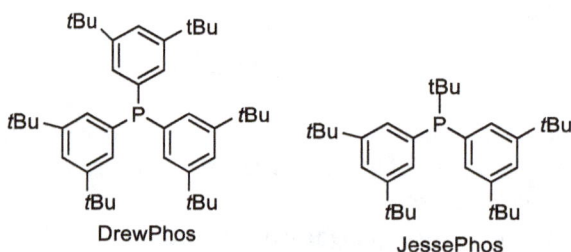

Fig. 21.30: Silyl-Negishi reaction between primary and secondary zinc organometallics and silicon electrophiles.

High yields are obtained with a low catalyst loading in short reaction times for various substituents on both reactants and, most important, provides unprecedented access to secondary silanes using abundant silyl electrophiles [37].

The chemoselective Negishi cross-coupling reactions of bis[(pinacolato)boryl] methylzinc halides with aryl (pseudo)halides catalyzed by palladium complexes leads to benzylic 1,1-diboronate esters (Fig. 21.31), important intermediates for further transformations with relevance in preparation of pharmaceutical analogues. The mild reaction conditions are compatible with a variety of functional groups. The best results were obtained with P(o-tolyl)$_3$ (L1) dicyclohexylphosphino-2′,4′,6′-triisopropyl biphenyl (X-Phos, L2) as ligands [38].

Promising results were obtained using trans-dichloro (1,3-bis-(2,6-diisopropyl-phenyl)imidazolylidinium)(3-chloro-pyridine)palladium [39] and other PEPPSI catalysts [40] in Negishi cross-coupling reactions.

R = H, Y = Br, L1, (93 %); Y = OTf, L2, (92 %)
R = OMe, Y = Br, L1, (82 %); Y = OTf, L2, (83 %)
R = Me₂N, Y = Br, L1 (76 %)
R = CN, Y = Br, L2, (60 %); Y = OTf, L2, (57 %)

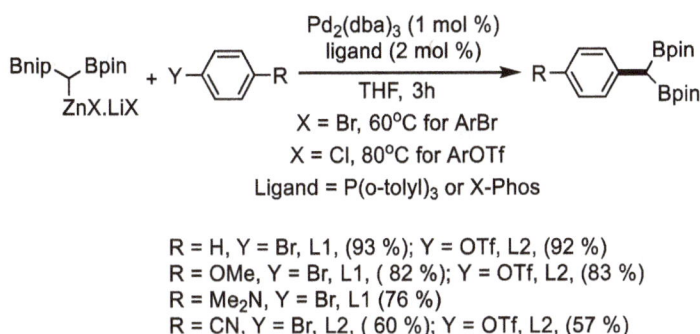

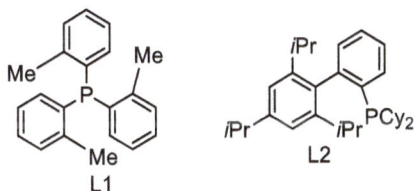

Fig. 21.31: Cross-coupling of bis[(pinacolato)boryl]methylzinc halides with aryl bromides.

The in situ preparation of the aryl, heteroaryl, alkyl or benzylic polyfunctional zinc reagents by the addition of zinc and LiCl to the corresponding organic halides or triflates (Fig. 21.32) undergo smooth Pd(0)-catalyzed Negishi cross-coupling reactions with aryl bromides in a one-pot procedure in high yields using PEPPSI-iPr (Fig. 21.20) ligand. This procedure avoids the manipulation of water and air of sensitive organozinc reagents [41]:

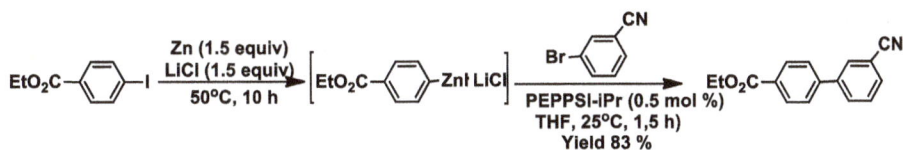

Fig. 21.32: In situ preparation of functionalized organozinc halides and cross-coupling with arylbromide.

The stereoselective cross-coupling of chiral secondary alkylzinc reagents with alkenyl and aryl halides using Pd-PEPPSI-iPent (Fig. 21.20) catalyst afforded α-chiral alkenes and arenes with high retention of configuration (dr up to 98:2) and yields up to 76% for three reaction steps (Fig. 21.33). These chiral mixed dialkylzincs are configurationally stable at room temperature for several hours [42].

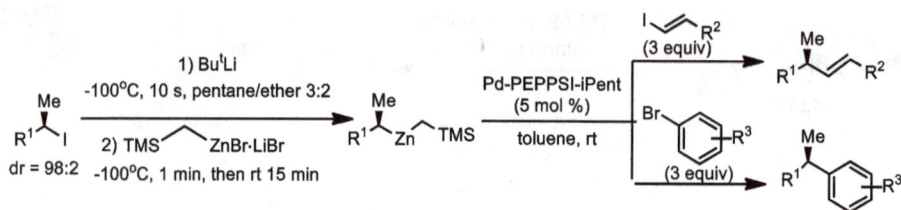

Fig. 21.33: Cross-coupling of chiral organozinc with alkenyl- and arylhalides.

This method was applied in the total synthesis of the sesquiterpenes (*S*)- and (*R*)-curcumene (Fig. 21.34) [42].

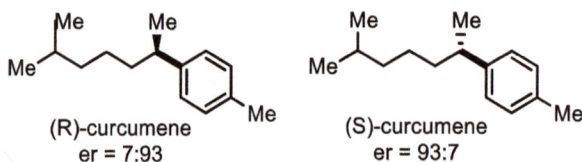

(R)-curcumene
er = 7:93

(S)-curcumene
er = 93:7

Fig. 21.34: (*R*)- and (*S*)-curcumene prepared by cross-coupling reaction.

To increase the catalytic activity of the resulting palladium *N*-heterocyclic carbene (NHC) complexes, a new, robust acenaphthoimidazol-ylidene palladium complex was prepared from the corresponding acenaphthoimidazolium chlorides by heating with $PdCl_2$ and K_2CO_3 in neat 3-chloropyridine (Fig. 21.35). Low catalyst loadings exhibited high catalytic activity toward Negishi cross-coupling reactions of alkylzinc reagents complexed with lithium chloride, R-ZnBr.LiCl, with a wide range of (hetero)aryl halides (including less reactive heterocyclic chloroarenes) under mild reaction conditions within 30 min (Fig. 21.35) [43]. Several sensitive functional groups are tolerated, and no β-hydride elimination was observed:

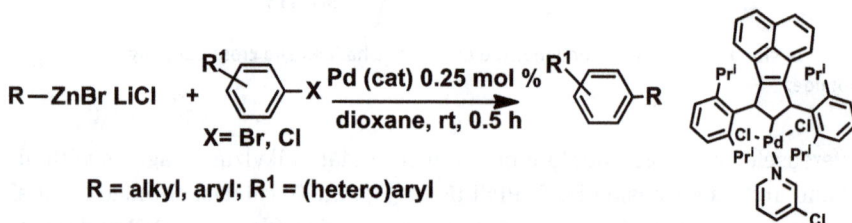

Fig. 21.35: Negishi cross-coupling of R-ZnBr.LiCl with aryl and heteroaryl halides.

Air-stable solid zinc pivalates of sensitive aromatics and heteroaromatics prepared using $TMPZn(OPiv)_3Mg(OPiv)Cl.LiCl$ react in mild conditions with a wide variety of electrophiles (Fig. 21.36) [44].

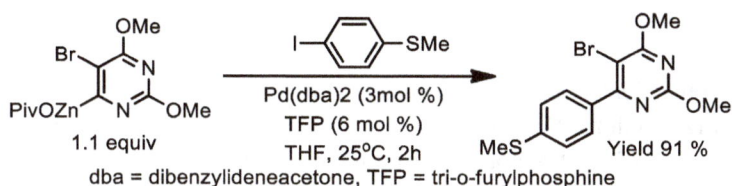

Fig. 21.36: Cross-coupling of solid zinc pivalate with functionalized aryl iodide.

Solid allylic zinc derivatives decorated with important functional groups such as esters or nitriles are tolerated in these couplings [45] with a broad range of electrophiles, exemplified in Fig. 21.37 [46].

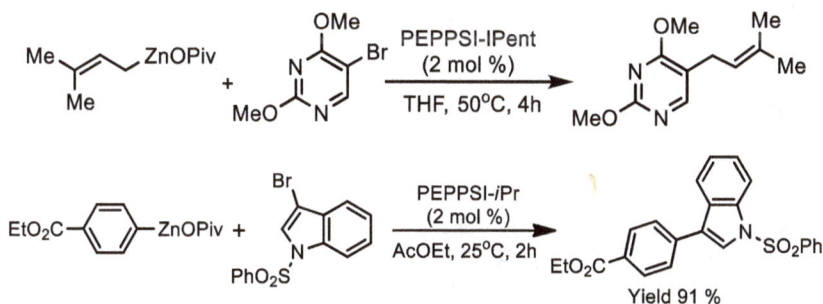

Fig. 21.37: Cross-coupling of solid zinc pivalates with functionalized heterocycles.

The solid, air- and moisture-stable organozinc pivalates (RZnOPiv) were proved to be good nucleophiles for Negishi cross-coupling using not only the traditional nickel and palladium catalyst but also the less expensive cobalt catalysts (Fig. 21.38) [47].

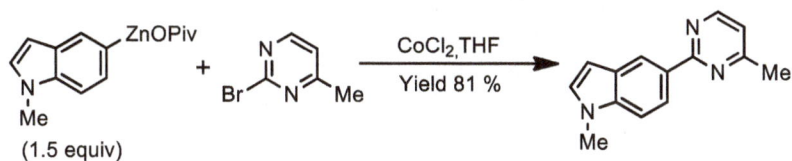

Fig. 21.38: Cobalt-catalyzed Negishi cross-coupling of zinc pivalate with heterocyclic bromide.

Negishi cross-coupling between functionalized aryl and heteroaryl zinc pivalates and various electron-poor aryl and heteroaryl halides (X = Cl, Br, I) as well as (E)- or (Z)-bromo- or iodo-alkenes proceed in mild condition in the presence of $CoCl_2$. Also, alkynyl bromides react with arylzinc pivalates providing arylated alkynes.

21.2.4 The Stille cross-coupling reactions

The cross-coupling reactions of organotin reagents with electrophiles are known as Stille cross-coupling reactions. Various classes of organotin compounds proved to be suitable for this type of reaction: organodistannanes, homo- and hetero-tetraorganostannanes. The palladium-catalyzed cross-couplings of organotin reagents with aryl bromides were reported in 1977 (Fig. 21.39) [48].

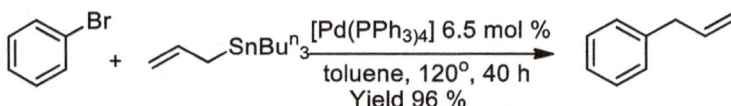

Fig. 21.39: Cross-coupling of phenylbromide with $C_3H_5SnBu^n_3$.

The synthesis of ketones (Fig. 21.40) by cross-coupling of aroyl chlorides with organostannanes under significantly milder reaction conditions was reported in 1978 [49]:

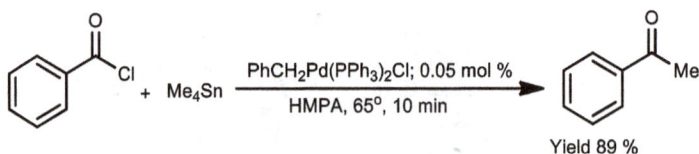

Fig. 21.40: Preparation of unsymmetrical ketones by Stille cross-coupling.

The cross-coupling reaction using organotin nucleophiles was explored and improved by Stille [50]. Due to the versatile methodology and the broad functional group compatibility, the Stille reaction became one of the important ways to build new CC bonds [51, 52]. Although the nucleophiles are, usually, tetraorganotin compounds during the cross-coupling process, only one of the organic groups at the tin atom is transferred. From economic reasons, in most of the cases the organostannane contains one group to be transferred (sometimes difficult to synthesize or expensive) and three simpler organic groups like methyl or n-butyl. This approach works because different groups are transferred with different rates, and the slowest transfer rate being noticed for the alkyl groups. An important advantage of the Stille cross-coupling is the mild conditions required which tolerate several functional groups (i.e., CO_2R, CN, OH or CHO) and the high yields in most of the cases. As it can be seen in Fig. 21.41, a wide variety of electrophiles and organotin compounds can be coupled [50].

The retention of configuration of the double bond is observed, regardless of the reactant containing the double bond. The reaction is regioselective in coupling reactions of allyl partners and occurs stereospecifically with inversion of configuration at sp^3 carbon centers bound to tin and/or halogen [50].

Organotin reagents

H-SnR$_3$ Alkyl-SnR$_3$ R'-C≡C-SnR$_3$ Aryl-CH$_2$-SnR$_3$ Aryl-SnR$_3$

Electrophiles

Aryl-CH$_2$-X
X = Cl, Br

X = Cl, Br

X = I, OTf

Aryl-X
X = Br, I

COOH
R-C-X
H
X = Br, I

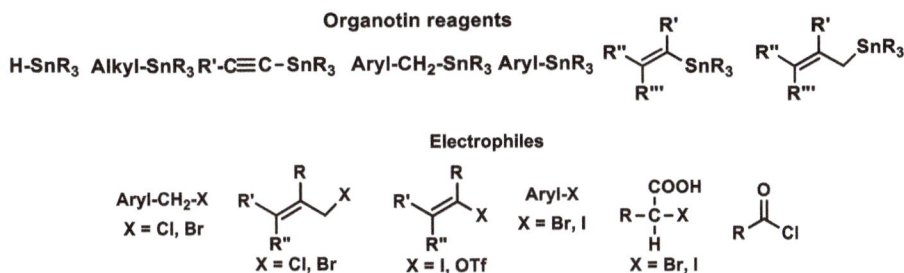

Fig. 21.41: Examples of organotin reagents and electrophiles suited for Stille cross-coupling.

Aryl halides, including aryl iodides, aryl bromides and activated aryl chlorides, were efficiently coupled with organotin compounds to afford the corresponding biaryls, alkenes and alkynes (Fig. 21.42) in good to excellent yields using Pd(OAc)$_2$/ diazabicyclooctane-catalytic system in the presence of KF or [Bu$^n_4$N]F [53].

$$\text{Ar-X} + \text{RSnBu}^n_3 \xrightarrow[\substack{\text{3 eq. KF or Bu}^n_4\text{NF}\\ \text{Dioxane, 80°C}}]{\substack{\text{Pd(OAc)}_2\ 3\ \text{mol \%}\\ \text{DABCO 6 mol \%}}} \text{Ar-R}$$

MeO—⟨ ⟩—X + PhSnBu$^n_3$ ⟶ MeO—⟨ ⟩—⟨ ⟩ (1)

X = Br, KF, 24 h, Yield 75 %
X = I, Bu$^n_4$NF. 12 h, Yield 90 %

O$_2$N—⟨ ⟩—Br + ⟍SnBu$^n_3$ ⟶ O$_2$N—⟨ ⟩—⟍ (2)

16 h, Yield 99 %

O$_2$N—⟨ ⟩—Br + Ph—≡—SnBu$^n_3$ ⟶ O$_2$N—⟨ ⟩—≡—Ph (3)

16 h, Yield 100 %

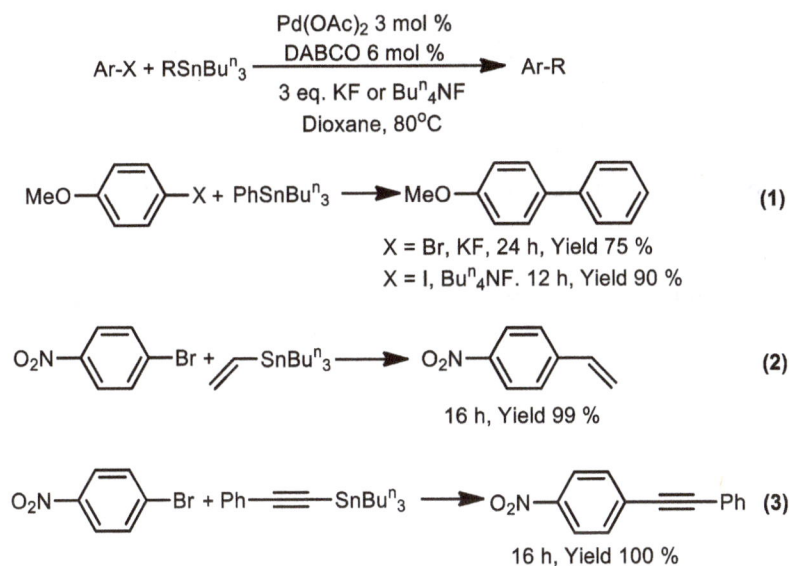

Fig. 21.42: Synthesis of biaryls (1), substituted styrene (2) and functionalized alkyne by Stille cross-coupling.

A useful alternative to Friedel–Crafts acylations for the synthesis of ketones is provided by chemoselective cross-coupling of aliphatic and aromatic acyl chlorides with organostannanes. A range of ketones are obtained in high yield using bis(di-*tert*-butylchlorophosphine)palladium(II) dichloride as precatalyst (Fig. 21.43). The catalyst tolerates various functional groups including aryl chlorides and bromides that usually undergo oxidative addition to palladium complexes [54].

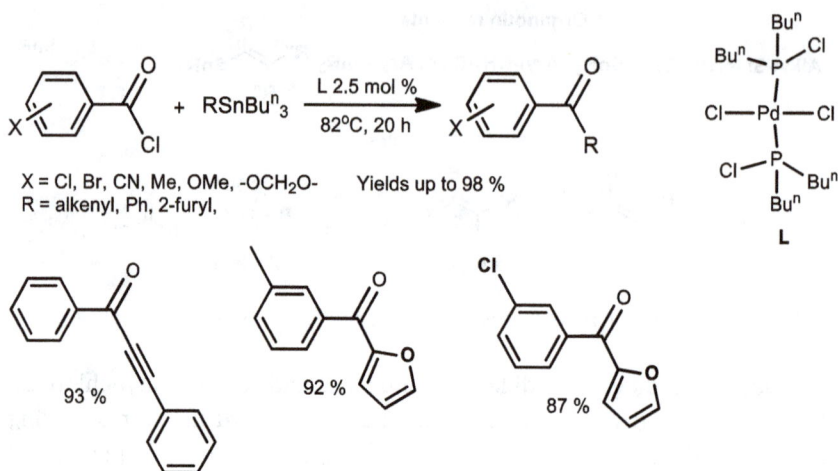

X = Cl, Br, CN, Me, OMe, -OCH$_2$O-   Yields up to 98 %
R = alkenyl, Ph, 2-furyl,

93 %   92 %   87 %

Fig. 21.43: Chemoselective cross-coupling of aromatic acyl chlorides with organostannanes.

Conventional palladium(II) acetate/PCy$_3$ (Pd(OAc)$_2$/PCy$_3$) under air and using CsF as a base are effective for stannylation of aryl halides and for one-pot two-step stannylation/Stille cross-coupling conducted without solvent (Fig. 21.44). The procedures were applied for cross-coupling with the (het)aryl bromide or iodide bearing acceptor, donor as well as the sterically demanding substituents. After completion of the stannylation step, a new portion of a catalyst, base and (het)aryl halide is added to the same flask. Nitro- and fluoro-substituted aryl halides readily participate in cross-coupling to furnish the corresponding biaryls in high yields. No evident disproportionation or competitive deprotonation were observed during the reaction of aldehyde or methyl ketone-substituted aryls [55].

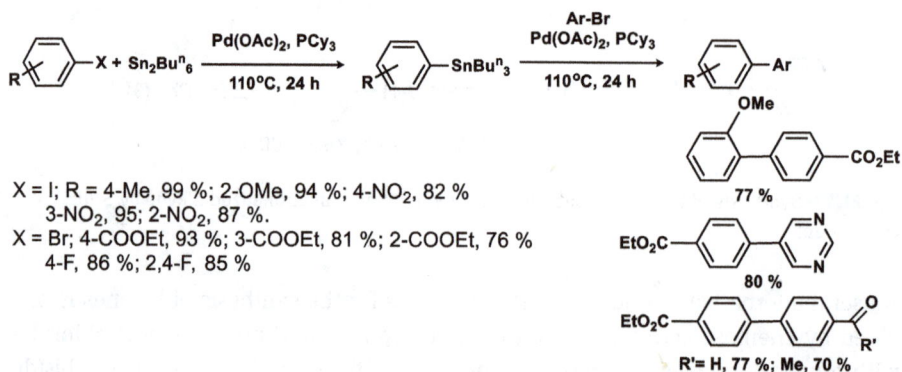

X = I; R = 4-Me, 99 %; 2-OMe, 94 %; 4-NO$_2$, 82 %
    3-NO$_2$, 95; 2-NO$_2$, 87 %.
X = Br; 4-COOEt, 93 %; 3-COOEt, 81 %; 2-COOEt, 76 %
    4-F, 86 %; 2,4-F, 85 %

77 %

80 %

R'= H, 77 %; Me, 70 %

Fig. 21.44: One-pot two-step stannylation/Stille cross-coupling for the synthesis of functionalized biaryls.

21.2.5 The Suzuki–Miyaura cross-coupling reactions

The promising results reported by Negishi in 1977 in the cross-coupling using boron derivatives as nucleophiles were followed by the extensive research of Suzuki [8, 9, 56]. In 1979, Suzuki and Miyaura reported the preparation in high yields and high regio- and stereospecificity-conjugated (E,Z)-, (Z,E)- or (Z,Z)-alkadienes (Suzuki–Miyaura cross-coupling reactions. The cross-coupling of (E)-1-alkenyldisiamylboranes and (E)-1-alkenyl-1,3,2-benzodioxaboroles with 1-alkenyl- or 1-alkynylhalides using Pd(PPh₃)₄ as catalyst is effective in the presence of a base such as sodium alkoxide, phenoxide or hydroxide. The steps in the catalytic cycle are (Fig. 21.45) oxidative addition (1), ligand exchange (2), transmetallation (3), trans/cis-isomerization (4) and reductive elimination (5).

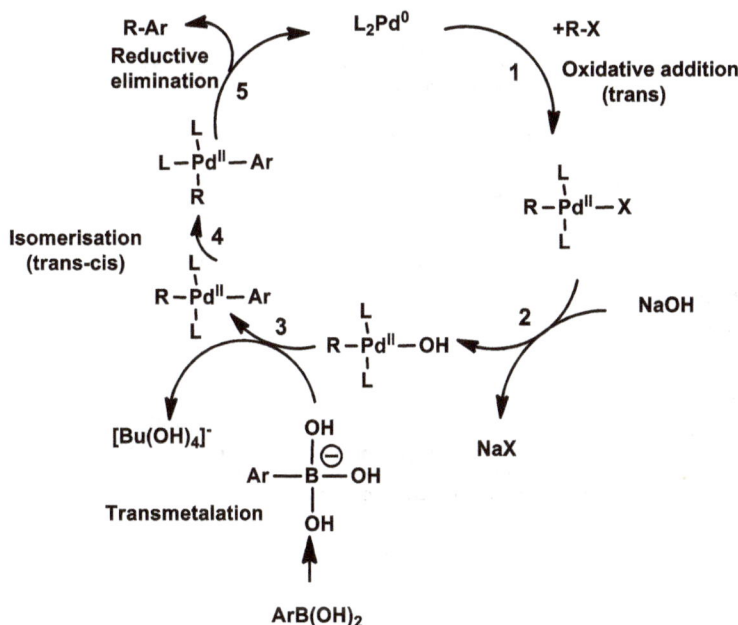

Fig. 21.45: Suzuki cross-coupling catalytic cycle.

The mechanism of Suzuki cross-coupling reactions involves the transmetallation between 1-alkenylborane and alkoxypalladium(II) complex generated through the metathetical displacement of a halogen atom from RPdL$_n$X with sodium alkoxide (Fig. 21.45).

The reactions take place with the retention of the configurations of the starting alkenylboranes and alkenyl bromides (Fig. 21.46) [57–59].

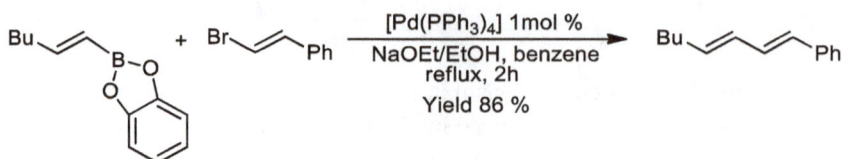

Fig. 21.46: Suzuki cross-coupling of alkenylborane and β-bromostyrene.

In the same conditions, the reaction of (2)-1-alkenyldisiamylboranes with 1-bromoalkynes provides conjugated (E)- and (2)-alkenynes (there was no reaction of (E)-1-alkenyl-1,3,2-benzodioxaboroles with haloalkynes) (Fig. 21.47). The reflux of a benzene solution of (E)-1-hexenyldisiamylboranes and 1-bromooctyne gave (5E)-tetradecen-7-yne in a 98% yield with an isomeric purity of 99% [57, 59].

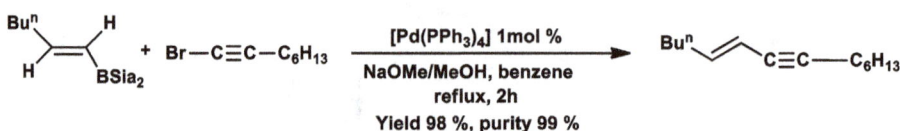

Fig. 21.47: Synthesis of conjugated (E)- and (2)-alkenynes.

Arylated (E)-alkenes (Fig. 21.48) can be obtained by the cross-coupling reaction of aryl halides with alk-1-enylboranes in the presence of $Pd(PPh_3)_4$ and bases such as sodium ethoxide [60].

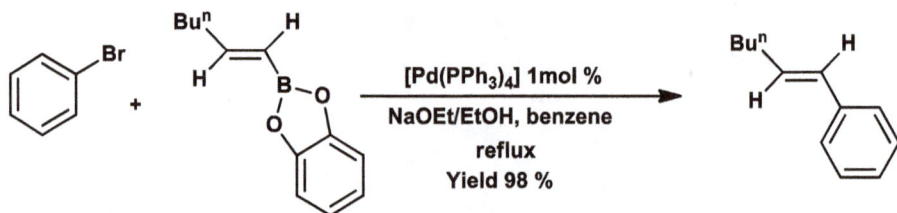

Fig. 21.48: Preparation of E-β-n-butyl styrene.

As the organoboron reagents are readily prepared by monohydroboration of acetylenes, the reactions provide a new regio- and stereoselective synthetic procedure for arylated (E)-alkenes in good yield from aryl halides and acetylenes. More advantages of Suzuki–Miyaura cross-coupling reactions that are worth mentioning are the mild reaction conditions and high product yields under both aqueous and heterogeneous conditions, toleration of a broad range of functional groups, application in one-pot synthesis, easy separation of inorganic boron compound, nontoxic reaction, hence, environmentally friendly.

21.2.6 The Hiyama–Denmark cross-coupling reactions

The next class of nucleophiles reported on the search of new reagents for cross-coupling was organosilicon. Hiyama used the same transition metals, nickel and palladium, as catalysts in the coupling of organosilicon (activated by a fluoride source) with aryl halides and triflates [61, 62]. The mechanism of Hiyama cross-coupling reactions (Fig. 21.49) follows the general steps of cross-coupling reactions. The nucleophile for the transmetallation step is generated in the reaction with fluorine derivatives, mainly tetraammonium fluorides.

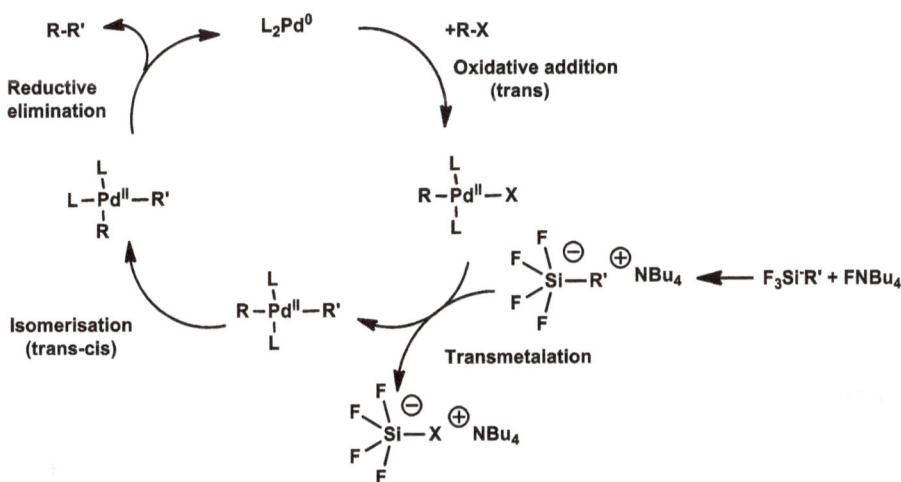

Fig. 21.49: Hiyama cross-coupling mechanism.

The first experiments used tris(dimethylamino)sulfonium difluorotrimethylsilicate (TAFS) as fluorine source and allylpalladium chloride dimer as catalyst in the reaction of aryl-, vinyl- and allyl-halides and iodides with vinyl-, ethynyl- and allyltrimethylsilane with high stereospecificity and chemoselectivity (Fig. 21.50) [61].

Functionalized styrene, conjugated dienes and enynes were prepared in moderate to high yield by one-pot procedure as the reaction conditions are mild, and a wide variety of organic functionality on both substrates – ester, ketone, carbonyls, ethoxy, hydroxy or aldehyde carbonyl – are tolerated.

Very good results were obtained in the coupling of alkenylsilacyclobutanes (like (E)- and (Z)-1-(1-heptenyl)-1-methysilacyclobutane) with organic halides (Fig. 21.51) [63].

It was found that along with the wise choice of the palladium catalyst and the fluoride source, the order of mixing of reagents is important. The influence on the reaction rate of the palladium catalysts follows the order: $Pd(dba)_2 Pd_2$-$(dba)_3 > Pd$ $(OAc)_2 Pd(OTf)_2 > (COD)PdBr_2 > [allylPdCl]_2 \gg (PhCN)_2 PdCl_2 \sim (Ph_3P)_2 PdCl_2$. The best

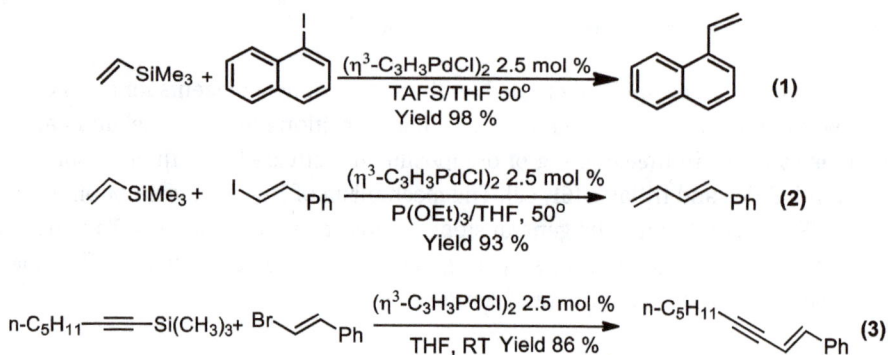

Fig. 21.50: Hiyama cross-coupling reactions of vinylsilanes with aryl halide (1) and vinylhalide (2) and of ethynyltrimethylsilane with vinylhalide (3).

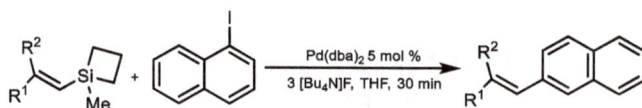

Fig. 21.51: Hiyama cross-coupling of alkenylsilacyclobutanes with naphthyliodide.

fluoride source was proved to be tetrabutylammonium fluoride (TBAF): when 3.0 equiv of TBAF was used, the reactions proceeded to completion within minutes.

Hiyama cross-coupling was proved to be effective for the Pd-catalyzed regioselective remote sp²–sp³ coupling reaction of chloromethylarenes with allyltrimethoxysilane or substituted allytrimetoxysilanes to form *para*-allyl-substituted methylarenes (Fig. 21.52) [64]. The reaction proceeds at room temperature with moderate to excellent yields.

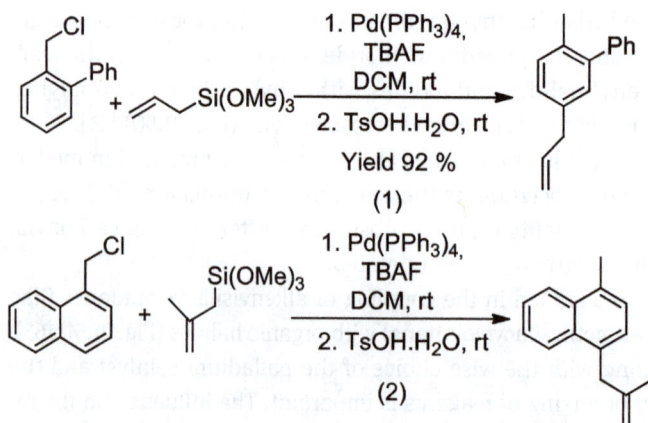

Fig. 21.52: Hiyama remote sp²–sp³ cross-coupling of 2-phenylbenzylchloride with allyltrimethoxysilane (1) and 1-(chloromethyl)naphthalene with (2-methylallyl)trimethoxysilane (2).

The use of fluoride for the formation of the pentacoordinated silicon species is preventing the compatibility with substrates bearing silyl-protecting groups; therefore, a search for an alternative led to fluoride-free Hiyama–Denmark cross-coupling reactions [65, 66]. According to this approach, the transmetallation occurs from a tetracoordinate species containing an Si–O–Pd linkage, as in the arylpalladium(II) silanolate complexes in Fig. 21.53 [67, 68], which are active in the cross-couplings.

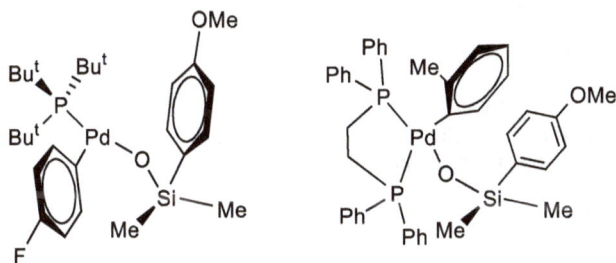

Fig. 21.53: Arylpalladium(II) silanolate complexes.

The metal silanolate precursors can be prepared using Brønsted bases such as KOSiMe$_3$, Cs$_2$CO$_3$, NaOBu^t, NaH, NaHMDS (HMDS = hexamethyldisilazane) or KH. Each activator is promoting a specific type of reaction and mechanism [67, 68].

The cross-coupling of alkenyl metal silanolates with functionalized aryl bromide and iodide in the presence of KOSiMe$_3$ in DME at ambient temperature (Fig. 21.54) led to high stereoselectivity and very good yield [69].

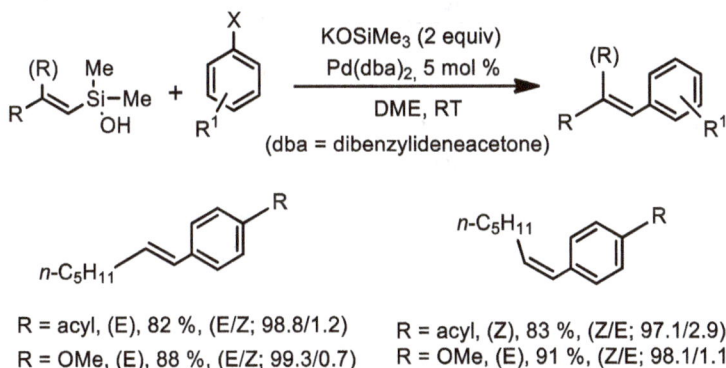

R = acyl, (E), 82 %, (E/Z; 98.8/1.2)   R = acyl, (Z), 83 %, (Z/E; 97.1/2.9)
R = OMe, (E), 88 %, (E/Z; 99.3/0.7)   R = OMe, (E), 91 %, (Z/E; 98.1/1.1)

Fig. 21.54: Fluoride-free cross-coupling of alkenyl silanolates with aryl iodides and bromides.

Aryl silanolates are less reactive than alkenyl- and alkynyl-silanolates, but using Cs$_2$CO$_3$ in toluene at 90 °C, the cross-coupling of dimethyl(4-methoxyphenyl)silanol with ethyl 4-iodobenzoate as well as other reaction partners are effective with high yields (Fig. 21.55). The hydration of Cs$_2$CO$_3$ with 3.0 equiv of water per equiv of

Cs_2CO_3 in reaction catalyzed by [allylPdCl]$_2$ with 1,4-bis(diphenylphosphino)butane or Ph$_3$As prevents the homocoupling of the aryl halides [70, 71].

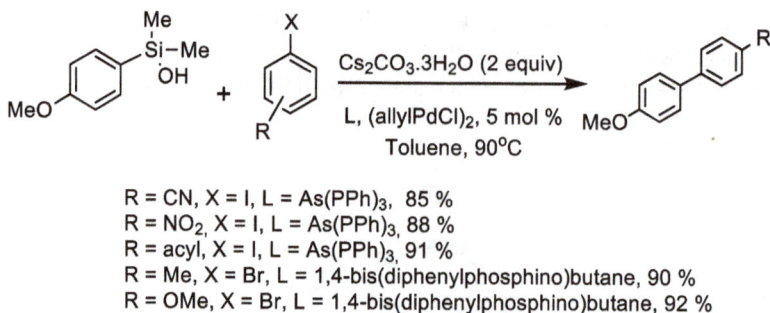

R = CN, X = I, L = As(PPh)$_3$, 85 %
R = NO$_2$, X = I, L = As(PPh)$_3$, 88 %
R = acyl, X = I, L = As(PPh)$_3$, 91 %
R = Me, X = Br, L = 1,4-bis(diphenylphosphino)butane, 90 %
R = OMe, X = Br, L = 1,4-bis(diphenylphosphino)butane, 92 %

Fig. 21.55: Reaction of aryl silanolates with functionalized aryl bromides and iodides.

Reaction conditions for cross-coupling of the much less reactive chloride imply heating 60 °C with 1.3 equiv of (E)- and (Z)-alkenylsilanolates. Functionalities like nitrile, ester, nitro, ketone and TBS-protected benzyl alcohol are tolerated (Fig. 21.56 (I)) [72]. Under the same conditions, 2- and 3-chloropyridine, mono- and di-*ortho*-substituted aryl chlorides react smoothly. The reactions are highly stereospecific. The coupling of (E)- and (Z)-styrylsilanolates (Fig. 21.56 (II)), prone to isomerization, proceeded with complete retention of the double bond geometry in dioxane at 90 °C.

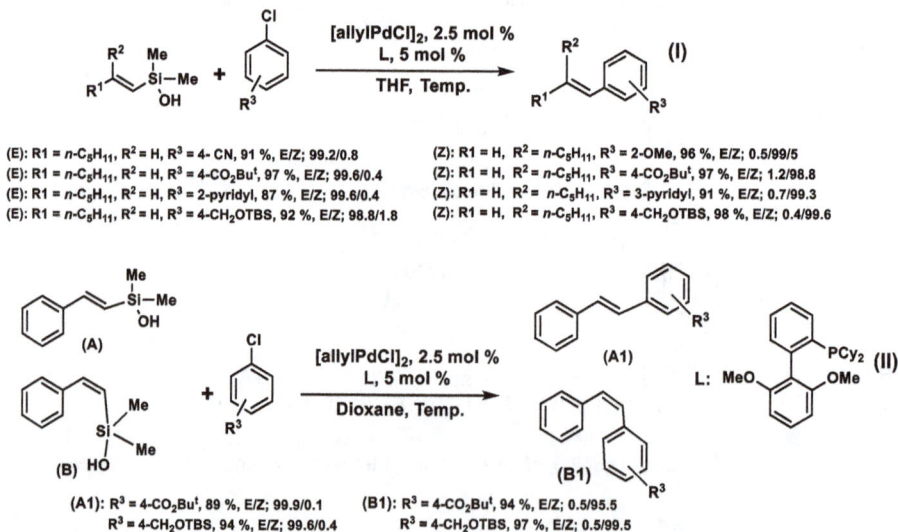

(E): R1 = n-C$_5$H$_{11}$, R^2 = H, R^3 = 4- CN, 91 %, E/Z; 99.2/0.8
(E): R1 = n-C$_5$H$_{11}$, R^2 = H, R^3 = 4-CO$_2$Bu^t, 97 %, E/Z; 99.6/0.4
(E): R1 = n-C$_5$H$_{11}$, R^2 = H, R^3 = 2-pyridyl, 87 %, E/Z; 99.6/0.4
(E): R1 = n-C$_5$H$_{11}$, R^2 = H, R^3 = 4-CH$_2$OTBS, 92 %, E/Z; 98.8/1.8

(Z): R1 = H, R^2 = n-C$_5$H$_{11}$, R^3 = 2-OMe, 96 %, E/Z; 0.5/99/5
(Z): R1 = H, R^2 = n-C$_5$H$_{11}$, R^3 = 4-CO$_2$Bu^t, 97 %, E/Z; 1.2/98.8
(Z): R1 = H, R^2 = n-C$_5$H$_{11}$, R^3 = 3-pyridyl, 91 %, E/Z; 0.7/99.3
(Z): R1 = H, R^2 = n-C$_5$H$_{11}$, R^3 = 4-CH$_2$OTBS, 98 %, E/Z; 0.4/99.6

(A1): R^3 = 4-CO$_2$Bu^t, 89 %, E/Z; 99.9/0.1
R^3 = 4-CH$_2$OTBS, 94 %, E/Z; 99.6/0.4

(B1): R^3 = 4-CO$_2$Bu^t, 94 %, E/Z; 0.5/95.5
R^3 = 4-CH$_2$OTBS, 97 %, E/Z; 0.5/99.5

Fig. 21.56: Cross-coupling of arylchloride with (E)- and (Z)-alkenylsilanolates (I) and (E)- and (Z)-styrylsilanolates (II).

For the cross-coupling of five-membered heterocyclic silanolates, their sodium salts formed in situ under the action of $NaOBu^t$ were used in reaction with electron-rich as well as with electron-poor arylhalides (Fig. 21.57) [73].

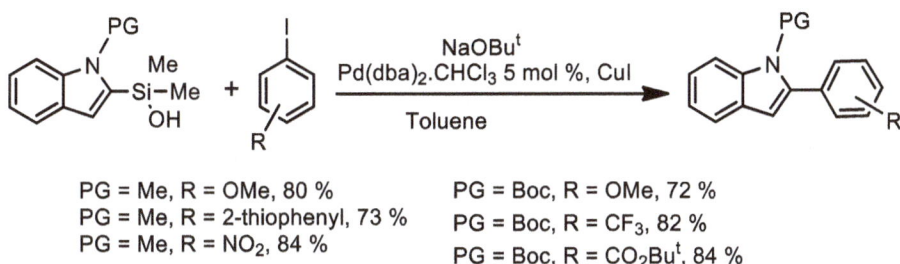

PG = Me, R = OMe, 80 %   PG = Boc, R = OMe, 72 %
PG = Me, R = 2-thiophenyl, 73 %   PG = Boc, R = CF_3, 82 %
PG = Me, R = NO_2, 84 %   PG = Boc, R = CO_2Bu^t, 84 %

Fig. 21.57: Cross-coupling of heterocyclic silanolates with functionalized aryl iodides.

Alkynylsilanols activated by potassium trimethylsilanolate ($KOSiMe_3$) are effective coupling partners with electron-rich and electron-deficient aryl iodides bearing a variety of functional groups (Fig. 21.58) [74].

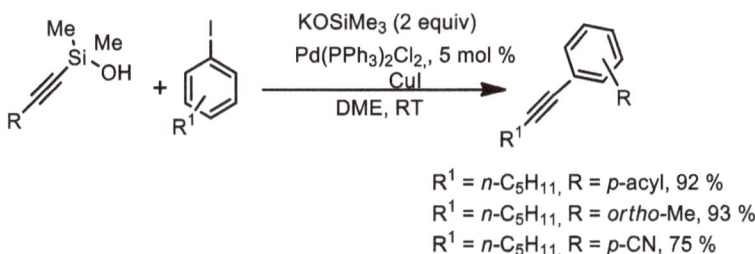

R^1 = n-C_5H_{11}, R = p-acyl, 92 %
R^1 = n-C_5H_{11}, R = $ortho$-Me, 93 %
R^1 = n-C_5H_{11}, R = p-CN, 75 %

Fig. 21.58: Cross-coupling of alkynylsilanols with aryliodides.

21.2.7 The Murahashi–Feringa cross-coupling reactions

The use of organolithium reagents as nucleophiles in transition metals catalyzed cross-couplings (Murahashi–Feringa cross-coupling reactions), although unexpected due to their high reactivity, water and air sensitivity, started along with the reactions already described in 1975. In time, Feringa (Nobel Laureate in Chemistry, 2016) succeeded to overcome many of the significant challenges by careful selection of catalysts and ingenious reaction design, contributing to the metal-catalyzed C–C bond-forming reactions using organolithium reagents.

The first attempts to cross-coupling organolithium derivatives with alkenyl and aryl halides in the presence of Pd(0) compounds, such as $Pd(PPh_3)_4$, to form alkenes stereoselectively under both stoichiometric [23] and catalytic conditions [75] gave promising results (yields were generally good to excellent) Fig. 21.59 [75].

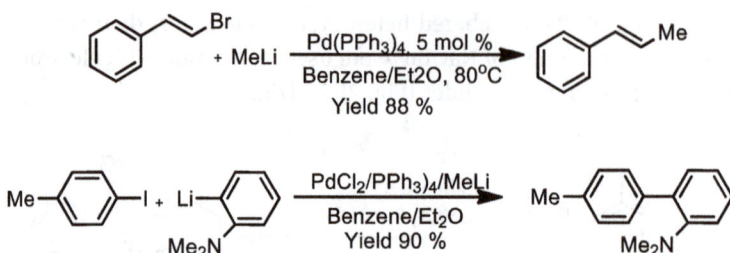

Fig. 21.59: The reaction of (E)-β-bromostyrene and p-iodotoluene with organolithium compounds in the presence of palladium catalyst.

As mentioned earlier, improvements have been made to the cross-coupling re-actions of organolithium compound with organic substrates [76, 77]. A selection of examples is presented below.

The experiments published by Feringa in 2013 show effective cross-couplings of a wide range of alkyl-, aryl- and heteroaryl-lithium reagents with aryl-bromides bearing halide, alcohol, acetal and ether functionalities, electron-withdrawing chlorides and electron-donating methoxy and dimethylamino substituents in the presence of palladium-phosphine complexes as catalysts. The process proceeds quickly under mild conditions (room temp.) and avoids the lithium halogen ex-change and homocoupling (Fig. 21.60). The preparation of alkyl-, aryl- and hetero-biaryl intermediates highlights the potential of these cross-coupling reactions for the synthesis of complex organic molecules [78].

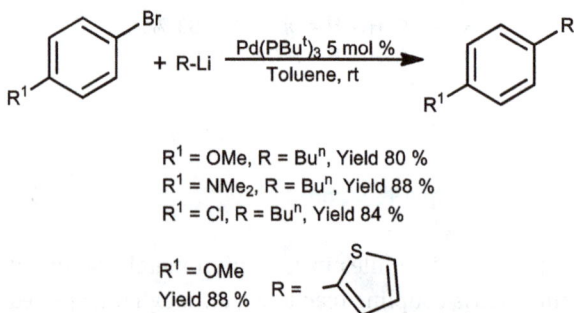

R^1 = OMe, R = Bu^n, Yield 80 %
R^1 = NMe$_2$, R = Bu^n, Yield 88 %
R^1 = Cl, R = Bu^n, Yield 84 %

R^1 = OMe
Yield 88 %   R =

Fig. 21.60: Palladium-catalyzed cross-coupling using organolithium nucleophiles.

The synthesis of 9,9-di-n-octyl-2,7-bis-thienylfluorene (Fig. 21.61) [78], an important building block in the preparation of optoelectronic organic materials, was achieved by Pd-catalyzed twofold arylation of bis-bromo-dialkylfluorene with 2-thienyllithium in high yield. The reaction conditions (see Fig. 21.61) recommend this approach com-pared to the synthesis of the same compound using tributyl(thiophen-2-yl)stannane [79] or Suzuki coupling [80].

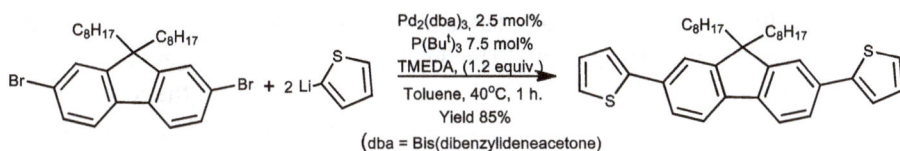

Fig. 21.61: Synthesis of 9,9-di-*n*-octyl-2,7-bis-thienylfluorene by catalytic cross-coupling using organolithium reagent.

Cross-coupling of both activated and deactivated aryl chlorides with aryl and heteroaryl lithium compounds was successful using Pd-PEPPSI-IPent (**A**) in toluene at room temperature or Pd₂(dba)₃/XPhos palladium(0) (dba = bis(dibenzylideneacetone) (**B**) in toluene at 40 °C, as catalysts (Fig. 21.62) [81]. Biaryl and heterobiaryl compounds can be prepared in high yields with short reaction times.

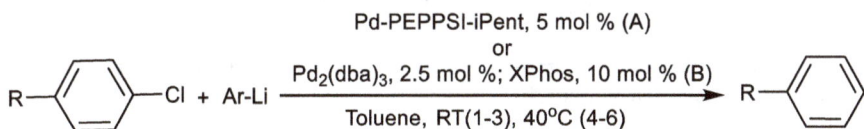

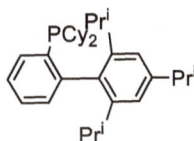

1. R = CF₃, Ar = Ph; A: 89 %, B: 95 %   4. R = Buⁿ, Ar = Ph; A: 86 %, B: 84 %

2. R = CF₃, Ar = (furan) A: 77 %   5. R = OMe, Ar = Ph; A: 94 %, B: 62 %

3. R = CF₃, Ar = (thiophene) A: 83 %   6. R = OMe, Ar = (furan) A: 80 %

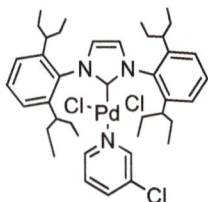

Pd-PEPPSI-iPent
dba = Bis(dibenzylideneacetone)palladium(0)

XPhos

Fig. 21.62: Pd-catalyzed cross-coupling of aryl lithium reagents with activated aryl chlorides (1–3) and reagents with deactivated aryl chlorides (4–6).

The lithium–halogen exchange of appropriately *ortho*-substituted aryl bromide was applied to Pd-catalyzed direct cross-coupling of two distinct aryl bromides (Fig. 21.63). The *ortho*-substituted aryl bromide should react faster than the other with alkyl lithium, providing the organolithium reagent for the cross-coupling. The catalysts

[Pd-PEPPSI-IPr] or [Pd-PEPPSI-IPent] led to an efficient one-pot synthesis of unsymmetrical biaryls at room temperature. The Bu^{t}Li was the lithiation agent of choice as other lithium alkyls such as Bu^{n}Li, Bu^{s}Li or Pr^{i}Li can react with the formation of alkylated products by palladium-catalyzed cross-coupling with aryl bromides [82].

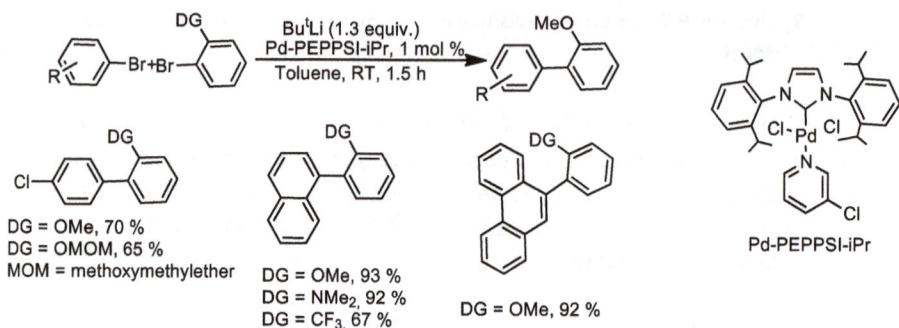

DG = OMe, 70 %
DG = OMOM, 65 %
MOM = methoxymethylether

DG = OMe, 93 %
DG = NMe$_2$, 92 %
DG = CF$_3$, 67 %

DG = OMe, 92 %

Pd-PEPPSI-iPr

Fig. 21.63: One-pot synthesis of unsymmetrical biaryls by cross-coupling of arylbromides with in situ prepared organolithium reagents.

The procedure gives good results for electron-donating and electron-withdrawing substituents on the electrophiles and for a variety of *ortho*-substituents on the in situ prepared nucleophile, -OMe, -NMe$_2$, -CF$_3$, -F, -OMOM, benzyl or benzofuran (Fig. 21.64) [82].

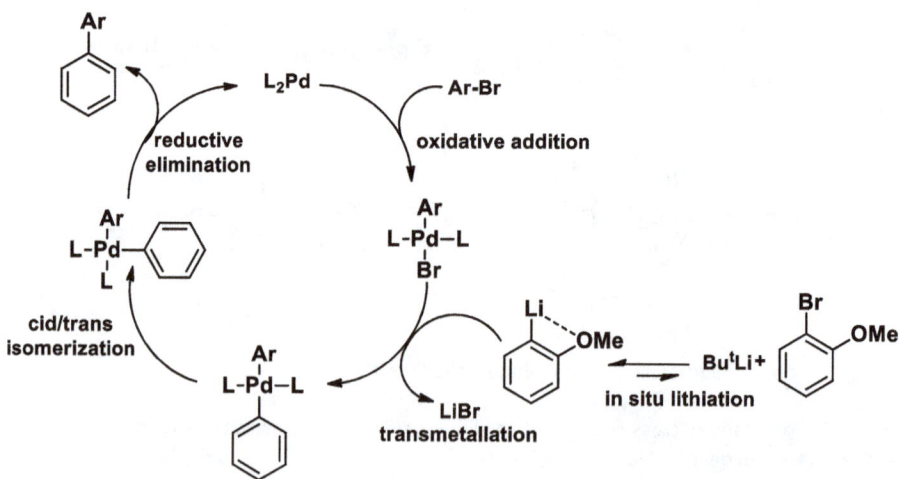

Fig. 21.64: Cross-coupling of organolithium reagents with arylhalides assisted by directing groups.

The organolithium reagents are able to cleave the inert C–O bond catalyzed by [Ni(cod)₂] (cod = 1,5-cyclooctadiene) catalysts with NHC ligands, Fig. 21.65 (**A**), and C–N bond in the presence of a [Pd(PPh₃)₂Cl₂] catalyst Fig. 21.65 (**B**), in one-pot procedure [83].

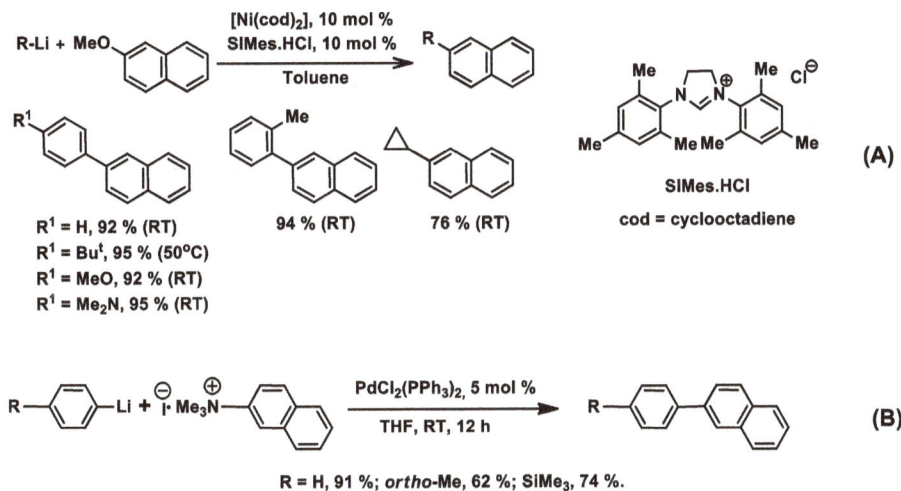

Fig. 21.65: Ni-catalyzed cross-coupling of 2-methoxynaphthalene with organolithium reagents.

The C–O bond cleavage is chemoselective as the OR groups such as OMe, OTBS, OMOM on the phenyl ring remained intact during the reaction. The very bulky *ortho*-substituted phenyllithium reacted smoothly, under heating up to 70 °C with slightly decreased yields. Good yields are obtained in the reaction of Ar-Li with π-rich heteroaromatic compounds, whereas thiophenyllithium, furanyllithium and π-deficient heteroaromatic compounds such as pyridinyllithium show a rather low reactivity [83].

An important class of compounds, allenes, used as synthons in the synthesis of complex organic molecules can be functionalized via cross-coupling between in situ-generated allenyl/propargyl-lithium species and aryl bromides using SPhos- or XPhos-based Pd catalysts. Both allenes and propargyl compounds are good precursors for the cross-coupling using this methodology (Fig. 21.66(**A**)). The procedure prevents the formation of the corresponding isomeric propargylic products, the allenic products being selectively obtained [84]. The treatment of propargyl derivatives, freshly distilled with alkyllithiums, *tert*-BuLi or *n*-BuLi at −78 °C in freshly distilled solvents generates an equilibrium of the allene lithium and propargyl lithium (Fig. 21.66(**B**)) [21]. A mixture of (I)/(II) in a 65/35 ratio was obtained and subsequently reacted with the arylhalides in the presence of different Pd catalysts.

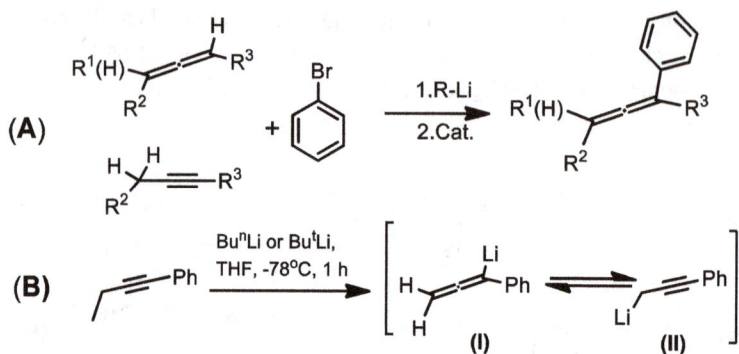

Fig. 21.66: Synthesis of tri- and tetra-substituted allenes by cross-coupling of the in situ generation of organolithium species (A); lithiation of 1-phenylpropargyl (B).

The reaction of allenes with aryl halides and organolithium reagents is straight-forward (Fig. 21.67), without the need of prior transmetallation or prefunctionaliza-tion of substrates with leaving groups [84].

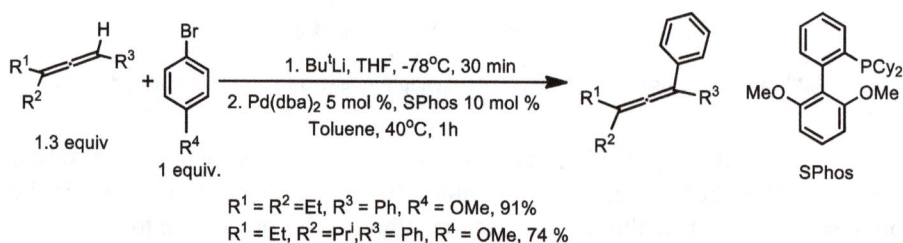

$R^1 = R^2 =$ Et, $R^3 =$ Ph, $R^4 =$ OMe, 91%
$R^1 =$ Et, $R^2 =$Pr^i,$R^3 =$ Ph, $R^4 =$ OMe, 74 %

Fig. 21.67: Direct arylation of allenyl-lithium species.

The direct coupling of alkynes bearing aliphatic or aromatic substituents and func-tionalized aryl bromides in the synthesis of tri-substituted allenes proceeds preserv-ing the selective formation of the allene derivative. Electron-withdrawing and electron-donating substituents on aryl bromides or extended aromatic systems did not influence the conversion or selectivity of the reaction (Fig. 21.68).

Another class with a significant role as synthons in the synthesis of complex organic molecules, pharmaceuticals, natural products and alkynes can be prepared using cross-coupling of the lithium alkynes with various electrophiles (Fig. 21.69) [85]. This approach is complementary to the Sonogashira reaction (see Section 21.2.8). The reactions take place in good to excellent yields under ambient conditions and short reaction times. The mild conditions make possible the tolerance of functional groups on aryls (Fig. 21.69(**A**)), functionalization of heterocycles (Fig. 21.69(**B**)), the presence of a variety of organolithium-sensitive functionalities (carbonyls, active

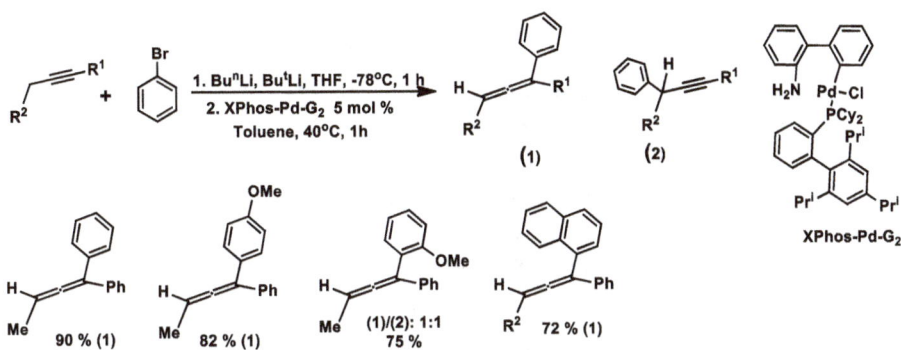

Fig. 21.68: Direct synthesis of arylated allene from alkynes.

methylenes) (Fig. 21.69(**C**)), and the preparation of compound containing more alkyl substituents (Fig. 21.69(**D**)) [85].

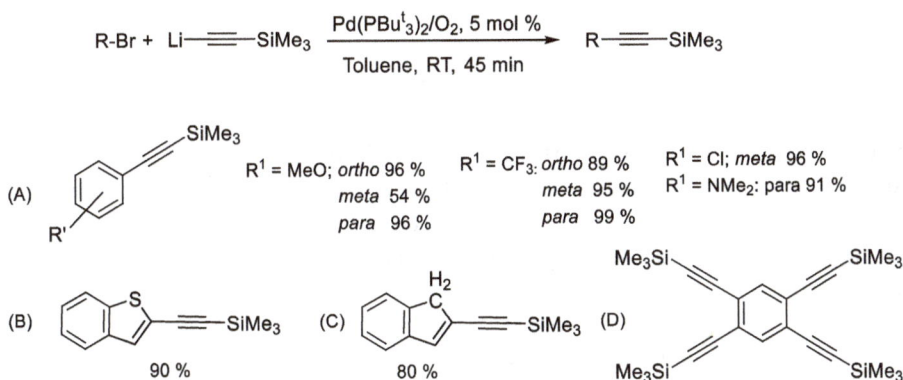

Fig. 21.69: Cross-coupling of lithium acetylides with (A) aryl bromides, (B) heterocycles, (C) substrates with active hydrogens and (D) polybromides.

An alternative to the organic solvents as reaction media is a palladium-catalyzed cross-couplings between organolithium reagents and (hetero)aryl halides (Br, Cl) procedure at room temperature in air, with water as the only reaction medium in the presence of NaCl. This was experimented with good results. Cross-coupling products involving $C(sp^3)–C(sp^2)$, $C(sp^2)–C(sp^2)$ and $C(sp)–C(sp^2)$ can be obtained with no side products, in about 20 s, and yields of up to 99% (Fig. 21.70) [86].

The concentration of NaCl solution, the rate of addition of the organolithium reagent and the presence of dissolved oxygen in water are very important. Aliphatic and aromatic organolithium reagents, MeLi to EtLi, HexylLi, Bu^sLi, Pr^iLi, PhLi and Me_3SiCH_2Li in reaction with (hetero)aryl halides led to the coupling products with high selectivities [86].

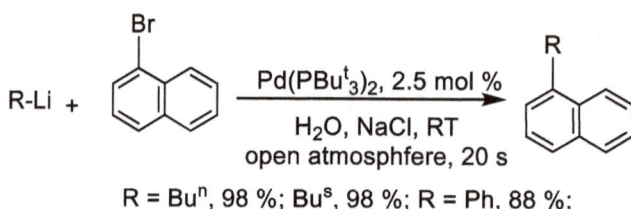

R = Bu^n, 98 %; Bu^s, 98 %; R = Ph, 88 %;

Other exemples:

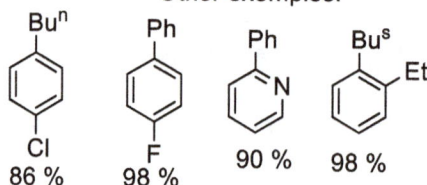

Fig. 21.70: Cross-couplings of organolithium reagents with arylbromides in water and open atmosphere.

21.2.8 Sonogashira cross-coupling reactions

Sonogashira reaction refers to the palladium-catalyzed cross-coupling of a terminal sp-hybridized carbon from an alkyne with a sp^2 (rarely sp^3) carbon of an aryl or vinyl halide (or triflate) (Fig. 21.71) [87, 88]. The first experiments reported by Sonogashira in 1975 used the copper salts as a cocatalyst [89].

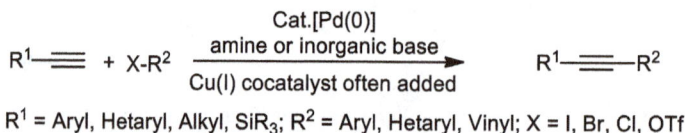

R^1 = Aryl, Hetaryl, Alkyl, SiR$_3$; R^2 = Aryl, Hetaryl, Vinyl; X = I, Br, Cl, OTf

Fig. 21.71: Sonogashira cross-coupling.

The Sonogashira cross-coupling reactions proceed in milder conditions compared to the non-cocatalyzed reactions (Fig. 21.72) [90, 91].

Precautions are usually needed in case of using the copper salts to avoid oxygen in order to diminish or eliminate the alkyne homocoupling through a copper-mediated reaction [92]. The search for "copper free" Sonogashira cross-coupling led to rich chemistry still known as "Sonogashira reactions.

The mechanism of the palladium/copper-catalyzed Sonogashira reaction can be described, taking into account two independent catalytic cycles (Fig. 21.73).

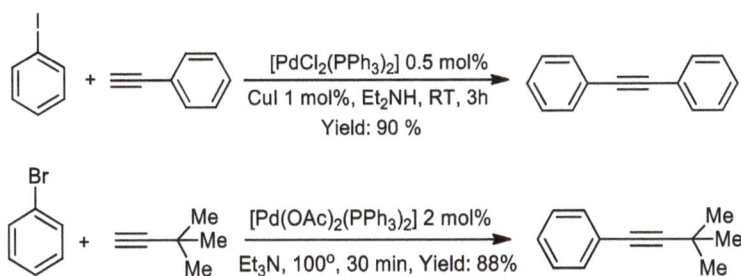

Fig. 21.72: Sonogashira cross-coupling of aryl- or alkylalkyne with arylhalides.

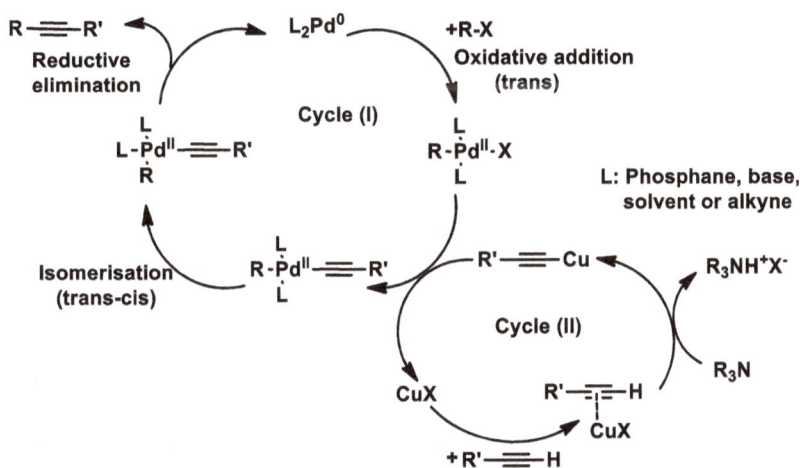

Fig. 21.73: Sonogashira cross-coupling catalytic cycle.

The first "'palladium cycle" (I) is the classical C–C cross-coupling formations [8] and starts with the catalytically active species Pd(0)L$_2$ formed by the dissociation of two ligands from Pd(0) complexes such as Pd(PPh$_3$)$_4$ or from Pd(II) complexes such as PdCl$_2$ (PPh$_3$)$_2$ in a reduction reaction with amines. The first step in the catalytic cycle is the oxidative addition of the aryl or vinyl halide, which is considered to be the rate-limiting step of the Sonogashira reaction, the barriers of oxidative addition increasing in the order of ArI < ArBr < ArCl [93]. The electron-withdrawing groups facilitate the oxidative addition reaction. The next step is the transmetallation with the copper acetylide formed in the "copper cycle" (cycle II). After the *cis/trans*-isomerization (common to all cross-coupling reactions), the final product is released by reductive elimination, regenerating the catalyst [Pd(0)L$_2$]. In the "copper cycle" (cycle II), prior coordination of the alkyne via a π-bond to the copper salt activates the terminal proton which is trapped by the base.

In the "copper-free" Sonogashira alternative, the transmetallation reaction is preceded by an exchange of ligands (Fig. 21.74): the akyne is π-coordinated to the palladium atom replacing a ligand from the starting catalytic species. The acetylenic proton is acidified by coordination and therefore easily removed by the base.

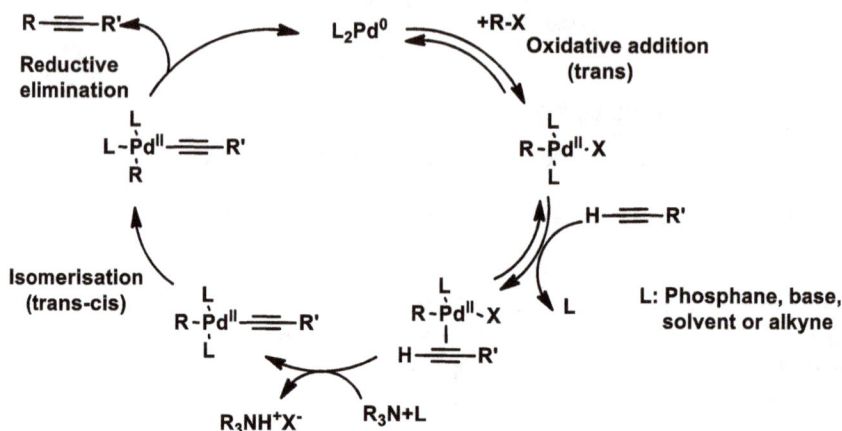

Fig. 21.74: The "copper-free" Sonogashira alternative catalytic cycle.

Another mechanism is considering a more complex role for the amines [92].

As already mentioned, palladium is the metal of choice when the copper cocatalyst was used. Both the generation of the catalytic-active form of the palladium complexes and the oxidative addition step are common for previously described C–C cross-coupling and are valid for Sonogashira reactions too. Increasing the steric bulk of the ligands and the electron richness will work well for Sonogashira couplings, the first favoring the formation of the law coordinate and highly active palladium complexes, and the second by favoring the oxidative addition [94].

Using NHC ligands, it was possible to apply Sonogashira for cross-coupling of functionalized, inactivated, β-hydrogen-containing primary alkyl bromides and iodides that contain a wide range of functional groups like esters, nitriles, olefins, acetals and unprotected alcohols with terminal alkynes under mild conditions (Fig. 21.75) [95].

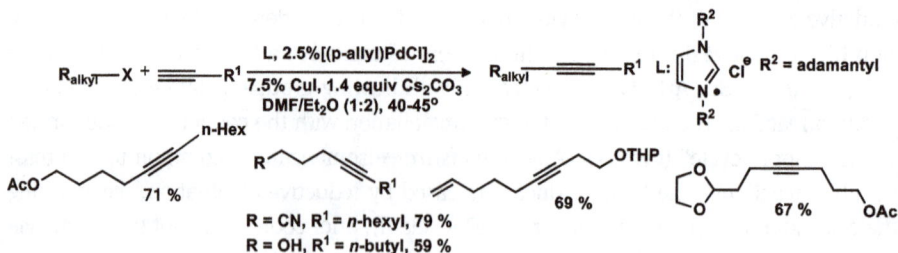

Fig. 21.75: Sonogashira cross-coupling of inactivated alkylhalides with alkynes.

The cross-coupling of secondary inactivated alkyl bromides with alkynes was realized using other classes of palladium–NHC complexes, with or without an amine as additive (Fig. 21.76) [86].

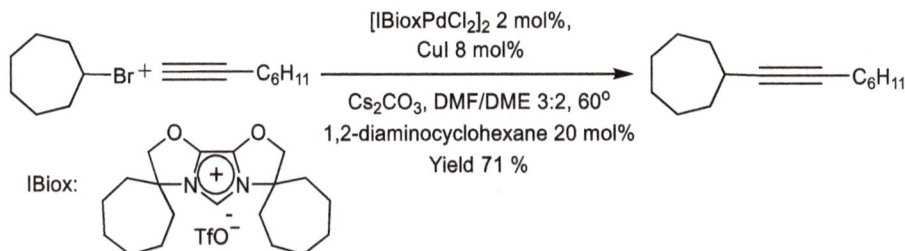

Fig. 21.76: Cross-coupling of secondary inactivated alkyl bromides with alkynes.

The use of the preformed complex, [IBioxPdCl$_2$]$_2$, led to a yield of up to 70%. In the same reaction conditions, the phosphine ligands were not suitable as ligands [96]. Functional groups like chloride, ester or epoxide in the alkylbromides are well tolerated due to the mild reaction conditions.

21.3 The Mizoroki–Heck reaction

The Mizoroki–Heck reaction became known at the time as the catalytic arylation and alkenylation of olefins, Fig. 21.77:

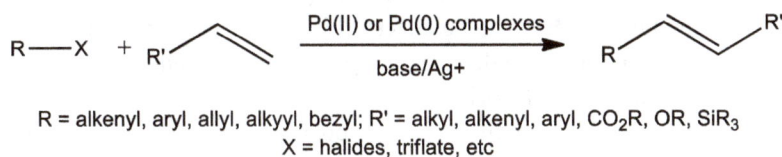

R = alkenyl, aryl, allyl, alkyyl, bezyl; R' = alkyl, alkenyl, aryl, CO$_2$R, OR, SiR$_3$
X = halides, triflate, etc

Fig. 21.77: Heck reaction.

The first results were published independently by Mizoroki (Fig. 21.78(**A**)) [97] and Heck (Fig. 21.78(**B**)) [98].

The contribution of Heck to the development of the organic synthesis mediated by transition metal catalysts was acknowledged with the Nobel Prize in 2010.

The reaction is catalyzed by palladium complexes, with or without phosphine ligands (phosphine-assisted vs. phosphine-free catalysis) [99]. A primary role of phosphine ligands is to support palladium in its zero-oxidation state in the form of stable PdL$_4$ or PdL$_3$ species. The phosphine-assisted approach is the classical and well-established method which gives excellent results in most cases. The catalytically active species, Pd(0)L$_2$ stabilized by the ligands present, can be formed by the

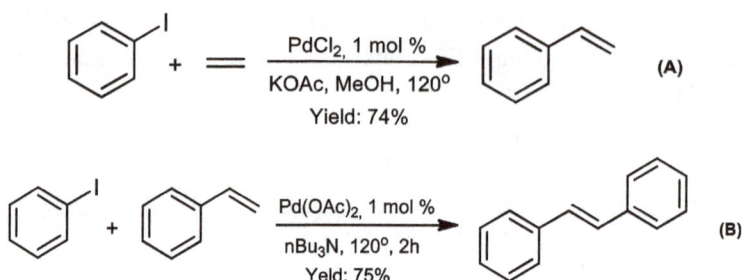

Fig. 21.78: Reactions of olefins with aryl iodides.

dissociation of two ligands from Pd(0) complexes such as Pd(PPh$_3$)$_4$, or can be formed in situ from Pd(II) complexes. The reduction of Pd(II) complexes to Pd(0) and the generation of active species through multiple ligand exchange equilibria is, in some cases, a latent (inductive) period due to the labile character of Pd(0) complexes [100]. The primary reduction of Pd(II) to Pd(0) is most likely accomplished by phosphines:

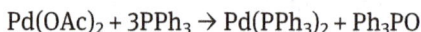

$$Pd(OAc)_2 + 3PPh_3 \rightarrow Pd(PPh_3)_2 + Ph_3PO$$

The reduction is assisted by amines or hard nucleophiles like hydroxide and alkoxide ions, water and water and acetate ion. The reduction of the palladium(II) takes place on the expense of the phosphine oxidation. The rate of reaction of the phosphines containing aryl groups substituted with electron-withdrawing groups are higher than those containing unsubstituted aryls. This behavior is related to the more efficient nucleophilic attack at electrophilic phosphorus atom. In phosphine-free systems, the reduction of Pd(II) can be effected by amines, if these are used as base, or olefins but they do not have detectable influence on the reduction rate in the presence of a phosphine [12, 101, 102].

There are two routes for Heck couplings: the nonpolar Heck reactions (Fig. 21.79) and the cationic (polar) Heck reactions (Fig. 21.80) both catalyzed by palladium complexes.

The nonpolar pathway (Fig. 21.79) involves coordination of the olefin via dissociation of one neutral ligand while the cationic (polar) (Fig. 21.80) involves coordination of the olefin via dissociation of the anionic ligand.

The first step of the catalytic cycles is the oxidative addition. The palladium(0) should be coordinated by two strongly bound ligands. The molar ratio Pd(0):ligand is very important for the concentration of active species. Both the choice and the concentration of the ligand in the reaction mixture will be taken into account to provide the required amount of reactive dicoordinated Pd(0) complex. An excess of the ligand will decrease the concentration of active species, which sometimes can lead to the inhibition of catalytic process while a low concentration will determine the disproportionation of the dicoordinated complex to a stable tricoordinate complex

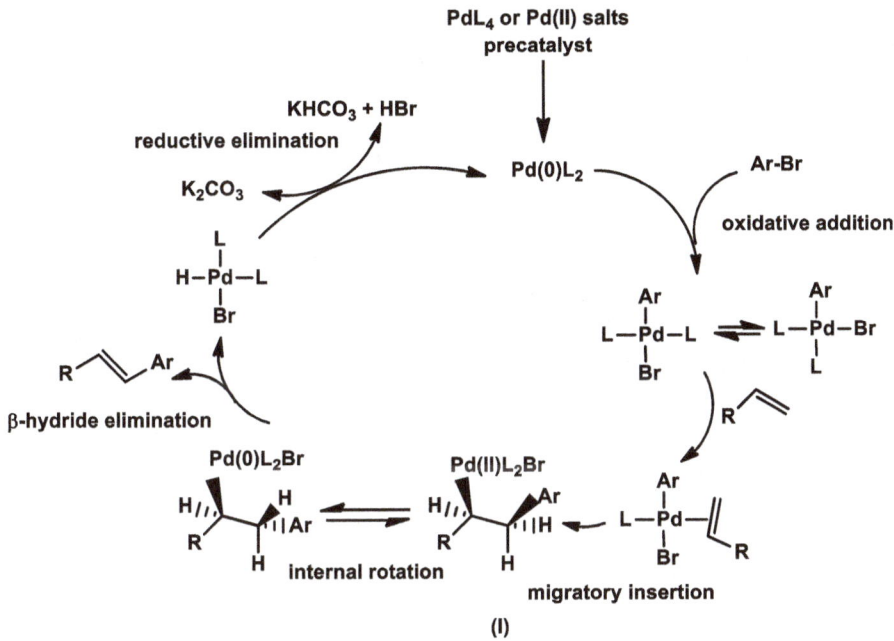

Fig. 21.79: Heck coupling catalytic cycle – nonpolar.

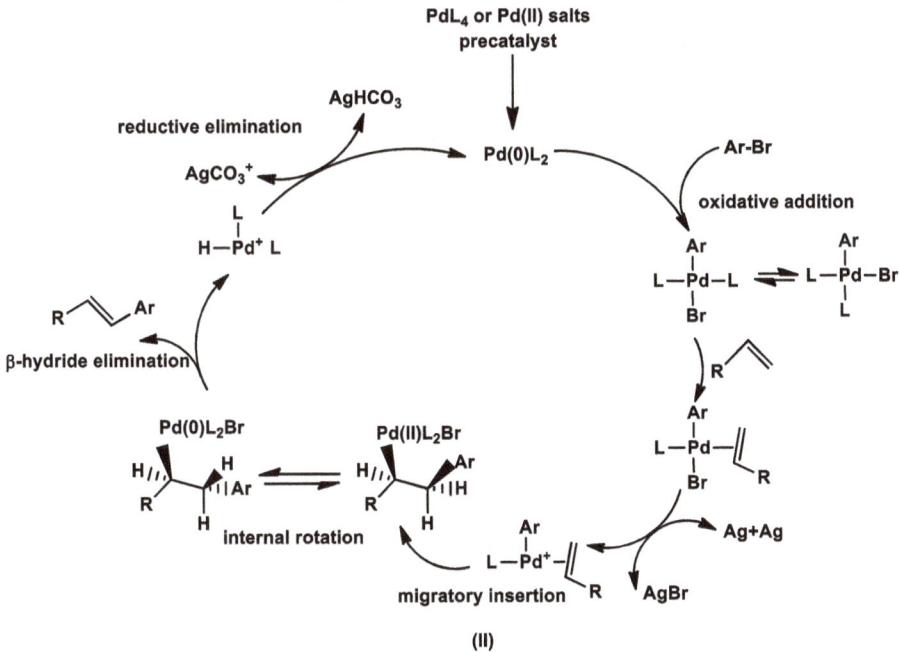

Fig. 21.80: Heck coupling catalytic cycle – cationic.

and unstable low-ligated complexes which undergo a fast aggregation to clusters, and finally the inactive metallic particles are formed:

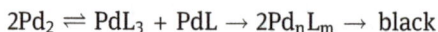

$$2Pd_2 \rightleftharpoons PdL_3 + PdL \rightarrow 2Pd_nL_m \rightarrow black$$

The product of the oxidative addition is a Pd(II) tetracoordinated complex. The relative orientation of the two components of the electrophile, *cis* or *trans*, depends on the polarity of the carbon–halide bond (see Oxidative addition in Section 21.2.1). The oxidative addition is less sensitive to the substituents in the electrophile but is sensitive to the nature of nucleofuge and the strength of C–X and M–X bonds. The order of reactivity is I > OTf > Br > Cl. The next two steps, not always marked as independent, are the coordination of the olefin to palladium and the migratory insertion. Two types of complexes can be taken into account for the coordination of the olefin to the metal: a tetracoordinated one formed by ligand–substrate exchange or a pentacoordinated structure. The energy barrier for the generation of the reactive configuration in a tetracoordinated complex is lower compared to the pentacoordinated one. Two coordination sites of the square planar complex are occupied by the olefin and the fragment that have to migrate onto the π-system. As a consequence, the control exerted by the catalyst depends on the remaining two ligands, which are one neutral and one anionic in the neutral complex, or both neutral in the cationic complex. For bidentate phosphine complexes, the polar path is the preferred one. The migratory insertion process requires a coplanar assembly of the metal, ethylene and the hydride. Therefore, the insertion process is stereoselective and occurs in a *syn* manner, according to experimental observations by Heck and theoretical calculations [103]. The β-hydride elimination is stereoselective and occurs in a *syn* manner, and its efficiency being related to the dissociation of the olefin from the palladium(II)–hydride complex (48). The elimination goes through a rather strong agostic interaction of palladium with hydrogen atom and thus proceeds as a concerted *syn*-process without the involvement of the base.

The regioselectivity can be affected by side reaction to be considered after elimination. The new alkene can coordinate to the palladium hydride and the next step is not fast enough; therefore, the migratory insertion may occur, leading to the formation of several products including the isomerization of the double bond (Fig. 21.81).

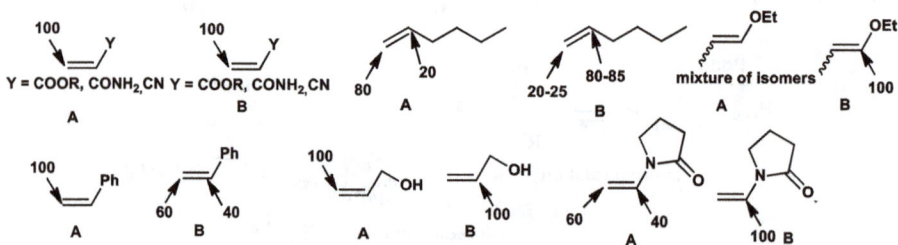

Fig. 21.81: Heck reaction regioselectivity in coordination–insertion process: nonpolar path (A) and cationic (polar) path (B).

Another way to the isomerization of the alkene leading to a product with the wrong stereochemistry is the scavenging of the palladium hydride by the starting alkene, which is always more reactive than the Heck product due to its smaller size.

Heck reactions of aryl bromides and activated aryl chlorides with a range of mono- and disubstituted olefins at room temperature afford arylated product with high E/Z stereoselection (Fig. 21.82). The corresponding reactions of a broad spectrum of electron-neutral and electron-rich aryl chlorides proceed at elevated temperature, also with high selectivity. In terms of scope and mildness, Pd/P(Bu^t)$_3$/Cy$_2$NMe represents an advance over previously reported catalysts for these Heck coupling processes [104].

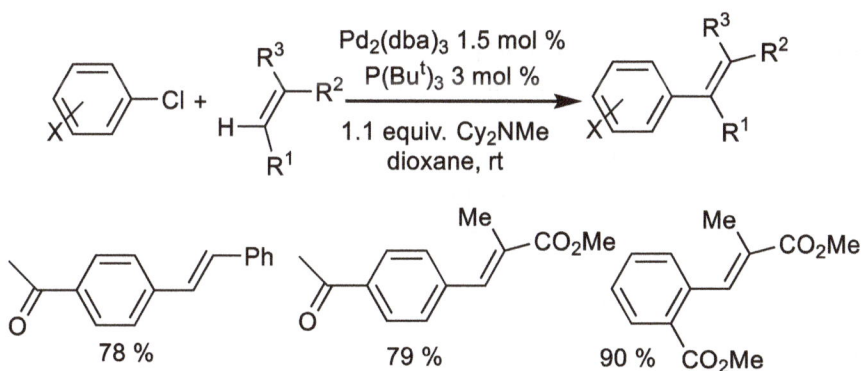

Fig. 21.82: Reactions of activated aryl chlorides with mono- and disubstituted olefins.

A palladium-catalyzed Heck reaction of substituted styrene derivatives with a variety of tertiary, secondary and primary alkyl bromides proceeds smoothly at room temperature upon irradiation with blue light-emitting diodes (blue light-emitting diodes) in the presence of a dual-phosphine ligand system (Fig. 21.83). The use of a dual-phosphine ligand system is crucial for the success of this transformation. The palladium source, Pd(PPh$_3$)$_4$ and Pd(PPh$_3$)$_2$Cl$_2$, is effective in the presence of Xantphos (both PPh$_3$ and Xantphos play an essential role in the reaction). Only bisphosphine ligands with a conjugated backbone similar to that of Xantphos proved to be effective [105].

A possible explanation for the suppression of the undesired β-hydride elimination is the enhancing of oxidative addition by the photoexcited state reactivity of the palladium complex.

The vinylation of electron-rich olefins by β-halostyrenes using hemilabile 1,3-bis (diphenylphosphino)propane monoxide as a ligand and palladium acetate in dimethylsulfoxide as solvent led to fast reactions even of the challenging 2-substituted vinyl ethers (Fig. 21.84). The 1:3 ratio of bromostyrene:vinylether has to be used. The

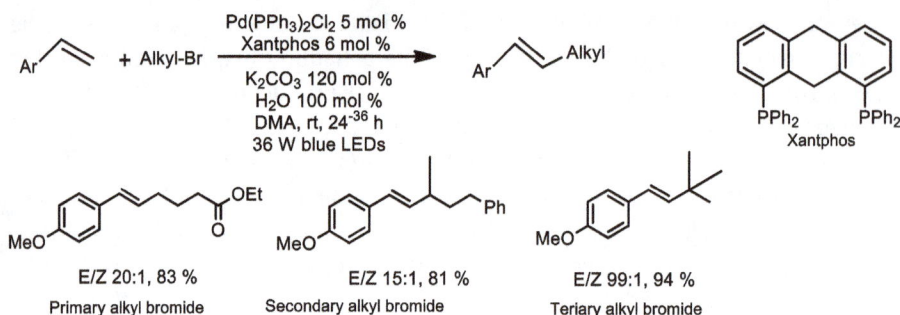

Fig. 21.83: Heck reaction of styrene derivatives with alkyl bromide activated by irradiation.

ketone (**I**) was separated after hydrolysis, but in some cases the separation of the aldol (**II**) was possible. The reactions are highly regioselective [106]:

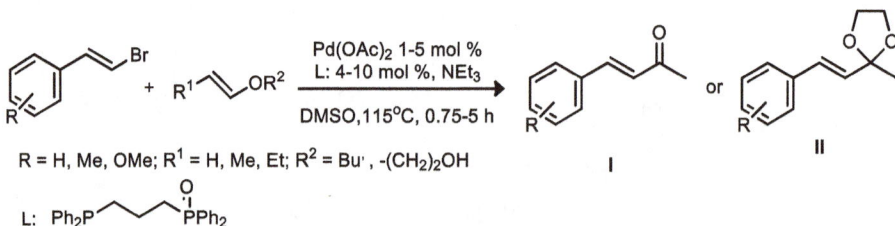

Fig. 21.84: Heck reaction of 2-substituted vinyl ethers and β-halostyrenes.

Reactions of electron-rich bromides such as *p*-methoxy-β-bromostyrene were complete in slightly shorter times. The reaction of *p*-acetyl-β-bromostyrene led to the aldol compound which is difficult to prepare via conventional aldol methodology. In addition to *p*-methoxy-β-bromostyrene, both *n*-butyl vinyl ether and 2-hydroxyethyl vinyl ether allowed access to cyclic ketals under identical conditions. The formation of a protected ketone in this way could be potentially useful when chemoselectivity would otherwise be a problem. The vinyl chlorides also react, even if over longer reaction times, affording exclusively branched products hydrolyzed and isolated as the ketones in good yields [106].

An important contribution to the organic synthesis, especially to the synthesis of natural products or pharmaceuticals, is the intramolecular Heck reaction, including the asymmetric synthesis. A variety of mono- and bidentate ligands have been experimented not only for Heck reaction but also for other reaction types [107]. Successful experiments affording a 96% ee (diequatorial) and good conversions in the intramolecular asymmetric Heck reaction used a Taddol-based monodentate ligand (Fig. 21.85) [108, 109].

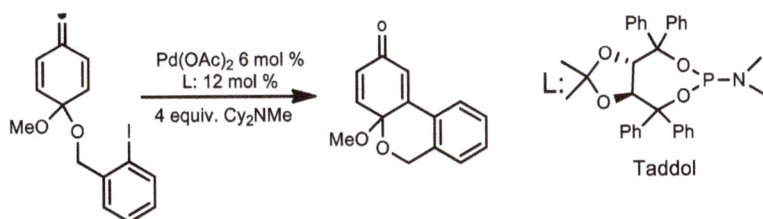

Fig. 21.85: Intramolecular Heck reaction using palladium taddol catalyst.

21.4 Hydroformylation

The hydroformylation of olefins, addition of CO and H_2 to an alkene function providing a new carbon–carbon and a new carbon–hydrogen bond, was discovered in 1938 by Otto Roelen. Hydroformylation is one of the largest industrially applied processes which relies on homogenous catalysis providing not only aldehydes but also alcohols as subsequent hydrogenation reactions [110].

There are two possible reaction products (Fig. 21.86): the linear (*normal*) and the branched (*iso*) aldehydes with the formation of a new stereocenter (asymmetric hydroformylation).

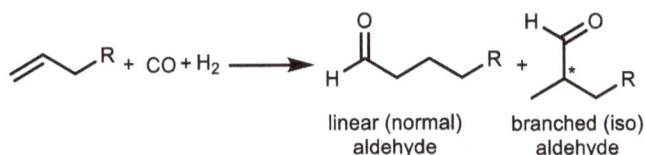

linear (normal)
aldehyde

branched (iso)
aldehyde

Fig. 21.86: Hydroformylation reaction.

The side reactions, alkene isomerization or alkene hydrogenation, have also been observed. As all the atoms are finally present in the product, this reaction is a prototype of an atom economic transformation with significant environmental advantages [111, 112].

For the application in organic synthesis, there are challenges regarding chemo-, regio-, diastereo- and enantioselectivity control in the course of the hydroformylation [113]. The catalyst used in the original research experiments was $HCo(CO)_4$ prepared in situ from $Co_2(CO)_8$ under hydrogen/carbon monoxide pressure. The mechanism of the cobalt-catalyzed hydroformylation was proposed by Heck and Breslow in 1960 and 1961 [114]. In the early 1960s, phosphine-modified cobalt catalysts were introduced [115].

When $HCo(CO)_4$ is used as a catalyst, the temperatures for reasonable reaction rates (110–180 °C) require rather high CO partial pressure; therefore, a total H_2/CO

pressures of 200–300 bar are needed. The search for better catalytic systems obtained with suitable metal and donor ligand modifications resulted in the use of rhodium-, platinum- and palladium-based catalysts [116]. Higher activity and chemoselectivity running the hydroformylation in the low to medium pressure range (5–100 bar) was obtained using phosphine-modified rhodium complexes (1968 with Wilkinson) [117–119]. The first mechanism for cobalt-catalyzed hydroformylation proposed in the early 1960s by Breslow and Heck [114] is widely accepted, with little modifications, for the phosphine-modified cobalt catalysts and the phosphine- or phosphite-modified rhodium-catalyzed hydroformylation (Fig. 21.87) [120–122].

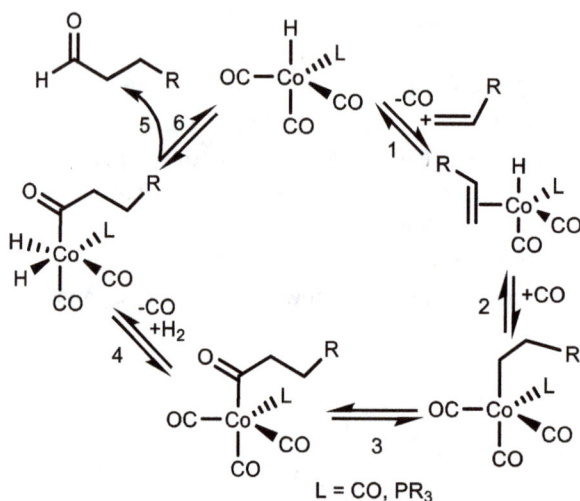

Fig. 21.87: Cobalt-catalyzed hydroformylation catalytic cycle.

The first step in the catalytic cycle of hydroformylation (Fig. 21.87) is the ligand exchange of a CO with the olefine (1) to produce a π-complex, a process without change in the oxidation state of the metal. The next step (2) is the migratory insertion of the olefine in the Co–H bond combined with the coordination of a CO molecule. Another migratory insertion (3) is providing the acyl complex. The oxidative addition of a hydrogen molecule (4) takes place with the increase of the formal oxidation state of the cobalt from +1 to +3. The final step (5) is the reductive elimination of the aldehyde and regenerates the catalytically active species.

For synthetic organic chemistry, the tolerance of functional groups is crucial. A wide range of sensitive and reactive functional groups – aldehydes, ketones, acetals, ketals nitriles, free alcohols, carboxylic acids, alkyl halides, nitro compounds, pyridine derivatives, *tert*-amines or tosylates – are compatible with hydroformylation conditions [113].

The effect of the ligands on reactivity and selectivity of rhodium catalysts led to the discovery of better catalytic systems. Good results were obtained using bidentate ligands, diphosphines and diphosphites [107, 123–127]. Examples of diphosphine ligands for regioselective hydroformylation of terminal alkene are presented in Fig. 21.81.

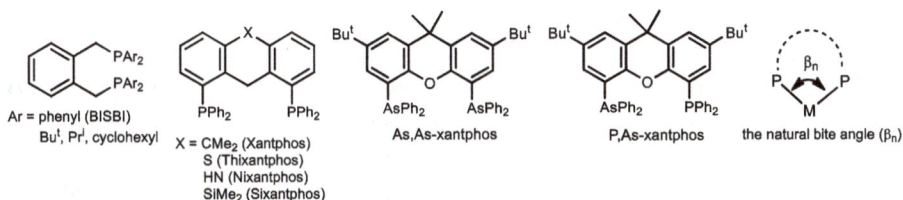

Ar = phenyl (BISBI)
Bu^t, Pr^i, cyclohexyl

X = CMe$_2$ (Xantphos)
S (Thixantphos)
HN (Nixantphos)
SiMe$_2$ (Sixantphos)

As,As-xantphos

P,As-xantphos

the natural bite angle (β_n)

Fig. 21.88: Bidentate ligands for hydroformylation reactions.

The activity and selectivity in the rhodium–diphosphine-catalyzed hydroformylation is influenced by the natural bite angle (β_n) of the diphosphine ligands (Fig. 21.88). The concepts of the natural bite angle (β_n) and the flexibility range are useful in the prediction of chelating properties of diphosphine ligands. The natural bite angle (β_n) of a diphosphine ligand is defined as the preferred chelation angle determined by ligand backbone only and not by metal valence angles. The flexibility range is defined as the accessible range of bite angles within less than 3 kcal/mol excess strain energy from the calculated natural bite angle [128].

The n-selectivity is a result of the dynamic equilibria between ee and ea (equatorial–apical) coordination isomers of the diphosphine ligands: the equilibria shift to the ee isomer for ligands with wider bite angles (high n-selectivity) and to ea isomer for ligands with smaller bite angles (which still give reasonably high n-selectivity) [129]. The hydroformylation of 1-octene or styrene using (diphosphine)Rh(CO)$_2$H affords the linear aldehyde with increased selectivity and activity, following the increase of natural bite angle. The CO dissociation rates of (diphosphine)Rh(CO)$_2$H complexes are orders of magnitude higher than the hydroformylation rates and it is not correlated with the natural bite angle. The bite angle affects the selectivity in the steps of alkene coordination and hydride migration.

The substitution at the ninth position on xanthene-type-based diphosphines results in tuning the electronic and steric properties covering a range of natural bite angles from 102° to 120.6° [129]. The *steric bite angle effect* and *the electronic angle effect* are associated with the properties of the abovementioned substituents and with the specific coordination to a given metal. In most of the cases, the linear-to-branched ratio of the aldehyde product increases with increase of the bite angle of the diphosphine. In the rhodium complexes of the rigid Xantphos-type ligands, the chelation mode, partially imposed by the natural bite angle, is also influenced by the phosphine basicity [124, 130].

High levels of *n*-selectivity, 66:1 linear-to-branched aldehyde ratio, was obtained for the rhodium catalysts derived from the bidentate ligand BISBI (bis((diphenylphosphino)methyl)-1,1′-biphenyl, Fig. 21.81, with natural bite angle of 113°) [124], in the hydroformylation of 1-hexene. The substituents on the aromatic groups in the structure of diphosphines influence both the reactivity and selectivity, as found when the phenyl groups in BISBI were replaced by strongly electron-withdrawing 3,5-$(CF_3)_2C_6H_3$ substituents: a fivefold rate increase and a 123:1.61 linear-to-branched ratio.

The in situ-prepared catalyst starting the bis-organophosphite ligand (Fig. 21.82) and $Rh(CO)_2(acac)$ (acac = acetylacetone) is active in hydroformylation of a wide variety of carbonyl-containing substrates like ketones (Fig. 21.89 **(1)**), esters, carboxylic acids (Fig. 21.89 **(2)**) and amides produced dicarbonyls in good-to-excellent yield.

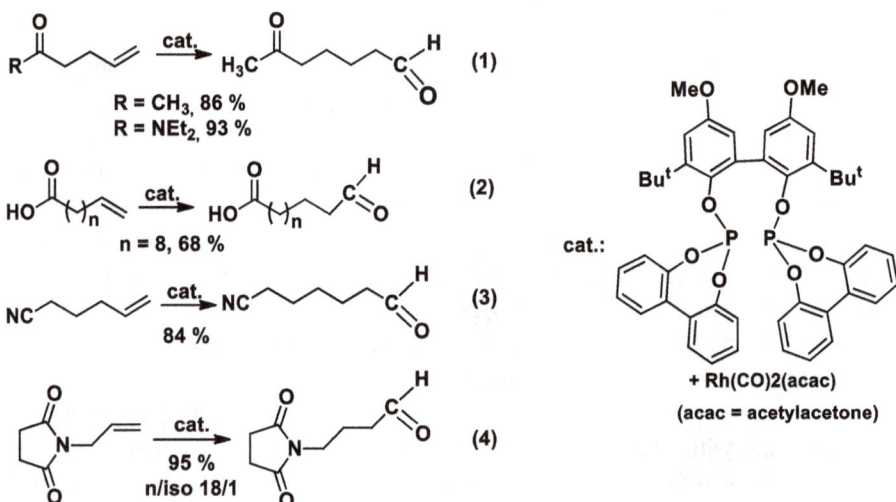

Fig. 21.89: Hydroformylation of carbonyl-containing substrates (1,2), nitrile (3) and *N*-allylsuccinimide (4).

Olefinic alcohols (even unprotected), nitriles (Fig. 21.89 **(3)**), and halides were also hydroformylated with very good results. Unsaturated acetals underwent hydroformylation to produce the monoprotected di-aldehydes, which have been shown to be synthetically useful intermediates. *N*-Allylsuccinimide was hydroformylated to the imide aldehyde in excellent yield (Fig. 21.89 **(4)**) [125].

The platinum complex (Sixantophos)$PtCl(SnCl_3)$ (natural bite angle of the ligand, Fig. 21.88, is 106.2°) was proved to be remarkably active catalyst for the hydroformylation of methyl 3-pentenoate. A prior step to get the desired linear aldehyde, a nylon-6 precursor, is the selective isomerization to methyl 4-pentenoate in the reaction mixture. At low CO concentration, isomerization to the more stable *n*-alkyl complex is faster than CO insertion, and the overall result of the hydroformylation is the linear aldehyde [131].

The selectivity for the linear aldehyde in the hydroformylation of 1-octene was moderate. Remarkably, the selectivity and reaction rate increased enormously if one of the phosphorus donor atoms was replaced by arsenic.

In the platinum/tin-catalyzed hydroformylation of 1-octene using the xantarsine and xantphosarsine ligands (Fig. 21.88, having the natural bite angle approximately the same as Xantphos), a very good selectivity was obtained.

These are the first examples of arsine-based ligands superior to phosphine ligands in the platinum/tin-catalyzed hydroformylation and can be related to the enhanced preference for the formation of *cis*-coordinated platinum/tin complexes [132].

21.4.1 Asymmetric hydroformylation

The enantioselective hydroformylation is a way to obtain chiral aldehydes, important intermediates in organic synthesis [133]. The general mechanism of asymmetric hydroformylation using rhodium complexes is presented in Fig. 21.90.

The catalytically active 16 valence electron species is formed by dissociation of one ligand L from the trigonal bipyramidal 18 valence electron species formed under syngas pressure and in the presence of donor ligands L such as phosphines, phosphites or carbon monoxide. The main catalytic cycle starts with the coordination of the alkene preferably in the equatorial position, thus furnishing a trigonal bipyramidal hydrido olefin complex (1). Alkene insertion into the Rh–H bond (2) takes place to form isomeric tetragonal alkyl rhodium complexes. The coordination of carbon monoxide (3) yields the trigonal bipyramidal complexes which is transformed by migratory insertion of the alkyl group to one of the coordinated carbon monoxide ligands (4) in a tetragonal acyl complex. Oxidative addition of molecular hydrogen (5) forms a tetragonal bipyramidal rhodium(II) complex. Subsequent reductive elimination (6) liberates the isomeric aldehydes and regenerates the catalytically active species (7).

As mentioned earlier, the selective hydroformylation of terminal and internal, including functionalized alkenes, toward the formation of the linear product using chelating ligands with high natural bond angles was well documented.

The role of the ligand in asymmetric hydroformylation, as in asymmetric synthesis mediated by transition metals in general, is more complex. A variety of chelating phosphorus ligand with the potential of inducing chirality has been described. The source of chirality is also diverse: bidentate C2-symmetric ligands with additional options for chirality on P-substituents (auxiliaries) or P-stereogenic atoms, C1-symmetric (backbone chirality), hybrid diphosphorus (chirality on phosphorus substituents) or mixed donor classes (chirality on P/N, O or S substituents) [134]. The hybrid ligands are combining the high activity displayed by one donor and the high selectivity

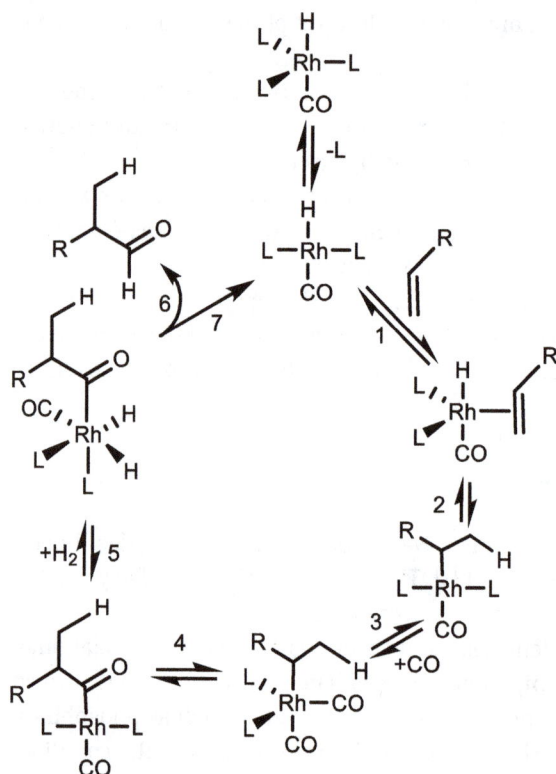

Fig. 21.90: Rhodium-catalyzed asymmetric hydroformylation catalytic cycle.

induced by using the second donor. The coordination to the metal center of the two donor atoms allows the tuning of the catalytic properties. As in the case of the chelating ligands already described for n-selectivity, the bite angle of the ligand is an important parameter.

Many chiral diphosphine, diphosphite and hybrid phosphine–phosphite ligands have been evaluated with regard to induce enantioselectivity in the course of the hydroformylation reaction (see for exemplification [134–136]). The first ligand used in asymmetric hydroformylation was (R,S)-BINAPHOS, a phosphine–phosphite compound, (R)-2-diphenylphosphino-1,1′-binaphtalen-2′-yl(S)-1,1′-binaphtalen-2,2′-diyl phosphite (Fig. 21.91). The major isomer significant in the catalytic process is ea-coordinated complex [137].

The coordination of the hybrid bidentate ligand to the rhodium in ee or ea relative positions in the trigonal bipyramidal complex is more relevant in the case of symmetric ligands. The position of the ligand has a direct influence on the hydride ligand usually residing in the apical position. The two possible isomers in case of the ea coordination of the ligand have different trans-relationship with the hydride

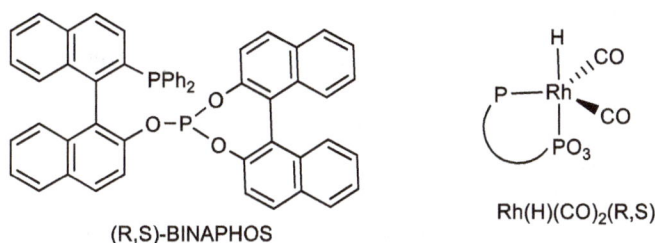

Fig. 21.91: (*R,S*)-BINAPHOS and the *ea* coordination to rhodium.

and as a result an important influence on reactivity of the two isomers. Theoretical studies suggest that alkene insertion into Rh–H bond, mostly irreversible, is the enantioselectivity-determining step [138, 139]. A preference for the phosphine moiety for the equatorial and the phosphite for the apical site was observed for BINAHOS as a consequence of the nonbonding steric effects induced by the bulky naphthyl and phenyl groups and by the larger electronic distortion induced by the better π-acceptor phosphite moiety at the equatorial position. This balance explains the exceptional behavior of this ligand. The ligand coordination has important implications in the mechanism of enantioinduction. In the stereoselectivity-determining transition state, the key ligand–substrate interactions occur between the styrene and the apical ligand moiety. All of the evidence indicated that the coordination preference in the resting state was transferred to the transition states. The control of the vacant sites available for substrate coordination plays a crucial role.

The asymmetric hydroformylation of styrene and *p*-substituted styrenes (CH_3, OCH_3, Cl, $CH_2CH(CH_3)_2$) using Rh(acac)(CO)$_2$ and (R,S)- or (S,R)-BINAPHOS as catalysts (prepared in advance or in situ from Rh(acac)(CO)$_2$ and 4.0 equiv of the ligand) resulted in >99% conversion, a branched/linear ratio in the range of 86/14–92/8 and an enantiomeric excess higher than 85%. Hydroformylation of internal terminal olefins and functionalized olefins (vinyl carboxylates) using the same catalytic system gave the desired products with very good yield and very high enantiomeric excess [140].

21.5 Hydrogenation

Hydrogenation is the addition of H_2 to a multiple bond (C = C, C ≡ C, C = O, C = N, C ≡ N, N = O, N = N, N ≡ N, etc.) to reduce it to a lower bond order. For industrial purposes, the heterogeneous catalytic hydrogenation is still very important. The exception is the homogeneous asymmetric hydrogenation, a way to prepare chiral compounds. The most common and simple type of hydrogenation is the reduction of a C = C double bond to a saturated alkane.

The key step in hydrogenation is the activation of hydrogen molecule. Oxidative addition is the most common method of activating H_2 on a transition metal complex with empty coordination site in order to coordinate the H_2 molecule prior to the oxidative addition (see oxidative addition in Section 21.2.1). The early transition metals with d^0 counts can activate H_2 by hydrogenolysis. The metal center needs to have empty orbitals to bind both the H_2 and the anionic ligand to be protonated. Metals like Ru(II) form the active catalytic species by the heterolytic cleavage. The proton is produced by the reaction with an external base and then transferred to the metal center: the actual catalyst, RuHCl species, is formed by the heterolysis of H_2 by the $RuCl_2$ precatalyst. During the last two processes mentioned, there is no change in oxidation state of the metal.

The early studies of the homogeneous catalytic hydrogenation is related to Wilkinson catalyst, $RhCl(PPh_3)_3$ [117]. The catalytic cycle is presented in Fig. 21.92.

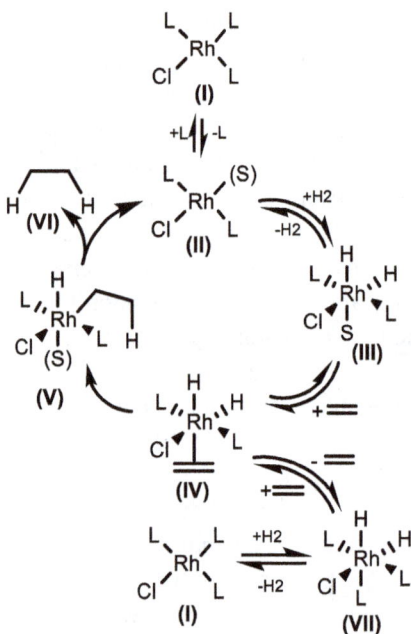

Fig. 21.92: Homogeneous hydrogenation catalytic cycle.

There are two possible catalytically active species starting from the Wilkinson's catalyst, $RhCl(PPh_3)_3$: one resulted by the loss of a PPh_3 ligand followed by coordination of a solvent molecule on the vacated site, $RhCl(S)(PPh_3)_2$ (II) and one formed by oxidative addition of hydrogen to $RhCl(PPh_3)_3$, $RhCl(H)_2(PPh_3)_3$ (VII). The formation of (II) is about 1,000 times faster than the formation of (VII). The next steps for (II) are the oxidative addition of hydrogen (to form (III) and the coordination of the olefin to form (IV). The coordination of the olefin to (VII) results in the formation of the same complex (IV). The migratory insertion affords the alkyl complex (V). The cycle is completed by the reductive elimination of the alkane (VI) and the release of the active catalytic species (II) (Fig. 21.92).

The cationic complexes like [Rh(cod)(PPh$_3$)$_2$]$^+$ [141] or [Ir(cod)(PCy$_3$)(Py)]$^+$ (py = pyridine) – the "super unsaturated" complexes (the Crabtree catalysts) [142] were proved to be more active than Wilkinson's catalyst. Exceptional results were reported in the hydrogenation of internal olefins (e.g., Me$_2$C = CMe$_2$). The Ru catalyst, HRuCl(PPh$_3$)$_3$, [143] is even more active for 1-alkenes.

An important role in the coordination step of the alkenes to the metal is played by the position of the double bond(s) and the substituents therein:

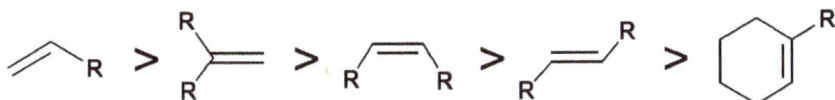

21.5.1 Asymmetric hydrogenation

Asymmetric hydrogenation became an important tool in the synthesis of chiral compounds and was recognized as such when the Nobel Prize in Chemistry was awarded to W. S. Knowles [144] and R. Noyory [145]. The first successful attempt of Knowles was the synthesis of L-DOPA using the chiral ligand DIPAMP (Fig. 21.93) [146], followed by the developments reported for Kagan's ligand DIOP (**A**) [147], Noyori's ligand BINAP (**B**) and the ruthenium complexes (**C**) [148] and a plethora of other chiral ligands (Fig. 21.94) [133].

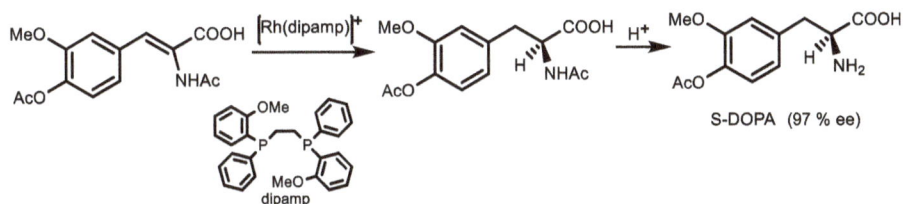

Fig. 21.93: Asymmetric synthesis of L-DOPA.

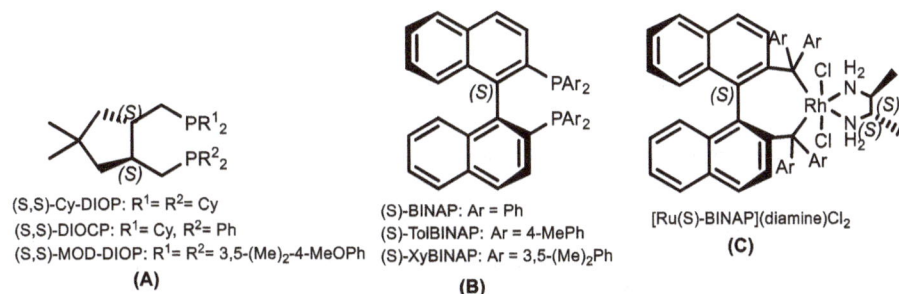

(S,S)-Cy-DIOP: R^1= R^2= Cy
(S,S)-DIOCP: R^1= Cy, R^2= Ph
(S,S)-MOD-DIOP: R^1= R^2= 3,5-(Me)$_2$-4-MeOPh
(**A**)

(S)-BINAP: Ar = Ph
(S)-TolBINAP: Ar = 4-MePh
(S)-XyBINAP: Ar = 3,5-(Me)$_2$Ph
(**B**)

[Ru(S)-BINAP](diamine)Cl$_2$
(**C**)

Fig. 21.94: Chiral ligands active for asymmetric hydrogenation.

There are many homogeneous catalytic asymmetric hydrogenation reactions of substrates of interest in the synthesis of complex organic molecules (pharmaceuticals, natural products, etc.), most providing over 90% enantiomeric excess [149].

The asymmetric hydrogenation of olefins using cationic iridium catalysts with bis(*ortho*-tolyl)phosphino-*tert*-butyloxazoline ligand and its analogues proved to be very efficient (Figs. 21.95–21.97). Weakly coordinated counterions are preferred to avoid the competition with the substrate in the coordination step. Unfunctionalized substrates including tri- or tetra-substituted olefins have been hydrogenated with good-to-excellent enantiomeric excess (*ee*) [150, 151]. The asymmetric hydrogenation of methyl stilbene with two of the iridium catalysts is illustrated in Fig. 21.95.

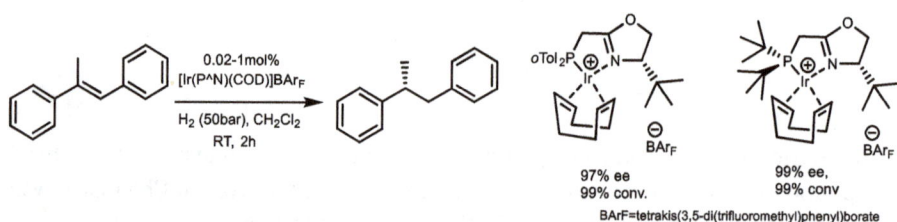

Fig. 21.95: Asymmetric hydrogenation of stilbene.

Two adjacent stereogenic centers can be introduced in a single step by asymmetric hydrogenation of notoriously unreactive substrates such as tetra-substituted unfunctionalized olefins with high efficiency and excellent enantioselectivity using the chiral Ir catalysts (BarF = tetrakis(3,5-di(trifluoromethyl)phenyl)borate, o-Tol = *ortho*-tolyl) (Fig. 21.96) [152].

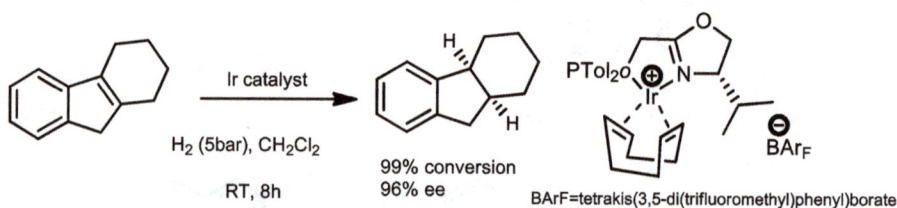

Fig. 21.96: Asymmetric hydrogenation to introduce two stereogenic centers.

High enantioselectivities for the hydrogenation of both (*E*)- and (*Z*)-2-(4-methoxyphenyl)-2-butane was achieved using phosphinite-oxazoline ligands (Fig. 21.97). The hydrogenated compounds result in the opposite configurations with the catalyst [153].

The dehydroamino acids, ketones and imines are hydrogenated successfully resulting, many cases, in chiral intermediates for multistep synthesis of important products, sometimes at industrial scale.

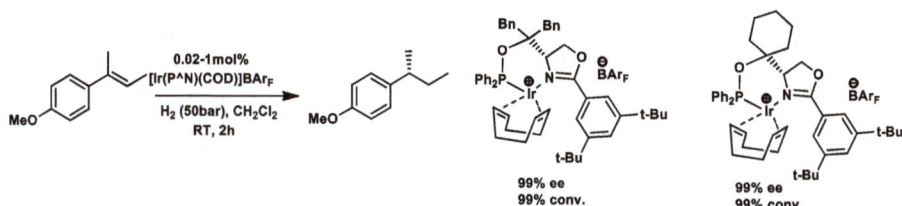

Fig. 21.97: Asymmetric hydrogenation of (*E*)- and (*Z*)-2-(4-methoxyphenyl)-2-butane.

Asymmetric hydrogenation of α- and β-dehydroamino acids was the very first achievements in the field reported by Knowles. The most studied substrates for asymmetric hydrogenation are (*Z*)-2-(acetamido)cinnamic acid, 2-(acetamido)acrylic acids and their methyl esters, frequently used as standards for the evaluation of new catalysts.

Cationic Rh(I) complexes of optically active diphosphine ligands such as CHIRA-PHOS ((2S,3S)-(−)-bis(diphenylphosphino)butane) [154], Et-DuPhos (1,2-bis(phospho-lano)ethanes) and DuPHOS (1,2-bis-(phospholano)benzenes) [155], Ph-BPE (1,2-bis (2,5-phenylphospholano)ethane) [156], TangPhos [157], DIPAMP [158] and BasPhos [159] have achieved very good enantioselectivities (*ee* from 96 to >99.9%) in asymmet-ric hydrogenation of α-dehydroamino acids and their derivatives. The preparation of these and more ligands is reviewed in [160], and their reactions are illustrated further in this chapter.

The mechanism of asymmetric hydrogenation of these substrates implies the bi-dentate coordination of the substrate to the metal through the carbon–carbon dou-ble bond and the amide oxygen. The two diastereoisomeric complexes thus formed react further, usually, with substantially different rates. The oxidative addition of the hydrogen to the Rh(I) complex, the rate-limiting step, is much faster for the iso-mer leading to the desired configuration, and the rapid equilibration is the key to a successful asymmetric hydrogenation.

DuPhos is highly selective for both protected α-dehydroamino acid isomers (Fig. 21.98). The products are formed with the same configuration as in the Rh catalyst used for hydrogenation [155]. With either Et-DuPHOS-Rh or Pr-DuPHOS-Rh catalyst, a variety of α-aminoacid derivatives have been obtained with enantioselectivities >99% ee. Using (*R,R*)-Pr-DuPHOS-Rh) (*R*)-*N*-acetylalanine methyl ester as well as alkyl- (Me, *n*- and *iso*-Pr, Bu), aryl- (Ph, 1- and 2-naphthyl), heteroaryl- (2-thienyl) and ferrocenyl-alanine derivatives were obtained with very high enantiomeric excess (>99%) [155]. The hindered *tert*-butylalanine was hydrogenated with 96.2% *ee*.

The absolute configurations of the hydrogenation products showed that for the methyl-, ethyl- and *n*-propyl-substituted ligands, the (*R,R*)-catalysts provided the R products, while for *iso*-propyl-substituted ligand, the (*R,R*)-catalyst yielded the S product. As expected, the catalysts of opposite absolute configuration (*S,S*) afforded products of opposite absolute configuration and with identical enantiomeric excess.

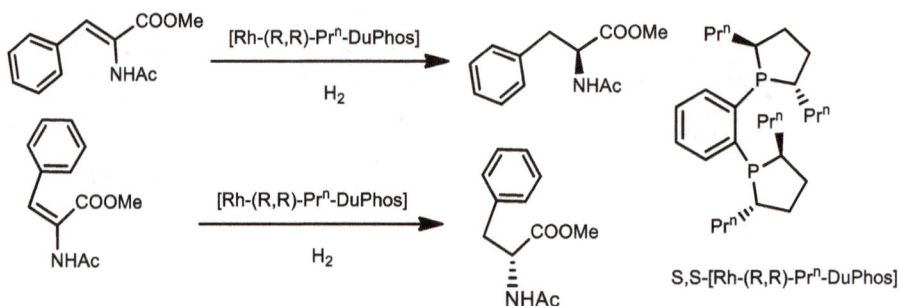

Fig. 21.98: Asymmetric hydrogenation of protected α-dehydroamino acids.

The same relationship holds for BPE ligand. The apparent inconsistency of change in absolute configuration of the products when the *iso*-propyl-substituted ligands are used is a consequence due to the fact that the Cahn–Ingold–Prelog designation changes from *R* to *S* (or vice versa) on moving from the Me-, Et- and *n*-Pr-substituted ligands to the *iso*-Pr-substituted analogue [155].

The cationic BINAP–Ru complexes (BINAP (2,2-bis(diphenylphosphanyl)-1,1– binaphthyl [148]) catalyze hydrogenation of α-(acylamino)-acrylic acids or esters to give the corresponding amino acid derivatives in high *ee* values (Fig. 21.99) [161].

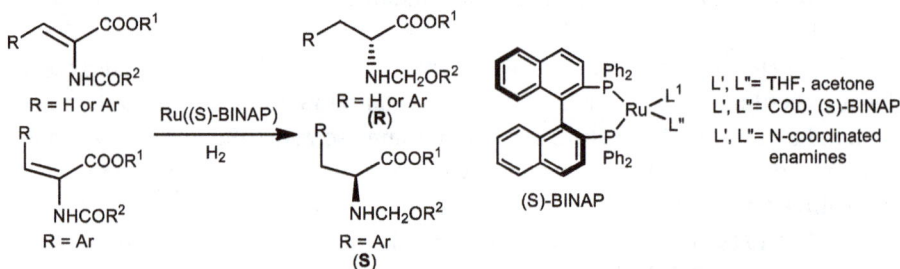

Fig. 21.99: Asymmetric hydrogenation of α-dehydroamino esters.

Asymmetric hydrogenation of β-dehydroamino acids is effective by using a variety of catalytic systems. Since many desirable β-dehydroaminoacids are unavoidably synthesized as a mixture of (*Z*)- and (*E*)-isomers, a separation step is generally required prior to hydrogenation. The Rh complexes of chiral diphosphine ligands BPE and DuPhos [162], BasPhos [159, 163] or Ru complex of BINAP have been reported to effectively catalyze the asymmetric hydrogenation of (*E*)-(β-acylamino)acrylates with high enantioselectivities (Fig. 21.100). Although most of the studies were carried out with relatively high catalyst loadings for screening purposes, an S/C ratio of 10,000 was achieved by Rh-BINAPINE on a series of (*Z*)-3-aryl-3-(acylamino)acrylic acid derivatives [130].

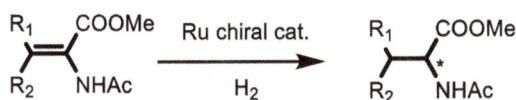

Fig. 21.100: Asymmetric hydrogenation of β-dehydroamino acids.

Excellent enantioselectivities (88–97%) were obtained using [Rh(NBD)$_2$]BF$_4$ and Josiphos ligands in the asymmetric hydrogenation of N-acetyl and N-methoxy-carbonyl β,β-diaryldehydroamino acids (highly congested tetra-substituted olefins) to prepare nonsymmetrically substituted β,β-diaryl α-amino acid esters, independent of electronic effects of the substituents on the phenyl ring. The hydrogenation works equally well with E- or Z-alkenes (Fig. 21.101). Along with substituted phenyl rings, benzofurans, azaindoles and pyridines are tolerated affording high enantioselectivities. Methoxycarbonyl-protected substrates also gave good results. In all cases, no diastereomeric products were observed, indicating that there was no isomerization of the olefin during the hydrogenation [164].

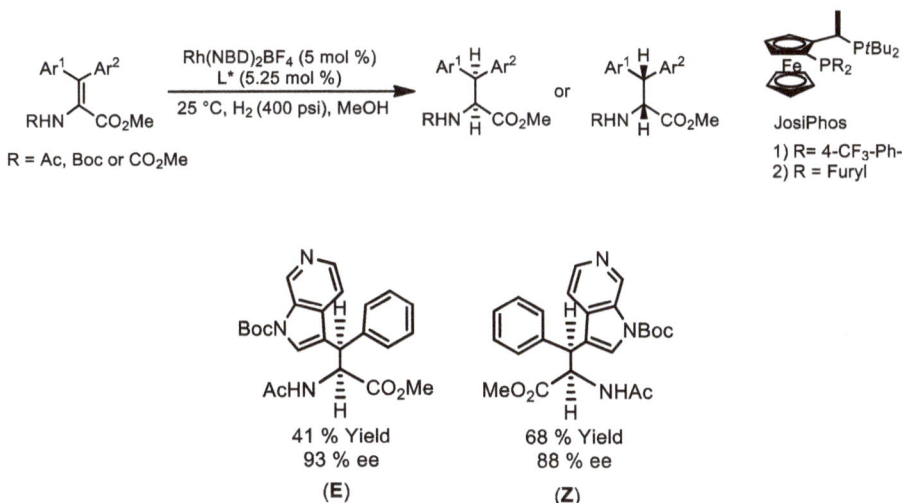

Fig. 21.101: Asymmetric hydrogenation of β,β-diaryl α-amino acid esters.

Unprotected (Z)-β-dehydroamino acid derivatives, challenging substrates for homogeneous hydrogenation, are effectively reacting (up to 96.1% ee) under catalytic action of Rh-JosiPhos complex in trifluoroethanol (Fig. 21.102) [165].

The mechanistic studies revealed that the hydrogenation went through the imine tautomer of the dehydroamino acid ester or amide (Fig. 21.102 (**A**)). These results show that the directing N-acyl group are not necessarily a prerequisite for asymmetric hydrogenation of the substrates, making this reaction mechanistically analogous to the hydrogenation of β-ketoesters and -amides.

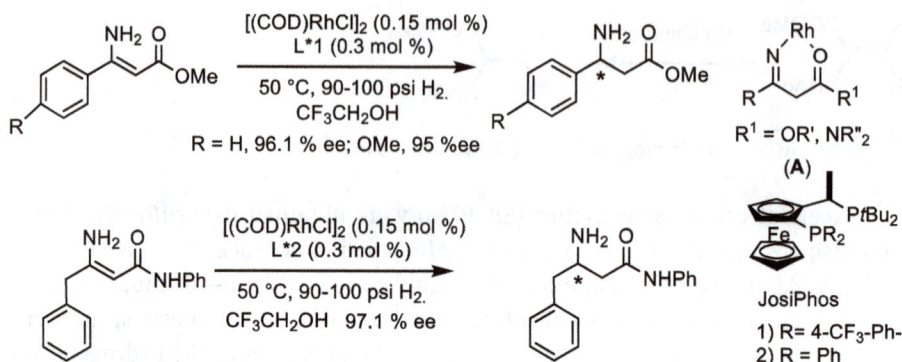

Fig. 21.102: Asymmetric hydrogenation of (Z)-β-dehydroamino acid.

Asymmetric hydrogenation of enamides affords chiral amines. Among many systems, Rh-BPE and Rh-TangPhos complexes ((1S,1S,2R,2R)-TangPhos (1S,1S,2R,2R)-1, 1-di-*tert*-butyl-[2,2]-diphospholanyl) form chiral amines with excellent enantioselectivities (*ee* > 99%) (Fig. 21.103). The β-substituted substrates, *E/Z* mixtures, are also reduced with high ee values regardless of the various β-substituents or the electronic properties of the 1-aryl group with Rh-TangPhos complexes in the presence of [Rh(nbd)]SbF$_6$ (nbd = norbornadiene) [157].

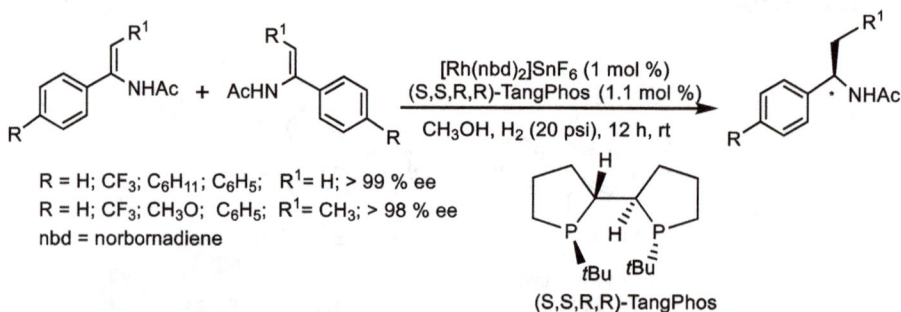

Fig. 21.103: TangPhos asymmetric hydrogenation of enamides.

Excellent enantioselectivities (over 99% ee) are obtained in the asymmetric hydrogenation of enamides using air-stable, highly unsymmetrical ferrocene-based phosphine-phosphoramidite ligands and rhodium complexes. Enantioselectivity was achieved using the ligand L* (Fig. 21.104) having (Sc)-central, (Rp)-planar and (Sa)-axial absolute configurations. The binaphthyl moiety is crucial for reactivity and enantioselectivity, and its absolute configuration plays a dominant role in determining the chirality of the hydrogenation products [166].

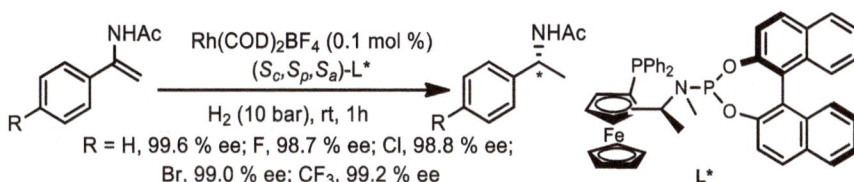

Fig. 21.104: Asymmetric hydrogenation of enamides.

Cyclic enamides can be hydrogenated with various Rh-L* complexes. A ligand that is very effective is the *ortho*-substituted BIPHEP ligand, o-Ph-hexaMeO-BIPHEP (3,3'-diphenyl-4,4',5,5',6,6'-hexamethoxybiphenyl-2,2'-diyl)bis(diphenylphosphine) (Fig. 21.105).

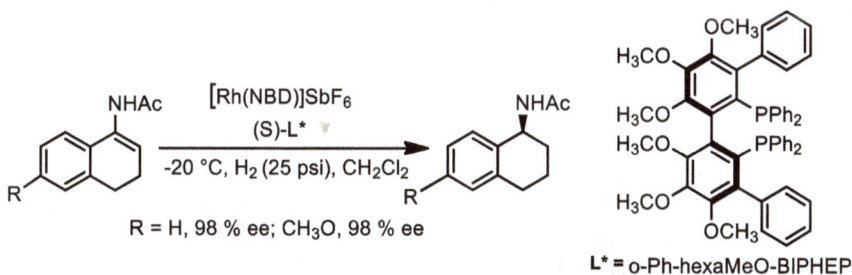

Fig. 21.105: Selective asymmetric hydrogenation of cyclic enamides.

The introduction of phenyl groups at 3,3'-positions have a strong influence on the conformation of P-aryl rings and a high enantioselectivity is achieved [167].

A useful method to prepare chiral alcohols is through asymmetric hydrogenation of the corresponding unsaturated enol esters. The ester-protected unsaturated alcohols have similar structures as the enamides but are less reactive due to the weaker coordinating ability of the ester groups compared to that of the amide in enamides. A few catalysts that have shown superior performance for the hydrogenation of enamides have also provided good selectivities and activities for unsaturated alcohols.

The asymmetric hydrogenation of β-branched enol esters – a new route for the synthesis of β-chiral primary alcohols – has been developed using a rhodium complex of ligand with a large bite angle. The enol ester substrates possessing an O-formyl directing group (Fig. 21.106) afforded the alcohols in quantitative yields and with excellent enantioselectivities [168].

The size of the alkyl chain has no influence on the reactivity and selectivity. The stereoselectivity is predominantly controlled by the chelating direction of the enol ester to the bisphosphine– rhodium–dihydride complex and not because of the difference between the two β-substituents. The (E)- and (Z)-isomers generate

R² = C₂H₅; R₁ = H, 95% yield, 97 % ee   R² = i-C₃H₇; R₁ = H, 94% yield, 98 % ee
R² = CH3; R¹ = H, 96 % yield, 97 % ee; CH₃, 96 % yield, 94 % ee; CF₃, 94 % yield, 96 % ee; Cl, 96 % yield, 97. 5 ee

Fig. 21.106: Preparation of β-chiral primary alcohols by asymmetric hydrogenation of enol esters.

products with the opposite enantioselectivity as in the case of the asymmetric hydrogenation of enamides [168].

The asymmetric hydrogenation of methoxymethylether-protected β-hydroxyl enamides using pure [Rh(DIOP*)(NBD)]SbF₆ as catalyst precursor afforded the chiral aminoalcohol quantitatively and with high enantioselectivity (98.1–99.6% ee) under very mild conditions within 1 h (Fig. 21.107).

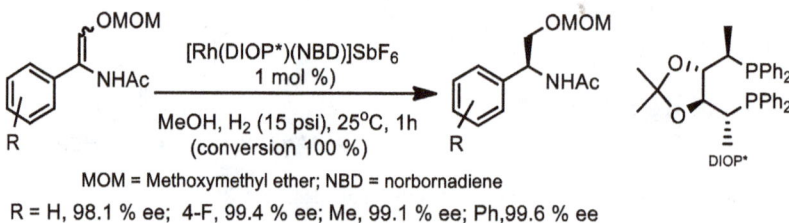

MOM = Methoxymethyl ether; NBD = norbornadiene
R = H, 98.1 % ee; 4-F, 99.4 % ee; Me, 99.1 % ee; Ph,99.6 % ee

Fig. 21.107: Preparation of aminoalcohols by asymmetric hydrogenation of β-hydroxyl enamides.

The conformational rigidity of the chiral ligand is responsible for both the reactivity and the enantioselectivity. The two coordinating sites can form a more rigid metallocycle with fewer available conformations, and thus enhance the enantioface differentiation leading to a chiral product with higher ee value.

Homoallylic alcohols can be hydrogenated with high ee. Early results were obtained using the Ru-BINAP catalysts to hydrogenate pure geraniol ((2E)-3,7-dimethyl-2,6-octadien-1-ol) and nerol ((Z)-3,7-dimethyl-2,6-octadien-1-ol) (Fig. 21.108).

The synthesis of (3R,7R)-3,7,11-trimethyldodecanol, an intermediate for the synthesis of α-tocopherol, was achieved using the same catalytic system. When racemic allylic alcohols are subjected to asymmetric hydrogenation, highly efficient kinetic resolution is achieved with a BINAP–Ru complex as the catalyst [169].

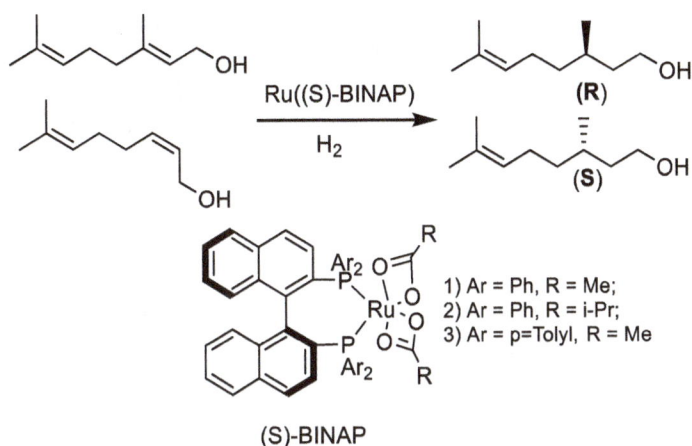

Fig. 21.108: Asymmetric hydrogenation of geraniol and nerol using Ru-BINAP catalysts.

21.6 Carbonylation of methanol

The preparation of acetic acid was developed by using various feedstocks – acetalde-hyde oxidation, hydrocarbon oxidation, direct oxidation of ethylene, ethane oxidation, direct syngas conversion and methanol carbonylation [170, 171]. The carbonylation of methanol in homogeneous catalysis with either Rh-based (developed as Monsanto pro-cess, 1970) or Ir-based (developed as Cativa process, 1996) catalytic systems (first men-tioned in 1968 [172]) became commercial production methods. The Monsanto process replaced the Wacker process (based on olefin oxidation) [173]. Acetica processes are a heterogeneous methanol carbonylation based on a supported rhodium catalyst [174]. Besides the catalysts used in the three mentioned processes, there are systems built using various ligands in search for milder reaction conditions [175].

Basically, the methanol (generated from the synthesis gas, Fig. 21.109 (**I**)) re-acted with carbon monoxide in the presence of a catalyst (carbonylation of metha-nol) to afford acetic acid (Fig. 21.109 (**II**)).

Fig. 21.109: Preparation of methanol (I) and carbonylation of methanol to acetic acid (II).

The catalyst system in Monsanto process have two components: iodide (the pro-moter), and the transition metal, rhodium or iridium. The role of iodide is simply to promote the conversion of methanol into the better electrophile, methyl iodide, the species which then undergoes oxidative addition with the transition metal catalyst.

The catalytic cycle in the Monsanto process (Fig. 21.110) begins with oxidative addition of methyl iodide to the 16-electron $[Rh(CO)_2I_2]^-$ complex (almost any source of Rh and I^- will work in this reaction as they will be converted to the actual catalyst, $[Rh(CO)_2I_2]^-$, under the reaction conditions). Coordination and insertion of carbon monoxide lead to a 18-electron acyl intermediate. The next step, reductive elimination, yields acetyl iodide and regenerates the active catalytic species.

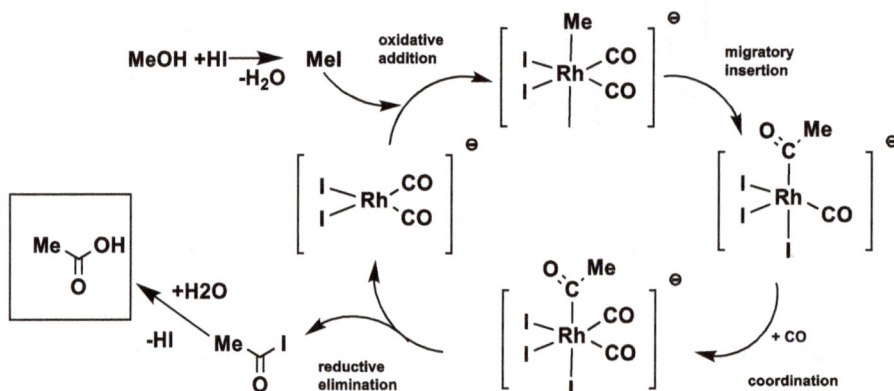

Fig. 21.110: The catalytic cycle in the Monsanto process of methanol carbonylation.

There are two catalytic cycles interconnected, one involving the transition metal and one involving the iodide. The acetyl iodide produced in the transition metal catalytic cycle is then hydrolyzed in the iodide one to give acetic acid. This hydrolysis produces HI which can then convert methanol to iodide and continue the cycle. At 180 °C, 30–40 bars and 10^{-3} M, the conversion of the catalyst up to 99% selectivity (on methanol) is obtained. It was demonstrated that concentrations of the reactants and products have no kinetic influence. The extremely facile formation of an acetyl complex observed during the process and the finding that oxidative addition is the rate-determining step are related to the high selectivity mentioned earlier [173].

The Monsanto process was further improved during the 1980s by adding a lithium or sodium iodide promoter to enable operation in a reduced water environment to improve raw material conversion and lower downstream separation costs.

The mechanism of iridium-catalyzed carbonylation of methanol (Cativa process) is more complex [176–178]. A range of compounds can enhance the activity of the iridium catalyst: carbonyl or halocarbonyl complexes of W, Re, Ru and Os or simple iodides of Zn, Cd, Hg, Ga and In [177, 179]. The iridium catalyst is more stable than the rhodium one, behavior associated with the stronger metal–ligand bonding which inhibits CO loss.

The iridium-based cycle is similar to the rhodium-based cycle but operates with different kinetics, responsible for the advantages of the Cativa over the Monsanto process. The oxidative addition of methyl iodide to the metal center – the rate-

determining step in case of rhodium – is about 150 times faster for the iridium catalyst [176]. The rate of iridium-catalyzed carbonylation displays a rather complicated
dependence on a range of process variables such as pCO, [MeI], [MeOAc] and [H$_2$O]
(maximum activity is achieved at ca. 5% w/w H$_2$O). The catalytic rate displays a
strong positive dependence on [MeOAc] but is zero order in [MeI] above a limiting
threshold and independent of CO partial pressure above ca. 10 bar [176, 179, 180].

The heterogeneous catalyst commercialized for the Acetica TM process is based
on heterogenized homogeneous catalyst, rhodium complex immobilized on polyvinyl
pyridine resin, compatible with elevated temperature and pressure. Under reaction
conditions, the Rh is converted to its catalytically active anion form [Rh(CO)$_2$I$_2$]$^-$ already described for the Monsanto process. The terminal nitrogen atoms of the resin
pyridine groups become positively charged after quaternization with methyl iodide
and efficiently incorporate the rhodium catalytic anionic complex by strong electrostatic interactions [174].

21.7 Metathesis reactions

The olefin metathesis reaction was developed as an important method in organic
synthesis, as mentioned in 2005 when the Nobel Prize in Chemistry was awarded to
Yves Chauvin [181], Robert H. Grubbs [182] and Richard R. Schrock [183]. During the
metathesis reaction, the carbon–carbon double bonds in an olefin (alkene) are cut
and then rearranged in a statistical fashion (Fig. 21.111).

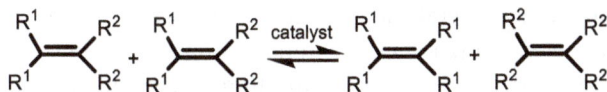

Fig. 21.111: Metathesis reactions.

The equilibrium is shifted to the right if one of the products, alkenes, is volatile
(ethylene is the best example) or easily removed. The role of ethylene is, however,
more important in the metathesis reactions (i.e., using a high pressure of ethylene
internal olefins can be converted to terminal olefins [184]. The concept was applied
for a wide variety of reactions:
Cross-metathesis (CM):

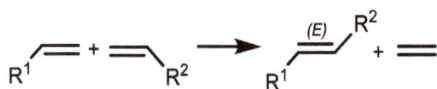

Ring-closing metathesis (RCM): \

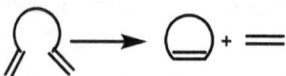

Ring-opening CM:

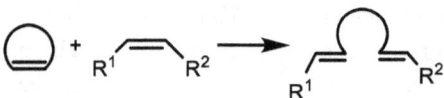

Ring-opening metathesis polymerization (ROMP):

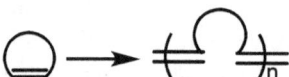

Acyclic diene metathesis polymerization (ADMEP):

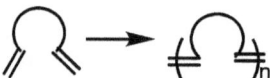

-Enyne metathesis:

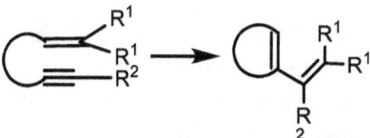

The mechanism of the olefin metathesis reaction was first proposed by Chauvin [185]. In time, the mechanism was experimentally supported and widely accepted (Fig. 21.112). The catalyst containing a metal–carbon double bond (a metal alkylidene complex) reacts with the olefin (a [2 + 2] cycloaddition) forming a metallacyclobutane which breaks up in the opposite fashion to afford a new alkylidene complex and the new olefin. The new olefin contains one methylene from the catalyst and one from the starting olefin, and the new metal alkylidene contains the alkylidene from the substrate olefin. The new metal alkylidene reacts with a new molecule of the substrate olefin to yield another metallacyclobutane intermediate. The cleavage of the metallacyclobutane intermediate in the forward direction yields the internal alkene and the metal alkylidene is ready to enter another catalytic cycle. Thus, each step in the catalytic cycle involves exchange of alkylidenes – metathesis.

Metathesis is a reversible reaction but, in this case, removal of ethene drives the reaction to completion.

The cycloaddition reactions of two alkenes to give cyclobutanes is symmetry forbidden and occurs only photochemically but, in this case, the presence of d-orbitals on the metal alkylidene fragment breaks this symmetry and favor the reaction.

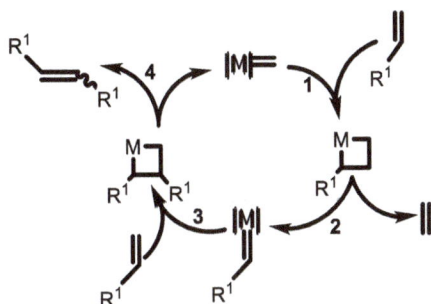

Fig. 21.112: The mechanism of the olefin metathesis.

21.7.1 Cross-metathesis (CM)

The first catalysts active for metathesis reactions were high-valent transition metal halide, oxide or oxo-halide with an alkylating cocatalyst such as an alkyl zinc or alkyl aluminum placed on an alumina or silica support, like $WCl_6/SnMe_4$ or Re_2O_7/Al_2O_3. Their use is limited due to their low tolerance for functional groups.

The formation of a metalacyclobutane during the metathesis reaction was supported by the experiment using Tebbe's reagent, $Cp_2Ti(\mu\text{-}Cl)(\mu\text{-}CH_2)AlMe_2$. Tebbe's reagent forms $Cp_2Ti = CH_2$, which undergoes stoichiometric Wittig-like reactions with ketones, aldehydes and other carbonyls, in the presence of strong bases, leading to the corresponding methylene derivatives but, in addition, exhibits metathesis activity. The mechanism of this reaction is identical to that of the olefin metathesis reaction except that the final step is not reversible. Grubb's studies showed the formation of titanacyclobutane, the important intermediate in olefin metathesis (Fig. 21.113) [186].

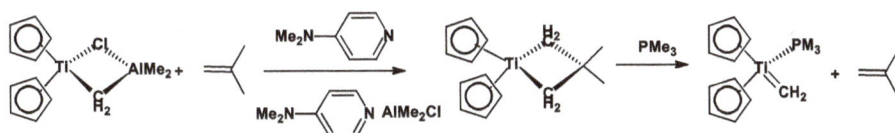

Fig. 21.113: Generation of titanacyclobutane as intermediate in olefin metathesis.

The development of well-defined catalysts was the key to the impressive contribution of olefin metathesis in organic chemistry. Some of the catalysts (both Schrock and Grubbs) became commercially available and the uses of olefin metathesis in the synthesis of complex structures are impressive. Several reviews on the topic are describing the synthesis of catalytic systems and their application in the metathesis reactions [184, 187–193].

Schrock's alkylidene complexes proved to be excellent catalysts for olefin metathesis (Fig. 21.114). Very good results were obtained using arylimido complexes of tungsten and molybdenum (Ar′N)(RO)$_2$Mo = CHR′ where Ar′ is typically 2,6-diisopropylphenyl, R′ can be virtually anything and R is neopentyl or neophyl (CMe$_2$Ph) [189].

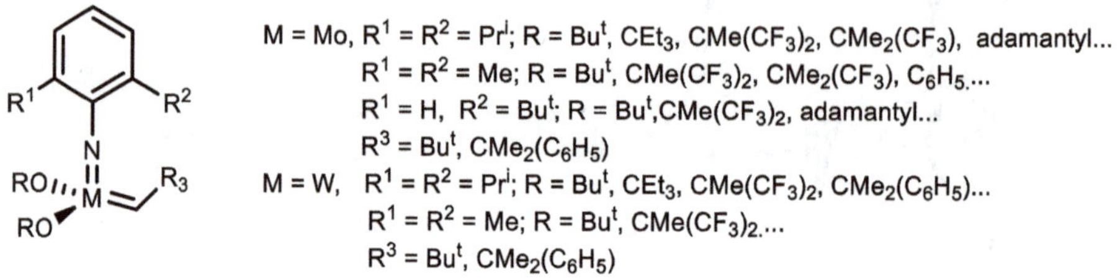

M = Mo, R^1 = R^2 = Pr^i; R = Bu^t, CEt$_3$, CMe(CF$_3$)$_2$, CMe$_2$(CF$_3$), adamantyl...
R^1 = R^2 = Me; R = Bu^t, CMe(CF$_3$)$_2$, CMe$_2$(CF$_3$), C$_6$H$_5$...
R^1 = H, R^2 = Bu^t; R = Bu^t,CMe(CF$_3$)$_2$, adamantyl...
R^3 = Bu^t, CMe$_2$(C$_6$H$_5$)
M = W, R^1 = R^2 = Pr^i; R = Bu^t, CEt$_3$, CMe(CF$_3$)$_2$, CMe$_2$(C$_6$H$_5$)...
R^1 = R^2 = Me; R = Bu^t, CMe(CF$_3$)$_2$...
R^3 = Bu^t, CMe$_2$(C$_6$H$_5$)

Fig. 21.114: Schrock's alkylidene complexes.

These catalysts are exceedingly active, metathesizing over 1,000 equivalents of cis-2-pentene to equilibrium in less than 1 min for R = CMe(CF$_3$)$_2$. The reactivity of these catalysts can be tuned very easily by changing the nature of the alkoxide ligands. For example, when R = tert-butyl, the complex reacts only with strained cyclic olefins, making it an ideal ROMP catalyst. Although these catalysts are air and water sensitive, they have a high tolerance for functionality and are 100% active. The source of their success comes from their coordinative and electronic unsaturation (making them electrophilic) and their bulky ligands (prevents bimolecular decomposition).

The results are depending on the electronic properties of the olefin: better yields are obtained for the electron-rich alkenes (Fig. 21.115) [194].

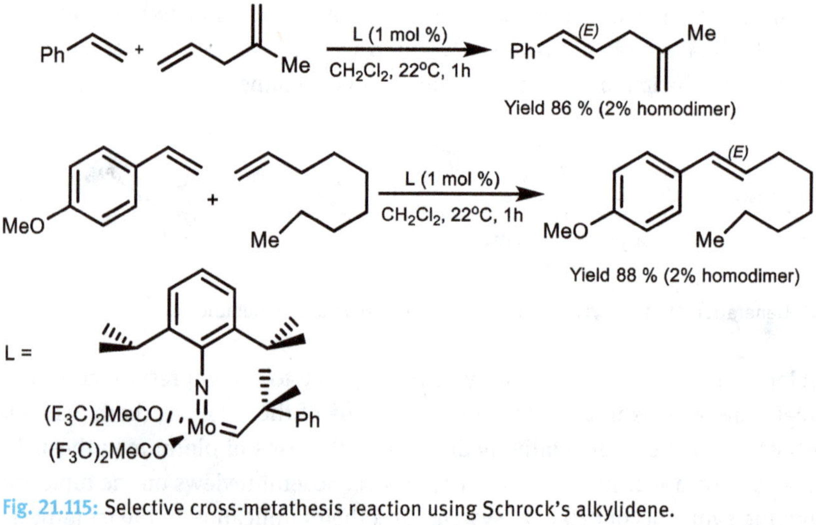

Yield 86 % (2% homodimer)

Yield 88 % (2% homodimer)

Fig. 21.115: Selective cross-metathesis reaction using Schrock's alkylidene.

Grubbs developed a series of Ru catalysts with high tolerance to air and water, exploiting the preference of ruthenium for soft Lewis bases and π-acids, such as olefins, over hard bases, such as oxygen-based ligands (Fig. 21.116). In the ruthenium catalysts, the metal is not in its highest oxidation state as in the case of the titanium-based or Schrock's alkylidene catalysts [193].

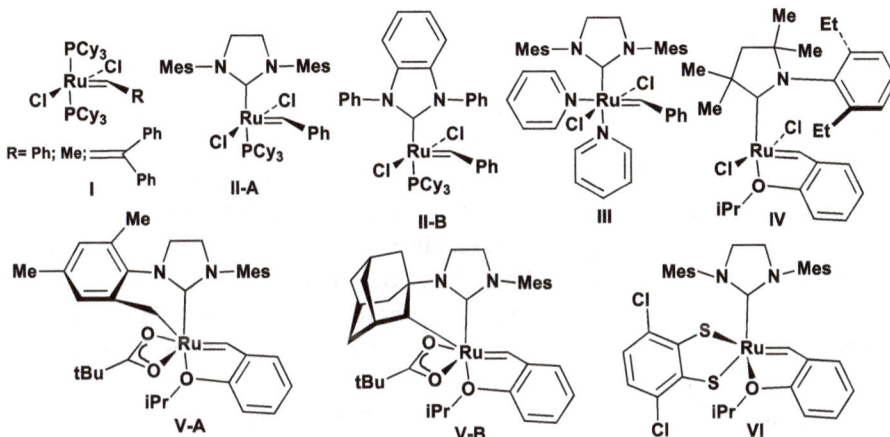

Fig. 21.116: Grubbs' catalysts.

For the first generation of Grubbs' catalysts the phosphine ligands proved to be the choice for high catalytic activity [195].

In the first two generations of Grubbs' catalytic precursors, the ruthenium is in a five-coordinated environment. In the next generations of Grubbs' catalysts, the replacement of one phosphine ligand by an NHC ligand resulted in a series of highly active catalysts with enhanced catalytic initiation rate, turnover number, stereoselectivity or lifetime. The favorable electron donation and steric bulk of the NHC ligand is the source of the increase in metathesis activity and was related with an increased rate of catalyst turnover. The replacement of another phosphine ligand with a weaker ligand, such as pyridine, resulted in an increase in activity up to a factor of 10^4 relative to catalyst.

The active catalytic species are four-coordinated and are formed by dissociation of one of the neutral ligands. This is the reason why the complexes of bulky tricyclohexylphosphine ligands are more effective than those containing triphenylphosphine ligands in the ruthenium-catalyzed metathesis of cyclooctene. The remaining neutral ligand of the resulting 14-electron complex is responsible for catalyst turnover.

The simple representation of CM reactions is presented in Fig. 21.117.

The homodimerization in CM of an olefin (Fig. 21.117 (**A**)) has its significance for some processes, but, when two olefins are reacted (Fig. 21.117 (**B**)), the expected product is (a), while (b) and (c) are the results of homodimerization secondary reactions.

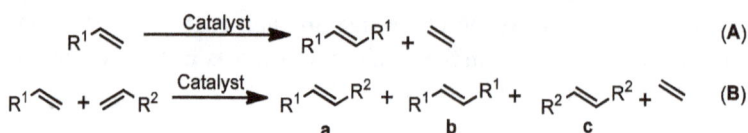

Fig. 21.117: Cross-metathesis reactions.

CM became useful for organic synthesis after a general model for the prediction of product selectivity, and stereoselectivity was elaborated based on investigation of the reactivity of several classes of olefins, including substituted and functionalized styrenes, secondary allylic alcohols, tertiary allylic alcohols and olefins with α-quaternary centers in the presence of a given catalyst. According to the relative abilities of the olefins to undergo homodimerization via CM and the susceptibility of their homodimers toward secondary metathesis reactions, the olefins were included in one of the four types: Type I – very reactive olefins toward homodimerization and whose homodimers can participate in CM as well as with their terminal olefin counterpart (i.e., terminal olefins, allyl silanes, irrespective the catalyst); Type II – slow homodimerization with homodimers sparingly reactive in subsequent metathesis reactions (allylic alcohols, ortho-substituted styrene with first- and second-generation Grubbs' catalysts); Type III – no homodimerization but reactive with Type I and II olefins (1,1-dissubstituded olefins, trisubstituted olefins with non-bulky substituents catalyzed by second-generation Grubbs' catalyst); Type IV – olefins inert to CM but not deactivate the catalyst (spectator) (protected trisubstituted allyl alcohols or vinyl nitro olefins) in the presence of second-generation Grubbs' catalysts or 1,1-disubstituted olefins in the presence of first-generation Grubbs' catalyst (Fig. 21.114), or Schrock catalyst (Fig. 21.114). There are also olefins that deactivate the catalyst. Steric and electronic effects can influence the behavior of an olefin and change from one type to another.

The homodimerization of α,β-unsaturated carbonyl compounds by a metathesis mechanism afforded only the E-isomer (Fig. 21.118) [196].

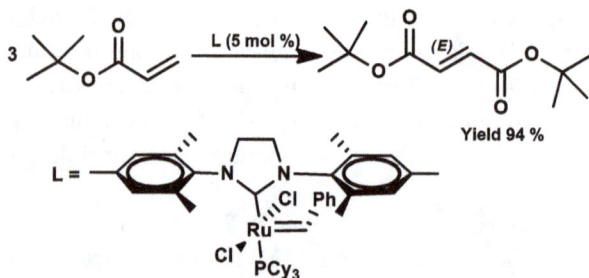

Fig. 21.118: Selective homodimerization of α,β-unsaturated carbonyl compounds.

An important reaction for the success of the CM is the secondary metathesis, the reaction of a product olefin with the propagating catalyst. The secondary metathesis is the way to selective CM reaction. The accessibility of all products of reaction **B** (Fig. 21.117), to the catalyst, including the homodimers is the key to an efficient secondary metathesis. It is also important that the desired cross-product is not redistributed into a statistical product mixture by these same secondary metathesis events [197].

Nonselective CM occurs when two Type I olefins are reacting with similar rates of homodimerization, and the reactivities of the homodimers and cross-products toward secondary metathesis events are high. In these reactions, the desired cross-product will be equilibrated with the various homodimers through secondary metathesis reactions resulting in a statistical product mixture.

The reaction between allylbenzene and protected allylic alcohols (1:2 equivalent) using ruthenium catalysts is an example of nonselective CM between two Type I olefins (Fig. 21.120) [193].

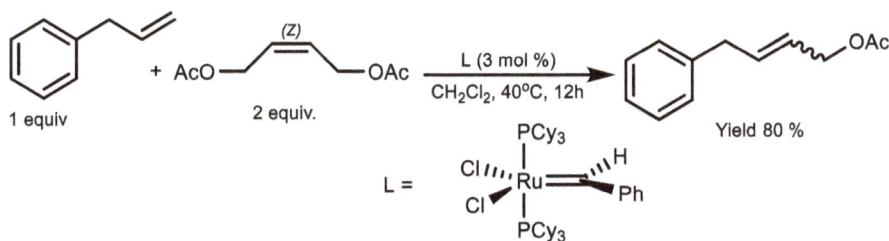

Fig. 21.120: Nonselective cross-metathesis between two Type I olefins.

In other cases, an excess of nearly 10 equiv of one CM partner is needed to provide 90% of the CM product. Nonselective product mixtures are usually obtained even in the CM reaction of two Type II olefins reacting with each other, although with lower yield [196, 197].

The selective CM can be achieved in the reaction of two olefins from two different types, with significantly different rates of dimerization and/or slower than CM the product formation. It is the case of the reaction of a Type I olefin with a less reactive Type II or Type III olefin (the last two undergoes homodimerization at a significantly lower rate or not at all). The product distribution, including the homodimers formed by Type I olefin, is driven toward the desired cross-product as ethylene is removed from the system preventing the regeneration of terminal olefins. Type I homodimer readily undergoes secondary metathesis with Type II/III olefins.

Selective CM of Type I terminal olefins with Type II olefins such as α,β-unsaturated carbonyl olefins, including acrylates, acrylamides, acrylic acid and vinyl ketones, result in highly selective CM reactions with high ($E/Z > 20{:}1$) stereoselectivity (Fig. 21.121) [198].

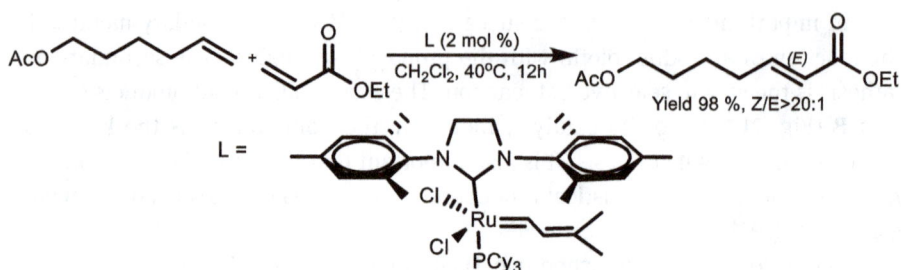

Fig. 21.121: Ruthenium complex-catalyzed selective cross-metathesis of Type I terminal olefins with Type II olefins.

Secondary allylic alcohols, Type II olefins, can be utilized in CM with moderate-to-high cross-product yields and good stereoselectivity [199].

Grubbs' catalysts for CM homodimerization of terminal olefins were proved to be effective to prepare Z-isomers. The removal of the ethylene from the reaction mixture prevents the decomposition of the catalyst (Fig. 21.122). Substrates like allylbenzene, methyl undecenoate, allyl acetate, 1-hexene, allyl trimethylsilane, 1-octene and allyl pinacol borane are reacting with very good yields and high selectivities [200].

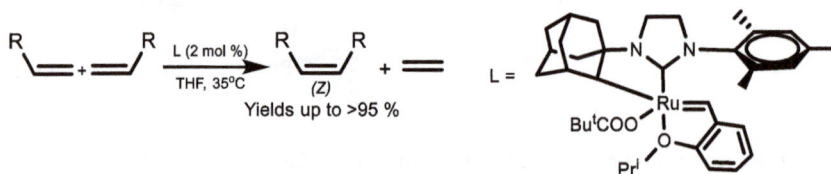

Fig. 21.122: CM homodimerization of terminal olefins.

Cyclometalated ruthenium complexes with bulky NHC ligands are efficient catalysts for Z-selective CM between acrylamides and common terminal olefins. The kinetic preference for CM is related to the presence of the pivalate anionic ligand (Fig. 21.123) [201].

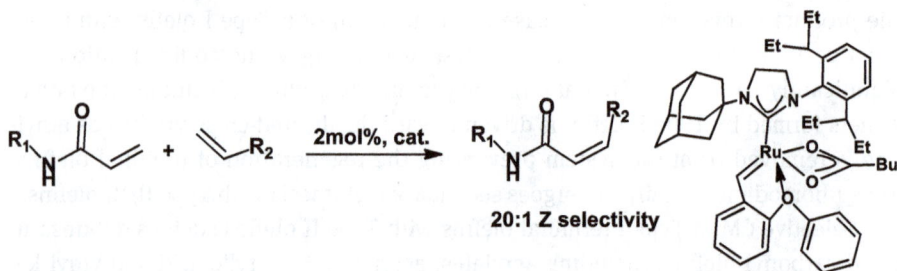

Fig. 21.123: Cyclometalated ruthenium complexes active for Z-selective cross-metathesis.

High *E*-selective (>98%) olefin CM reactions kinetically controlled occur between two *trans*-olefins and between a *trans*-olefin and a terminal olefin using dithiolate-ligated ruthenium complexes Fig. 21.124 [202].

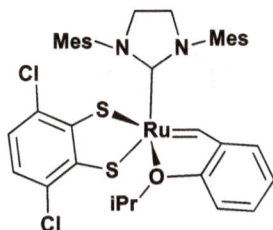

Fig. 21.124: Dithiolate-ligated ruthenium complexes active in olefin cross-metathesis.

Steric bulk in the allylic position, as well as alkyl substitution directly on the double bond, greatly reduces the rate of homodimerization and such olefins are classified as Type 2 or 3. For example, the ketal of methylvinylketone gives a near-quantitative yield of the cross-product, Fig. 21.125. Steric bulk also favors the *E*-isomer [190].

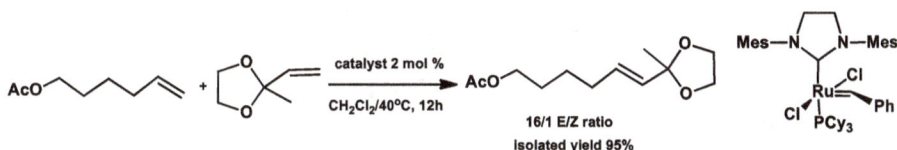

Fig. 21.125: Cross-metathesis of the methylvinylketone ketal with 6-acetyl-1-hexene.

21.7.2 Ring-closing metathesis (RCM)

RCM is a method to prepare rings without appreciable strain. The RCM reaction involves equilibria; therefore, running the experiment are usually conducted at low dilution. Most of the reactions are intra- rather than intermolecular. Removal of the volatile by-product drives the equilibrium to the ring-closed product. The catalysts are selected to have good reactivity with terminal olefins and low reactivity with internal ones.

The preparation of heterocycles is possible using both molybdenum or ruthenium catalysts. Early experiments afforded various functionalized heterocycles with very good yields (Fig. 21.126) [203].

Seven- and eight-membered heterocycles, important intermediates for the synthesis of medicinally significant and structurally complex molecules, are accessible through [2,6-Pr$^i_2$ C$_6$H$_5$NMo{OC(CF$_3$C$_6$)$_2$Me}$_2$CHCMe$_2$Ph] or the Grubbs' catalyst [Cl$_2$ {(c-C$_6$ H$_{11}$)$_3$ P}$_2$RuCHPh]-catalyzed RCM (Fig. 21.127). The efficiency of RCM does vary depending on the substitution pattern of the substrate alkene and enol ether moieties [204, 205].

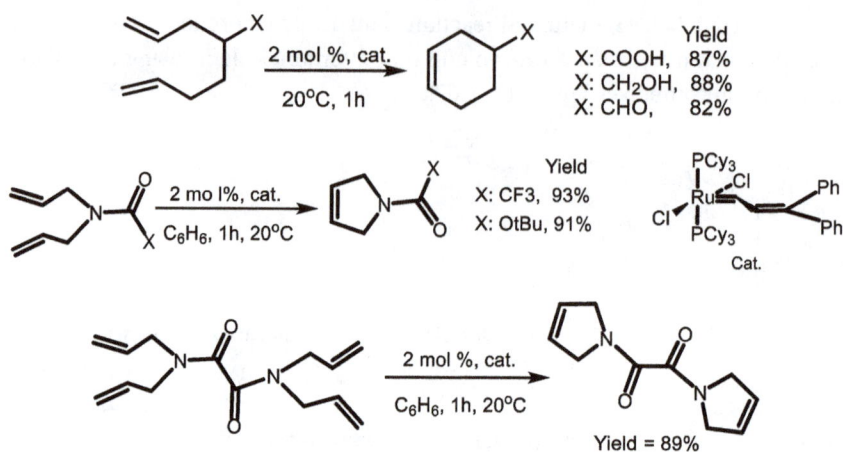

Fig. 21.126: Ruthenium complexes catalyzed the preparation of heterocycles.

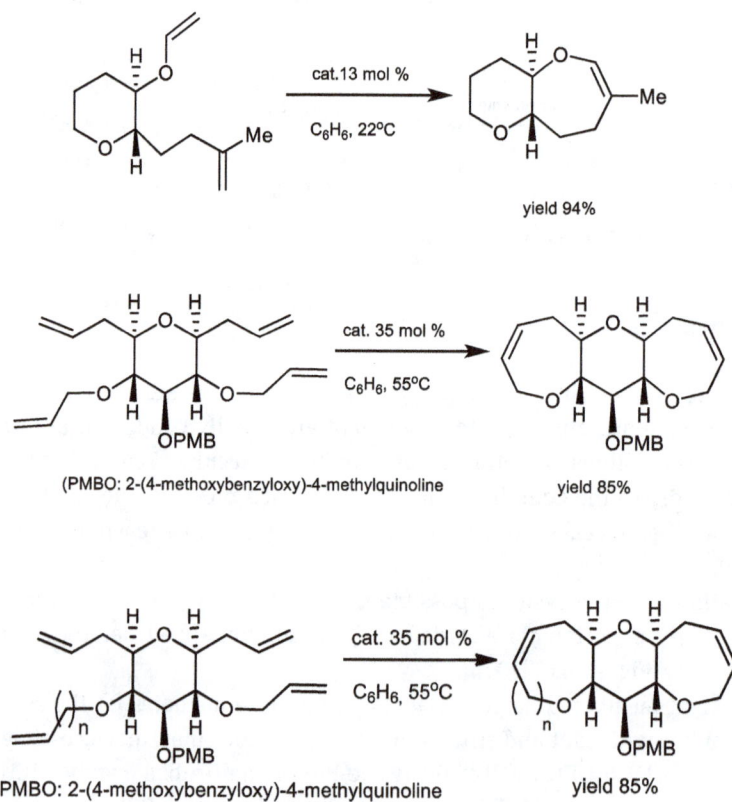

yield 94%

(PMBO: 2-(4-methoxybenzyloxy)-4-methylquinoline

yield 85%

PMBO: 2-(4-methoxybenzyloxy)-4-methylquinoline

yield 85%

Fig. 21.127: Seven- and eight-membered heterocycles prepared by RCM.

Cyclic tetra-substituted systems can be prepared using ruthenium-catalyzed RCM (Fig. 21.128). A wide range of functionalities such as nitrogen-, oxygen-, sulfur-, silicon- and carbon-tethered groups, as well as very challenging fluorine and boron atoms are tolerated. The heterocycles thus prepared are important intermediates for the synthesis of compounds containing morpholine moiety [206].

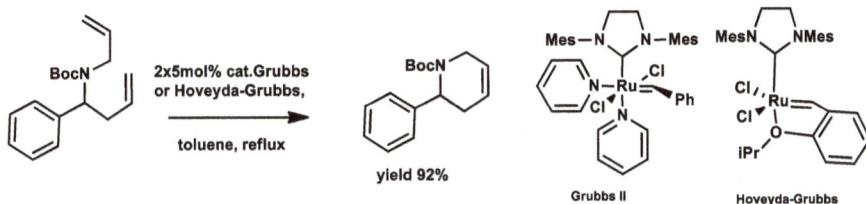

Fig. 21.128: Ruthenium-catalyzed ring-closing metathesis.

Examples of enantioselective catalysis using a chiral metathesis catalyst are reported for both ruthenium and high-oxidation-state alkylidene complexes. The ruthenium RCM is illustrated in Fig. 21.129 [207].

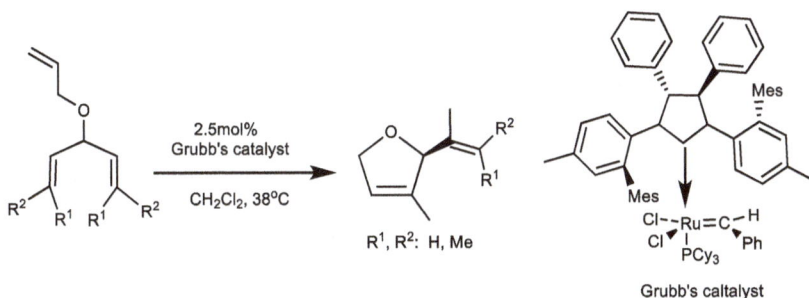

Fig. 21.129: Ruthenium complex-catalyzed ring-closing enantioselective metathesis.

Asymmetric RCM using high-oxidation molybdenum alkylidene can be conducted with solvent (Fig. 21.130 (**1**)) [208] or without a solvent, in green chemistry conditions (Fig. 21.130 (**2**)) [191].

A combination of transformations of ring-opening/CM affords new unsaturated systems. An important application is the preparation of chiral compounds with a good control of stereochemistry of the new double bonds. Both chiral ruthenium and molybdenum catalysts promote highly selective asymmetric ring-opening metathesis/CM. The choice of the catalyst is in close relation with the structure of the substrates. The reaction of oxabicyclic and azabicyclic substrates with styrene (Fig. 21.131) was used to exemplify the way different catalysts are acting for a given substrate [209, 210].

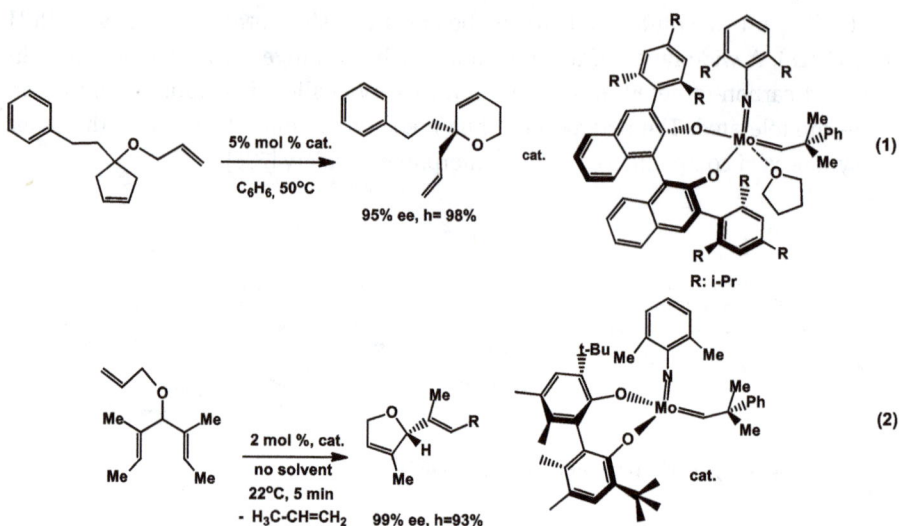

Fig. 21.130: Asymmetric ring-closing metathesis run in solvent (1) or without a solvent (2).

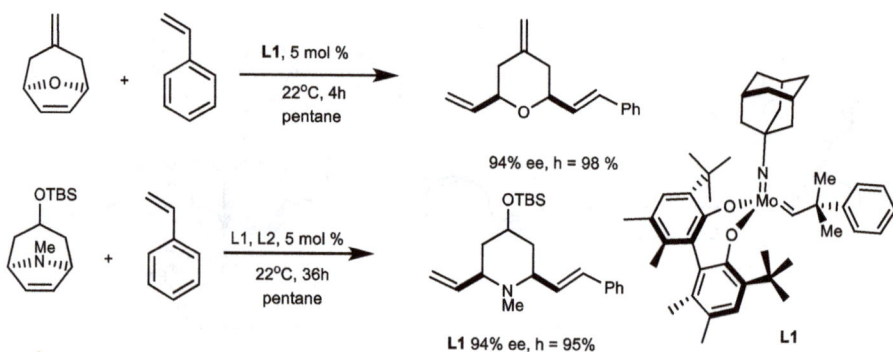

Fig. 21.131: Chiral molybdenum alkylidene-catalyzed ring-opening/cross-metathesis.

21.7.3 Ring-opening metathesis polymerization (ROMP)

Some of the earliest commercial applications of olefin metathesis involved ROMP of monomers containing strained, unsaturated rings to produce stereoregular and monodisperse polymers and copolymers. The polymerization of dicyclopentadiene (Fig. 21.132) is one of the best-known examples of ring-opening polymerization [211].

The mechanism of the ROMP reaction (Fig. 21.133) is identical with the mechanism of olefin metathesis and involves the same type of catalysts.

Initiation begins with coordination of a transition metal alkylidene complex to a cyclic olefin. Subsequent [2 + 2]-cycloaddition affords a four-membered metallacyclobutane

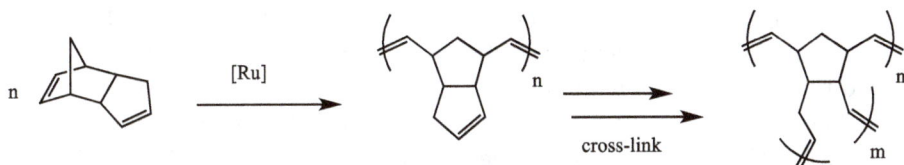

Fig. 21.132: Ring-opening polymerization of dicyclopentadiene.

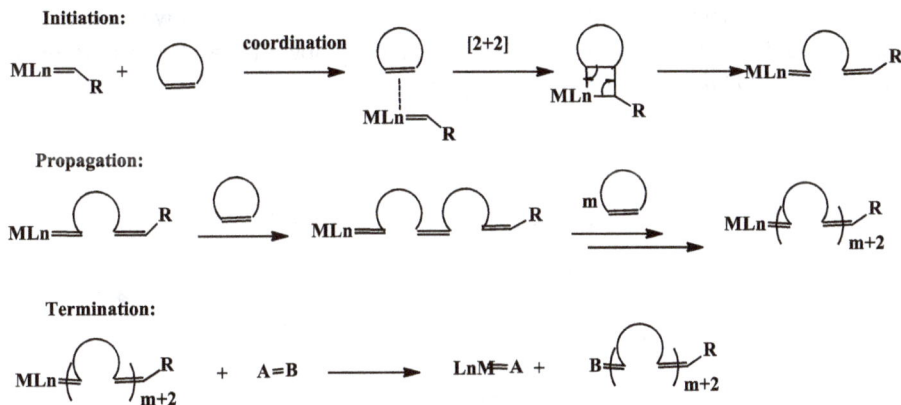

Fig. 21.133: Ring-opening polymerization mechanism.

intermediate, which effectively forms the beginning of a growing polymer chain. The metallacyclobutane intermediate undergoes a cycloreversion reaction to afford a new metal alkylidene with the same reactivity toward cyclic olefins like the initiator. The propagation process takes place until all monomer is consumed, a reaction equilibrium is reached or the reaction is terminated. If another monomer is added in the system, the result is the formation of block copolymers [212].

The ROMP falls in the category of living polymerization [213]. The generally accepted definition of a living polymerization is as that of a chain polymerization proceeding without termination or transfer. The release of the polymer in living ROMP reactions is achieved by the addition of a specialized reagent. Along with the selective removal and deactivation of the transition metal from the end of the growing polymer chain, the reagent is introducing functional group in place of the metal. In processes that are not living, ROMP products can deliver mixtures that contain other cyclic or linear olefins formed by secondary metathesis processes.

The driving force for the ROMP reaction is the relief of ring strain. Olefins with little or no ring strain cannot be polymerized because there is no thermodynamic preference for polymer versus monomer. Strained cyclic olefins such as those shown below have sufficient ring strain to make this process possible. Monomers based on norbornene and norbornadiene derivatives are good candidates for ROMP reactions. There are few examples of monocyclic olefins that have been polymerized successfully in a living manner via ROMP although three-, four-, and eight-membered rings have the right strain for this purpose. ROMP of a wide variety of 3-functionalized cyclobutenes containing ether, ester, alcohol, amine, amide and carboxylic acid substituents have been conducted using $(PCy_3)_2Cl_2Ru = CHCH = CPh_2$ and $(PCy_3)_2Cl_2Ru = CHPh$ [214].

ROMP of 3-methyl-3-phenylcyclopropene and 3-(2-methoxyethyl)-3-methylcyclopropene,using molybdenum alkylidenes as initiators (Fig. 21.134), led to the corresponding polymers with very good yield (94–97%). The reactions were quenched through addition of excess benzaldehyde, which is known to react in a Wittig-like manner to yield benzylidene-capped polymers [215, 216].

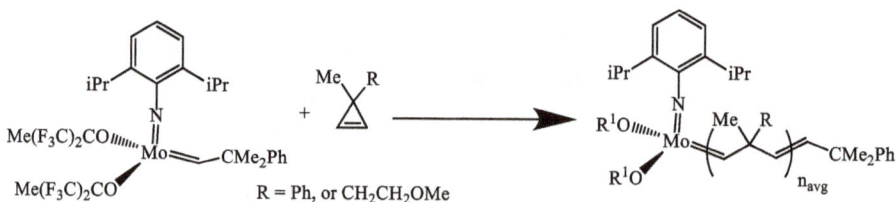

Fig. 21.134: ROMP of 3,3-disubstituted cyclopropane using Schrock's catalyst.

The Grubbs'-type ruthenium initiator $(H_2IMes)(PCy_3)Cl_2RuCHC_6H_5$ ($H_2IMes = 1,3$-di-mesitylimidazolidine) was proved to be a good initiator for polymerization of 3-methyl-3-phenylcyclopropene.

An important property of an ROMP catalyst is to react with high *cis*-selectivity (up to >98%) and high tacticity control (up to >98% syndiotactic). The preparation of highly microstructurally controlled norbornene-, norbornadiene- and cyclopropene-derived polymers is thus possible. Molybdenum and tungsten alkylidene were used successfully [217, 218].

Ring-opening polymerization of bistrifluoromethylnorbornadiene via enantiomorphic site control was achieved using molybdenum complex (Fig. 21.135) [216].

Norbornene ROMP (Fig. 21.136 (**1**)), with the addition of 1-octene as a chain transfer agent (CTA), generated the polymer with high *cis*-content (>98%, *Z*-selectivity) and high syndiotacticity (>98%). The norbornadiene ROMP (Fig. 21.136 (**2**)) afforded the polymeric product in high yield with the same level of selectivity control (97% *cis*, > 98% syndiotactic). Significant stereogenic metal control was observed in the ROMP reaction of 3-methyl-3-phenylcyclopropene (Fig. 21.136 (**3**)), with very high *cis, syndioselectivity* [219].

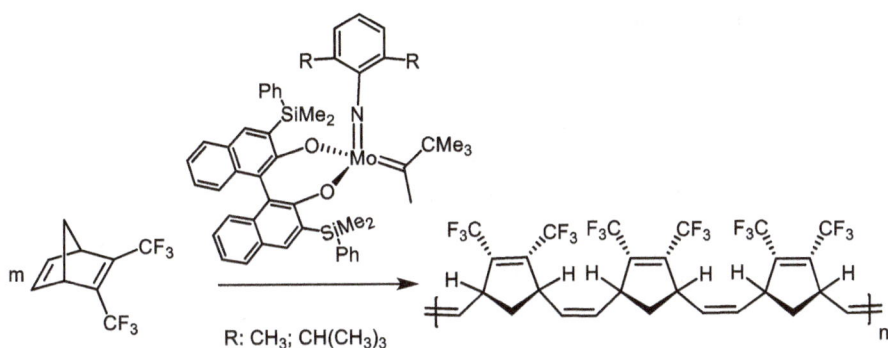

Fig. 21.135: ROMP of bistrifluoromethylnorbornadiene.

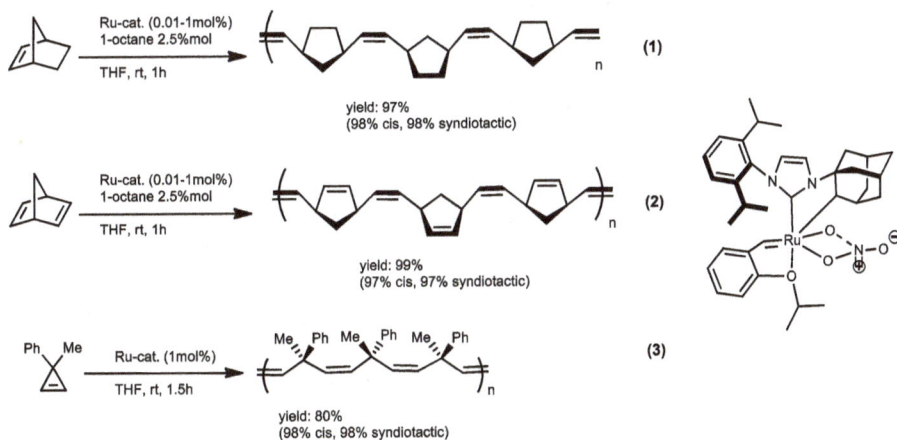

Fig. 21.136: ROMP of norbornene (1), norbornadiene (2) and 3-methyl-3-phenylcyclopropene (3) using Grubbs' catalyst.

21.7.4 Acyclic diene metathesis polymerization (ADMEP)

The more suitable systems for the ADMET are α,ω-dienes. The general reaction is described as follows (Fig. 21.137).

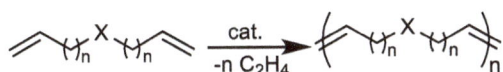

Fig. 21.137: Acyclic diene metathesis polymerization.

The formation and permanent removal of ethylene is the actual driving force of these reactions. When functional dienes are the substrate for ADMEP, other elimination products are formed and must be removed to avoid polymerization–depolymerization equilibria.

The well-defined Mo-, W-, as well as Ru-based carbenes can act as catalysts for the preparation of a wide variety of homo- and copolymers by ADMET process.

21.8 Polymerization

The transition metal-catalyzed polymerization of olefins is the source of the plastic materials and one of very important industrial processes. The polyolefin products have targeted applications in various fields such as health and medical, food packaging, pipes and fittings, consumer and durable goods, and rigid packaging. The properties of ethylene-based polymers are defined by polymer's molecular weight, molecular weight distribution, short-chain branching (type and amount), short-chain branching distribution, long-chain branching level and block structure. To obtain the appropriate polymeric material, specific catalytic systems and reaction conditions/parameters are applied.

The breakthrough in the field of polymer chemistry marked by the simultaneous discovery of Ziegler's [220, 221] and Natta's [222] groups in 1955 was the observation that a mixture of $TiCl_4/AlEt_3$ is a good catalytic system for the polymerization of alkenes. Their discovery was rewarded with the Nobel Prize in 1963. The activation of a metal halide by aluminum alkyls are generally referred to as Ziegler–Natta catalysis. A representation of the polymerization reaction and the structure of various polymers are presented in Fig. 21.138.

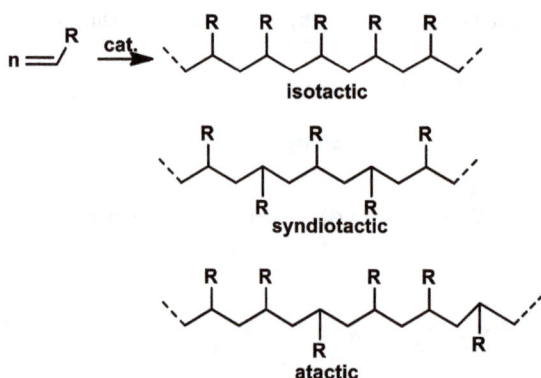

Fig. 21.138: Structure of polymers.

The most familiar plastics made via early transition metal-catalyzed polymeri-zation include high-density polyethylene (HDPE, R = H), linear low-density polyeth-ylene (LLDPE, R = mostly H with some Et, Bu or Hx), polypropylene (R = Me) and ethylene–propylene–diene-modified rubber (EPDM, R = H, Me and alkenyl).

Mechanism. The commonly accepted mechanism for the olefin polymerization re-action is described in Fig. 21.139.

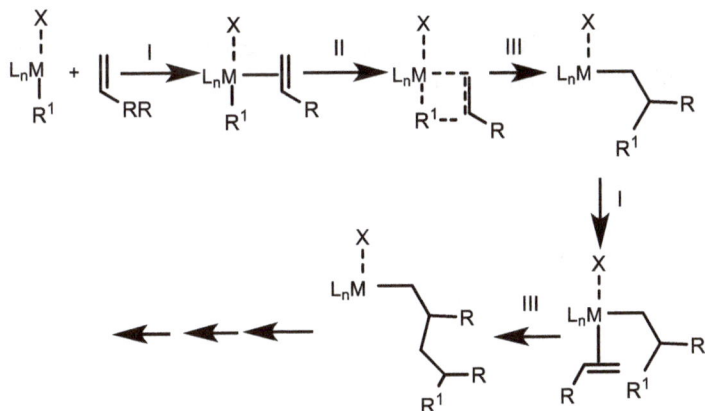

Fig. 21.139: Mechanism of the olefin polymerization.

The first step (I) is the coordination of an olefin via a π-bond to an electron-defi-cient metal center already containing a σ-transition metal–carbon bond to form a weakly bound π-olefin complex (no or very weak back-donation). The next step is the migratory insertion of the olefin into transition metal–carbon bond olefin via a four-center transition state (II) forming new σ-metal–carbon and carbon–carbon bonds (III). This step is exothermic by ~ 20 kcal/mol, the energy difference between the carbon–carbon π- and σ-bonds. For typical early transition metal, d^0 catalysts, the activation energy is ~10 kcal/mol, making this a very facile reaction. The pro-cess is repeated extending the chain.

Chain growth can be terminated by β-hydride elimination, hydrogenolysis, the incorporation of functional groups or adding main group organometallics in the po-lymerization mixture. The first, β-hydride elimination is thermodynamically favor-able (the activation energy is ~30 kcal/mol plus the energy difference between the metal–carbon and metal–hydrogen bonds).

β-Hydride elimination (Fig. 21.140) leads to polymers with vinyl (R = H) or vinyl-idene (R = Me, Et, Pr, etc.) end groups which, given the appropriate conditions, may also be incorporated into the growing polymer chains giving long-chain branched products affording branched copolymers (not always desirable).

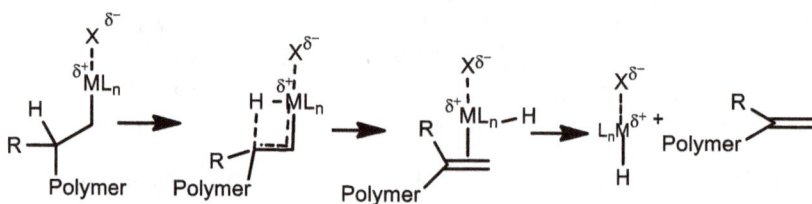

Fig. 21.140: Termination process by β-hydride elimination.

Another termination process of great practical use is hydrogenolysis (Fig. 21.141), leading to polymers with saturated end groups:

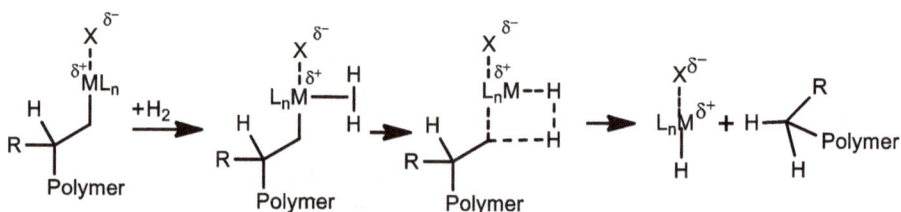

Fig. 21.141: Termination process by hydrogenolysis.

The resulting metal hydride is prone to start a new fast-growing polymer chain, increasing the overall activity of the catalyst.

The incorporation of functional groups at the end of polyolefin chains offers an opportunity to prepare polyolefin building blocks as starting point for polymers with designed properties. The reactivity of the carbon–metal bond formed during the polymerization enables the fast and reversible chain transfer reactions between the active metal center and a main group metal center (Fig. 21.142). Main group organometallics used in the catalytic systems, that is, AlR_3, MgR_2 $ZnEt_2$, BR_3, GaR_3, PbR_4 or SnR_4 act as a reversible CTA according to the mechanism of degenerative chain transfer:

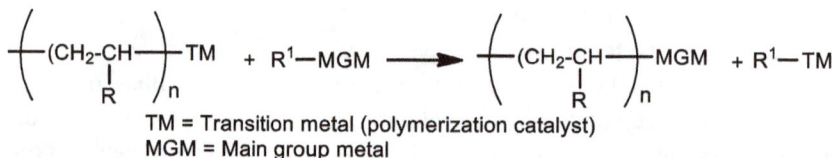

TM = Transition metal (polymerization catalyst)
MGM = Main group metal

Fig. 21.142: Reversible chain transfer reactions between the active metal center and a main group metal center.

Catalyzed chain growth on a main group metal became an efficient tool to functionalize polyolefins. The growth of the chains is looking like occurring on the main group organometallic compound while being catalyzed by the active metal center [223, 224].

Catalysts. The first olefin polymerization processes described by Ziegler and Natta were conducted in heterogeneous systems consisting of a high-valent transition metal halide, oxide or oxo-halide with alkyl aluminum as alkylating cocatalyst, often prepared on supports: $TiCl_4/MgCl_2/AlEt_3$, $CrO_3/Al_2O_3/AlEt_3$ and $VOCl_3/AlEt_3$. As little is known about the nature of the actual catalytic species in these systems, they are referred to as "Black box." Although very active as catalysts they have a very low tolerance for functional groups because of their Lewis acidic nature.

An important improvement was the discovery of the catalytic properties of the high-valent transition metal complex Cp_2ZrCl_2 in combination with methylalumoxane $[MeAlO]_n$, (MAO) (a hydrolysis product of $AlMe_3$) by Kaminsky [225]. The catalytic system was active for the homopolymerization of ethylene to high-density polyethylene. Rationally designed zirconium systems soon followed, that is, Et$(Ind)_2ZrCl_2$/MAO (Ind = indenyl) for the synthesis of isotactic polypropylene and $Pr^i(Cp)(Flr)MCl_2$/MAO (M = Zr, Hf, Flr = fluorenyl), for the preparation of syndiotactic polypropylene [226, 227]. The activity of C_2-symmetric ethanobridged-indenyl-titanocene/MAO system (Fig. 21.143), as the first example of an effective homogeneous isospecific propylene polymerization catalyst, was demonstrated by Ewen [226].

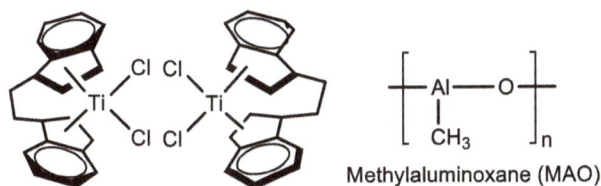

Methylaluminoxane (MAO)

Fig. 21.143: Ethanobridged-indenyl-titanocene/MAO catalytic system.

Besides the good activity, these systems allowed a remarkable tunability via ligand changes in the precatalyst complex. For example, the chiral-active site of the Kaminsky–Brintzinger catalytic system, in the presence of MAO (huge excess, Fig. 21.144), generates isotactic polypropylene [228].

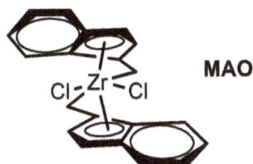

MAO

Fig. 21.144: Kaminsky–Brintzinger catalytic system.

A plethora of stereospecific olefin polymerization with chiral metallocene catalysts was reported in the same period [229].

The catalytic systems are formed from the precatalyst and a cocatalyst. The typical metallocene precatalysts are neutral dichloride complexes of group IV transition metals (L_nMCl_2; L = ancillary ligand, M = metal). The role of the cocatalyst is to abstract one halogen ion (Cl-abstraction) and to transform the resulting cation into the catalytic-active species $[L_nM\text{-}R]^+$ (R = alkyl) by alkylation. The cocatalyst also provides the counterion to the active species, determining the ion-pairing interactions crucial for this type of catalysis. The activators and the activation processes have a great influence on the structure of the polymers [230, 231].

Methylaluminoxane (MAO), although still ill-defined oligomeric compound, is fulfilling the three roles, activator/alkylator/scavenger, effectively and simultaneously. This excellent activator is comprised of a dynamic mixture of $(MeAlO)_n$ cages and trimethylaluminum, which reacts with the precatalyst likely via transient $[AlMe_2]^+$ species. The synthesis of polymers with a highly defined microstructure, tacticity and stereoregularity as well as new cycloolefin, long-chain branched or blocky copolymers with excellent properties was possible by using MAO as cocatalyst [232].

The borate salt $Al\text{-}H\text{-}Al^+[B(C_6F_5)_4]^-$ ($Al\text{-}H\text{-}Al^+ = [Bu^i_2(DMA)Al]_2(\mu\text{-}H)^+$), stable at room temperature, is a molecular activator able to completely activate dichloride metallocene and prototypical post-metallocene precatalysts unlike the simpler $[AlBu^i_2]^+[B(C_6F_5)_4]^-$. As little as 50 equiv of $Al\text{-}H\text{-}Al^+[B(C_6F_5)_4]^-$ are required for efficient catalyst activation and impurity scavenging, the orders of magnitude below the amounts are usually required with MAO or $AlBu^i_3$ [233].

After the discovery that the actual active species in catalytic process is the cation $[L_nM\text{-}R]^+$ (R = alkyl), some cationic homogeneous catalysts like $[Cp_2Zr(CH_2Ph)(THF)]$ $[BPh_4]$ with weakly coordinating anions were described. Their synthesis started from the reaction of early transition metal alkyl complexes, such as Cp'_2ZrMe_2 (Cp′ = methylcyclopentadienyl) with oxidizing tetraphenylborate salts, including $AgBPh_4$ and $(Cp_2Fe)BPh_4$ (Fig. 21.145) [234].

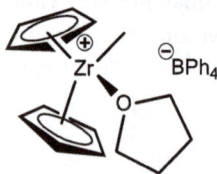

Fig. 21.145: Cationic catalyst for polymerization.

This type of catalysts have also exceptional tunability by the choice of various ligands [234].

The stereochemistry of the polymerization by the single-site polymerization catalysts is strongly related to the structure/symmetry of precatalyst. It was shown that the five main symmetry categories are generally producing specific polymeric structures: C_{2v} and C_s-symmetric catalysts that have mirror planes containing the two diastereotopic coordination sites typically produce atactic polymers or moderately

stereoregular polymers by chain-end control mechanisms [235]. The C_s-symmetric catalysts that have a mirror plane reflecting two enantiotopic coordination sites frequently
produce syndiotactic polymers. C_2-symmetric complexes, both racemic mixtures and
enantiomerically pure ones, typically produce isotactic polymers via a site control
mechanism (Fig. 21.146).

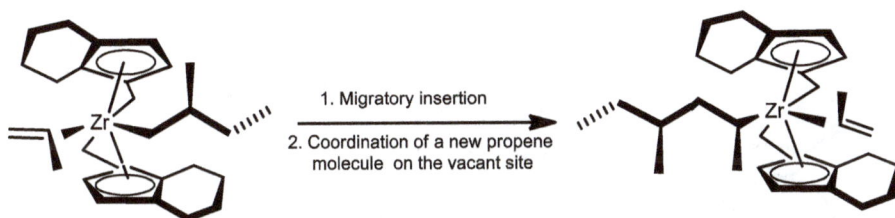

Fig. 21.146: C_2-symmetric complex to produce isotactic polypropylene by migratory insertion/
coordination of the propene into Zr-polymer bond followed by coordination of a new propene
molecule on the vacant site.

The asymmetric C_1 complexes produce polymer architectures ranging from highly isotactic, to atactic, including isotactic–atactic stereoblock and hemiisotactic [235]. The
particular behavior of each catalyst is not always falling within this scheme.

Nonmetallocene ligands with catalytic systems with improved comonomer incorporation, molecular weight and even variable tacticity have been developed
[223, 236, 237]. The scandium(III) complex, 21.147 (**1**), containing a ligand framework composed of cyclopentadienyl linked *via* a silicon dimethyl fragment to a *tert*-
butylamido unit functioned as α-olefin oligomerization catalysts without the need
for an activator [238, 239].

The complexes depicted in Fig. 21.147 belong to the so-called "constrained geometry complex" (**CGC**) catalysts. The term "constrained geometry complex" was originally
used for complexes in which a bidentate ligand built from a π-bonded moiety (e.g.,
cyclopentadienyl or a derivative) linked to a donor atom by a bridge is coordinated to a
metal center in a chelate manner in such a way that the angle at the metal (between
the centroid of the π-system and the additional donor atom) is smaller compared to
unbridged complexes. In the same class of compound, other systems that are included
are as follows: (i) other *ansa*-complexes with η^5:η^1 coordination, where at least one of
the coordinating fragments of bridged cyclopentadienyl-amido complexes is replaced
by an isolobal fragment; (ii) other *ansa*-complexes with η^5:η^1 coordination, where at
least one of the coordinating fragments is not isolobal to the formally replaced fragments of bridged cyclopentadienyl-amido complexes; and (iii) other *ansa*-complexes
with a coordination mode different from η^5:η^1 coordination [243].

The CGCs used for copolymerization of ethylene and α-olefins gave better results compared to metallocenes and metallocenophanes, probably, due to a more
Lewis acidity of the transition metal center. The decreased tendency of the bulk

(1)

(2) R^1=R^2=R^4=H; R^3= Bu^t
(3) R^1=R^2=R^3=R^4= Me
(4) R^1=R^4=H, R^2=R^3= Ph

(5) R^1=R^2=R^3=R^4= H
(6) R^1= OEt, R^2= H
(7) R^1=NMe$_2$,R^2= H
(8) R^1=H,   R^2= OMe
(9) R^1=H, R^2 = N-pirolidine

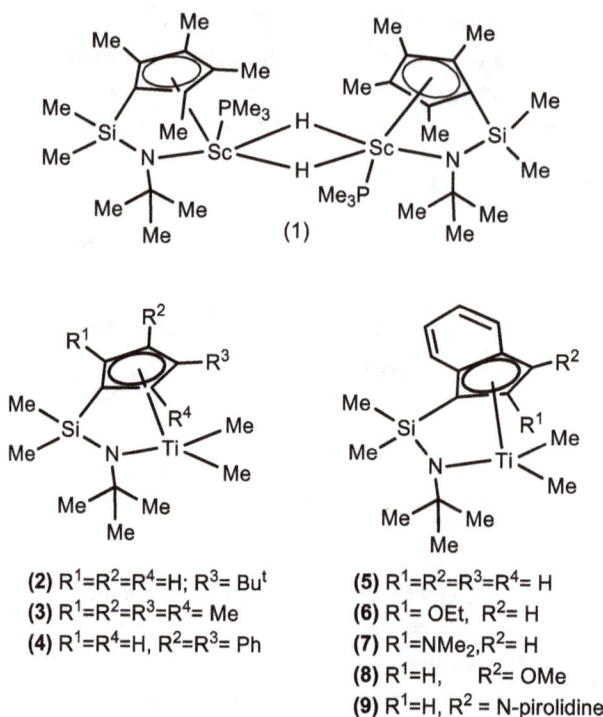

Fig. 21.147: Constrained geometry complex (CGC) catalysts: (1) [239], (2) [240] (3–4) [241] and (5–9) [242].

polymer chain to undergo chain transfer reactions is one of the advantages. The high thermal stability of alkyl and dialkyl CGCs allows higher polymerization temperatures compared with the metallocene catalysts.

Activation of titanium(II) CGC diene complexes can be achieved with common activators such as MAO and B(C$_6$F$_5$)$_3$ [244] or with other boron derivatives like carboranes [245]. The reaction of both metallocene and CGCs with dienes and B(C$_6$F$_5$)$_3$ led to zwitterionic species containing a cationic transition metal center with only weak stabilization by the counter-ion, acting as single-component olefin polymerization catalysts with no additional cocatalyst required [246–248]. To exemplify, the reaction of Cp$_2$M(C$_4$H$_6$) (M = Zr, Hf) with B(C$_6$F$_5$)$_3$ afforded Cp$_2$Zr(+)(μ-C$_4$H$_6$)B(−)(C$_6$F$_5$)$_3$ [248] and [(C$_5$H$_4$)SiMe$_2$(N-t-Bu)]M(C$_4$H$_6$) (M = Ti, Zr) with B(C$_6$F$_5$)$_3$ afforded [(C$_5$H$_4$)SiMe$_2$(N-t-Bu)]M(+)(μ-C$_4$H$_6$)B(−)(C$_6$F$_5$)$_3$ [247]. The structure of the Zr compounds features the butadiene bound as an η^3-allyl, a dative Zr←F-C(ortho) interaction and a single Zr(+) . . . H'-CB(−) agostic interaction (Fig. 21.148) [247].

Stoichiometric reaction between constrained geometry precatalyst [(η^5-C$_5$Me$_4$) (SiMe$_2$-N-t-Bu)]TiMe$_2$ and [Ph$_3$C][HCB$_{11}$Cl$_{11}$] with ortho-difluorobenzene (o-C$_6$H$_4$F$_2$) led to the formation of a complex containing a cationic μ-CH$_3$ dimer and an unassociated carborane anion, as shown by single-crystal X-ray analysis (Fig. 21.149) [245].

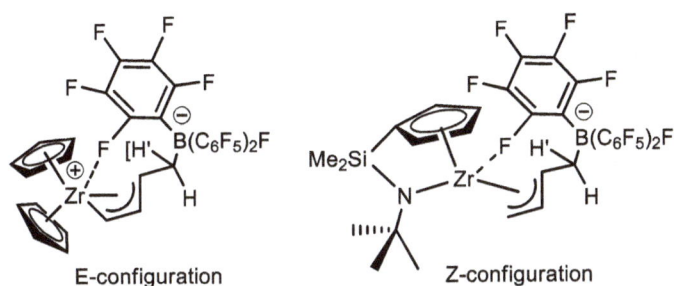

Fig. 21.148: Zwitterionic species acting as single-component olefin polymerization catalysts [248].

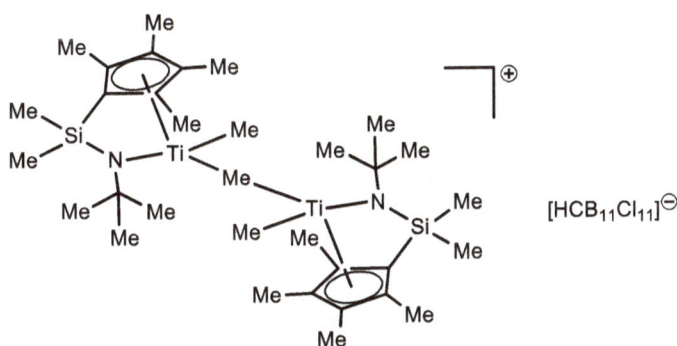

Fig. 21.149: Cationic μ-CH$_3$ dimer.

The reaction of [Me$_3$NH][HCB$_{11}$Cl$_{11}$] in a 2:1 mixture of C$_6$D$_6$ and o-C6H4F2 with [(η^5-C$_5$Me$_4$)(SiMe$_2$-N-t-Bu)]TiMe$_2$ led to a cationic titanium complex and the anionic carborane (Fig. 21.150) [245].

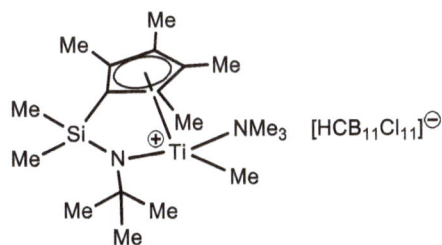

Fig. 21.150: Titanium cationic complex with anionic carborane.

In search of more efficient catalyst systems affording tailored (co)polymers and technologies meeting the demands of green chemistry, the combination of already known different single-site catalysts in a "single" multisite catalyst seems to be a very good solution. The development of microreactors, embedded in the catalyst, to create a virtual nanometer-scale cascade within the catalyst and the polyolefin particles is one of the choices [237]. Interaction of sites via chain shuttling or molecular switching of sites enables the formation of a wide variety of segmented polyolefins. High-throughput screening enables identification of complementary single-site catalysts with matched compatibilities and polymerization kinetics.

The development of catalytic chain transfer polymerization, already mentioned before, also became known as chain shuttling [249, 250] and molecular switching of sites by catalytic group transfer polymerization on a single-site catalyst [251–254] high precision in block copolymer synthesis. This progress opens a new dimension for tailoring polyolefin block copolymers and also many other hydrocarbon materials.

References

[1] Stille JK, Lau KSY. Mechanisms of oxidative addition of organic halides to Group 8 Transition-metal complexes. Acc Chem Res 1977, 10, 434–42.
[2] Collman JP. Patterns of organometallic reactions related to homogeneous catalysis. Acc Chem Res 1968, 1, 136–43.
[3] Tolman CA. Steric effects of phosphorus ligands in organometallic chemistry and homogeneous catalysis. Chem Rev 1977, 77, 313–48.
[4] Calderazzo F. Synthetic and mechanistic aspects of inorganic insertion reactions insertion of carbon monoxide. Angew Chem Int Ed Eng 1977, 16, 299–311.
[5] Bai W, Chen J, Jia G. Reactions of Nucleophiles with Coordinated Alkynes, Alkenes, and Allenes. Knochel P, Molander G A Comprehensive Organic Synthesis II. Elsevier Publ, 2014, 580–647.
[6] Negishi EI, Wang G, Rao H, Xu Z. Alkyne elementometallation-pd-catalyzed cross-coupling. Toward synthesis of all conceivable types of acyclic alkenes in high yields, efficiently, selectively, economically, and safely: "green" way. J Org Chem 2010, 75, 3151–82.
[7] Knappke CEI, Von Wangelin AJ. 35 years of palladium-catalyzed cross-coupling with Grignard reagents: How far have we come? Chem Soc Rev 2011, 40, 4948–62.
[8] Johansson Seechurn CCC, Kitching MO, Colacot TJ, Snieckus V. Palladium-catalyzed cross-coupling: A historical contextual perspective to the 2010 Nobel Prize. Angew Chem Int Ed 2012, 51, 5062–85.
[9] Suzuki A. Cross-coupling reactions of organoboranes: An easy way to construct C-C bonds (Nobel Lecture). Angew Chem Int Ed 2011, 50, 6722–37.
[10] González-Sebastián L, Morales-Morales D. Cross-coupling reactions catalysed by palladium pincer complexes. A review of recent advances. J Organomet Chem 2019, 893, 39–51.
[11] Biffis A, Centomo P, Del Zotto A, Zecca M. Pd Metal catalysts for cross-couplings and related reactions in the twenty-first century: A critical review. Chem Rev 2018, 118, 2249–95.
[12] Beletskaya IP, Cheprakov AV. The Heck reaction as a sharpening stone of palladium catalysis. Chem Rev 2000, 100, 3009–66.
[13] Hazra S, Johansson Seechurn CCC, Handa S, Colacot TJ. The resurrection of Murahashi coupling after four decades. ACS Catal 2021, 11, 13188–202.
[14] Negishi E. Magical power of transition metals: Past, present, and future (Nobel Lecture). Angew Chem Int Ed 2011, 50, 6738–64.
[15] Ziegler DS, Karaghiosoff K, Knochel P. Generation of aryl and heteroaryl magnesium reagents in toluene by Br/Mg or Cl/Mg exchange. Angew Chem Int Ed 2018, 57, 6701–04.

[16] Vechorkin O, Proust V, Hu X. Functional group tolerant Kumada-Corriu-Tamao coupling of nonactivated alkyl halides with aryl and heteroaryl nucleophiles: Catalysis by a pincer complex permits the coupling of functionalized Grignard reagents. J Am Chem Soc 2009, 131, 9756–66.

[17] Clososki GC, Rohbogner CJ, Knochel P. Direct magnesiation of polyfunctionalized arenes and heteroarenes using (tmp)$_2$Mg·2LiCl. Angew Chem Int Ed 2007, 46, 7681–84.

[18] Krasovskiy A, Straub BF, Knochel P. Highly efficient reagents for Br/Mg exchange. Angew Chem Int Ed 2006, 45, 159–62.

[19] Martin R, Buchwald SL. Pd-catalyzed Kumada-Corriu cross-coupling reactions at low temperatures allow the use of Knochel-type Grignard reagents. J Am Chem Soc 2007, 129, 3844–45.

[20] Tamao K, Sumitani K, Kumada M. Selective carbon-carbon bond formation by cross-coupling of Grignard reagents with organic halides. Catalysis by nickel-phosphine complexes. J Am Chem Soc 1972, 94, 4374–76.

[21] Tamao K. Discovery of the cross-coupling reaction between Grignard reagents and C(sp2) halides catalyzed by nickel-phosphine complexes. J Organomet Chem 2002, 653, 23–26.

[22] Corriu RJP, Masse JP. Activation of Grignard reagents by transition-metal complexes. A new and simple synthesis of trans-stilbenes and polyphenyls. J Chem Soc Chem Commun 1972, 144a–144a.

[23] Yamamura M, Moritani I, Murahashi SI. The reaction of σ-vinylpalladium complexes with alkyllithiums. Stereospecific syntheses of olefins from vinyl halides and alkyllithiums. J Organomet Chem 1975, 91, 3–6.

[24] Kalvet I, Magnin G, Schoenebeck F. Rapid room-temperature, chemoselective c–c coupling of poly(pseudo)halogenated arenes enabled by palladium(I) catalysis in air. Angew Chem Int Ed 2017, 56, 1581–85.

[25] O'Brien CJ, Kantchev EAB, Hadei N, Chass GA, Lough A, Hopkinson AC, Organ MG. Easily prepared air- and moisture-stable Pd-NHC (NHC = N-heterocyclic carbene) complexes: a reliable, user-friendly, highly active palladium precatalyst for the Suzuki-Miyaura reaction. Chem Eur J 2006, 12, 4743–48.

[26] Sinha N, Champagne PA, Rodriguez MJ, Lu Y, Kopach ME, Mitchell D, Organ MG. One-pot sequential Kumada–Tamao-Corriu couplings of (hetero)aryl polyhalides in the presence of Grignard-sensitive functional groups using Pd-PEPPSI-IPent Cl. Chem Eur J 2019, 25, 6508–12.

[27] Krasovskiy AL, Haley S, Voigtritter K, Lipshutz BH. Stereoretentive Pd-catalyzed Kumada–Corriu couplings of alkenyl halides at room temperature. Org Lett 2014, 16, 4066–69.

[28] Negishi E ichi. Palladium- or nickel-catalyzed cross coupling. A new selective method for carbon-carbon bond formation. Acc Chem Res 1982, 15, 340–48.

[29] Negishi E, Baba S. Novel stereoselective alkenyl–aryl coupling via nickel-catalysed reaction of alkenylanes with aryl halides. J Chem Soc Chem Commun 1976, 596b–597b.

[30] Negishi EI, Ao K, Okukado N. Selective carbon-carbon bond formation via transition metal catalysis. 3. A highly selective synthesis of unsymmetrical biaryls and diarylmethanes by the nickel- or palladium-catalyzed reaction of aryl- and benzylzinc derivatives with aryl halides. J Org Chem 1977, 42, 1821–23.

[31] King AO, Okukado N, Negishi EI. Highly general stereo-, regio-, and chemo-selective synthesis of terminal and internal conjugated enynes by the Pd-catalysed reaction of alkynylzinc reagents with alkenyl halides. J Chem Soc Chem Commun 1977, 683–84.

[32] King AO, Negishi EI, Villani FJ, Silveira A. A general synthesis of terminal and internal arylalkynes by the palladium-catalyzed reaction of alkynylzinc reagents with aryl halides. J Org Chem 1978, 43, 358–60.

[33] Krasovskiy A, Duplais C, Lipshutz BH. Stereoselective Negishi-like couplings between alkenyl and alkyl halides in water at room temperature. Org Lett 2010, 12, 4742–44.

[34] Krasovskiy A, Lipshutz BH. Highly selective reactions of unbiased alkenyl halides and alkylzinc halides: Negishi-plus couplings. Org Lett 2011, 13, 3822–25.

[35] Krasovskiy A, Lipshutz BH. Ligand effects on Negishi couplings of alkenyl halides. Org Lett 2011, 13, 3818–21.

[36] Huang CY, Doyle AG. Nickel-catalyzed Negishi alkylations of styrenyl aziridines. J Am Chem Soc 2012, 134, 9541–44.

[37] Cinderella AP, Vulovic B, Watson DA. Palladium-catalyzed cross-coupling of silyl electrophiles with alkylzinc halides: A Silyl-Negishi reaction. J Am Chem Soc 2017, 139, 7741–44.

[38] Lee H, Lee Y, Cho SH. Palladium-catalyzed chemoselective Negishi cross-coupling of bis [(pinacolato)boryl]methylzinc halides with aryl (pseudo)halides. Org Lett 2019, 21, 5912–16.

[39] Organ MG, Avola S, Dubovyk I, Hadei N, Kantchev EAB, O'Brien CJ, Valente G. A user-friendly, all-purpose Pd-NHC (NHC = N-heterocyclic carbene) precatalyst for the Negishi reaction: A step towards a universal cross-coupling catalyst. Chem Eur J Eur J 2006, 12, 4749–55.

[40] Çalimsiz S, Organ MG. Negishi cross-coupling of secondary alkylzinc halides with aryl/ heteroaryl halides using Pd–PEPPSI–IPent. Chem Commun 2011, 47, 5181–83.

[41] Sase S, Jaric M, Metzger A, Malakhov V, Knochel P. One-pot Negishi cross-coupling reactions of in situ generated zinc reagents with aryl chlorides, bromides, and triflates. J Org Chem 2008, 73, 7380–82.

[42] Skotnitzki J, Kremsmair A, Keefer D, Gong Y, de Vivie-riedle R, Knochel P. Stereoselective Csp^3–Csp^2 cross-couplings of chiral secondary alkylzinc reagents with alkenyl and aryl halides. Angew Chem Int Ed 2020, 59, 320–24.

[43] Liu Z, Dong N, Xu M, Sun Z, Tu T. Mild Negishi cross-coupling reactions catalyzed by acenaphthoimidazolylidene palladium complexes at low catalyst loadings. J Org Chem 2013, 78, 7436–44.

[44] Stathakis CI, Manolikakes SM, Knochel P. TMPZnOPiv·LiCl: A new base for the preparation of air-stable solid zinc pivalates of sensitive aromatics and heteroaromatics. Org Lett 2013, 15, 1302–05.

[45] Ellwart M, Knochel P. Preparation of solid, substituted allylic zinc reagents and their reactions with electrophiles. Angew Chem Int Ed 2015, 54, 10662–65.

[46] Bernhardt S, Manolikakes G, Kunz T, Knochel P. Preparation of solid salt-stabilized functionalized organozinc compounds and their application to cross-coupling and carbonyl addition reactions. Angew Chem Int Ed 2011, 50, 9205–09.

[47] Hammann JM, Lutter FH, Haas D, Knochel PA. Robust and broadly applicable cobalt-catalyzed cross-coupling of functionalized bench-stable organozinc pivalates with unsaturated halides. Angew Chem Int Ed 2017, 56, 1082–86.

[48] Kosugi M, Kurino K, Takayama K, Migita T. The reaction of organic halides with allyltrimethyltin. J Organomet Chem 1973, 56, 1–3.

[49] Milstein D, Stille JK. A general, selective, and facile method for ketone synthesis from acid chlorides and organotin compounds catalyzed by palladium. J Am Chem Soc 1978, 100, 3636–38.

[50] Stille JK. The palladium-catalyzed cross-coupling reactions of organotin reagents with organic electrophiles. Angew Chem Int Ed Eng 1986, 25, 508–24.

[51] Farina V. New perspectives in the cross-coupling reactions of organostannanes. Pure Appl Chem 1996, 68, 73–78.

[52] Beletskaya IP. The cross-coupling reactions of organic halides with organic derivatives of tin, mercury and copper catalyzed by palladium. J Organomet Chem 1983, 250, 551–64.

[53] Li J-H, Liang Y, Wang D-P, Liu W-J, Xie Y-X, Yin D-L. Efficient Stille cross-coupling reaction catalyzed by the Pd(OAc) 2/Dabco catalytic system. J Org Chem 2005, 70, 2832–34.

[54] Lerebours R, Camacho-Soto A, Wolf C. Palladium-catalyzed chemoselective cross-coupling of acyl chlorides and organostannanes. J Org Chem 2005, 70, 8601–04.

[55] Gribanov PS, Golenko YD, Topchiy MA, Minaeva LI, Asachenko AF, Nechaev MS. Stannylation of aryl halides, Stille cross-coupling, and one-pot, two-step stannylation/Stille cross-coupling reactions under solvent-free conditions. Eur J Org Chem 2018, 2018, 120–25.

[56] Chemler SR, Trauner D, Danishefsky SJ. The β-alkyl Suzuki-Miyaura cross-coupling reaction: Development, mechanistic study, and applications in natural product synthesis. Angew Chem Int Ed 2001, 40, 4544–68.

[57] Miyaura N, Yamada K, Suzuki A. A new stereospecific cross-coupling by the palladium-catalyzed reaction of 1-alkenylboranes with 1-alkenyl or 1-alkynyl halides. Tetrahedron Lett 1979, 20, 3437–40.

[58] Miyaura N, Yamada K, Suzuki A. A new stereospecific cross-coupling by the palladium-catalyzed reaction of 1-alkenylboranes with 1-alkenyl or l-alkynyl halides. Tetrahedron Lett 1979, 20, 3437–40.

[59] Miyaura N, Yamada K, Suginome H, Suzuki A. Novel and convenient method for the stereo- and regiospecific synthesis of conjugated alkadienes and alkenynes via the palladium-catalyzed cross-coupling reaction of 1-alkenylboranes with bromoalkenes and bromoalkynes. J Am Chem Soc 1985, 107, 972–80.

[60] Miyaura N, Suzuki A. Stereoselective synthesis of arylated (E)-alkenes. by the reaction of alk-1 -enylboranes with aryl halides in the presence of palladium catalyst. J Chem Soc Chem Commun 1979, 866–67.

[61] Hatanaka Y, Hiyama T. Cross-coupling of organosilanes with organic halides mediated by palladium catalyst and tris(diethylamino)sulfonium difluorotrimethylsilicate. J Org Chem 1988, 53, 918–20.

[62] Hiyama T. How I came across the silicon-based cross-coupling reaction. J Organomet Chem 2002, 653, 58–61.

[63] Denmark SE, Choi JY. Highly stereospecific, cross-coupling reactions of lkenylsilacyclobutanes. J Am Chem Soc 1999, 121, 5821–22.

[64] Zhang S, Cai J, Yamamoto Y, Bao M. Palladium-Catalyzed sp^2-sp^3 coupling of chloromethylarenes with allyltrimethoxysilane: Synthesis of allyl arenes. J Org Chem 2017, 82, 5974–80.

[65] Denmark SE, Regens CS. Palladium-catalyzed cross-coupling reactions of organosilanols and their salts: Practical alternatives to boron- and tin-based methods. Acc Chem Res 2008, 41, 1486–99.

[66] Denmark SE, Ambrosi A. Why you really should consider using palladium-catalyzed cross-coupling of silanols and silanolates. Org Process Res Dev 2015, 19, 982–94.

[67] Denmark SE, Smith RC. Mechanistic duality in palladium-catalyzed cross-coupling reactions of aryldimethylsilanolates intermediacy of an 8-Si-4 arylpalladium(II) silanolate. J Am Chem Soc 2010, 132, 1243–45.

[68] Tymonko SA, Smith RC, Ambrosi A, Ober MH, Wang H, Denmark SE. Mechanistic significance of the Si–O–Pd bond in the palladium-catalyzed cross-coupling reactions of arylsilanolates. J Am Chem Soc 2015, 137, 6200–18.

[69] Denmark SE, Sweis RF. Fluoride-free cross-coupling of organosilanols. J Am Chem Soc 2001, 123, 6439–40.

[70] Denmark SE, Ober MH. Cross-coupling reactions of arylsilanols with substituted aryl halides. Org Lett 2003, 5, 1357–60.

[71]   Denmark SE, Ober MH. Palladium-catalyzed cross-coupling reactions of substituted aryl(dimethyl)silanols. Adv Synth Catal 2004, 346, 1703–14.

[72]   Denmark SE, Kallemeyn JM. Stereospecific palladium-catalyzed cross-coupling of (E)- and (Z)-alkenylsilanolates with aryl chlorides. J Am Chem Soc 2006, 128, 15958–59.

[73]   Denmark SE, Baird JD, Regens CS. Palladium-catalyzed cross-coupling of five-membered heterocyclic silanolates. J Org Chem 2008, 73, 1440–55.

[74]   Denmark SE, Tymonko SA. Cross-coupling of alkynylsilanols with aryl halides promoted by potassium trimethylsilanolate. J Org Chem 2003, 68, 9151–54.

[75]   Murahashi S, Yamamura M, Yanagisawa K, Mita N, Kondo K. Stereoselective synthesis of alkenes and alkenyl sulfides from alkenyl halides using palladium and ruthenium catalysts. J Org Chem 1979, 44, 2408–17.

[76]   Pace V, Luisi R. Expanding the synthetic portfolio of organolithiums: Direct use in catalytic cross-coupling reactions. ChemCatChem 2014, 6, 1516–19.

[77]   Firth JD, O'Brien P. Cross-coupling knows no limits: Assessing the synthetic potential of the palladium-catalysed cross-coupling of organolithiums. ChemCatChem 2015, 7, 395–97.

[78]   Giannerini M, Fañanás-Mastral M, Feringa BL. Direct catalytic cross-coupling of organolithium compounds. Nat Chem 2013, 5, 667–72.

[79]   Woon KL, Aldred MP, Vlachos P, Mehl GH, Stirner T, Kelly SM. et al., Electronic charge transport in extended nematic liquid crystals. Chem Mater 2006, 18, 2311–17.

[80]   Liu L, Ho CL, Wong WY, Cheung KY, Fung MK, Lam WT. et al., Effect of oligothienyl chain length on tuning the solar cell performance in fluorene-based polyplatinynes. Adv Funct Mater 2008, 18, 2824–33.

[81]   Hornillos V, Giannerini M, Vila C, Fañanás-Mastral M, Feringa BL. Catalytic direct cross-coupling of organolithium compounds with aryl chlorides. Org Lett 2013, 15, 5114–17.

[82]   Vila C, Cembellín S, Hornillos V, Giannerini M, Fañanás-Mastral M, Feringa BL. t BuLi-mediated one-pot direct highly selective cross-coupling of two distinct aryl bromides. Chem Eur J 2015, 21, 15520–24.

[83]   Yang ZK, Wang DY, Minami H, Ogawa H, Ozaki T, Saito T. et al., Cross-coupling of organolithium with ethers or aryl ammonium salts by C–O or C–N bond cleavage. Chem Eur J 2016, 22, 15693–99.

[84]   Mateos-Gil J, Mondal A, Castiñeira Reis M, Feringa BL. Synthesis and functionalization of allenes by direct Pd-catalyzed organolithium cross-coupling. Angew Chem Int Ed 2020, 59, 7823–29.

[85]   Helbert H, Visser P, Hermens JGH, Buter J, Feringa BL. Palladium-catalysed cross-coupling of lithium acetylides. Nat Catal 2020, 3, 664–71.

[86]   Dilauro G, Francesca Quivelli A, Vitale P, Capriati V, Perna FM. Water and sodium chloride: Essential ingredients for robust and fast Pd-catalysed cross-coupling reactions between organolithium reagents and (hetero)aryl halides. Angew Chem Int Ed 2019, 58, 1799–802.

[87]   Chinchilla R, Nájera C. The Sonogashira reaction: A booming methodology in synthetic organic chemistry. Chem Rev 2007, 107, 874–922.

[88]   Chinchilla R, Nájera C. Recent advances in Sonogashira reactions. Chem Soc Rev 2011, 40, 5084.

[89]   Sonogashira K, Tohda Y, Hagihara N. A convenient synthesis of acetylenes: Catalytic substitutions of acetylenic hydrogen with bromoalkenes, iodoarenes and bromopyridines. Tetrahedron Lett 1975, 16, 4467–70.

[90]   Cassar L. Synthesis of aryl- and vinyl-substituted acetylene derivatives by the use of nickel and palladium complexes. J Organomet Chem 1975, 93, 253–57.

[91]   Dieck HA, Heck FR. Palladium catalyzed synthesis of aryl, heterocyclic and vinylic acetylene derivatives. J Organomet Chem 1975, 93, 259–63.

[92] Evano G, Blanchard N, Toumi M. Copper-mediated coupling reactions and their applications in natural products and designed biomolecules synthesis. Chem Rev 2008, 108, 3054–131.

[93] Gottardo C, Kraft TM, Hossain MS, Zawada PV, Muchall HM. Linear free-energy correlation analysis of the electronic effects of the substituents in the Sonogashira coupling reaction. Can J Chem 2008, 86, 410–15.

[94] An Der Heiden M, Plenio H. The effect of steric bulk in Sonogashira coupling reactions. Chem Commun 2007, 972–74.

[95] Eckhardt M, Fu GC. The first applications of carbene ligands in cross-couplings of alkyl electrophiles: Sonogashira reactions of unactivated alkyl bromides and iodides. J Am Chem Soc 2003, 125, 13642–43.

[96] Altenhoff G, Würtz S, Glorius F. The first palladium-catalyzed Sonogashira coupling of unactivated secondary alkyl bromides. Tetrahedron Lett 2006, 47, 2925–28.

[97] Mizoroki T, Mori K, Ozaki A. Arylation of olefin with aryl iodide catalyzed by palladium. Bull Chem Soc Jpn 1971, 44, 581.

[98] Heck RF, Nolley JP. Palladium-catalyzed vinylic hydrogen substitution reactions with aryl, benzyl, and styryl halides. J Org Chem 1972, 37, 2320–22.

[99] Jutand A. The Mizoroki–Heck Reaction. Oestreich M ed Mechanisms of the Mizoroki–Heck reaction. Chichester, UK: John Wiley & Sons, Ltd, 2009, 1–50.

[100] Amatore C, Jutand A. Anionic Pd(0) and Pd(II) intermediates in palladium-catalyzed Heck and cross-coupling reactions. Acc Chem Res 2000, 33, 314–21.

[101] Farina V. High-Turnover palladium catalysts in cross-coupling and Heck chemistry: A critical overview. Adv Synth Catal 2004, 346, 1553–82.

[102] Littke AF, Fu GC. Palladium-catalyzed coupling reactions of aryl chlorides. Angew Chem Int Ed 2002, 41, 4176–211.

[103] Thorn DL, Hoffmann R. The olefin insertion reaction. J Am Chem Soc 1978, 100, 2079–90.

[104] Littke AF, Fu GC. A versatile catalyst for Heck reactions of aryl chlorides and aryl bromides under mild conditions. J Am Chem Soc 2001, 123, 6989–7000.

[105] Wang GZ, Shang R, Cheng WM, Fu Y. Irradiation-induced Heck reaction of unactivated alkyl halides at room temperature. J Am Chem Soc 2017, 139, 18307–12.

[106] McConville M, Saidi O, Blacker J, Xiao J. Regioselective Heck vinylation of electron-rich olefins with vinyl halides: Is the neutral pathway in operation? J Org Chem 2009, 74, 2692–98.

[107] Leeuwen PWNMV, Kamer PCJ, Claver C, Pàmies O, Diéguez M. Phosphite-containing ligands for asymmetric catalysis. Chem Rev 2011, 111, 2077–118.

[108] Imbos R, Minnaard AJ, Feringa BL. A highly enantioselective intramolecular Heck reaction with a monodentate ligand. J Am Chem Soc 2002, 124, 184–85.

[109] Imbos R, Minnaard AJ, Feringa BL. Monodentate phosphoramidites; versatile ligands in catalytic asymmetric intramolecular Heck reactions. Dalton Trans. 2003, 2017–23.

[110] Franke R, Selent D, Börner A. Applied hydroformylation. Chem Rev 2012, 112, 5675–732.

[111] Trost BM. The atom economy – A search for synthetic efficiency. Science 1991, 254, 1471–77.

[112] Trost BM. Atom Economy – A challenge for organic synthesis: Homogeneous catalysis leads the way. Angew Chem Int Ed Eng 1995, 34, 259–81.

[113] Breit B, Seiche W. Recent Advances on chemo-, regio- and stereoselective hydroformylation. Synthesis (Stuttg). 2001, 1–36.

[114] Heck RF, Breslow DS. The reaction of cobalt hydrotetracarbonyl with olefins. J Am Chem Soc 1961, 83, 4023–27.

[115] Slaugh LH, Mullineaux RD. Novel hydroformylation catalysts. J Organomet Chem 1968, 13, 469–77.

[116] Beller M, Cornils B, Frohning CD, Kohlpaintner CW. Progress in hydroformylation and carbonylation. J Mol Catal A Chem 1995, 104, 17–85.

[117] Osborn JA, Jardine FH, Young JF, Wilkinson G. The preparation and properties of tris(triphenylphosphine) halogenorhodium(I) and some reactions thereof including catalytic homogeneous hydrogenation of olefins and acetylenes and their derivatives. J Chem Soc A Inorg Phys Theor 1966, 1711–32.

[118] Evans D, Osborn JA, Wilkinson G. Hydroformylation of alkenes by use of rhodium complex catalysts. J Chem Soc A Inorg Phys Theor 1968, 3133–42.

[119] Brown CK, Wilkinson G. Homogeneous hydroformylation of alkenes with hydridocarbonyltris-(triphenylphosphine)rhodium(I) as catalyst. J Chem Soc A Inorg Phys Theor Chem 1970, 53, 2753–64.

[120] Jardine FH. Carbonylhydrido tris(triphenylphosphine)rhodium(I). Polyhedron 1982, 1, 569–605.

[121] Pwnm VL, Kamer PCJ, Reek JNH, Dierkes P. Ligand bite angle effects in metal-catalyzed C-C bond formation. Chem Rev 2000, 100, 2741–69.

[122] Torrent M, Solà M, Frenking G. Theoretical studies of some transition-metal-mediated reactions of industrial and synthetic importance. Chem Rev 2000, 100, 439–93.

[123] van Leeuwen PWNM, Kamer PCJ, Reek JNH, Dierkes P. Ligand bite angle effects in metal-catalyzed C–C bond formation. Chem Rev 2000, 100, 2741–70.

[124] Casey CP, Whiteker GT, Melville MG, Petrovich LM, Gavney JA, Powell DR. Diphosphines with natural bite angles near 120° increase selectivity for n-aldehyde formation in rhodium-catalyzed hydroformylation. J Am Chem Soc 1992, 114, 5535–43.

[125] Cuny GD, Buchwald SL. Practical, high-yield, regioselective, rhodium-catalyzed hydroformylation of functionalized α-olefins. J Am Chem Soc 1993, 115, 2066–68.

[126] Kamer PCJ, Van Leeuwen PWNM, Reek JNH. Wide bite angle diphosphines: Xantphos ligands in transition metal complexes and catalysis. Acc Chem Res 2001, 34, 895–904.

[127] Carbó JJ, Maseras F, Bo C, van Leeuwen PWNM. Unraveling the origin of regioselectivity in rhodium diphosphine catalyzed hydroformylation. A DFT QM/MM study. J Am Chem Soc 2001, 123, 7630–37.

[128] Casey CP, Whiteker GT. The natural bite angle of chelating diphosphines. Isr J Chem 1990, 30, 299–304.

[129] Van Der Veen LA, Keeven PH, Schoemaker GC, Reek JNH, Kamer PCJ, Van Leeuwen PWNM, Lutz M, Spek AL. Origin of the bite angle effect on rhodium diphosphine catalyzed hydroformylation. Organometallics 2000, 19, 872–82.

[130] Van Der Veen LA, Boele MDK, Bregman FR, Kamer PCJ, Van Leeuwen PWNM, Goubitz K, Fraanje J, Schenk H, Bo C. Electronic effect on rhodium diphosphine catalyzed hydroformylation: The bite angle effect reconsidered. J Am Chem Soc 1998, 120, 11616–26.

[131] Meessen P, Vogt D, Keim W. Highly regioselective hydroformylation of internal, functionalized olefins applying Pt/Sn complexes with large bite angle diphosphines. J Organomet Chem 1998, 551, 165–70.

[132] Van Der Veen LA, Keeven PK, Kamer PCJ, Van Leeuwen PWNM. Wide bite angle amine, arsine and phosphine ligands in rhodium-and platinum/tin-catalysed hydroformylation 1. J Chem Soc Dalton Trans 2000, 2105–12.

[133] Ojima I. Catalytic Asymmetric Synthesis. Ojima I ed Catalytic Asymmetric Synthesis: Third Edition. Hoboken, NJ, USA: John Wiley & Sons, Inc., 2010.

[134] Chikkali SH, Van Der Vlugt JI, Reek JNH. Hybrid diphosphorus ligands in rhodium catalysed asymmetric hydroformylation. Coord Chem Rev 2014, 262, 1–15.

[135] Agbossou F, Carpentier J-F, Mortreux A. Asymmetric hydroformylation. Chem Rev 1995, 95, 2485–506.

[136] Czauderna CF, Slawin AMZ, Cordes DB, van der Vlugt JI, Kamer PCJ. P-stereogenic wide bite angle diphosphine ligands. Tetrahedron 2019, 75, 47–56.

[137] Sakai N, Mano S, Nozaki K, Takaya H. Highly enantioselective hydroformylation of olefins catalyzed by new phosphinephosphite-Rh(I) complexes. J Am Chem Soc 1993, 115, 7033–34.

[138] Gleich D, Schmid R, Herrmann WA. A molecular model to explain and predict the stereoselectivity in rhodium-catalyzed hydroformylation. Organometallics 1998, 17, 2141–43.

[139] Aguado-Ullate S, Saureu S, Guasch L, Carbó JJ. Theoretical studies of asymmetric hydroformylation using the Rh-(R,S)-BINAPHOS catalyst – Origin of coordination preferences and stereoinduction. Chem Eur J 2012, 18, 995–1005.

[140] Nozaki K, Sakai N, Nanno T, Higashijima T, Mano S, Horiuchi T. et al. Highly enantioselective hydroformylation of olefins catalyzed by rhodium (I) complexes of new chiral phosphine-phosphite ligands. J Am Chem Soc 1997, 119, 4413–23.

[141] Osborn JA, Schrock RR. Coordinatively unsaturated cationic complexes of rhodium (I), iridium (I), palladium(II), and platinum(II). Generation, synthetic utility, and some catalytic studies. J Am Chem Soc 1971, 93, 3089–91.

[142] Crabtree R. I. Compounds in catalysis. Acc Chem Res 1979, 12, 331–37.

[143] James BR, Markham LD, Wang DKW. Stoicheiometric hydrogenation of olefins using HRuCl(PPh₃)₃ and formation of an ortho-metallated ruthenium(II) complex. J Chem Soc Chem Commun 1974, 439.

[144] Knowles WS. Asymmetric hydrogenations (Nobel Lecture). Angew Chem Int Ed 2002, 41, 1998–2007.

[145] Noyori R. Asymmetric catalysis: Science and opportunities (Nobel Lecture). Angew Chem Int Ed 2002, 41, 2008–22.

[146] Knowles WS. Asymmetric hydrogenation. Acc Chem Res 1983, 16, 106–12.

[147] Dang TP, Kagan HB. The asymmetric synthesis of hydratropic acid and amino-acids by homogeneous catalytic hydrogenation. J Chem Soc D Chem Commun 1971, 7, 481.

[148] Miyashita A, Yasuda A, Takaya H, Toriumi K, Ito T, Souchi T. et al. Synthesis of 2,2'-bis (diphenylphosphino)-1,1'-binaphthyl (BINAP), an atropisomeric chiral bis(triaryl)phosphine, and its use in the rhodium (I)-catalyzed asymmetric hydrogenation of α-(acylamino)acrylic acids. J Am Chem Soc 1980, 102, 7932–34.

[149] Seo CSG, Morris RH. Catalytic homogeneous asymmetric hydrogenation: Successes and opportunities. Organometallics 2019, 38, 47–65.

[150] Pfaltz A, Blankenstein J, Hilgraf R, Hörmann E, McIntyre S, Menges F, Schönleber M, Smidt SP, Wüstenberg B, Zimmermann N. Iridium-catalyzed enantioselective hydrogenation of olefins. Adv Synth Catal 2003, 345, 33–43.

[151] Roseblade SJ, Pfaltz A. Iridium-catalyzed asymmetric hydrogenation of olefins. Acc Chem Res 2007, 40, 1402–11.

[152] Schrems MG, Neumann E, Pfaltz A. Indium-catalyzed asymmetric hydrogenation of unfunctionalized tetrasubstituted olefins. Angew Chem Int Ed 2007, 46, 8274–76.

[153] Menges F, Pfaltz A. Threonine-derived phosphinite-oxazoline ligands for the ir-catalyzed enantioselective hydrogenation. Adv Synth Catal 2002, 344, 40–44.

[154] Fryzuk MD, Bosnich B. Asymmetric synthesis production of optically active amino acids by catalytic hydrogenation. J Am Chem Soc 1977, 99, 6262–67.

[155] Burk MJ, Feaster JE, Nugent WA, Harlow RL. Preparation and use of C2-symmetric bis (phospholanes): Production of α-amino acid derivatives via highly enantioselective hydrogenation reactions. J Am Chem Soc 1993, 115, 10125–38.

[156] Pilkington CJ, Zanotti-Gerosa A. Expanding the family of phospholane-based ligands: 1,2-bis (2,5-diphenylphospholano)ethane. Org Lett 2003, 5, 1273–75.

[157] Tang W, Zhang X, Chiral A. 1,2-bisphospholane ligand with a novel structural motif: Applications in highly enantioselective Rh-catalyzed hydrogenations. Angew Chem Int Ed 2002, 41, 1612–14.

[158] Xie Y, Lou R, Li Z, Mi A, Jiang Y. DPAMPP in catalytic asymmetric reactions: Enantioselective synthesis of L-homophenylalanine. Tetrahedron Asymmetry 2000, 11, 1487–94.

[159] Heller D, Holz J, Komarov I, Drexler H-J, You J, Drauz K. et al., On the enantioselective hydrogenation of isomeric β-acylamido β-alkylacrylates with chiral Rh(I) complexes – comparison of phosphine ligands and substrates. Tetrahedron Asymmetry 2002, 13, 2735–41.

[160] Benessere V, Del Litto R, De Roma A, Ruffo F. Carbohydrates as building blocks of privileged ligands. Coord Chem Rev 2010, 254, 390–401.

[161] Miyashita A, Takaya H, Souchi T, Noyori R. 2, 2′-bis(diphenylphosphino)-1, 1′-binaphthyl (binap). Tetrahedron 1984, 40, 1245–53.

[162] Jerphagnon T, Renaud JL, Demonchaux P, Ferreira A, Bruneau C. Enantioselective hydrogenation of isomeric β-acetamido β-alkylacrylates: Crucial influence of temperature. Tetrahedron Asymmetry 2003, 14, 1973–77.

[163] Heller D, Drexler HJ, You J, Baumann W, Drauz K, Krimmer HP. et al., On the enantioselective hydrogenation of isomeric methyl 3-acetamidobutenoates with RhI complexes. Chem Eur J Eur J 2002, 8, 5196–202.

[164] Molinaro C, Scott JP, Shevlin M, Wise C, Ménard A, Gibb A, Junker EM, Lieberman D. Catalytic, asymmetric, and stereodivergent synthesis of non-symmetric β,β-Diaryl-α-Amino Acids. J Am Chem Soc 2015, 137, 999–1006.

[165] Hsiao Y, Rivera NR, Rosner T, Krska SW, Njolito E, Wang F, Sun Y, Armstrong JD, Grabowski EJJ, Tillyer RD, Spindler F, Malan C. Highly efficient synthesis of β-amino acid derivatives via asymmetric hydrogenation of unprotected enamines. J Am Chem Soc 2004, 126, 9918–19.

[166] Hu XP, Zheng Z. Unsymmetrical hybrid ferrocene-based phosphine-phosphoramidites: A new class of practical ligands for Rh-catalyzed asymmetric hydrogenation. Org Lett 2004, 6, 3585–88.

[167] Tang W, Chi Y, Zhang X. An ortho-substituted BIPHEP ligand and its applications in Rh-catalyzed hydrogenation of cyclic enamides. Org Lett 2002, 4, 1695–98.

[168] Liu C, Yuan J, Zhang J, Wang Z, Zhang Z, Zhang W. Rh-catalyzed asymmetric hydrogenation of β-branched enol esters for the synthesis of β-chiral primary alcohols. Org Lett 2018, 20, 108–11.

[169] Takaya H, Ohta T, Sayo N, Kumobayashi H, Akutagawa S, Inoue S, Kasahara I, Noyori R. Enantioselective hydrogenation of allylic and homoallylic alcohols. J Am Chem Soc 1987, 109, 1596–97.

[170] Yoneda N, Kusano S, Yasui M, Pujado P, Wilcher S. Recent advances in processes and catalysts for the production of acetic acid. Appl Catal A Gen 2001, 221, 253–65.

[171] Budiman AW, Nam JS, Park JH, Mukti RI, Chang TS, Bae JW, Choi MJ. Review of Acetic Acid Synthesis from Various Feedstocks Through Different Catalytic Processes. Catal Surv from Asia 2016, 20, 173–93.

[172] Paulik FE, Roth JF. Novel catalysts for the low-pressure carbonylation of methanol to acetic acid. Chem Commun 1968, 11, 2427.

[173] Forster D. Mechanistic pathways in the catalytic carbonylation of methanol by rhodium and iridium complexes. Advances Organometallic Chemistry. 1979, 255–67.

[174] Yoneda N, Minami T, Weiszmann J, Spehlmann B. The Chiyoda/UOP AceticaTM process: A novel acetic acid technology. Stud Surf Sci Catal 1999, 121, 93–98.

[175] Thomas CM, Süss-Fink G. Ligand effects in the rhodium-catalyzed carbonylation of methanol. Coord Chem Rev 2003, 243, 125–42.

[176] Maitlis PM, Haynes A, Sunley GJ, Howard MJ. Methanol carbonylation revisited: Thirty years on. J Chem Soc – Dalton Trans 1996, 2187–96.

[177] Forster D. Kinetic and spectroscopic studies of the carbonylation of methanol with an iodide-promoted iridium catalyst. J Chem Soc Dalton Trans 1979, 1639–45.

[178] Sunley GJ, Watson DJ. High productivity methanol carbonylation catalysis using iridium. The CativaTM process for the manufacture of acetic acid. Catal Today 2000, 58, 293–307.

[179] Haynes A, Maitlis PM, Morris GE, Sunley GJ, Adams H, Badger PW, Bowers CM, Cook DB, Elliott PIP, Ghaffar T, Green H, Griffin TR, Payne M, Pearson JM, Taylor MJ, Vickers PW, Watt RJ. Promotion of Iridium-catalyzed methanol carbonylation: mechanistic studies of the Cativa process. J Am Chem Soc 2004, 126, 2847–61.

[180] Ghaffar T, Adams H, Maitlis PM, Sunley GJ, Baker MJ, Haynes A. Spectroscopic identification and reactivity of [Ir(CO)$_3$I$_2$Me], a key reactive intermediate in iridium catalysed methanol carbonylation. Chem Commun 1998, 2, 1023–24.

[181] Chauvin Y. Olefin metathesis: The early days (Nobel Lecture). Angew Chem Int Ed 2006, 45, 3740–47.

[182] Grubbs RH. Olefin-metathesis catalysts for the preparation of molecules and materials (Nobel Lecture). Angew Chem Int Ed 2006, 45, 3760–65.

[183] Schrock RR. Multiple metal-carbon bonds for catalytic metathesis reactions (Nobel Lecture). Angew Chem Int Ed 2006, 45, 3748–59.

[184] Hoveyda AH, Liu Z, Qin C, Koengeter T, Mu Y. Impact of ethylene on efficiency and stereocontrol in olefin metathesis: When to add it, when to remove it, and when to avoid it. Angew Chem 2020, 59, 22324–48.

[185] Jean-Louis Hérisson P, Chauvin Y. Catalyse de transformation des oléfines par les complexes du tungstène II. Télomérisation des oléfines cycliques en présence d'oléfines acycliques. Die Makromol Chemie 1971, 141, 161–76.

[186] Howard TR, Lee JB, Grubbs RH. Titanium metallacarbene-metallacyclobutane reactions: Stepwise metathesis. J Am Chem Soc 1980, 102, 6876–78.

[187] Trnka TM, Grubbs RH. The development of L$_2$X$_2$Ru=CHR olefin metathesis catalysts: An organometallic success story. Acc Chem Res 2001, 34, 18–29.

[188] Schrock RR. High oxidation state multiple metal-carbon bonds. Chem Rev 2002, 102, 145–79.

[189] Schrock RR, Hoveyda AH. Molybdenum and tungsten imido alkylidene complexes as efficient olefin-metathesis catalysts. Angew Chem Int Ed 2003, 42, 4592–633.

[190] Grubbs RH. Olefin metathesis. Tetrahedron 2004, 60, 7117–40.

[191] Hoveyda AH, Schrock RR. Catalytic asymmetric olefin metathesis. Org Synth Set 2008, 210–29.

[192] Kotha S, Dipak MK. Strategies and tactics in olefin metathesis. Tetrahedron 2012, 68, 397–421.

[193] Ogba OM, Warner NC, O'Leary DJ, Grubbs RH. Recent advances in ruthenium-based olefin metathesis. Chem Soc Rev 2018, 47, 4510–44.

[194] Crowe WE, Zhang ZJ. Highly selective cross-metathesis of terminal olefins. J Am Chem Soc 1993, 115, 10998–99.

[195] Forcina V, Garcia-Dominguez A, Lloyd-Jones G. Kinetics of initiation of the third generation grubbs metathesis catalyst: Convergent associative and dissociative pathways. Faraday Discuss 2019, 220, 179–95.

[196] Choi T-L, Lee CW, Chatterjee AK, Grubbs RH. Olefin metathesis involving ruthenium enoic carbene complexes. J Am Chem Soc 2001, 123, 10417–18.

[197] Chatterjee AK, Choi T, Sanders DP, Grubbs RH. A general model for selectivity in olefin cross metathesis. J Am Chem Soc 2003, 125, 11360–70.

[198] Chatterjee AK, Morgan JP, Scholl M, Grubbs RH. Synthesis of functionalized olefins by cross and ring-closing metatheses. J Am Chem Soc 2000, 122, 3783–84.

[199] Blackwell HE, O'Leary DJ, Chatterjee AK, Washenfelder RA, Bussmann DA, Grubbs RH. New approaches to olefin cross-metathesis. J Am Chem Soc 2000, 122, 58–71.

[200] Keitz BK, Endo K, Herbert MB, Grubbs RH. Z-Selective homodimerization of terminal olefins with a ruthenium metathesis catalyst. J Am Chem Soc 2011, 133, 9686–88.

[201] Xu Y, Wong JJ, Samkian AE, Ko JH, Chen S, Houk KN, Grubbs RH. Efficient Z-selective olefin-acrylamide cross-metathesis enabled by sterically demanding cyclometalated ruthenium catalysts. J Am Chem Soc 2020, 142, 20987–93.

[202] Johns AM, Ahmed TS, Jackson BW, Grubbs RH, Pederson RL. High trans kinetic selectivity in ruthenium-based ole fi n cross- metathesis through stereoretention. Org Lett 2016, 18, 772–75.

[203] Fu GC, Nguyen ST, Grubbs RH. Catalytic ring-closing metathesis of functionalized dienes by a ruthenium carbene complex. J Am Chem Soc 1993, 115, 9856–57.

[204] Clark JS, Kettle JG. Synthesis of brevetoxin sub-units by sequential ring-closing metathesis and hydroboration. Tetrahedron Lett 1997, 38, 123–26.

[205] Clark JS, Hamelin O. Synthesis of polycyclic ethers by two- directional double ring-closing metathesis. Angew Chem Int Ed Engl 2000, 39, 372–74.

[206] Heinrich F, Durand D, Starck J, Michelet V. Ruthenium metathesis: A Key step to access a new cyclic tetrasubstituted olefin platform. Org Lett 2020, 22, 7064–67.

[207] Seiders TJ, Ward DW, Grubbs RH. Enantioselective ruthenium-catalyzed ring-closing metathesis. Org Lett 2001, 2247–50.

[208] Cefalo DR, Kiely AF, Wuchrer M, Jamieson JY, Schrock RR, Hoveyda AH. Enantioselective synthesis of unsaturated cyclic tertiary ethers by mo-catalyzed olefin metathesis. J Am Chem Soc 2001, 123, 3139–40.

[209] Ibrahem I, Yu M, Schrock RR, Hoveyda AH, Highly Z. And enantioselective ring-opening/cross-metathesis reactions catalyzed by stereogenic-at-mo adamantylimido complexes. J Am Chem Soc 2009, 131, 3844–45.

[210] Cotez GA, Baxter CA, Schrock RR, Hoveyda AH. Comparison of Ru- and Mo-based chiral olefin metathesis catalysts Complementarity in asymmetric ring-opening/cross-metathesis reactions of oxa- and azabicycles. Org Lett 2007, 9, 2871–74.

[211] Davidson TA, Wagener KB, Priddy DB. Polymerization of dicyclopentadiene : A tale of two mechanisms. Macromolecules 1996, 29, 786–88.

[212] Bielawski CW, Grubbs RH. Living ring-opening metathesis polymerization. Prog Polym Sci 2007, 32, 1–29.

[213] Szwarc M. "Living" Polymers. Nature 1956, 178, 1168.

[214] Maughon BR, Grubbs RH. Ruthenium alkylidene initiated living ring-opening metathesis polymerization (ROMP) of 3-substituted cyclobutenes. Macromolecules 1997, 30, 3459–69.

[215] Singh R, Czekelius C, Schrock RR. Living ring-opening metathesis polymerization of cyclopropenes. Macromolecules 2006, 39, 1316–17.

[216] Flook MM, Gerber LCH, Debelouchina GT, Schrock RR. Z-selective and syndioselective ring-opening metathess polymerization (ROMP) initiated by monoaryloxidepyrrolide (MAP) catalysts. Macromolecules 2010, 43, 7515–22.

[217] Bazan GC, Oskam JH, Cho H, Park LY, Schrock RR. Living ring-opening metathesis pPolymerization of 2,3-difunctionalized 7-oxanorbornenes and 7-oxanorbornadienes by Mo(CHCMe₂R)(N-2,6-C₆H₃-i-Pr₂)(0-t-Bu), and Mo(CHCMe₂R)(N-2,6-C₆H₃-i-Pr₂)(OCMe₂CF₃)₂. J Am Chem Soc 1991, 113, 6899–907.

[218] Bazan GC, Schrock RR, O'Regan MB, Thomas JK, Davis WM, Khosravi E, Feast WJ, Gibson VC. Living ring-opening metathesis polymerization of 2,3-difunctionalized norbornadienes by Mo(CH-*t*-Bu)(N-2,6-C₆H₃-i-Pr₂)(O-t-Bu)₂. J Am Chem Soc 1990, 112, 8378–87.

[219] Dumas A, Tarrieu R, Vives T, Roisnel T, Dorcet V, Baslé O, Mauduit M. A versatile and highly Z-Selective olefin metathesis ruthenium catalyst based on a readily accessible N-heterocyclic carbene. ACS Catal 2018, 8, 3257–62.

[220] Ziegler K, Holzkamp E, Breil H, Martin H. Polymerisation von Äthylen und anderen Olefinen. Angew Chem 1955, 67, 426.

[221] Ziegler K, Holzkamp E, Breil H, Martin H. DM. Normaldruck-Polyäthylen-Verfahren. Angew Chem 1955, 67, 541–47.

[222] Natta G, Pino P, Corradini P, Danusso F, Moraglio G, Mantica E, Mazzanti G. Crystalline high polymers of α-olefins. J Am Chem Soc 1955, 77, 1708–10.

[223] Gibson VC, Spitzmesser SK. Advances in non-metallocene olefin polymerization catalysis. Chem Rev 2003, 103, 283–315.

[224] Gibson VC. Shuttling polyolefins to a new materials dimension. Science 2006, 312, 703–04.

[225] Sinn H, Kaminsky W. Ziegler-Natta catalysis. Adv Organomet Chem 1980, 99–149.

[226] Ewen JA. Mechanisms of stereochemical control in propylene polymerizations with soluble Group 4B metallocene/methylalumoxane catalysts. J Am Chem Soc 1984, 106, 6355–64.

[227] Ewen JA, Jones RL, Razavi A, Ferrara JD. Syndiospecific propylene polymerizations with Group 4 metallocenes. J Am Chem Soc 1988, 110, 6255–56.

[228] Kaminsky W, Külper K, Brintzinger HH, Wild FRWP. Polymerization of propene and butene with a chiral zirconocene and methylalumoxane as cocatalyst. Angew Chem Int Ed Eng 1985, 24, 507–08.

[229] Brintzinger HH, Fischer D, Mülhaupt R, Rieger B, Waymouth RM. Stereospecific olefin polymerization with chiral metallocene catalysts. Angew Chem Int Ed Eng 1995, 34, 1143–70.

[230] Bochmann M. The chemistry of catalyst activation: The case of Group 4 polymerization catalysts. Organometallics 2010, 29, 4711–40.

[231] Chen EYX, Marks TJ. Cocatalysts for metal-catalyzed olefin polymerization: Activators, activation processes, and structure-activity relationships. Chem Rev 2000, 100, 1391–434.

[232] Kaminsky W. Discovery of methylaluminoxane as cocatalyst for olefin polymerization. Macromolecules 2012, 45, 3289–97.

[233] Zaccaria F, Zuccaccia C, Cipullo R, Budzelaar PHM, Vittoria A, Macchioni A. et al., Methylaluminoxane's molecular cousin: A Well-defined and "Complete" Al-activator for molecular olefin polymerization catalysts. ACS Catal 2021, 11, 4464–75.

[234] Jordan RF, LaPointe RE, Bajgur CS, Echols SF, Willett R. Chemistry of cationic zirconium(IV) benzyl complexes. One-electron oxidation of d⁰ organometallics. J Am Chem Soc 1987, 109, 4111–13.

[235] Coates GW. Precise control of polyolefin stereochemistry using single-site metal catalysts. Chem Rev 2000, 100, 1223–52.

[236] Klosin J, Fontaine PP, Figueroa R. Development of Group IV molecular catalysts for high temperature ethylene-α-olefin copolymerization reactions. Acc Chem Res 2015, 48, 2004–16.

[237] Stürzel M, Mihan S, Mülhaupt R. From multisite polymerization catalysis to sustainable materials and all-polyolefin composites. Chem Rev 2016, 116, 1398–433.

[238] Shapiro PJ, Bunel E, Schaefer WP, Bercaw JE. A unique example of a single component a-olefin polymerization catalyst. Organometallics 1990, 9, 867–69.

[239] Shapiro PJ, Cotter WD, Schaefer WP, Labinger JA, Bercaw JE. Model Ziegler-Natta α-olefin polymerization catalysts derived from [{(η⁵-C₅Me₄)SiMe₂(η¹-NCMe₃)KPMe₃)Sc(μ₂-H)]₂ and [{(η⁵-C₅Me₄)SiMe₂(η¹-NCMe₃)}Sc(M₂-CH₂CH₂CH₃)]₂ Synthesis, Structures, and kinetic and equilibrium investigations of the catalytically activ. J Am Chem Soc 1994, 116, 4623–40.

[240] Okuda J. Functionalized cyclopentadienyl ligands, IV Synthesis and Complexation of Linked Cyclopentadienyl-amido. Ligands Chem Ber 1990, 123, 1649–51.

[241] Arriola DJ, Bokota M, Campbell RE, Klosin J, LaPointe RE, Redwine OD, Shankar RB, Timmers FJ, Abboud KA. Penultimate effect in ethylene-styrene copolymerization and the discovery of highly active ethylene-styrene catalysts with increased styrene reactivity. J Am Chem Soc 2007, 129, 7065–76.

[242] Klosin J, Kruper WJ, Nickias PN, Roof GR, Abboud KA. Complexes. Dramatic substituent effect on catalyst efficiency and polymer molecular weight. 2001, 2663–65.

[243] Braunschweig H, Breitling FM. Constrained geometry complexes-Synthesis and applications. Coord Chem Rev 2006, 250, 2691–720.

[244] Devore DD, Timmers FJ, Hasha DL, Rosen RK, Marks TJ, Deck PA, Stern CL. Constrained-Geometry Titanium(II) Diene Complexes. Structural Diversity and Olefin Polymerization Activity. Organometallics 1995, 14, 3132–34.

[245] Gunther SO, Lai Q, Senecal T, Huacuja R, Bremer S, Pearson DM, DeMott JC, Bhuvanesh N, Ozerov OV, Klosin J. Highly efficient carborane-based activators for molecular olefin polymerization catalysts. ACS Catal 2021, 11, 3335–42.

[246] Piers WE. Zwitterionic metallocenes. Chem Eur J 1998, 4, 13–18.

[247] Hannig F, Fröhlich R, Bergander K, Erker G, Petersen JL. Structural and spectroscopic evidence for intramolecular agostic M···H-C and dative Zr←-F-C(*ortho*) interactions in the zwitterionic metal complexes [(C$_5$H$_4$)SiMe$_2$(N-t-Bu)]M(+)(μ-C$_4$H$_6$)B(-)(C$_6$F$_5$)$_3$, M = Ti, Zr. Organometallics 2004, 23, 4495–502.

[248] Temme B, Erker G, Karl J, Luftmann H, Fröhlich R, Kotila S. Reaction of (butadiene) zirconocene with tris(pentafluorophenyl)borane – A novel way of generating methylalumoxane-free homogeneous Ziegler-Type cCatalysts. Angew Chem Int Ed Eng 1995, 34, 1755–57.

[249] Arriola DJ, Carnahan EM, Hustad PD, Kuhlman RL, Wenzel TT. Catalytic production of olefin block copolymers via chain shuttling polymerization. Science 2006, 312, 714–19.

[250] Hustad PD, Kuhlman RL, Carnahan EM, Wenzel TT, Arriola DJ. An exploration of the effects of reversibility in chain transfer to metal in olefin polymerization. Macromolecules 2008, 41, 4081–89.

[251] Sita LR. Ex Uno Plures ("Out of One, Many"): New paradigms for expanding the range of polyolefins through reversible group transfers. Angew Chem Int Ed 2009, 48, 2464–72.

[252] Zhang W, Wei J, Sita LR. Living coordinative chain-transfer polymerization and copolymerization of ethene, α-olefins, and α,ω-nonconjugated dienes using dialkylzinc as "surrogate" chain-growth sites. Macromolecules 2008, 41, 7829–33.

[253] Wei J, Zhang W, Sita LR. Aufbaureaktion Redux: Scalable production of precision hydrocarbons from AlR$_3$ (R = Et or iBu) by dialkyl zinc mediated ternary living coordinative chain-transfer polymerization. Angew Chem, Int Ed 2010, 49, 1768–72.

[254] Zhang W, Sita LR. Highly Efficient, Living coordinative chain-transfer polymerization of propene with ZnEt$_2$: practical production of ultrahigh to very low molecular weight amorphous atactic polypropenes of extremely narrow polydispersity. J Am Chem Soc 2008, 130, 442–43.

Index

https://doi.org/10.1515/9783110695274-023

www.ingramcontent.com/pod-product-compliance
Lightning Source LLC
Chambersburg PA
CBHW080136220326
41598CB00032B/5083